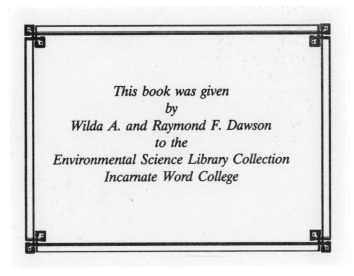

ILLUSTRATED HANDBOOK
OF
PHYSICAL-CHEMICAL PROPERTIES
AND
ENVIRONMENTAL FATE
FOR
ORGANIC CHEMICALS

Volume II
Polynuclear Aromatic Hydrocarbons,
Polychlorinated Dioxins,
and Dibenzofurans

Donald Mackay
Wan Ying Shiu
Kuo Ching Ma

LEWIS PUBLISHERS
Boca Raton   Ann Arbor   London   Tokyo

Library of Congress Cataloging-in-Publication Data

Mackay, Donald, Ph.D.
Illustrated handbook of physical-chemical properties and environmental fate for organic
chemicals/Donald Mackay, Wan Ying Shiu, and Kuo Ching Ma.

p. cm.
Includes bibliographical references.
Contents: v. 1. Monoaromatic hydrocarbons, chlorobenzenes, and PCBs.
1. Organic compounds - Environmental aspects - Handbooks, manuals, etc.
2. Environmental chemistry - Handbooks, manuals, etc.
I. Shiu, Wan Ying. II. Ma, Kuo Ching. III. Title.
TD196.073M32 1991
628.5'2 - dc20   91-33888
ISBN 0-87371-513-6 (Volume I)
ISBN 0-87371-583-7 (Volume II)

LEWIS PUBLISHERS
121 South Main Street, Chelsea, Michigan 48118

Printed in the United States of America   5 6 7 8 9 0
Printed on acid-free paper

PREFACE

This series of Handbooks brings together physical-chemical data for similarly structured groups of chemical substances, which influence their fate in the multimedia environment of air, water, soils, sediments and their resident biota. The task of assessing chemical fate locally, regionally and globally is complicated by the large (and increasing) number of chemicals of potential concern, by uncertainties in their physical-chemical properties, and by lack of knowledge of prevailing environmental conditions such as temperature, pH and deposition rates of solid matter from the atmosphere to water, or from water to bottom sediments. Further, reported values of properties such as solubility are often in conflict. Some are measured accurately, some approximately and some are estimated by various correlation schemes from molecular structure. In some cases units or chemical identity are wrongly reported. The user of such data thus has the difficult task of selecting the "best" or "right" values. There is justifiable concern that the resulting deductions of environmental fate may be in substantial error. For example, the potential for evaporation may be greatly underestimated if an erroneously low vapor pressure is selected.

To assist the environmental scientist and engineer in such assessments, this Handbook contains compilations of physical-chemical property data for series of chemicals such as the aromatic hydrocarbons. It has long been recognized that within such series, properties vary systematically with molecular size, thus providing guidance about the properties of one substance from those of its homologs. Plots of these systematic property variations are provided to check the reported data and provide an opportunity for interpolation and even modest extrapolation to estimate unreported properties of other homologs. Most handbooks treat chemicals only on an individual basis, and do not contain this feature of chemical-to-chemical comparison which can be valuable for identifying errors and estimating properties.

The data are taken a stage further and used to estimate likely environmental partitioning tendencies, i.e., how the chemical is likely to become distributed between the various media which comprise our biosphere. The results are presented numerically and pictorially to provide a visual impression of likely environmental behavior. This will be of interest to those assessing environmental fate by confirming the general fate characteristics or behavior profile. It is, of course, only possible here to assess fate in a "typical" or "generic" or "evaluative" environment, thus no claim is made that a chemical will behave in this manner in all situations, but this assessment should reveal the broad characteristics of behavior. These evaluative fate assessments are generated using simple fugacity models which flow naturally from the compilations of data on physical-chemical properties of relevant chemicals.

It is hoped that this series of Handbooks will be of value to environmental scientists and engineers and to students and teachers of environmental science. Its aim is to contribute to better assessments of chemical fate in our multimedia environment by serving as a reference source for environmentally relevant physical-chemical property data of classes of chemicals and by illustrating the likely behavior of these chemicals as they migrate throughout our biosphere.

Donald Mackay, born and educated in Scotland, received his degrees in Chemical Engineering from The University of Glasgow. After a period of time in the petrochemical industry he joined The University of Toronto, where he is now a Professor in the Department of Chemical Engineering and Applied Chemistry, and in the Institute for Environmental Studies. Professor Mackay's primary research is the study of organic environmental contaminants, their sources, fates, effects, and control, and particularly in understanding and modeling their behavior with the aid of the fugacity concept. His work has focused especially on the Great Lakes Basin and on cold northern climates.

Wan-Ying Shiu is a Research Associate in the Department of Chemical Engineering and Applied Chemistry, and the Institute for Environmental Studies, University of Toronto. She received her Ph.D. in Physical Chemistry from the Department of Chemistry, University of Toronto, M.Sc. in Physical Chemistry from St. Francis Xavier University and B.Sc. in Chemistry from Hong Kong Baptist College. Her research interest is in the area of physical-chemical properties and thermodynamics for organic chemicals of environmental concern.

Kuo-Ching Ma obtained his Ph.D. from The Florida State University, M.Sc. from The University of Saskatchewan and B.Sc. from The National Taiwan University; all in Physical Chemistry. After working many years in the Aerospace, Battery Research, Fine Chemicals and Metal Finishing industries in Canada as Research Scientist, Technical Supervisor/Director, he is now dedicating his time and interests to environmental research.

TABLE OF CONTENTS

ILLUSTRATED HANDBOOK

OF

PHYSICAL-CHEMICAL PROPERTIES

AND

ENVIRONMENTAL FATE

FOR

ORGANIC CHEMICALS

Volume II

Polynuclear Aromatic Hydrocarbons,
Polychlorinated Dioxins,
and Dibenzofurans

1. INTRODUCTION
1.1 THE INCENTIVE

It is alleged that there are some 60,000 chemicals in current commercial production, with approximately 1000 being added each year. Most are organic chemicals. Of these, perhaps 500 are of environmental concern because of their presence in detectable quantities in various components of the environment, their toxicity, their tendency to bioaccumulate, or their persistence. A view is emerging that some of these chemicals are of such extreme environmental concern that all production and use should be ceased, i.e., as a global society we should elect not to synthesize or use these chemicals. They should be "sunsetted". PCBs, "dioxins" and freons are examples. A second group consists of chemicals which are of concern because they are used or discharged in large quantities, or they are toxic or persistent. They are, however, of sufficient value to society that their continued use is justified, but only under conditions in which we fully understand their sources, fate and effects. This understanding is essential if society is to be assured that there are no adverse ecological or human health effects. Other groups of increasingly benign chemicals can presumably be treated with less rigor.

A key feature of this "cradle to grave" approach is that society must improve its skills in assessing chemical fate in the environment. We must better understand where chemicals originate, how they migrate in, and between, the various media of air, water, soils, sediments and their biota which comprise our biosphere. We must understand how these chemicals are transformed by chemical and biochemical processes and thus how long they will persist in the environment. We must seek a fuller understanding of the effects which they will have on the multitude of interacting organisms which occupy these media, including ourselves.

It is now clear that the fate of chemicals in the environment is controlled by a combination of two groups of factors. First are the prevailing environmental conditions such as temperatures, flows and accumulations of air, water and solid matter and the composition of these media. Second are the properties of the chemicals which influence partitioning and reaction tendencies, i.e., whether the chemical evaporates or associates with sediments, and how the chemical is eventually destroyed by conversion to other chemical species.

In recent decades there has emerged a discipline within environmental science concerned with increasing our understanding of how chemicals behave in our multimedia environment. It has been termed "chemodynamics". Practitioners of this discipline include scientists and engineers, students and teachers who attempt to measure, assess and predict how this large number of chemicals will behave in laboratory, local, regional and global environments. These individuals need data on physical-chemical and reactivity properties, as well as information on how these properties translate into environmental fate. This Handbook provides a compilation of such data and uses them to estimate the broad features of environmental fate. It does so for classes or groups of chemicals, instead of the usual approach of treating chemicals on an individual basis. This has the advantage that systematic variations in properties with molecular size can be revealed and used to check reported values, interpolate and even extrapolate to other chemicals of similar structure.

With the advent of inexpensive and rapid computation there has been a remarkable growth in interest in this general area of Quantitative Structure-Property Relationships (QSPRs). The ultimate goal is to use information about chemical structure to deduce physical-chemical properties, environmental partitioning and reaction tendencies, and even uptake and effects on biota. The goal is far from being realized, but considerable progress has been made, as is briefly reviewed in a following section. In this Handbook we adopt a simple, and well tried, approach of using molecular structure to deduce a molar volume, which in turn is related to physical-chemical properties. Undoubtedly, other molecular descriptors such as surface area or topological indices have the potential to give more accurate correlations and will be used in the future, but at this stage we believe that the improvements in accuracy obtained by using these more complex descriptors do not justify the computational effort of generating them, at least for the purposes of routine, general assessments. In some cases the fundamental causes of the relationships remain obscure.

A major benefit of this simple QSPR analysis is that it reveals likely errors in reported data. Regrettably, the scientific literature contains a great deal of conflicting data with reported values often varying over several orders of magnitude. There are some good, but more not-so-good reasons for this lack of accuracy. Many of these properties are difficult to measure because they involve analyzing very low concentrations of 1 part in 10^9 or 10^{12}. For many purposes an approximate value, for example, that a solubility is less than 1 mg/L, is adequate. There has been a mistaken impression that if a vapor pressure is low, as is the case with DDT, it is not important. DDT evaporates appreciably from solution in water despite its low vapor pressure, because of its low solubility in water. In some cases the units are reported incorrectly or there are uncertainties about temperature or pH. In other cases the chemical is wrongly identified. One aim of this Handbook is to assist the user to identify such problems and provide guidance when selecting appropriate values.

The final aspect of chemical fate treated in this Handbook is the depiction or illustration of likely chemical fate. This is done using a series of multimedia "fugacity" models as is described in a later section. The authors' aim is to convey an impression of likely environmental partitioning and transformation characteristics, i.e., we seek to generate a "behavior profile". A fascinating feature of chemodynamics is that chemicals differ so greatly in their behavior. Some, such as chloroform, evaporate rapidly and are dissipated in the atmosphere. Others, such as DDT, partition into the organic matter of soils and sediments and the lipids of fish, birds and mammals. Phenols tend to remain in water subject to fairly rapid transformation processes such as biodegradation and photolysis. By entering the physical-chemical data into a model of chemical fate in a generic or evaluative environment, it is possible to estimate the likely general features of the chemical's behavior and fate. The output of these calculations is presented numerically and pictorially.

In total, the aim of this series of Handbooks is to provide a useful reference work for those concerned with the assessment of the fate of existing and new chemicals in the environment.

1.2 PHYSICAL-CHEMICAL PROPERTIES
1.2.1 The key physical-chemical properties

The major differences between behavior profiles of organic chemicals in the environment are attributable to physical-chemical properties. The key properties are believed to be solubility in water, vapor pressure, octanol-water partition coefficient, dissociation constant in water (when relevant) and susceptibility to degrading or transformation reactions. Other essential molecular descriptors are molecular mass and molar volume, with properties such as critical temperature and pressure and molecular area being occasionally useful for specific purposes.

Chemical identity may appear to present a trivial problem, but many chemicals have several names, and subtle differences between isomers (e.g., cis and trans) may be ignored. The most commonly accepted identifiers are the IUPAC name and the Chemical Abstracts System (CAS) number. More recently methods have been sought of expressing the structure in line notation form so that computer entry of a series of symbols can be used to define a three-dimensional structure. The Wiswesser Line Notation is quite widely used, but it appears that for environmental purposes it will be superceded by the SMILES (Simplified Molecular Identification and Line Entry System, Anderson et al. 1987).

Molecular mass is readily obtained from structure. Also of interest are molecular volume and area, which may be estimated by a variety of methods.

Solubility in water and vapor pressure are both "saturation" properties, i.e., they are measurements of the maximum capacity which a phase has for dissolved chemical. Vapor pressure P (Pa) can be viewed as a "solubility in air", the corresponding concentration C (mol/m^3) being P/RT where R is the ideal gas constant (8.314 J/mol $\cdot$ K) and T is absolute temperature (K). Although most chemicals are present in the environment at concentrations well below saturation, these concentrations are useful for estimating air-water partition coefficients as ratios of saturation values. It is usually assumed that the same partition coefficient applies at lower sub-saturation concentrations. Vapor pressure and solubility thus provide estimates of air-water partition coefficients K_{AW} or Henry's law constants H (Pa $\cdot$ m^3/mol), and thus the relative air-water partitioning tendency.

The octanol-water partition coefficient K_{OW} provides a direct estimate of hydrophobicity or of partitioning tendency from water to organic media such as lipids, waxes and natural organic matter such as humin or humic acid. It is invaluable as a method of estimating K_{OC}, the organic carbon-water partition coefficient, the usual correlation invoked being that of Karickhoff (1981)

$$K_{OC} = 0.41 \ K_{OW}$$

It is also used to estimate fish-water bioconcentration factors K_B or BCF using a correlation similar to that of Mackay (1982)

$$K_B = 0.05 \ K_{OW}$$

3

where the term 0.05 corresponds to a 5% lipid content of the fish.

For ionizing chemicals it is essential to quantify the extent of ionization as a function of pH using the dissociation constant pKa. The parent and ionic forms behave and partition quite differently, thus pH and the presence of other ions may profoundly affect chemical fate.

Characterization of chemical reactivity presents a severe problem in Handbooks. Whereas radioisotopes have fixed half-lives, the half-life of a chemical in the environment depends not only on·the intrinsic properties of the chemical, but also on the nature of the surrounding environment. Factors such as sunlight intensity, hydroxyl radical concentration and the nature of the microbial community, as well as temperature, affect the chemical's half-life so it is impossible (and misleading) to document a single reliable half-life. The compilation by Howard et al. (1991) provides an excellent review of the existing literature for a large number of chemicals. It is widely used as a source document in this work. The best that can be done is to suggest a semi-quantitative classification of half-lives into groups, assuming average environmental conditions to apply. Obviously a different class will generally apply in air and bottom sediment. In this compilation we use the following class ranges for chemical reactivity in a single medium such as water.

Class	Mean half-life (hours)	Range (hours)
1	5	< 10
2	17 (~ 1 day)	10-30
3	55 (~ 2 days)	30-100
4	170 (~ 1 week)	100-300
5	550 (~ 3 weeks)	300-1,000
6	1700 (~ 2 months)	1,000-3,000
7	5500 (~ 8 months)	3,000-10,000
8	17000 (~ 2 years)	10,000-30,000
9	55000 (~ 6 years)	> 30,000

These times are divided logarithmically with a factor of approximately 3 between adjacent classes. With the present state of knowledge it is probably misleading to divide the classes into finer groupings; indeed, a single chemical may experience half-lives ranging over three classes, depending on season.

A recurring problem in compilation of this type is the criteria which should be used for selecting the "best" value of a property when several values are reported. An element of judgement is necessary. The usual considerations are:
(1) the age of the data and acknowledgment of previous conflicting or supporting values,
(2) the method of determination,

4

(3) the perception of the objectives of the authors, not necessarily as an indication of competence, but often as an indication of the need of the authors for accurate values.

In this Handbook we have used these considerations as well as information derived from the QSPR analyses.

It is appropriate, therefore, to review briefly the experimental methods which are commonly used for property determinations and comment on their accuracy.

1.2.2. Experimental methods
Solubility in water

The conventional method of preparing saturated solutions for the determination of solubility is batch equilibration. An excess amount of solute chemical is added to water and equilibrium is achieved by shaking gently (generally referred as the "shake flask method") or slow stirring with a magnetic stirrer. The aim is to prevent formation of emulsions or suspensions and thus avoid extra experimental procedures such as filtration or centrifuging which may be required to ensure that a true solution is obtained. Experimental difficulties can still occur because of the formation of emulsion or microcrystal suspensions with the sparingly soluble chemicals such as higher normal alkanes and polycyclic aromatic hydrocarbons (PAHs). An alternative approach is to coat a thin layer of the chemical on surface of the equilibration flask before water is added. An accurate "generator column" method has also been developed (Weil et al. 1974, May et al. 1978a,b) in which a column is packed with an inert solid support, such as glass beads or Chromosorb, and then coated with the solute chemical. Water is pumped through the column at a controlled, known flow rate to achieve saturation.

The method of concentration measurement of the saturated solution depends on the solute solubility and its chemical properties. Some common methods used for solubility measurement are listed below.

1. Gravimetric or volumetric methods (Booth and Everson 1948)

    An excess amount of solid compound is added to a flask containing water to achieve saturation solution by shaking, stirring, centrifuging until the water is saturated with solute and undissolved solid or liquid residue appears, often as a cloudy phase. For liquids, successive known amounts of solute may be added to water and allowed to reach equilibrium, and the volume of excess undissolved solute is measured.

2. Instrumental methods

    a. UV spectrometry (Andrews and Keffer 1950, Bohon and Claussen 1951, Yalkowsky et al. 1976);

    b. Gas chromatographic analysis with FID, ECD or other detectors (McAuliffe 1968, Mackay et al. 1975, Chiou et al. 1982);

    c. Fluorescence spectrophotometry (Mackay and Shiu 1977);

    d. Interferometry (Gross and Saylor 1931);

    e. High-pressure liquid chromatography (HPLC) with R.I., UV or fluorescence detection (May et al. 1978a,b, Wasik et al. 1983, Shiu et al. 1988, Doucette and Andren 1988a);

    f. Nephelometric methods (Davis and Parke 1942, Davis et al. 1942, Hollifield 1979).

For most organic chemicals the solubility is reported at a defined temperature in distilled water. For substances which ionize (e.g. phenols, carboxylic acids and amines) it is essential to report the pH of the determination because the extent of ionization affects the solubility. It is common to maintain the desired pH by buffering with an appropriate electrolyte mixure. This raises the complication that the presence of electrolytes modifies the water structure and changes the solubility. The effect is usually "salting-out". For example, many hydrocarbons have solubilities in seawater about 75% of their solubilities in distilled water. Care must thus be taken to interpret and use reported data properly when electrolytes are present.

The most common problem encountered with reported data is inaccuracy associated with very low solubilities, i.e., those less than 1.0 mg/L. Such solutions are difficult to prepare, handle and analyze, and reported data are often contain appreciable errors.

Octanol-water partition coefficient K_{ow}

The experimental approaches are similar to those for solubility, i.e., employing shake flask or generator-column techniques. Concentrations in both the water and octanol phases may be determined after equilibration. Both phases can then be analyzed by the instrumental methods discussed above and the partition coefficient is calculated from the concentration ratio C_o/C_w. This is actually the ratio of solute concentration in octanol saturated with water to that in water saturated with octanol.

As with solubility, K_{ow} is a function of the presence of electrolytes and for dissociating chemicals it is a function of pH. Accurate values can generally be measured up to about 10^6, but accurate measurement beyond this requires meticulous technique. A common problem is that the presence of small quantities of emulsified octanol in the water phase could create a high concentration of chemical in that emulsion which would cause an erroneously high apparent water phase concentration.

Considerable success has been achieved by calculating K_{ow} from molecular structure; thus, there has been a tendency to calculate K_{ow} rather than measure it, especially for "difficult" hydrophobic chemicals. These calculations are, in some cases, extrapolations and can be in serious error. Any calculated log K_{ow} value above 7 should be regarded as suspect, and any experimental or calculated value above 8 should be treated with extreme caution.

Details of experimental methods are described by Fujita et al. (1964), Leo et al. (1971); Hansch and Leo (1979), Rekker (1977), Chiou et al. (1977), Miller et al. (1984), Bowman and Sans (1983), Woodburn et al. (1984), Doucette and Andren (1987), and De Bruijn et al. (1989).

Vapor pressure

In principle, the determination of vapor pressure involves the measurement of the saturation concentration or pressure of the solute in a gas phase. The most reliable methods involve direct determination of these concentrations, but convenient indirect methods are also available based on evaporation rate measurement or chromatographic retention times. Some methods and approaches are listed below.
   a. Direct measurement by use of pressure gauges: diaphram gauge (Ambrose et al. 1975), Rodebush gauge (Sears & Hopke 1947), inclined-piston gauge (Osborn & Douslin 1975);
   b. Comparative ebulliometry (Ambrose 1981);

c. Effusion methods, torsion and weight-loss (Balson 1947, Bradley and Cleasby 1953, Hamaker and Kerlinger 1969, De Kruif 1980);

d. Gas saturation or transpiration methods (Spencer and Cliath 1970, 1972, Sinke 1974, Macknick and Prausnitz 1979, Westcott et al. 1981, Rodorf 1985a,b, 1986);

e. Dynamic coupled-column liquid chromatographic method - a gas saturation method (Sonnefeld et al. 1983);

f. Calculation from evaporation rates and vapor pressures of reference compound (Gückel et al. 1974, 1982, Dobbs and Grant 1980, Dobbs and Cull 1982);

g. Calculation from GC retention time data (Hamiltan 1980, Westcott and Bidleman 1982, Bidleman 1984, Kim et al. 1984, Foreman and Bidleman 1985, Burkhard et al. 1985a, Hinckley et al. 1990).

The greatest difficulty and uncertainty arises when determining the vapor pressure of chemicals of low volatility, i.e., those with vapor pressures below 1.0 Pa. Vapor pressures are strongly dependent on temperature, thus accurate temperature control is essential. Data are often regressed against temperature and reported as Antoine or Clapeyron constants. Care must be taken when using the Antoine or other equations to extropolate data beyond the temperature range specified. It must be clear if the data apply to the solid or liquid phase of the chemical.

Henry's law constant

The Henry's law constant is essentially an air-water partition coefficient which can be determined by measurement of solute concentrations in both phases. This raises the difficulty of accurate analytical determination in two very different media which require different techniques. Accordingly, some effort has been devoted to devising techniques in which concentrations are measured in only one phase and the other concentration is deduced by a mass balance. These methods are generally more accurate. The principal difficulty arises with hydrophobic, low-volatility chemicals which can establish only very small concentrations in both phases.

Henry's law constant can be regarded as a ratio of vapor pressure to solubility, thus it is subject to the same effects which electrolytes have on solubility and temperature has on both properties. Some methods are as follows:

a. Equilibrium batch stripping (Mackay et al. 1979, Dunnivant et al. 1988);

b. EPICS (Equilibrium Partioning In Closed Systems) method (Lincoff and Gossett 1984; Gossett 1987, Ashworth et al. 1988);

c. Wetted-wall column (Fendinger and Glotfelty 1988, 1990);

d. Headspace analyses (Hussam and Carr 1985);

e. Calculation from vapor pressure and solubility (Mackay and Shiu 1981).

When using vapor pressure and solubility data it is essential to ensure that both properties apply to the same chemical phase, i.e. both are of the liquid, or of the solid. Occasionally, a solubility is of a solid while a vapor pressure is extrapolated from higher temperature liquid phase data.

1.3 QUANTITATIVE-STRUCTURE-PROPERTY RELATIONSHIPS (QSPRs)
1.3.1 Objectives

Because of the large number of chemicals of actual and potential concern, the difficulties and cost of experimental determinations, and scientific interest in elucidating the fundamental molecular determinants of physical-chemical properties, a considerable effort has been devoted to generating quantitative structure-activity relationships (QSARs). This concept of structure-property relationships or structure-activity relationships is based on observations of linear free-energy relationships, and usually takes the form of a plot or regression of the property or interest as a function of an appropriate molecular descriptor which can be obtained from merely a knowledge of molecular structure.

Such relationships have been applied to solubility, vapor pressure, K_{OW}, Henry's law constant, reactivities, bioconcentration data and several other environmentally relevant partition coefficients. Of particular value are relationships involving various manifestations of toxicity, but these are beyond the scope of this Handbook. These relationships are valuable because they permit values to be checked for "reasonableness" and (with some caution) interpolation is possible to estimate undetermined values. They may be used (with extreme caution!) for extrapolation.

A large number of descriptors have been, and are being, proposed and tested. Dearden (1990) and the compilation by Karcher and Devillers (1990) give comprehensive accounts of descriptors and their applications.

Among the most commonly used molecular descriptors are molecular weight and volume, the number of specific atoms (e.g., carbon or chlorine), surface areas (which may be defined in various ways), refractivity, parachor, steric parameters, connectivities and various topological parameters. Several quantum chemical parameters can be calculated from molecular orbital calculations including charge, electron density and superdelocalizability.

It is likely that existing and new descriptors will be continued to be tested, and that eventually a generally preferred set of readily accessible parameters will be adopted of routine use for correlating purposes. From the viewpoint of developing quantitative correlations it is very desirable to seek a linear relationship between descriptor and property, but a nonlinear or curvilinear relationship is quite adequate for illustrating relationships and interpolating purposes. In this Handbook we have elected to use the simple descriptor of molar volume at the normal boiling point as estimated by the LeBas method (Reid et al. 1987). This parameter is very easily calculated and proves to be adequate for the present purposes of plotting property versus relationship without seeking linearity.

The LeBas method is based on a summation of atomic volumes with adjustment for the volume decrease arising from ring formation. The full method is described by Reid et al. (1987), but for the purposes of this compilation the volumes and rules as listed in Table 1.1 are used.

9

Table 1.1   LeBas Molar Volume

	increment, cm^3 /mol
carbon	14.8
hydrogen	3.7
oxygen	7.4
in methyl esters and ethers	9.1
in ethyl esters and ethers	9.9
join to S, P, or N	8.3
nitrogen	
doubly bonded	15.6
in primary amines	10.5
in secondary amines	12.0
bromine	27.0
chlorine	24.6
fluorine	8.7
iodine	37.0
sulfur	25.6
ring	
three-membered	-6.0
four-membered	-8.5
five-membered	-11.5
six-membered	-15.0
naphthalene	-30.0
anthracene	-47.5

Example: The experimental molar volume of chlorobenzene is 115 cm^3/mol. From the above rules, the LeBas molar volume for chlorobenzene (C_6H_5Cl) is:

$$V = 6 \times 14.8 + 5 \times 3.7 + 24.6 - 15 = 117 \ cm^3/mol$$

Accordingly, plots are presented at the end of each chapter for solubility, vapor pressure, K_{OW}, and Henry's law constant versus LeBas molar volume.

A complication arises in that two of these properties (solubility and vapor pressure) are dependent on whether the solute is in the liquid or solid state. Solid solutes have lower solubilities and vapor pressures than they would have if they had been liquids. The ratio of the (actual) solid to the (hypothetical subcooled) liquid solubility or vapor pressure is termed the fugacity ratio and can be estimated from the melting point and the entropy of fusion ΔS_{fus} as discussed by Mackay and Shiu (1981). For solid solutes, the correct property to plot is the calculated or extrapolated subcooled liquid solubility. This is calculated in this Handbook using the relationship suggested by Yalkowsky (1979) which implies an entropy of fusion of 56 J/mol K or 13.5 cal/mol · K

$$C^S_S/C^S_L = P^S_S/P^S_L = \exp\{6.79(1 - T_M/T)\}$$

where C^S is solubility, P^S is vapor pressure, subcripts S and L referring to solid and liquid phases, T_M is melting point and T is the system temperature, both in absolute (K) units. The fugacity ratio is given in the data tables at 25 °C, the usual temperature at which physical-chemical property data are reported. For liquids, the fugacity ratio is 1.0.

1.3.2 Examples

Recently, there have been efforts to extend the long established concept of Quantitative Structure-Activity Relationships (QSARs) to Quantitative Structure-Property Relationships (QSPRs) to compute all relevant environmental physical-chemical properties (such as aqueous solubility, vapor pressure, octanol-water partition coefficient, Henry's law constant, bioconcentration factor (BCF)) and sorption coefficient from molecular structure. Examples are Burkhard (1984) and Burkhard et al. (1985a) who calculated solubility, vapor pressure, Henry's law constant, K_{OW} and K_{OC} for all PCB congeners. Hawker and Connell (1988) also calculated log K_{OW}; Abramowitz and Yalkowsky (1990) calculated melting point and solubility for all PCB congeners based on the correlation with total surface area (planar TSAs). Doucette and Andren (1988b) used six molecular descriptors to compute the K_{OW} of some chlorobenzenes, PCBs and PCDDs. Mailhot and Peters (1988) employed seven molecular descriptors to compute physical-chemical properties of some 300 compounds. Isnard and Lambert (1988, 1989) correlated solubility, K_{OW} and BCF for a large number of organic chemicals. Nirmalakhandan and Speece (1988a,b, 1989) used molecular connectivity indices to predict aqueous solubility and Henry's law constants for 300 compounds over 12 logarithmic units in solubility. Kamlet and coworkers (1987, 1988) have developed the solvatochromic parameters with the intrinsic molar volume to predict solubility, log K_{OW} and toxicity of organic chemicals. Warne et al. (1990) correlated solubility and K_{OW} for lipophilic organic compound with 39 molecular descriptors and physical-chemical properties. Other correlations are reviewed by Lyman et al. (1982). As Dearden (1990) has pointed out, "new parameters are continually being devised and tested, although the necessity of that may be questioned, given the vast number already available". It must be emphasized, however, that regardless of how accurate these predicted or estimated properties are claimed to be, utimately they have to be confirmed or verified by experimental measurement.

A fundamental problem encountered in these correlations is that the molecular descriptors can be calculated with relatively high precision, usually within a few percent. The accuracy may not always be high, but for empirical correlation purposes precision is more important than accuracy. The precision and accuracy of the experimental data are often poor, frequently ranging over a factor of two or more. Certain isomers may yield identical descriptors, but have different properties. There is thus an inherent limit to the applicability of QSPRs imposed by the quality of the experimental data, and further efforts to improve descriptors, while interesting and potentially useful, are unlikely to yield demonstrably improved QSPRs.

11

For correlation of **solubility** the correct thermodynamic quantities for correlation are the activity coefficient γ, or the excess Gibbs free energy ΔG, as discussed by Pierotti et al. (1959) and Tsonopoulos and Prausnitz (1971). Examples of such correlations are given below.

1. Carbon number or carbon plus chlorine number (Tsonopoulos and Prausnitz 1971, Mackay and Shiu 1977);
2. Molar volume cm^3/mol
   a. Liquid molar volume - from density (McAuliffe 1966, Lande and Banerjee 1981, Chiou et al. 1982, Abernethy et al. 1988);
   b. Molar volume by additive group contribution method, e.g., LeBas method, Schroeder method (Reid et al. 1987, Miller et al. 1985);
   c. Intrinsic molar volume, V_I, cm^3/mol - from van der Waals radius with solvatochromic parameters α and β (Leahy 1986, Kamlet et al. 1987, 1988);
   d. Characteristic molecular volume, m^3/mol (McGowan and Mellors 1986);
3. Molecular volume - $\mathring{A}^3/mol$ (cubic Angstrom per mole)
   a. van der Waals volume (Bondi 1964);
   b. Total Molecular Volume (TMV) (Pearlman et al. 1984, Pearlman 1986);
4. Total Surface Area (TSA) - $\mathring{A}^2/mol$ (Hermann 1971, Yalkowsky and Valvani 1976, Yalkowsky et al. 1979, Pearlman 1986, Andren et al. 1987, Hawker and Connell 1988);
5. Molecular Connectivity indices, χ (Kier and Hall 1976, Andren et al. 1987, Nirmalakhandan and Speece 1988b, 1989);
7. Boiling point (Almgren et al. 1979);
8. Melting point (Amidon and Williams 1982);
9. Melting point and TSA (Abramowitz and Yalkowsky 1990);
10. High Pressure Liquid Chromatography (HPLC) - retention data (Locke 1974, Whitehouse & Cooke 1982).

Several workers have explored the linear relationship between octanol-water partition coefficient and solubility as means of estimating solubility.

Hansch et al. (1968) established the linear free-energy relationship between aqueous and octanol-water partition of organic liquid. Others, such as Tulp and Hutzinger (1978), Yalkowsky et al. (1979), Mackay et al. (1980), Banerjee et al. (1980), Chiou et al. (1982), Bowman and Sans (1983), Miller et al. (1985), Andren et al. (1987) and Andren and Doucette (1988b) have all presented similar but modified relationships.

The UNIFAC (UNIQUAC Functional Group Activity Coefficient) group contribution (Fredenslund et al. 1975) is widely used for predicting the activity coefficient in nonelectrolyte liquid mixtures by using group-interaction parameters. This method has been used by Kabadi and Danner (1979), Banerjee (1985), Arbuckle (1983, 1986), Banerjee and Howard (1988) and Al-Sahhaf (1989).

HPLC retention time data have been used as a psuedo-molecular descriptor by Whitehouse and Cooke (1982), Hafkenscheid and Tomlinson (1981), Tomlinson and Hafkenscheid (1986) and Swann et al. (1983).

The **octanol-water partition coefficient** K_{ow} is widely used as a descriptor of hydrophobicity. Variation in K_{ow} is primarily attributable to variation in activity coefficient in the aqueous phase (Miller et al. 1985); thus, the same correlations used for solubility in water are applicable to K_{ow}. Most widely used is the Hansch-Leo compilation of data (Leo et al. 1971, Hansch and Leo 1979) and related predictive methods. Examples of K_{ow} correlations are

I. Molecular descriptors
1. Molar volumes: LeBas method; from density; intrinsic molar volume; characteristic molecular volume (Abernethy et al. 1988, Chiou 1985, Kamlet et al. 1988, McGowan and Mellors 1986);
2. TMV (De Bruijn and Hermens 1990);
3. TSA (Yalkowsky et al. 1979, Yalkowsky et al. 1983, Hawker and Connell 1988);
4. Molecular connectivity indices (Doucette and Andren 1988b);
5. Molecular weight (Doucette and Andren 1988b).
II. Group contribution methods
1. π-constant or hydrophobic substituent method (Hansch et al. 1968, Hansch & Leo 1979, Doucette and Andren 1988b);
2. Fragmental constants or f-constant (Rekker 1977, Yalkowsky et al. 1983);
3. Hansch & Leo's f-constant (Hansch & Leo 1979; Doucette and Andren 1988b).
III. From solubility - K_{ow} relationship
IV. HPLC retention data
1. HPLC-k' capacity factor (Könemann et al. 1979, McDuffie 1981);
2. HPLC-RT retention time (Veith et al. 1979, Rappaport and Eisenreich 1984, Doucette and Andren 1988b);
3. HPLC-RV retention volume (Garst 1984);
4. HPLC-RT/MS HPLC retention time with mass spectrometry (Burkhard et al. 1985c).
V. Reversed-phase thin-layer chromatography (TLC) (Bruggeman et al. 1982).
VI. Molar refractivity (Yoshida et al. 1983).

As with solubility and octanol-water partition coefficient, **vapor pressure** can be estimated with a variety of correlations as discussed in detail by Burkhard et al. (1985a) and summarized as follows:
1. Interpolation or extrapolation from equation for correlating temperature relationships, e.g., the Clausius-Clapeyron, Antoine equations (Burkhard et al. 1985a);
2. Carbon or chlorine numbers (Mackay et al. 1980, Shiu and Mackay 1986);
3. LeBas molar volume (Shiu et al. 1987, 1988);
4. Boiling point T_B and heat of vaporization ΔH_v (Mackay et al. 1982);
5. Group contribution method (Macknick and Prausnitz 1979);

6. UNIFAC group contribution method (Burkhard et al. 1985a, Banerjee et al.1990);
7. Molecular weight and Gibbs' free energy of vaporization ΔG_v (Burkhard et al. 1985a);
8. TSA and ΔG_v (Amidon and Anik 1981, Burkhard et al. 1985a, Hawker 1989);
9. Molecular connectivity indices (Kier and Hall 1976, 1986, Burkhard et al. 1985a);
10. Melting point T_M and GC retention index (Bidleman 1984, Burkhard et al. 1985a);
11. Solvatochromic parameters and intrinsic molar volume (Banerjee et al. 1990).

As described earlier, **Henry's law constants** can be calculated from the ratio of vapor pressure and aqueous solubility. Henry's law constants do not show a simple linear pattern as solublity, K_{OW} or vapor pressure when plotted against simple molecular descriptors, such as numbers of chlorine or LeBas molar volume, e.g. PCBs (Burkhard et al. 1985b), pesticides (Suntio et al. 1988), and chlorinated dioxins (Shiu et al. 1988). Henry's law constants can be estimated from:
1. UNIFAC-derived Activity coefficients (Arbuckle 1983);
2. Molecular connectivity indices (Nirmalakhandan and Speece 1988b, Sabljic and Güsten 1989);
3. Total surface area - planar TSA (Hawker 1989);
4. Bond contribution method (Meylan and Howard 1991).

Bioconcentration factors:
1. Correlation with K_{OW} (Neely et al. 1974, Könemann and van Leeuwen 1980, Veith et al. 1980, Chiou et al. 1977, Mackay 1982, Briggs 1981, Garten and Trabalka 1983, Davies and Dobbs 1984, Oliver and Niimi 1988, Isnard and Lambert 1988);
2. Correlation with solubility (Kenaga 1980, Kenaga and Goring 1980, Briggs 1981, Garten and Trabalka 1983, Davies and Dobbs 1984, Isnard and Lambert 1988);
3. Correlation with K_{OC} (Kenaga 1980, Kenaga and Goring 1980, Briggs 1981);
4. Calculation with HPLC retention data (Swann et al. 1983);
5. Calculation with solvatochromic parameters (Hawker 1989a, 1990).

Sorption coefficients:
1. Correlation with K_{OW} (Karickhoff et al. 1979, Schwarzenbach and Westall 1981, Mackay 1982, Oliver 1984);
2. Correlation with solubility (Karickhoff et al. 1979);
3. Molecular connectivity indices (Sabljic 1984, 1987, Sabljic et al. 1989);
4. From HPLC retention data (Swann et al. 1983, Szabo et al. 1990).

1.4 FATE MODELS
1.4.1 Evaluative Environmental Calculations
The nature of these calculations has been described in a series of papers, notably Mackay (1979), Paterson and Mackay (1985), Mackay and Paterson (1990, 1991), and a recent text (Mackay 1991). Only the salient features are presented here. Three calculations are completed for each chemical, namely the Level I, II and III fugacity calculations.

1.4.2 Level I Fugacity Calculation
The Level I calculation describes how a given amount of chemical partitions at equilibrium between six media: air, water, soil, bottom sediment, suspended sediment and fish. No account is taken of reactivity. Whereas most early evaluative environments have treated a one square kilometer region with about 70% water surface (simulating the global proportion of ocean surface), it has become apparent that a more useful approach is to treat a larger, principally terrestrial area similar to a jurisdictional region such as a U.S. state. The area selected is 100,000 km^2 or 10^{11} m^2 which is about the area of Ohio, Greece or England.

The atmospheric height is selected as a fairly arbitrary 1000 m reflecting that region of the troposphere which is most affected by local air emissions. A water surface area of 10% or 10,000 km^2 is used, with a water depth of 20 m. The water volume is thus $2x10^{11}$ m^3. The soil is viewed as being well mixed to a depth of 10 cm and is considered to be 2% organic carbon. It has a volume of $9x10^9$ m^3. The bottom sediment has the same area as the water, a depth of 1 cm and an organic carbon content of 4%. It thus has a volume of 10^8 m^3.

For the Level I calculation both the soil and sediment are treated as simple solid phases with the above volumes, i.e., the presence of air or water in the pores of these phases is ignored.

Two other phases are included for interest. Suspended matter in water is often an important medium when compared in sorbing capacity to that of water. It is treated as having 20% organic carbon and being present at a volume fraction in the water of $5x10^{-6}$, i.e., it is about 5 mg/L. Fishes are also included at an entirely arbitrary volume fraction of 10^{-6} and are assumed to contain 5% lipid, equivalent in sorbing capacity to octanol. These two phases are small in volume and rarely contain an appreciable fraction of the chemical present, but it is in these phases that the highest concentration of chemical often exists.

Another phase which is introduced later in the Level III model is aerosol particles with a volume fraction in air of $2x10^{-11}$, i.e., approximately 30 $\mu g/m^3$. Although negligible in volume, an appreciable fraction of the chemical present in the air phase may be associated with aerosols. Aerosols are not treated in Level I or II calculations because their capacity for chemical is usually negligible when compared with soil.

These dimensions and properties are summarized in Table 1.2. The user is encouraged to modify these dimensions to reflect conditions in a specific area of interest.

Table 1.2a Compartment Dimensions and Properties for Level I and II Calculations

Compartment	Air	Water	Soil	Sediment	Suspended Sediment	Fish
Volume, V (m³)	10^{14}	2×10^{11}	9×10^9	10^8	10^6	2×10^5
Depth, h (m)	1000	20	0.1	0.01	-	-
Area, A (m²)	100×10^9	10×10^9	90×10^9	10×10^9	-	-
Org. Fraction (ϕ_{OC})	-	-	0.02	0.04	0.2	-
Density, ρ (kg/m³)	1.2	1000	2400	2400	1500	1000
Adv. Residence Time, t (hours)	100	1000	-	50,000	-	-
Adv. flow, G (m³/h)	10^{12}	2×10^8	-	2000	-	-

Table 1.2b Bulk Compartment Dimensions and Volume Fraction (v) for Level III Calculations

Air	Total volume	10^{14} m³	(as above)
	Air phase	10^{14} m³	
	Aerosol phase	2000 m³	(v = 2×10^{-11})
Water	Total volume	2×10^{11} m³	
	Water phase	2×10^{11} m³	(as above)
	Suspended sediment phase	10^6 m³	(v = 5×10^{-6})
	Fish phase	2×10^5 m³	(v = 1×10^{-6})
Soil	Total volume	18×10^9 m³	
	Air phase	3.6×10^9 m³	(v = 0.2)
	Water phase	5.4×10^9 m³	(v = 0.3)
	Solid phase	9.0×10^9 m³	(v = 0.5) (as above)
Sediment	Total volume	500×10^6 m³	
	Water phase	400×10^6 m³	(v = 0.8)
	Solid phase	100×10^6 m³	(v = 0.2) (as above)

The amount of chemical introduced in the Level I calculation is an arbitrary 100,000 kg or 100 tonnes. If dispersed entirely in the air, this amount yields a concentration of 1 $\mu g/m^3$ which is not unusual for ubiquitous contaminants such as hydrocarbons. If dispersed entirely in the water, the concentration is a higher 500 $\mu g/m^3$ or 500 ng/L, which again is reasonable for a well-used chemical of commerce. The corresponding value in soil is about 0.0046 $\mu g/g$. It is believed that this amount is a reasonable common value for evaluative purposes. Clearly for restricted chemicals such as PCBs, this amount is too large, but it is preferable to adopt a common evaluative amount for all substances. No significance should, of course, be attached to the absolute values of the concentrations which are deduced from this arbitrary amount. Only the relative values have significance.

The Level I calculation proceeds by deducing the fugacity capacities, Z values for each medium (see Table 1.3), following the procedures described by Mackay (1991). These working equations show the necessity of having data on molecular mass, water solubility, vapor pressure, and octanol-water partition coefficient. The fugacity f (Pa) common to all media is deduced as

$$f = M/\Sigma \ V_i Z_i$$

where M is the total amount of chemical (mol), V_i is the medium volume (m^3) and Z_i is the corresponding fugacity capacity for the chemical in each medium.

The molar concentration C (mol/m^3) can then be deduced as Zf mol/m^3 or as WZf g/m^3 or 1000 WZf/ρ $\mu g/g$ where ρ is the phase density (kg/m^3) and W is the molecular mass (g/mol). The amount m_i in each medium is $C_i V_i$ mol, and the total in all media is M mol. The **BASIC** computer program for undertaking this calculation is appended. For those who prefer a spreadsheet format, an identical Lotus 123° program is also provided.

The information obtained from this calculation includes the concentrations, amounts and distribution. In the figures, a pie chart illustrates the distribution between the four primary compartments of air, water, soil and sediment, the amount in fish and suspended sediment being ignored. This information is useful as an indication of the relative concentrations.

Note that this simple treatment assumes that the soil and sediment phases are entirely solid, i.e., there are no air or water phases present to "dilute" the solids. Later in the Level III calculation these phases and aerosols are included.

Table 1.3a   Equations for Phase Z values used in Levels I and II and the Bulk Phase values used in Level III

Air	$Z_1 = 1/RT$
Water	$Z_2 = 1/H = C^S/P^S$
Soil	$Z_3 = Z_2 \cdot \rho_3 \cdot \phi_3 \cdot K_{OC}/1000$
Sediment	$Z_4 = Z_2 \cdot \rho_4 \cdot \phi_4 \cdot K_{OC}/1000$
Suspended Sediment	$Z_5 = Z_2 \cdot \rho_5 \cdot \phi_5 \cdot K_{OC}/1000$
Fish	$Z_6 = Z_2 \cdot \rho_6 \cdot L \cdot K_{OW}/1000$
Aerosol	$Z_7 = Z_1 \cdot 6\times10^6/P^S_L$
where	R = gas constant (8.314 J/mol K)
	T = absolute temperature (K)
	C^S = solubility in water (mol/m^3)
	P^S = vapor pressure (Pa)
	H = Henry's law constant (Pa $\cdot$ m^3/mol)
	P^S_L = liquid vapor pressure (Pa)
	K_{OW} = octanol-water partition coefficient
	K_{OC} = organic-carbon partition coefficient (= 0.41 K_{OW})
	ρ_i = density of phase i (kg/m^3)
	ϕ_i = mass fraction organic-carbon in phase i (g/g)
	L = lipid content of fish

Note for solids $P^S_L = P^S_S/\exp\{6.79(1 - T_M/T)\}$ where T_M is melting point (K) of the solute.

Table 1.3b Bulk Phase Z values, Z_{Bi} deduced as $\Sigma\ v_iZ_i$, in which the coefficients, e.g., 2×10^{-11}, are the volume fractions v_i of each pure phase as specified in Table 1.2b

Air	$Z_{B1} = Z_1 + 2\times10^{-11}\ Z_7$	(approximately 30 μg/m^3 aerosols)
Water	$Z_{B2} = Z_2 + 5\times10^{-6}\ Z_5 + 1\times10^{-6}\ Z_6$	(5 ppm solids, 1 ppm fish by volume)
Soil	$Z_{B3} = 0.2\ Z_1 + 0.3\ Z_2 + 0.5\ Z_3$	(20% air, 30% water, 50% solids)
Sediment	$Z_{B4} = 0.8\ Z_2 + 0.2\ Z_4$	(80% water, 20% solids)

1.4.3 Level II Fugacity Calculation

The Level II calculation simulates a situation in which chemical is continuously discharged into the multimedia environment and achieves a steady-state equilibrium condition at which input and output rates are equal. The task is to deduce the rates of loss by reaction and advection.

The reaction rate data developed for each chemical in the tables are used to select a reactivity class as described earlier, and hence a first-order rate constant for each medium. Often these rates are in considerable doubt, thus the quantities selected should be used with extreme caution because they may not be widely applicable. The rate constants k_i h^{-1} are used to calculate reaction D values for each medium D_{Ri} as $V_i Z_i k_i$. The rate of reactive loss is then $D_{Ri} f$ mol/h.

For advection, it is necessary to select flow rates. This is conveniently done in the form of advective residence times, t h, thus the advection rate G_i is V_i/t m^3/h for each medium. For air, a residence time of 100 hours is used (approximately 4 days), which is probably too long for the geographic area considered, but shorter residence times tend to cause air advective loss to be a dominant mechanism. For water, a figure of 1000 hours (42 days) is used, reflecting a mixture of rivers and lakes. For sediment burial (which is treated as an advective loss), a time of 50000 hours or 5.7 years is used. Only for very persistent, hydrophobic chemicals is this process important. No advective loss from soil is included. The D value for loss by advection D_{Ai} is $G_i Z_i$ and the rates are $D_{Ai} f$ mol/h. These rates are listed in Table 1.2.

There may thus be losses caused by both reaction and advection D values for the four primary media. These loss processes are not included for fish or suspended matter. At steady-state, equilibrium conditions the input rate E mol/h can be equated to the sum of the output rates, from which the common fugacity can be calculated as follows

$$E = f \cdot \Sigma \, D_{Ai} + f \cdot \Sigma \, D_{Ri}$$

thus,

$$f = E/(\Sigma \, D_{Ai} + \Sigma \, D_{Ri})$$

The common assumed emission rate is 1000 kg/h or 1 tonne/h. To achieve an amount equivalent to the 100 tonnes in the Level I calculation requires an overall residence time of 100 hours. Again the concentrations and amounts m_i and $\Sigma \, m_i$ or M can be deduced, as well as the reaction and advection rates. These rates obviously total to give the input rate E. Of particular interest are the relative rates of these loss processes, and the overall persistence or residence time which is calculated as

$$t_O = M/E$$

where M is the total amount present. It is also useful to calculate a reaction and an advection persistence t_R and t_A as

$$t_R = M/\Sigma \, D_{Ri} f \qquad t_A = M/\Sigma \, D_{Ai} f$$

19

Obviously,
$$1/t_O = 1/t_R + 1/t_A$$

These persistences indicate the likelihood of the chemical being lost by reaction as distinct from advection. The percentage distribution of chemical between phases is identical to that in Level I. A pie chart depicting the distribution of losses is presented.

1.4.4 Level III Fugacity Calculation

Whereas the Level I and II calculations assume equilibrium to prevail between all media, this is recognized as being excessively simplistic and even misleading. In the interests of algebraic simplicity only the four primary media are treated for this level. The task is to develop expressions for intermedia transport rates by the various diffusive and nondiffusive processes as described by Mackay (1991). This is done by selecting values for 12 intermedia transport velocity parameters which have dimensions of velocity (m/h or m/year), are designated as U_i m/h and are applied to all chemicals. These parameters are used to calculate seven intermedia transport D values.

It is desirable to calculate new "bulk phase" Z values for the four primary media which include the contribution of dispersed phases within each medium as described by Mackay and Paterson (1991) and as listed in Tables 1.2 and 1.3. The air is now treated as an air-aerosol mixture, water as water plus suspended particles and fish, soil as solids, air and water, and sediment as solids and porewater. The Z values thus differ from the Level I and Level II "pure phase" values. The necessity for introducing this complication arises from the fact that much of the intermedia transport of the chemicals occurs in association with the movement of chemical in these dispersed phases. To accommodate this change the same volumes of the soil solids and sediment solids are retained, but the total phase volumes are increasd. These Level III volumes are also given in Table 1.2. The reaction and advection D values employ the generally smaller bulk phase Z values but the same resisdence times, thus the G values are increased and the D values are generally larger.

Intermedia D values

The justisfication for each intermedia D value follows. It is noteworthy that, for example, air-to-water and water-to-air values differ because of the presence of one-way nondiffusive processes. A fuller description of the background to these calculations is given by Mackay (1991).

1. Air to water (D_{12})

Four processes are considered: diffusion (absorption), dissolution in rain of gaseous chemical, and wet and dry deposition of particle-associated chemical.

For diffusion, the conventional two-film approach is taken with water-side (k_W) and air-side (k_A) mass transfer coefficients (m/h) being defined. Values of 0.05 for k_W and 5 m/h for k_A are used. The absorption D value is then

$$D_{VW} = 1/(1/(k_A A_W Z_1) + 1/(k_W A_W Z_2))$$

where A_W is the air-water area (m^2) and Z_1 and Z_2 are the pure air and water Z values. The velocities k_A and k_W are designated as U_1 and U_2.

For rain dissolution, a rainfall rate of 0.876 m/year is used, i.e., U_R or U_3 is 10^{-4} m/h. The D value for dissolution D_{RW} is then

$$D_{RW} = U_R A_W Z_2 = U_3 A_W Z_2$$

For wet deposition, it is assumed that the rain scavenges Q (scavenging ratio) or about 200,000 times its volume of air. Using a particle concentration (volume fraction) v_Q of 2×10^{-11}, this corresponds to the removal of Qv_Q or 4×10^{-6} volumes of aerosol per volume of rain. The total rate of particle removal by wet deposition is then $Qv_Q U_R A_W$ m^3/h, thus the wet "transport velocity" $Qv_Q U_R$ is 4×10^{-10} m/h.

For dry deposition, a typical deposition velocity U_Q of 10 m/h is selected yielding a rate of particle removal of $U_Q v_Q A_W$ or $2 \times 10^{-10} A_W$ m^3/h corresponding to a transport velocity of 2×10^{-10} m/h. Thus,

$$U_4 = Qv_Q U_R + U_Q v_Q = v_Q(QU_R + U_Q)$$

The total particle transport velocity U_4 for wet and dry deposition is thus 6×10^{-10} m/h and the total D value D_{QW} is

$$D_{QW} = U_4 A_W Z_7$$

where Z_7 is the aerosol Z value.

The overall D value is given by

$$D_{12} = D_{VW} + D_{RW} + D_{QW}$$

2. Water to air (D_{21})

Evaporation is treated as the reverse of absorption thus D_{21} is simply D_{VW} as before.

3. Air to soil (D_{13})

A similar approach is adopted as for air to water transfer. Four processes are considered with rain dissolution (D_{RS}) and wet and dry deposition (D_{QS}) being treated identically except that the area term is now the air-soil area A_S.

For diffusion, the approach of Jury et al. (1983, 1984) is used as described by Mackay and Stiver (1991) and Mackay (1991) in which three diffusive processes are treated. The air boundary layer is characterized by a mass transfer coefficient k_S or U_7 of 5 m/h, equal to that of the air-water MTC coefficient k_A used in D_{12}.

For diffusion in the soil air-pores, a molecular diffusivity of 0.02 m²/h is reduced to an effective diffusivity using a Millington-Quirk type of relationship by a factor of about 20 to 10^{-3} m²/h. Combining this with a path length of 0.05 m gives an effective air to soil mass transfer coefficient k_{SA} of 0.02 m/h which is designated as U_5.

Similarly, for diffusion in water a molecular diffusivity of 2×10^{-6} m²/h is reduced by a factor of 20 to an effective diffusivity of 10^{-7} m²/h, which is combined with a path length of 0.05 m to give an effective soil to water mass transfer coefficient of k_{SW} 2×10^{-6} m/h.

It is probable that capillary flow of water contributes to transport in the soil. For example, a rate of 7 cm/year would yield an equivalent water velocity of 8×10^{-6} m/h which exceeds the water diffusion rate by a factor of four. For illustrative purposes we thus select a water transport velocity or coefficient U_6 in the soil of 10×10^{-6} m/h, recognizing that this may be in error by a substantial amount, and will vary with rainfall characteristics and soil type.

The soil processes are in parallel with boundary layer diffusion in series, so the final equation is

$$D_{VS} = 1/[1/D_S + 1/(D_{SW} + D_{SA})]$$

where

$$
\begin{aligned}
D_S &= U_7 A_S Z_1 & (U_7 &= 5 \text{ m/h}) \\
D_{SW} &= U_6 A_S Z_2 & (U_6 &= 10 \times 10^{-6} \text{ m/h}) \\
D_{SA} &= U_5 A_S Z_1 & (U_5 &= 0.02 \text{ m/h})
\end{aligned}
$$

where A_S is the soil horizontal area.

Air-soil diffusion thus appears to be much slower than air-water diffusion because of the slow migration in the soil matrix. In practice, the result will be a nonuniform composition in the soil with the surface soil (which is much more accessible to the air than the deeper soil) being closer in fugacity to the atmosphere.

The overall D value is given as

$$D_{13} = D_{VS} + D_{QS} + D_{RS}$$

4. Soil to air (D_{31})

Evaporation is treated as the reverse of absorption, thus the D value is simply D_{VS}.

5. Water to sediment (D_{24})

Two processes are treated, diffusion and deposition.

Diffusion is characterized by a mass transfer coefficient U_8 of 10^{-4} m/h which can be regarded as a molecular diffusivity of 2×10^{-6} m²/h divided by a path length of 0.02 m. In practice, bioturbation may contribute substantially to this exchange process, and in shallow water current-induced turbulence may also increase the rate of transport. Diffusion in association with organic colloids is not included.

The D value is thus given as $U_8 A_W Z_2$.

Deposition is assumed to occur at a rate of 5000 m³/h which corresponds to addition of a depth of solids of 0.438 cm/year; thus 43.8% of the solids resident in the accessible bottom sediment is added each year. This rate is about 12 cm³/m² • day which is high compared to values observed in large lakes. The velocity U_9, corresponding to the addition of 5000 m³/h over the area of 10^{10} m², is thus 5×10^{-7} m/h.

It is assumed that of this 5000 m³/h deposited, 2000 m³/h or 40% is buried (yielding the advective flow rate in Table 1.2), 2000 m³/h or 40% is resuspended (as discussed later) and the remaining 20% is mineralized organic matter. The organic carbon balance is thus only approximate.

The transport velocities are thus:

deposition U_9 5.0×10^{-7} m/h or 0.438 cm/years

resuspension U_{10} 2.0×10^{-7} m/h or 0.175 cm/year

burial U_B 2.0×10^{-7} m/h or 0.175 cm/year
                                 (included as an advective residence time of 50,000 h)

The water to sediment D value is thus

$$D_{24} = U_8 A_W Z_2 + U_9 A_W Z_5$$

where Z_5 is the Z value of the particles in the water column.

6. Sediment to water (D_{42})
This is treated similarly to D_{24} giving:

$$D_{42} = U_8 A_W Z_2 + U_{10} A_W Z_4$$

where U_{10} is the sediment resuspension velocity of 2.0×10^{-7} m/h and Z_4 is the Z value of the sediment solids.

7. Sediment advection (D_{A4})
This D value is $U_B A_W Z_4$ where U_B, the sediment burial rate, is 2.0×10^{-7} m/h. It can be viewed as $G_B Z_{B4}$ where G_B is the total burial rate specified as V_S / t_B where t_B (residence time) is 50,000 h, and V_S (the sediment volume) is the product of sediment depth (0.01 cm) and area A_W.

23

Z_4, Z_{B4} are the Z values of the sediment solids and of the bulk sediment respectively. Since there are 20% solids, Z_{B4} is about 0.2 Z_4. There is a slight difference between these approaches because in the advection approach (which is used here) there is burial of water as well as solids.

8. Soil to water (D_{32})

It is assumed that there is run-off of water at a rate of 50% of the rain rate, i.e., the D value is

$$D = 0.5\ U_3 A_s Z_2 = U_{11} A_s Z_2$$

thus the transport velocity term U_{11} is $0.5U_3$ or 5×10^{-5} m/h.

For solids run-off it is assumed that this run-off water contains 200 parts per million by volume of solids; thus the corresponding velocity term U_{12} is $200 \times 10^{-6} U_{11}$, i.e., 10^{-8} m/h. This corresponds to the loss of soil at a rate of about 0.1 mm per year. If these solids were completely deposited in the aquatic environment (which is about 1/10th the soil area), they would accumulate at about 0.1 cm per year, which is about a factor of four less than the deposition rate to sediments. The implication is that most of this deposition is of naturally generated organic carbon and from sources such as bank erosion.

Summary

The twelve intermedia transport parameters are listed in Table 1.4 and the equations are summarized in Table 1.5.

Table 1.4   Intermedia Transport Parameters

U		m/h	m/year
1	Air side, air-water MTC*, k_A	5	43800
2	Water side, air-water MTC, k_W	0.05	438
3	Rain rate, U_R	10^{-4}	0.876
4	Aerosol deposition	6×10^{-10}	5.256×10^{-6}
5	Soil-air phase diffusion MTC, k_{SA}	0.02	175.2
6	Soil-water phase diffusion MTC	10×10^{-6}	0.0876
7	Soil-air boundary layer MTC, k_S	5	43800
8	Sediment-water MTC	10^{-4}	0.876
9	Sediment deposition	5.0×10^{-7}	0.00438
10	Sediment resuspension	2.0×10^{-7}	0.00175
11	Soil-water run-off	5×10^{-5}	0.438
12	Soil-solids run-off	10^{-8}	0.0000876

* Mass transfer coefficient
with,
Scavenging ratio $Q = 2 \times 10^5$
Dry deposition velocity $U_Q = 10$ m/h
Sediment burial rate $U_B = 2.0 \times 10^{-7}$ m/h

Table 1.5 Intermedia Transport D Value Equations

Air-Water

$$D_{12} = D_{VW} + D_{RW} + D_{QW}$$
$$D_{VW} = A_W/(1/U_1Z_1 + 1/U_2Z_2)$$
$$D_{RW} = U_3A_WZ_2$$
$$D_{QW} = U_4A_WZ_7$$

Water-Air

$$D_{21} = D_{VW}$$

Air-Soil

$$D_{13} = D_{VS} + D_{RS} + D_{QS}$$
$$D_{VS} = 1/(1/D_S + 1/(D_W + D_A))$$
$$D_S = U_7A_SZ_1$$
$$D_{SA} = U_5A_SZ_1$$
$$D_{SW} = U_6A_SZ_2$$
$$D_{RS} = U_3A_SZ_2$$
$$D_{QS} = U_4A_SZ_7$$

Soil-Air

$$D_{31} = D_{VS}$$

Water-Sediment

$$D_{24} = U_8A_WZ_2 + U_9A_WZ_5$$

Sediment-Water

$$D_{42} = U_8A_WZ_2 + U_{10}A_WZ_4$$

Soil-Water

$$D_{32} = U_{11}A_SZ_2 + U_{12}A_SZ_3$$

Algebraic solution

Four mass balance equations can be written, one for each medium resulting in a total of four unknown fugacities, enabling simple algebraic solution as shown in Table 1.6. From the four fugacities, the concentration, amounts and rates of all transport and transformation processes can be deduced, yielding a complete mass balance.

The new information from the Level III calculations are the intermedia transport data, i.e., the extent to which chemical discharged into one medium tends to migrate into another. This migration pattern depends strongly on the proportions of the chemical discharged into each medium; indeed, the relative amounts in each medium are largely a reflection of the locations of discharge. It is difficult to interpret these mass balance diagrams because, for example, chemical depositing from air to water may have been discharged to air, or to soil from which it evaporated, or even to water from which it is cycling to and from air.

To simplify this interpretation, it is best to conduct three separate Level III calculations in which unit amounts (1000 kg/h) are introduced individually into air, soil and water. Direct discharges to sediment are unlikely and are not considered here. These calculations show clearly the extent to which intermedia transport occurs. If, for example, the intermedia D values are small compared to the reaction and advection values, the discharged chemical will tend to remain in the discharge or "source" medium with only a small proportion migrating to other media. Conversely, if the intermedia D values are relatively large the chemical becomes very susceptible to intermedia transport. This behavior is observed for persistent substances such as PCBs which have very low rates of reaction.

A direct assessment of multimedia behavior is thus possible by examining the proportions of chemical found at steady-state in the "source" medium and in other media. For example, when discharged to water, an appreciable fraction of the benzene is found in air, whereas for atrazine, only a negligible fraction of atrazine reaches air.

Table 1.6 Level III Solutions to Mass Balance Equations

Mass balance equations:

Air $\qquad$ $E_1 + f_2 D_{21} + f_3 D_{31} = f_1 D_{T1}$

Water $\qquad$ $E_2 + f_1 D_{12} + f_3 D_{32} + f_4 D_{42} = f_2 D_{T2}$

Soil $\qquad$ $E_3 + f_1 D_{13} = f_3 D_{T3}$

Sediment $\qquad$ $E_4 + f_2 D_{24} = f_4 D_{T4}$

where E_i is discharge rate, E_4 usually being zero.

$$D_{T1} = D_{R1} + D_{A1} + D_{12} + D_{13}$$

$$D_{T2} = D_{R2} + D_{A2} + D_{21} + D_{23} + D_{24}, \qquad (D_{23} = 0)$$

$$D_{T3} = D_{R3} + D_{A3} + D_{31} + D_{32}, \qquad (D_{A3} = 0)$$

$$D_{T4} = D_{R4} + D_{A4} + D_{42}$$

Solution:

$$f_2 = [E_2 + J_1 J_4/J_3 + E_3 D_{32}/D_{T3} + E_4 D_{42}/D_{T4}]/(D_{T2} - J_2 J_4/J_3 - D_{24} \cdot D_{42}/D_{T4})$$

$$f_1 = (J_1 + f_2 J_2)/J_3$$

$$f_3 = (E_3 + f_1 D_{13})/D_{T3}$$

$$f_4 = (E_4 + f_2 D_{24})/D_{T4}$$

where

$$J_1 = E_1/D_{T1} + E_3 D_{31}/(D_{T3} \cdot D_{T1})$$

$$J_2 = D_{21}/D_{T1}$$

$$J_3 = 1 - D_{31} \cdot D_{13}/(D_{T1} \cdot D_{T3})$$

$$J_4 = D_{12} + D_{32} \cdot D_{13}/D_{T3}$$

28

Linear additivity

Because these equations are entirely linear, the solutions can be scaled linearly. The concentrations resulting from a discharge of 2000 kg/h are simply twice those of 1000 kg/h. Further, if discharge of 1000 kg/h to air causes 500 kg in water and discharge of 1000 kg/h to soil causes 100 kg in water, then if both discharges occur simultaneously, there will be 600 kg in water. If the discharge to soil is increased to 3000 kg/h, the total amount in the water will rise to (500 + 300) or 800 kg. It is thus possible to deduce the amount in any medium arising from any combination of discharge rates by scaling and adding the responses from the unit inputs. This "linear additivity principle" is more fully discussed by Stiver and Mackay (1989).

In the diagrams presented later, these three-unit (1000 kg/h) responses are given. Also, an illustrative "three discharge" mass balance is given in which a total of 1000 kg/h is discharged, but in proportions judged to be typical of chemical use and discharge to the environment. For example, benzene is believed to be mostly discharged to air with minor amounts to soil and water.

Also given in the tables are the rates of reaction, advection and intermedia transport for each case.

The reader can deduce the fate of any desired discharge pattern by appropriate scaling and addition. It is important to re-emphasize that because the values of transport velocity parameters are only illustrative, actual environmental conditions may be quite different; thus, simulation of conditions in a specific region requires determination of appropriate parameter values as well as the site specific dimensions, reaction rate constants and the physical-chemical properties which prevail at the desired temperature.

In total, the aim is to convey an impression of the likely environmental behavior of the chemical in a readily assimilable form.

1.5. DATA SOURCES AND PRESENTATIONS

1.5.1 Data sources

Most physical properties such as molecular weight (MW, g/mol), melting point (M.P., °C), boiling point (B.P., °C), and density have been obtained from commonly used Handbooks such as the CRC Handbook of Physics and Chemistry (Weast 1972, 1984), Lange's Handbook of Chemistry (Dean 1979, 1985) and the Merck Index (1983, 1987). Other physical-chemical properties such as aqueous solubility, vapor pressure, octanol-water partition coefficient, Henry's law constant, bioconcentration factor and sorption coefficient have been obtained from scientific journals or other environmental Handbooks, notably Verschueren's Handbook of Environmental Data on Organic Chemicals (1977, 1983) and Howard et al.'s Handbook of Environmental Fate and Exposure Data, Vol. I and II (1989, 1990). Other important sources of vapor pressure are the CRC Handbook of Physics and Chemistry (Weast 1972), Lange's Handbook of Chemistry (Dean 1985), the Handbook of Vapor Pressures and Heats of Vaporization of Hydrocarbons and Related Compounds (Zwolinski and Wilhoit 1971), the Vapor Pressure of Pure Substances (Boublik et al. 1973, 1984), the Handbook of the Thermodynamics of Organic Compounds (Stephenson and Malanowski 1987). For aqueous solubilities, valuable sources include the IUPAC Solubility Data Series (1985, 1989a,b) and Horvath's Halogenated Hydrocarbons, Solubility-Miscibility with Water (Horvath 1982). Octanol-water partition coefficients are conveniently obtained from the compilation by Leo et al. (1971) and Hansch and Leo (1979), or can be calculated from molecular structure by the methods of Hansch and Leo (1979) or Rekker (1977). Lyman et al. (1982) also outline methods of estimating solubility, K_{ow}, vapor pressure, and bioconcentration factor for organic chemicals. The recent Handbook of Environmental Degradation Rates by Howard et al. (1991) is a valuable source of rate constants and half-lives for inclusion in subsequent fugacity calculations.

The most reliable sources of data are the original citations in the reviewed scientific literature. Particularly reliable are those papers which contain a critical review of data from a number of sources as well as independent experimental determinations. Calculated or correlated values are reported in the tables but are viewed as being less reliable. A recurring problem is that a value is frequently quoted, then requoted and the original paper may not be cited. The aim in this work has been to gather and list the citations, interpret them and select a "best" or "most likely" value. To assist in this process, plots are prepared of properties as a function of molar volume as the molecular descriptors. These are discussed at the end of each chapter.

1.5.2 Data Format

Each data sheet lists the following properties, although not all quantities are included for all chemicals. In all cases citations are provided.

Common Name:
Synonym:
Chemical Name:
CAS Registry No:
Molecular Formula:

Molecular Weight (g/mol):
Melting Point (°C):
Boiling Point (°C):
Density (g/cm^3 at 20 °C):
Molar Volume (cm^3/mol):
Molecular Volume (Å^3):
Total Surface Area, TSA (Å^2):
Heat of Fusion, ΔH_{fus}, kcal/mol:
Entropy of Fusion, ΔS_{fus}, cal/mol K (or e.u.):
Fugacity Ratio at 25 °C:
Water Solubility (g/m^3 or mg/L at 25°C):
Vapor Pressure (Pa at 25°C):
Henry's Law Constant (Pa m^3/mol) or Air-Water Partition Coefficient:
Octanol-Water Partition Coefficient K_{OW} or log K_{OW}:
Bioconcentration Factor K_B or BCF (or log K_B):
Sorption Partition Coefficient to Organic Carbon K_{OC} or to Organic Matter K_{OM}:
Half-lives in the Environment:
     Air:
     Surface water:
     Goundwater:
     Soil:
     Sediment:
     Biota:
Environmental Fate Rate Constants or Half-Lives:
     Volatilization/Evaporation:
     Photolysis:
     Oxidation or Photooxidation:
     Hydrolysis:
     Biotransformation/Biodegradation:
     Bioconcentration, Uptake (k_1) and Elimination (k_2) Rate Constants:

1.5.3 Explanation of Data Presentations
Example: Naphthalene (data sheets presented in Chapter 2)
1. Chemical Properties.
    The names, formula, melting and boiling point and density data are self-explanatory.

    The molar volumes are in some cases at the stated temperature and in others at the normal boiling point. Certain calculated molecular volumes are also used; thus the reader is cautioned to ensure that when using a molar volume in any correlation, it is correctly selected. In the case of polynuclear aromatic hydrocarbons, the LeBas molar volume is regarded as suspect because of the compact nature of the multiring compounds. It should thus be regarded as merely an indication of relative volume, not absolute volume.

The total surface areas (TSAs) are calculated in various ways and may contain the hydration shell, thus giving a much larger area. Again, the reader is cautioned to ensure that values are consistent.

Heats of fusion, ΔH_{fus}, are generally expressed in kcal/mol or kJ/mol and entropies of fusion, ΔS_{fus} in cal/mol·K (e.u. or entropy unit) or J/mol·K. In the case of liquids such as benzene, it is 1.0. For solids it is a fraction representing the ratio of solid to liquid solubility or vapor pressure. It is generally assumed that for a rigid organic molecule, the entropy of fusion is 13.5 e.u. or 56 J/mol·K, which is an average value of a number of organic compounds (Yalkowsky 1979, Miller et al. 1984). The fugacity ratio, F, given is calculated using ΔS_{fus} = 56 J/mol·K in the following expression

$$F = \exp\{(\Delta H_{fus}/RT)(1 - T_M/T)\} = \exp\{(\Delta S_{fus}/R)(1 - T_M/T)\}$$

where R is the ideal gas constant (8.314 J/mol·K or 1.987 cal/mol·K) and T_M is the melting point and T is the system temperature (K).

As is apparent, a wide variety of solubilities (in units of g/m³ or the equivalent mg/L) have been reported. Experimental data have the method of determinations indicated. In other compilations of data the reported value has merely been quoted from another secondary source. In some cases the value has been calculated. The abbreviations are generally self-explanatory and usually include two entries, the method of equilibration followed by the method of determination. From these values a single value is selected for inclusion in the summary data table. In the case of naphthalene the selected solubility value is 31.0 g/m³ at 25°C. From an examination of the data it is judged that the true value almost certainly lies between 28.0 and 34.0 g/m³.

The vapor pressure data are treated similarly with a value of 10.40 Pa being selected. The true value is judged to lie between 9.0 and 12.0 Pa. Vapor pressures are, of course very temperature dependent.

The Henry's law constant data are measured in some cases, but in other cases are calculated from the ratio of vapor pressure and solubility (in units of mol/m³). In this case a value of 45.0 Pa m³/mol is selected, the actual value probably lying between 30 and 70 Pa m³/mol. Care must be exercised when water is appreciably soluble in the chemical because the assumption that the Henry's law constant is the ratio of vapor pressure to solubility may be invalid.

The octanol-water partition coefficient data are similarly a combination of calculated and experimental values. A value of 3.37 is selected as being the most likely value of log K_{OW}, i.e., K_{OW} is 2344.

A number of (log) bioconcentration factors are listed. These generally range from 0.5 to 1.6 which corresponds to BCFs of 3 to 40. This range could be interpreted as lipid contents ranging from 0.13 to 1.7% but it is likely that metabolism occurs, giving lower-than-equilibrium

BCFs. If no metabolism occurred, it is expected that a 5% lipid fish would have a BCF of about 120. These larger maximum values are calculated later in the fugacity assessments.

The (log) organic-carbon partition coefficients listed range from 2.67 to 3.28. It is expected that K_{OC} usually lies in the range of 20 to 80% of K_{OW}, i.e., log K_{OC} will range from 470 to 1900. Organic matter partition coefficients are also reported. Since organic carbon accounts for some 50 to 60% of the content of organic matter, K_{OM} is expected to be about half K_{OC}.

The reader is advised to consult the original reference when using these values of BCF, K_{OC} and K_{OM}, to ensure that conditions are as close as possible to those of specific interest.

The "Half-life in the Environment" data reflect observations of the rate of disappearance of the chemical from a medium, without necessarily identifying the cause of mechanism of loss. For example, loss from water may be a combination of evaporation, biodegradation and photolysis. Clearly these times are highly variable and depend on factors such as temperature, meteorology and the nature of the media. Again, the reader is urged to consult the original reference.

The "Environmental Fate Rate Constants" refer to specific degradation processes rather than media. As far as possible the original numerical quantities are given and thus there is a variety of time units with some expressions being rate constants and others half-lives.

The conversion is

$$k = 0.693/t_{1/2}$$

where k is the first-order rate constant (h^{-1}) and $t_{1/2}$ is the half-life (h).

From these data a set of medium-specific degradation reaction half-lives was selected for use in Level II and III calculations. Emphasis was based on the fastest and the most plausible degradation process for each of the environmental compartment considered. Instead of assuming an equal half-life for both the water and soil compartment as suggested by Howard et al. (1991), a slower active class (in the reactivity table described earlier) was assigned for soil and sediment compared to that of the water compartment. This is in part because the major degradation processes are often photolysis (or photooxidation) and biodegradation. There is an element of judgement in this selection and it may be desirable to explore the implications of selecting other values. The selected values of the polynuclear aromatic hydrocarbons are given in Table 2.3 at the end of Chapter 2.

In summary, the physical-chemical and environmental fate data listed result in the selection of values of solubility, vapor pressure, K_{OW} and reaction half-lives which are used in the evaluative environmental calculations.

33

The physical-chemical data of PAHs are also plotted in the appropriate QSPR plots on Figures 1.1 to 1.6 (which are the same as Figures 2.1 to 2.6 later). These plots show that the naphthalene data are relatively "well-behaved" and are consistent with data obtained for homologous chemicals. In the case of naphthalene this QSPR plot is of little value because this is a well-studied chemical, but for other less-studied chemicals the plots are invaluable as a means of checking the reasonableness of data. The plots can also be used, with appropriate caution, to estimate data for untested chemicals.

Figure 1.1 shows the linear relationships among various molecular descriptors. Figures 1.2 to 1.5 show the dependence of the physical-chemical properties on LeBas molar volume. Figure 1.2 shows the solubilities of the PAHs decrease steadily with increasing molar volume. The vapor pressure data are similar but K_{OW} increases with increasing molar volume also in a linear fashion. The plot between Henry's law constant and molar volume (Figure 1.5) is more scattered. Figure 1.6 shows the often reported inverse relationship between octanol-water partition coefficient and the subcooled liquid solubility. Plots such as these are discussed in more detail at the conclusion of each chapter.

The QSPR plots show that an increase in molar volume by 25 cm^3/mol generally causes:
(i)   A decrease in log solubility by approximately 0.625 unit, i.e., a factor of 4.2;
(ii) A decrease in log vapor pressure by approximately 1.20 units, i.e., a factor of 18;
(iii) An increase in log Henry's law constant of approximately 0.625 unit or a factor of 4.2;
(iv) An increase in log K_{OW} by approximately 0.625 unit, i.e., a factor of 4.2.
The net result is a marked increase in hydrophobicity with molecular weight and also an increased tendency to partition from air to water. This latter behavior contrasts with chlorinated benzenes and biphenyls in which the Henry's law constant is relatively constant for the series.

1.5.4. Evaluative Calculations

The illustrative evaluative environmental calculations discussed here and presented later for a number of chemicals have been modified from those in Volume I of this series to give more data in a more spacious layout. Level I and II diagrams are assigned to separate pages and the physical-chemical properties are included in the Level I diagram. The Level III diagram is identical to that in Volume I, but additional pie-chart diagrams are included to show how the mass of chemical is distributed between compartments, and how loss processes are distributed as a function of the compartment of discharge. Generally, if discharge is into a compartment such as water, most chemical will be found in that compartment, and will react there but a quantity does migrate to other compartments and is lost from these media. Three pie-charts corresponding to discharges of 1000 kg/h into air, water and soil are included. A fourth pie-chart with discharges to all three compartments is also given. This latter chart is in principle the linear sum of the first three, but since the overall residence times differ, the diagram with the longer residence time, and greater resident mass, tends to dominate.

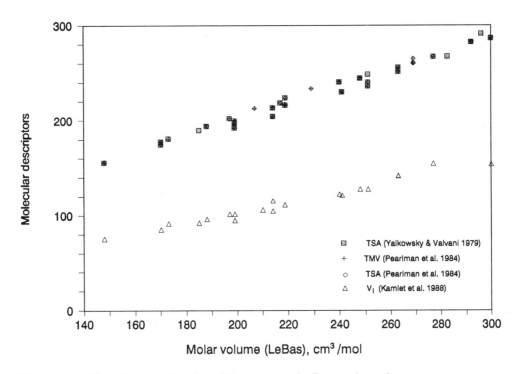

Figure 1.1 Plot of molecular descriptors versus LeBas molar volume.

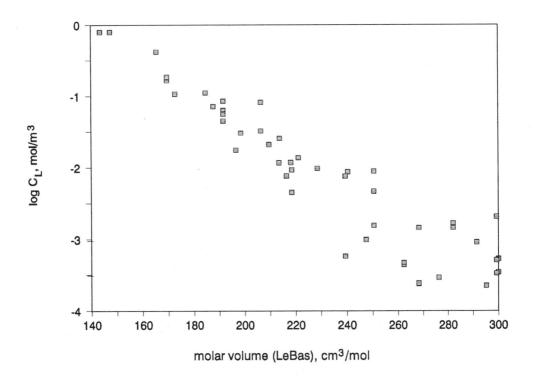

Figure 1.2 Plot of log C_L (liquid solubility) versus molar volume.

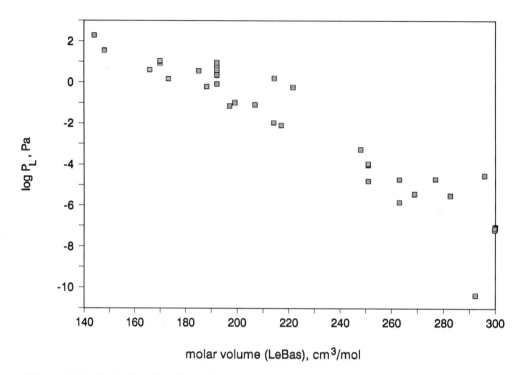

Figure 1.3   Plot of log P_L (liquid vapor pressure) versus molar volume.

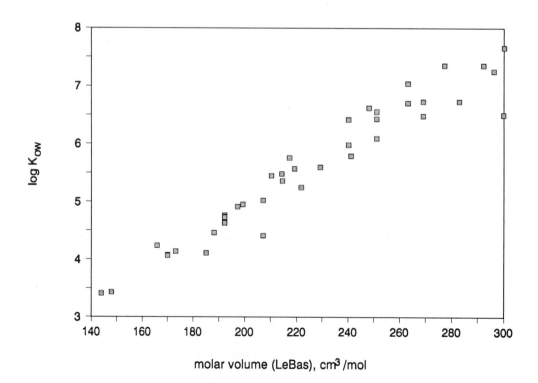

Figure 1.4   Plot of log K_{OW} versus LeBas molar volume.

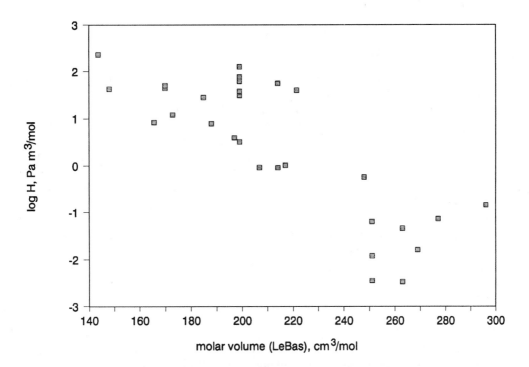

Figure 1.5   Plot of log H (Henry's law constant) versus molar volume.

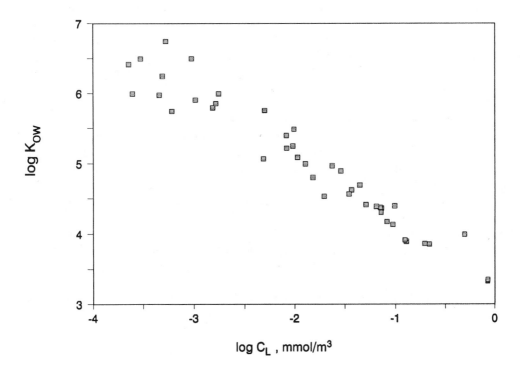

Figure 1.6   Plot of log K_{OW} versus log C_L (liquid solubility).

Level I

The Level I calculation suggests that if 100,000 kg (100 tonnes) of naphthalene are introduced into the 100,000 km^2 environment, 74% will partition into air at a concentration of 7.4 x 10^{-7} g/m^3 or about 0.7 μg/m^3. The water will contain 8.5% at a low concentration of 42 μg/m^3 or equivalently 42 ng/L. Soils would contain 18% of the naphthalene at 8 x 10^{-4} μg/g and sediments about 0.4% at 16 x 10^{-4} μg/g. These soil and sediment values would barely be detectable as a result of the moderate tendency of naphthalene to sorb to organic matter in these media. The fugacity is calculated to be 1.42 x 10^{-5} Pa. The dimensionless soil-water and sediment-water partition coefficients or ratios of Z values are 46 and 91 as a result of a K_{OC} of about 960 and a few percent organic carbon in these media. There is little evidence of bioconcentration with a very low fish concentration of 5 x 10^{-3} μg/g. The pie chart in Figure 1.7 (which is the same as the Level I diagram for naphthalene in Chapter 2) clearly shows that air is the primary medium of accumulation. The more hydrophobic and less volatile PAHs tend to partition less into air and more into soil and sediment. Note that only four media (air, water, soil and bottom sediment) are depicted in the pie chart, therefore the sum of % distribution is slightly less than 100%.

Level II

The Level II calculation includes the reaction half-lives of 17 h in air, 170 h in water, 1700 h in soil and 5500 h in sediment. No reaction is included for suspended sediment or fish. The input of 1000 kg/h results in an overall fugacity of 4 x 10^{-6} Pa which is about 26% of the Level I value. The concentrations and amounts in each medium are thus about 26% of the Level I values. The relative mass distribution is identical to Level I. The primary loss mechanism is reaction in air which accounts for 792 kg/h or 79% of the input. Most of the remainder is lost by advective outflow. The water, soil and sediment loss processes are unimportant largely because so little of the naphthalene is present in these media, and because of the slower reaction and advection rates. The overall residence time is 26.4 h; thus there is an inventory of naphthalene in the system of 26.4 x 1000 or 26400 kg. The pie chart in Figure 1.8 shows the dominance of air reaction and advection.

If the primary loss mechanism of atmospheric reaction is accepted as having a 17 h half-life, the D value is 1.6 x 10^9 mol/Pa$\cdot$h. For any other process to compete with this would require a value of at least 10^8 mol/Pa$\cdot$h. This is achieved by advection (4 x 10^8) but the other processes range in D value from 4300 (advection in bottom sediment) to 1.9 x 10^7 (reaction in water) and are thus a factor of over 20 or more smaller. The implication is that the water reaction rate constant would have to be increased by 20-fold to become significant. The soil rate constant would require an increase by 100 and the sediment by 10000. These are inconceivably large numbers corresponding to very short half-lives, thus the actual values of the rate constants in these media are relatively unimportant in this context. They need not be known accurately for Level II calculations. The most sensitive quantity is clearly the atmospheric reaction rate.

The amounts in the compartments can be calculated easily from the total amount and the percentages of mass distribution in Level I. For example, the amount in water is 8.5% of 26400 kg or 2240 kg.

Chemical name: Naphthalene

Level I calculation: (six compartment model)

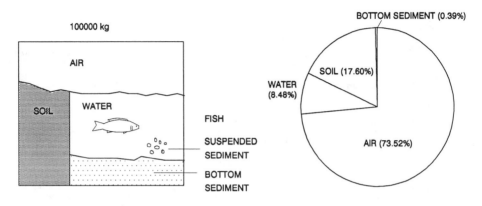

Distribution of mass

physical-chemical properties:

MW: 128.18

M.P.: 80.5°C

Fugacity ratio: 0.284

vapor pressure: 10.4 Pa

solubility: 31 mg/L

log K_{OW} : 3.37

Compartment	Z	Concentration				Amount	Amount
	mol/m3 Pa	mol/m3	mg/L (or g/m3)	ug/g		kg	%
Air	4.034E-04	5.736E-09	7.352E-07	6.202E-04		73524	73.524
Water	2.325E-02	3.306E-07	4.238E-05	4.238E-05		8475.7	8.476
Soil	1.073E+00	1.525E-05	1.955E-03	8.146E-04		17596.1	17.596
Biota (fish)	2.725E+00	3.875E-05	4.967E-03	4.967E-03		0.9935	9.93E-04
Suspended sediment	6.705E+00	9.532E-05	1.222E-02	8.146E-03		12.219	1.22E-02
Bottom sediment	2.146E+00	3.050E-05	3.910E-03	1.629E-03		391.024	0.3910
	Total					100000	100

f = 1.422E-05 Pa

Figure 1.7 Fugacity Level I calculation for naphthalene in a generic environment
(dimensions defined in Table 1.2).

39

Chemical name: Naphthalene

Level II calculation: (six compartment model)

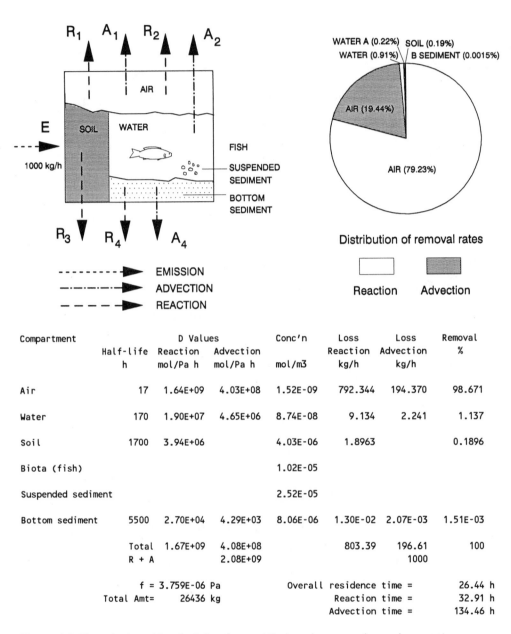

Compartment	Half-life h	D Values Reaction mol/Pa h	Advection mol/Pa h	Conc'n mol/m3	Loss Reaction kg/h	Loss Advection kg/h	Removal %
Air	17	1.64E+09	4.03E+08	1.52E-09	792.344	194.370	98.671
Water	170	1.90E+07	4.65E+06	8.74E-08	9.134	2.241	1.137
Soil	1700	3.94E+06		4.03E-06	1.8963		0.1896
Biota (fish)				1.02E-05			
Suspended sediment				2.52E-05			
Bottom sediment	5500	2.70E+04	4.29E+03	8.06E-06	1.30E-02	2.07E-03	1.51E-03
	Total	1.67E+09	4.08E+08		803.39	196.61	100
	R + A		2.08E+09			1000	

f = 3.759E-06 Pa
Total Amt= 26436 kg

Overall residence time = 26.44 h
Reaction time = 32.91 h
Advection time = 134.46 h

Figure 1.8 Fugacity Level II calculation for naphthalene in a generic environment (dimensions defined in Table 1.2).

40

Level III

The Level III calculation includes an estimation of intermedia transport. Examination of the magnitude of the intermedia D values given in the fate diagram (Figure 1.9 and Figure 1.10, which are the same as Figure 2.9 and Figure 2.10 in Chapter 2) suggests that air-water and air-soil exchange are most important with water-sediment and soil-water transport being slower in potential transfer rate. The magnitude of these larger intermedia transport D values (approximately 10^7 mol/Pa $\cdot$ h) compared to the atmospheric reaction and advection values of 10^8 to 10^9 suggests that reaction and advection will be very fast relative to transport.

The bulk Z values are similar for air and water to the values for the "pure" phases in Level I and II, but they are lower for soil and sediment because of the "dilution" of the solid phase with air or water.

The first row of figures at the tables at the foot of Figure 1.9 describes the condition if 1000 kg/h is emitted into the air. The result is similar to the Level II calculation with 19600 kg in air, 542 kg in water, 924 kg in soil and only 25 kg in sediment. It can be concluded that naphthalene discharged to the atmosphere has very little potential to enter other media. The rates of transfer from air to water is only about 4 kg/h. Even if the transfer coefficients were increased by a factor of 10, the rates would remain negligible. The reason for this is the value of the mass transfer coefficients which control this transport process. The overall residence time is 21 hours, similar to Level II.

If 1000 kg/h of naphthalene is discharged to water, as in the second row, there is predictably a much higher concentration in water (by a factor of nearly 300). There is reaction of 612 kg/h in water, advective outflow of 150 kg/h and transfer to air of 238 kg/h with negligible loss of 2 kg/h to sediment. The amount in the water is 150000 kg, thus the residence time in the water is 150 hours and the overall environmental residence time is a longer 162 hours. The key processes are thus reaction in water (half-life 170 h), evaporation (half-life 440 h) and advective outflow (residence time 1000 h). The evaporation half-life can be calculated as (0.693 x mass in water)/rate of transfer, i.e., (0.693 x 150000)/238 = 440 h. Clearly competition between reaction and evaporation in the water determines the overall fate. 93 % of the naphthalene discharged is now found in the water and the concentration is a fairly high, namely 7.5 x 10^{-4} g/m^3 or 750 ng/L.

The third row shows the fate if discharge is into soil. The amount in soil is 2 x 10^6 kg, reflecting an overall 2000 h residence time. The rate of reaction in soil is only 824 kg/h, there is no advection, thus the other loss mechanism is transfer to air (T_{31}) at a rate of 154 kg/h, with a relatively minor 22 kg/h to water by run-off. The soil concentration of 0.11 g/m^3 is controlled almost entirely by the rate at which the naphthalene reacts.

The net result is that naphthalene behaves entirely differently when discharged to the three media. If discharged to air it reacts rapidly and advects with a residence time of 21 h or about 1 day with little transport to soil or water. If discharged to water it reacts and evaporates to air

Fugacity Level III calculations: (four compartment model)

Chemical name: Naphthalene

Phase Properties and Rates:

Compartment	Bulk Z mol/m3 Pa	Half-life h	D Values Reaction mol/Pa h	Advection mol/Pa h
Air (1)	4.034E-04	17	1.64E+09	4.03E+08
Water (2)	2.329E-02	170	1.90E+07	4.66E+06
Soil (3)	5.434E-01	1700	3.99E+06	—
Sediment (4)	4.477E-01	5500	2.82E+04	4.48E+03

	E(1)=1000	E(2)=1000	E(3)=1000	E(1,2,3)
Overall residence time =	21.15	161.82	2029.01	264.13 h
Reaction time =	26.33	201.50	2101.70	322.38 h
Advection time =	107.38	821.81	58664.77	1461.91 h

EMISSION (E)
REACTION (R)
ADVECTION (A)
TRANSFER D VALUE mol/Pa h

Phase Properties, Compositions, Transport and Transformation Rates:

Emission, kg/h

E(1)	E(2)	E(3)	f(1)	f(2)	f(3)	f(4)	C(1)	C(2)	C(3)	C(4)	m(1)	m(2)	m(3)	m(4)	Total Amount, kg
1000	0	0	3.797E-06	9.074E-07	7.511E-07	8.554E-07	1.964E-07	2.709E-06	5.233E-05	4.909E-05	1.964E+04	5.418E+02	9.419E+02	2.455E+01	2.115E+04
0	1000	0	9.019E-07	2.514E-04	1.784E-07	2.370E-04	4.664E-08	7.507E-04	1.243E-05	1.360E-02	4.664E+03	1.501E+05	2.237E+02	6.802E+03	1.618E+05
0	0	1000	6.038E-07	5.629E-06	1.613E-03	5.307E-06	3.122E-08	1.680E-05	1.124E-01	3.045E-04	3.122E+03	3.361E+03	2.022E+06	1.523E+02	2.029E+06
600	300	100	2.609E-06	7.654E-05	1.618E-04	7.216E-05	1.349E-07	2.285E-04	1.127E-02	4.141E-03	1.349E+04	4.570E+04	2.029E+05	2.071E+03	2.641E+05

Intermedia Rate of Transport, kg/h

T12	T13	T21	T24	T31	T32	T42
air-water	air-soil	water-air	water-sed	soil-air	soil-water	sed-water
3.602E+00	4.658E-01	8.578E-01	6.604E-03	7.164E-02	1.017E-02	3.020E-03
8.554E-01	1.106E-01	2.377E+02	1.830E+00	1.702E-02	2.415E-03	8.369E-01
5.726E-01	7.405E-02	5.322E+00	4.097E-02	1.538E+02	2.183E+01	1.874E-02
2.475E+00	3.200E-01	7.236E+01	5.571E-01	1.543E+01	2.190E+00	2.548E-01

Emission, kg/h

E(1)	E(2)	E(3)	R(1)	R(2)	R(3)	R(4)	A(1)	A(2)	A(3)	A(4)
1000	0	0	8.005E+02	2.208E+01	3.84E-01	3.093E-03	1.964E+02	5.418E-01	4.909E-04	—
0	1000	0	1.901E+02	6.120E+02	9.12E-02	8.571E-01	4.664E+01	1.501E+02	1.360E-01	—
0	0	1000	1.273E+02	1.370E+01	8.24E+02	1.919E-02	3.122E+01	3.361E+00	3.045E-02	—
600	300	100	5.501E+02	1.863E+02	8.27E+01	2.609E-01	1.349E+02	4.570E+01	4.141E-01	—

Figure 1.9   Fugacity Level III calculation for naphthalene in a generic environment (dimensions defined in Table 1.2).

42

Level III Distribution

Chemical name: Naphthalene

Distribution of mass

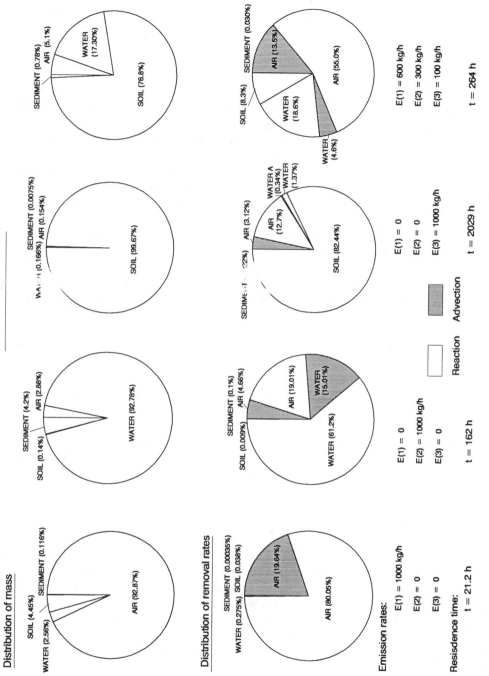

Distribution of removal rates

Emission rates:

E(1) = 1000 kg/h
E(2) = 0
E(3) = 0

Resisdence time:
t = 21.2 h

E(1) = 0
E(2) = 1000 kg/h
E(3) = 0

t = 162 h

E(1) = 0
E(2) = 0
E(3) = 1000 kg/h

t = 2029 h

E(1) = 600 kg/h
E(2) = 300 kg/h
E(3) = 100 kg/h

t = 264 h

Figure 1.10 Fugacity level III distributions of naphthalene for four emission scenarios in a generic environment.

43

with a residence time of 162 h or 1 week. If discharged to soil it mostly reacts with a residence time in soil of about 2000 h or 3 months.

The final scenario is a combination of discharges, 600 kg/h to air, 300 kg/h to water, and 100 kg/h to soil. The concentrations, amounts and transport and transformation rates are merely linearly combined versions of the three initial scenarios. For example, the rate of reaction in air is now 550 kg/h. This is 0.6 of the first (air emission) rate of 800 kg/h, i.e., 480 kg/h, plus 0.3 of the second (water emission) rate of 190 kg/h, i.e., 57 kg/h and 0.1 of the third (soil emission) rate of 127 kg/h, i.e., 13 kg yielding a total of (480 + 57 + 13) or 550 kg/h. It is also apparent that the amount in the air of 13500 kg causing a concentration of 0.135 $\mu g/m^3$ is attributable to emissions to air (0.6 x 0.196 or 0.118 $\mu g/m^3$), emissions to water (0.3 x 0.047 or 0.014 $\mu g/m^3$) and emissions to soil (0.1 x 0.031 or 0.003 $\mu g/m^3$). The concentration in water of 2.29 x 10^{-4} g/m^3 or 229 $\mu g/m^3$ or ng/L is largely attributable to the discharges to water which alone cause 0.3 x 751 or 225 $\mu g/m^3$. Although more is emitted to air it contributes about 1.6 $\mu g/m^3$ to the water with soil emissions accounting for about 1.6 $\mu g/m^3$. Similarly, the prevailing soil concentration is controlled by the rate of discharge to the soil.

In this multimedia discharge scenario the overall residence time is 264 hours, which can be viewed as 60% of the air residence time of 21 h, 30% of the water residence time of 162 h and 10% of the overall soil residence time of 2029 h. The overall amount in the environment of 264,100 kg is thus largely controlled by the discharges to soil which account for (0.1 x 264100) or 20,290 kg or 77% of the total.

Figure 1.10 shows the distributions of mass and removal process rates for these four scenarios. Clearly when naphthalene is discharged into a specific medium, most of the chemical is found in that medium.

Finally, it is interesting to note that the fugacities in this final case (in units of μPa) are for the four media 2.6 (air), 76.6 (water), 162 (soil) and 72 (sediment). The soil, sediment and water are fairly close to equilibrium, i.e., within a factor of about 2, with the air notably "undersaturated" by a factor of about 30. This is the result of the rapid loss processes from air.

It is believed that these three behavior profiles when combined in the fourth give a comprehensive illustration and explanation of the environmental fate characteristics of this and other chemicals. They show which intermedia transport processes are important and how levels in various media arise from discharges into other media. The same broad fate characteristics as described in the generic environment are believed to be generally applicable to other environments. Certainly this evaluation should, in most cases, identify the key physical-chemical properties, reactions and intermedia transport parameters. With a knowledge of the key parameters, more effort can be devoted to obtaining more accurate site-specific values, and sensitivity analyses can be conducted. In essence, the evaluation translates physical-chemical data into environmental fate information.

1.6 REFERENCES

Abernethy, S., Mackay, D., McCarty, L.S. (1988) "Volume fraction" correlation for narcosis in aquatic organisms: The key role or partitioning. *Environ. Toxicol. Chem.* 7, 469-481.

Abramowitz, R., Yalkowsky, S.H. (1990) Estimation of aqueous solubility and melting point of PCB congeners. *Chemosphere* 21, 1221-1229.

Almgren, M., Grieser, F., Powell, J.R., Thomas, J.K. (1979) A correlation between the solubility of aromatic hydrocarbons in water and micellar solutions, with their normal boiling points. *J. Chem. Eng. Data* 24, 285-287.

Al-Sahhaf, T.A. (1989) Prediction of the solubility of hydrocarbons in water using UNIFAC. *J. Environ. Sci. Health* A24, 49-56.

Ambrose, D. (1981) Reference value of vapor pressure. The vapor pressures of benzene and hexafluorobenzene. *J. Chem. Thermodyn.* 13, 1161-1167.

Ambrose, D., Lawrenson, L.J., Sprake, C.H.S. (1975) The vapour pressure of naphthalene. *J. Chem. Thermodyn.* 7, 1173-1176.

Amidon, G.L., Williams, N.A. (1982) A solubility equation for non-electrolytes in water. *Intl. J. Pharm.* 11, 249-156.

Amidon, G.L., Anik, S.T. (1981) Application of the surface area approach to the correction and estimation of aqueous solubility and vapor pressure. Alkyl aromatic hydrocarbons. *J. Chem. Eng. Data* 26, 28-33.

Anderson, E., Veith, G.D., Weininger, D. (1987) *SMILES: A Line Notation and Computerized Interpreter for Chemical Structures.* US EPA Environmental Research Brief, EPA/600/M-87/021.

Andren, A.W., Doucette, W.J., Dickhut, R.M. (1987) Methods for estimating solubilities of hydrophobic organic compounds: Environmental modeling efforts. In: *Sources and Fates of Aquatic Pollutants.* Hites, R.A., Eisenreich, S.J., Eds., pp. 3-26, Advances in Chemistry Series 216, American Chemical Society, Washington, D.C.

Andrews, L.J., Keffer, R.M. (1950a) Cation complexes of compounds containing carbon-carbon double bonds. IV. The argentation of aromatic hydrocarbons. *J. Am. Chem. Soc.* 72, 3644-3647.

Andrews, L.J., Keefer, R.M. (1950b) Cation complexes of compounds containing carbon-carbon double bonds. VII. Further studies on the argentation of substituted benzenes. *J. Am. Chem. Soc.* 72, 5034-5037.

Arbuckle, W.B. (1983) Estimating activity coefficients for use in calculating environmental parameters. *Environ. Sci. Technol.* 17, 537-542.

Arbuckle, W.B. (1986) Using UNIFAC to calculate aqueous solubilities. *Environ. Sci. Technol.* 20, 1060-1064.

Ashworth, R.A., Howe, G.B., Mullins, M.E., Roger, T.N. (1988) Air-water partitioning coefficients of organics in dilute aqueous solutions. *J. Hazard. Materials* 18, 25-36.

Balson, E.W. (1947) Studies in vapour pressure measurement. Part III. An effusion manometer sensitive to 5×10^{-6} millimetres of mercury: vapour pressure of D.D.T. and other slightly volatile substances. *Trans. Farad. Soc.* 43, 54-60.

Banerjee, S. (1985) Calculation of water solubility of organic compounds with UNIFAC-derived parameters. *Environ. Sci. Technol.* 19, 369-370.

Banerjee, S., Howard, P.H. (1988) Improved estimation of solubility and partitioning through correction of UNIFAC-derived activity coefficients. *Environ. Sci. Technol.* 22, 839-841.

Banerjee, S., Howard, P.H., Lande, S.S. (1990) General structure-vapor pressure relationships for organics. *Chemosphere* 21, 1173-1180.

Banerjee, S., Yalkowsky, S.H., Valvani, S.C. (1980) Water solubiltiy and octanol/water partition coefficients of organics. Limitations of the solubility-partition coefficient correlation. *Environ. Sci. Technol.* 14, 1227-1229.

Bidleman, T.F. (1984) Estimation of vapor pressures for nonpolar organic compounds by capillary gas chromatography. *Anal. Chem.* 56, 2490-2496.

Bohon, R.L., Claussen, W.F. (1951) The solubility of aromatic hydrocarbons in water. *J. Am. Chem. Soc.* 73, 1571-1576.

Bondi, A. (1964) van der Waals volumes and radii. *J. Phys. Chem.* 68, 441-451.

Booth, H.S., Everson, H.E. (1948) Hydrotropic solublities: solublities in 40 percent sodium xylenesulfonate. *Ind. Eng. Chem.* 40, 1491-1493.

Boublik, T., Fried, V., Hala, E. (1973) *The Vapor Pressure of Pure Substances*, Elsevier, Amsterdam.

Boublik, T., Fried, V., Hala, E. (1984) *The Vapor Pressure of Pure Substances*, 2nd revised ed., Elsevier, Amsterdam.

Bowman, B.T., Sans, W.W. (1983) Determination of octanol-water partitioning coefficient (K_{ow}) of 61 organophosphorus and carbamate insecticides and their relationship to respective water solubility (S) values. *J. Environ. Sci. Health* B18, 667-683.

Bradley, R.S., Cleasby, T.G. (1953) The vapour pressure and lattice energy of some aromatic ring compounds. *J. Chem. Soc.* 1953, 1690-1692.

Briggs, G.G. (1981) Theoretical and experimental relationships between soil adsorption, octanol-water partition coefficients, water solubilities, bioconcentration factors, and the Parachor. *J. Agric. Food Chem.* 29, 1050-1059.

Bruggeman, W.A., van der Steen, J., Hutzinger, O. (1982) Reversed-phase thin-layer chromatography of polynuclear aromatic hydrocarbons and chlorinated biphenyls. Relationship with hydrophobicity as measured by aqueous solubility and octanol-water partition coefficient. *J. Chromatogr.* 238, 335-346.

Budavari, S., Ed. (1989) *The Merck Index. An Encylopedia of Chemicals, Drugs and Biologicals.* 11th ed., Merck & Co. Inc., Rahway, New Jersey.

Burkhard, L.P. (1984) *Physical-Chemical Properties of the Polychlorinated Biphenyls: Measurement, Estimation, and Application to Environmental Systems.* Ph.D. Thesis, University of Wisconsin-Madison, Wisconsin.

Burkhard, L.P., Andren, A.W., Armstrong, D.E. (1985a) Estimation of vapor pressures for polychlorinated biphenyls: A comparison of eleven predictive methods. *Environ. Sci. Technol.* 19, 500-507.

Burkhard, L.P., Armstrong, D.E., Andren, A.W. (1985b) Henry's law constants for polychlorinated biphebyls. *Environ. Sci. Technol.* 590-595.

Burkhard, L.P., Kuehl, D.W., Veith G.D. (1985c) Evaluation of reversed phase liquid chromatograph/mass spectrophotometry for estimation of n-octanol/water partition coefficients of organic chemicals. *Chemosphere* 14, 1551-1560.

Chiou, C.T. (1981) Partition coefficient and water solubility in environmental chemistry. In: *Hazard Assessment of Chemicals Current Developments*. Vol. 1, pp. 117-153. Academic Press, New York.

Chiou, C.T. (1985) Partition coefficients of organic compounds in lipid-water systems and correlations with fish bioconcentration factors. *Environ. Sci. Technol.* 19, 57-62.

Chiou, C.T., Freed, V.H., Schmedding, D.W. (1977) Partition coefficient and bioaccumulation of selected organic chemicals. *Environ. Sci. Technol.* 11, 475-478.

Chiou, C.T., Schmedding, D.W., Manes, M. (1982) Partitioning of organic compounds in octanol-water system. *Environ. Sci. Technol.* 16, 4-10.

Davies, R.P., Dobbs, A.J. (1984) The prediction of bioconcentration in fish. *Water Res.* 18, 1253-1262.

Davis, W.W., Krahl, M.E., Clowes, G.H. (1942) Solubility of carcinogenic and related hydrocarbons in water *J. Am. Chem. Soc.* 64, 108-110.

Davis, W.W., Parke, Jr, T.V. (1942) A nephelometric method for determination of solubilities of extremely low order. *J. Am. Chem. Soc.* 64, 101-107.

Dean, J.D., Ed. (1979) *Lange's Handbook of Chemistry*. 12th ed., McGraw-Hill, New York, N.Y.

Dean, J.D., Ed. (1985) *Lange's Handbook of Chemistry*. 13th ed., McGraw-Hill, New York, N.Y.

Dearden, J.C. (1990) Physico-chemical descriptors. In: *Practical Applications of Quantitative Structure-Activity Relationships (QSAR) in Environmental Chemistry and Toxicology*. Karcher, W. and Devillers, J., Eds., pp. 25-60. Kluwer Academic Publisher, Dordrecht, Netherlands.

De Bruijn, J., Busser, G., Seinen, W., Hermens, J. (1989) Determination of octanol/water partition coefficient for hydrophobic organic chemicals with the "slow-stirring" method. *Environ. Toxicol. Chem.* 8, 499-512.

De Bruijn, J., Hermens, J. (1990) Relationships between octanol/water partition coefficients and total molecular surface area and total molecular volume of hydrophobic organic chemicals. *Quant. Struct.-Act. Relat.* 9, 11-21.

Dobbs, A.J., Grant, C. (1980) Pesticide volatilization rate: a new measurement of the vapor pressure of pentachlorophenol at room temperature. *Pestic. Sci.* 11, 29-32.

Dobbs, A.J., Cull, M.R. (1982) Volatilization of chemical relative loss rates and the estimation of vapor pressures. *Environ. Pollut. (Ser. B)* 3, 289-298.

Doucette, W.J., Andren, A.W. (1987) Correlation of octanol/water partition coefficients and total molecular surface area for highly hydrophobic aromatic compounds. *Environ. Sci. Technol.* 21, 521-524.

Doucette, W.J., Andren, A.W. (1988a) Aqueous solubility of selected biphenyl, furan, and dioxin congeners. *Chemosphere* 17, 243-252.

Doucette, W.J., Andren, A.W. (1988b) Estimation of octanol/water partition coefficients: Evaluation of six methods for highly hydrophobic aromatic hydrocarbons. *Chemosphere* 17, 345-359.

Dunnivant, F.M., Coate, J.T., Elzerman, A.W. (1988) Experimentally determined Henry's law constants for 17 polychlorobiphenyl congeners. *Environ. Sci. Technol.* 22, 448-453.

Fendinger, N.J., Glotfelty, D.E. (1988) A laboratory method for the experimental determination of air-water Henry's law constants for several pesticides. *Environ. Sci. Technol.* 22, 1289-1293.

Fendinger, N.J., Glotfelty, D.E. (1990) Henry's law constants for selected pesticides, PAHs and PCBs. *Environ. Toxicol. Chem.* 9, 731-735.

Foreman, W.T., Bidleman, T.F. (1985) Vapor pressure estimates of individual polychlorinated biphenyls and commercial fluids using gas chromatographic retention data. *J. Chromatogr.* 330, 203-216.

Fredenslund, A., Jones, R.L., Prausnitz, J.M. (1975) Group-contribution estimation of activity coefficients in nonideal liquid mixtures. *AIChE J.* 21, 1086-1099.

Fujita, T., Iwasa, J., Hansch, C. (1964) A new substituent constant, "pi" derived from partition coefficients. *J. Am. Chem. Soc.* 86, 5175-5180.

Garst, J.E. (1984) Accurate, wide-range, automated, high-performance chromatographic method for the estimation of octanol/water partition coefficients. II: Equilibrium in partition coefficient measurements, additivity of substituent constants, and correlation of biological data, *J. Pharm. Sci.* 73, 1623-1629.

Garten, C.T., Trabalka, J.R. (1983) Evaluation of models for predicting terrestrial food chain behavior of xenobiotics. *Environ. Sci. Technol.* 17, 590-595.

Gossett, R. (1987) Measurement of Henry's law constants for C_1 and C_2 chlorinated hydrocarbons. *Environ. Sci. Technol.* 21, 202-208.

Gross, P.M., Saylor, J.H. (1931) The solubilities of certain slightly soluble organic compounds in water. *J. Am. Chem. Soc.* 1931, 1744-1751.

Gückel, W., Rittig, R., Synnatschke, G. (1974) A method for determining the volatility of active ingredients used in plant protection. II. Application to formulated products. *Pestic. Sci.* 5, 393-400.

Gückel, W. Kästel, R., Lawerenz, J., Synnatschke, G. (1982) A method for determining the volatility of active ingredients used in plant protection. Part III: The temperature relationship between vapour pressure and evaporation rate. *Pestic. Sci.* 13, 161-168.

Hafkenscheid, T.L., Tomlinson, E. (1981) Estimation of aqueous solubilities of organic non-electrolytes using liquid chromatographic retention data. *J. Chromatogr.* 218, 409-425.

Hamaker, J.W. Kerlinger, H.O. (1969) Vapor pressures of pesticides. *Adv. Chem. Ser.* 86, 39-54.

Hamliton, D.J. (1980) Gas chromatographic measurement of volatility of herbicide esters. *J. Chromatogr.* 195, 75-83.

Hansch, C., Leo, A. (1979) *Substituent Constants for Correlation Analysis in Chemistry and Biology.* Wiley-Interscience, New York, N.Y.

Hansch, C., Quinlan, J.E., Lawrence, G.L. (1968) The linear-free energy relationship between partition coefficient and aqueous solubility of organic liquids. *J. Org. Chem.* 33, 347-350.

Hawker, D.W. (1989) The relationship between octan-1-ol/water partition coefficient and aqueous solubility in terms of solvatochromic parameters. *Chemosphere* 19, 1586-1593.

Hawker, D.W. (1990a) Vapor pressures and Henry's law constants of polychlorinated biphenyls. *Environ. Sci. Technol.* 23, 1250-1253.

Hawker, D.W. (1990b) Description of fish bioconcentration factors in terms of solvatochromic parameters. *Chemosphere* 20, 267-477.

Hawker, D.W., Connell, D.W. (1988) Octanol-water partition coefficients of polychlorinated biphenyl congeners. *Environ. Sci. Technol.* 22, 382-387.

Hermann, R.B. (1971) Theory of hydrophobic bonding. II. The correlation of hydrocarbon solubility in water with solvent cavity surface area. *J. Phys. Chem.* 76, 2754-2758.

Hinckley, D.A., Bidleman, T.F., Foreman, W.T. (1990) Determination of vapor pressures for nonpolar and semipolar organic compounds from gas chromatographic retention data. *J. Chem. Eng. Data* 35, 232-237.

Hollifield, H.C. (1979) Rapid nephelometric estimate of water solubility of highly insoluble organic chemicals of environmental interest. *Bull. Environ. Contam. Toxicol.* 23, 579-586.

Horvath, A.L. (1982) *Halogenated Hydrocarbons, Solubility - Miscibility with Water.* Marcel Dekker, Inc., New York, N.Y.

Howard, P.H., Ed. (1989) *Handbook of Fate and Exposure Data for Organic Chemicals. Vol. I. Large Production and Priority Pollutants.* Lewis Publishers, Chelsea, Michigan.

Howard, P.H., Ed. (1990) *Handbook of Fate and Exposure Data for Organic Chemicals. Vol. - II - Solvents.* Lewis Publishers, Inc., Chelsea, MI.

Howard, P.H., Boethling, R.S., Jarvis, W.F., Meylan, W.M., Michalenko, E.M. (1991) *Handbook of Environmental Degradation Rates.* Lewis Publishers, Inc., Chelsea, Michigan.

Hussam, A, & Carr, P.W. (1985) A study of a rapid and precise methodology for the measurement of vapor liquid equilibria by headspace gas chromatography. *Anal. Chem.* 57, 793-801.

Isnard, P., Lambert, S. (1988) Estimating bioconcentration factors for octanol-water partition coefficient and aqueous solubility. *Chemosphere* 17, 21-34.

Isnard, P., Lambert, S. (1989) Aqueous solubility/n-octanol-water partition coefficient correlations. *Chemosphere* 18, 1837-1853.

IUPAC Solubility Data Series (1989a) *Vol. 37: Hydrocarbons (C_5 - C_7) with Water and Seawater.* Shaw, D.G., Ed., Pergamon Press, Oxford, England.

IUPAC Solubility Data Series (1989b) *Vol. 38: Hydrocarbons (C_8 -C_{36}) with Water and Seawater.* Shaw, D.G., Ed., Pergamon Press, Oxford, England.

Jury, W.A., Spencer, W.F., Farmer, W.J. (1983) Behavior assessment model for trace organics in soil: I. Model description. *J. Environ. Qual.* 12, 558-566.

Jury, W.A., Farmer, W.J., Spencer, W.F. (1984a) Behavior assessment model for trace organics in soil: II. Chemical classification and parameter sensitivity. *J. Environ. Qual.* 13, 567-572.

Jury, W.A., Farmer, W.J., Spencer, W.F. (1984b) Behavior assessment model for trace organics in soil: III. Application of screening model. *J. Environ. Qual.* 13, 573-579.

Jury, W.A., Spencer, W.F., Farmer, W.J. (1984) Behavior assessment model for trace organics in soil: IV. Review of experimental evidence. *J. Environ. Qual.* 13, 580-587.

Kabadi, V.N., Danner, R.P. (1979) Nomograph solves for solubilities of hydrocarbons in water. *Hydrocarbon Processing* 68, 245-246.

Kamlet, M.J., Doherty, R.M., Abraham, M.H., Carr, P.W., Doherty, R.F., Raft, R.W. (1987) Linear solvation energy relationships. Important differences between aqueous solublity relationships for aliphatic and aromatic solutes. *J. Phy. Chem.* 91, 1996.

Kamlet, M.J., Doherty, R.M., Carr, P.W., Mackay, D., Abraham, M.H., Taft, R.W. (1988) Linear solvation energy relationships. 44. Parameter estimation rules that allow accurate prediction of octanol/water partition coefficients and other solubility and toxicity properties of polychlorinated biphenyls and polycyclic aromatic hydrocarbons. *Environ. Sci. Technol.* 22, 503-509.

Karcher, W., Devillers, J., Eds., (1990) *Practical Applications of Quantitative-Structure-Activity Relationships (QSAR) in Environmental Chemistry and Toxicology.* Kluwer Academic Publisher, Dordrecht, Netherlands.

Karickhoff, S.W. (1981) Semiempirical estimation of sorption of hydrophobic pollutants on natural sediments and soil. *Chemosphere* 10, 833-846.

Karickhoff, S.W., Brown, D.S., Scott, T.A. (1979) Sorption of hydrophobic pollutants on natural water sediments. *Water Res.* 13, 241-248.

Kenaga, E.E. (1980) Predicted bioconcentration factors and soil sorption coefficients of pesticides and other chemicals. *Ecotox. Environ. Saf.* 4, 26-38.

Kenaga, E.E., Goring, C.A.I. (1980) Relationship between water solubility, soil sorption, octanol-water partitioning, and concentration of chemicals in biota. In: *Aquatic Toxicology.* Eaton, J.G., Parrish, P.R., Hendrick, A.C., Eds., pp. 78-115, Am. Soc. for Testing and Materials, STP 707, Philadelphia.

Kier, L.B., Hall, L.H. (1976) Molar properties and molecular connectivity. In: *Molecular Connectivity in Chemistry and Drug Design.* Medicinal Chem. Vol. 14, pp. 123-167, Academic Press, New York.

Kier, L.B., Hall, L.H. (1986) *Molecular Connectivity in Structure-Activity Analysis.* Wiley, New York.

Kim, Y.-H., Woodrow, J.E., Seiber, J.N. (1984) Evaluation of a gas chromatographic method for calculating vapor pressures with organophosphorus pesticides. *J. Chromotagr.* 314, 37-53.

Könemann, H., van Leeuewen, K. (1980) Toxicokinetics in fish: accumulation of six chlorobenzenes by guppies. *Chemosphere* 9, 3-19.

Könemann, H., Zelle, R., Busser, F. (1979) Determination of log P_{oct} values of chloro-substituted benzenes, toluenes and anilines by high-performance liquid chromatography on ODS-silica. *J. Chromatogr.* 178, 559-565.

Lande, S.S., Banerjee, S. (1981) Predicting aqueous solubility of organic nonelectrolytes from molar volume. *Chemosphere* 10, 751-759.

Leahy, D.E. (1986) Intrinsic molecular volume as a measure of the cavity term in linear solvation energy relationships: octanol-water partition coefficients and aqueous solubilities. *J. Pharm. Sci.* 75, 629-636.

Leo, A., Hansch, C., Elkins, D. (1971) Partition coefficients and their uses. *Chem. Rev.* 71, 525-616.

Lincoff, A.H., Gossett, J.M. (1984) The determination of Henry's law constants for volatile organics by equilibrium partitioning in closed systems. In: *Gas Transfer at Water Surfaces*. Brutsaert, W., Jirka, G.H., Eds., pp. 17-26, D. Reidel Publishing Co., Dordrecht, Holland.

Locke, D. (1974) *J. Chromatogr. Sci.* 12, 433.

Lyman, W.J., Reehl, W.F., Rosenblatt, D.H. (1982) *Handbook of Chemical Property Estimation Methods*. McGraw-Hill, New York.

Mackay, D. (1979) Finding fugacity feasible. *Environ. Sci. Technol.* 13, 1218-1223.

Mackay, D. (1982) Correlation of bioconcentration factors. *Environ. Sci. Technol.* 16, 274-278.

Mackay, D. (1991) *Multimedia Environmental Models. The Fugacity Approach.* Lewis Publishers, Inc., Chelsea, Michigan.

Mackay, D., Bobra, A.M., Shiu. W.Y., Yalkowsky, S.H. (1980) Relationships between aqueous solubility and octanol-water partition coefficient. *Chemosphere* 9, 701-711.

Mackay, D., Bobra, A.M., Chan, D.W., Shiu, W.Y. (1982) Vapor pressure correlation for low-volatility environmental chemicals. *Environ. Sci. Technol.* 16, 645-649.

Mackay, D., Paterson, S. (1990) Fugacity models. In: *Practical Applications of Quantitative Structure-Activity Relationships (QSAR) in Environmental Chemistry and Toxicology.* Karcher, W., Devillers, J., Eds., pp. 433-460, Kluwer Academic Publishers, Dordrecht, Holland.

Mackay, D., Paterson, S. (1991) Evaluating the multimedia fate of organic chemicals: A Level III fugacity model. *Environ. Sci. Technol.* 25, 427-436.

Mackay, D., Shiu, W.Y. (1977) Aqueous solubility of polynuclear aromatic hydrocarbons. *J. Chem. Eng. Data* 22, 339-402.

Mackay, D., Shiu, W.Y. (1981) A critical review of Henry's law constants for chemicals of environmental interest. *J. Phys. Chem. Ref. Data* 11, 1175-1199.

Mackay, D., Shiu, W.Y., Sutherland, R.P. (1979) Determination of air-water Henry's law constants for hydrophobic pollutants. *Envrion. Sci. Technol.* 13, 333-337.

Mackay, D., Shiu, W.Y., Wolkoff, A.W. (1975) Gas chromatographic determination of low concentration of hydrocarbons in water by vapor phase extration. *ASTM STP 573*, pp. 251-258, American Society for Testing and Materials, Philadelphia, Pa.

Mackay, D., Stiver, W.H. (1991) Predictability and environmental chemistry. In: *Environmental Chemistry of Herbicides*. Vol. II. Grover, R., Lessna, A.J., Eds., pp. 281-297. CRC Press, Boca Raton, FL.

Macknick, A.B., Prausnitz, J.M. (1979) Vapor pressure of high-molecular weight hydrocarbons. *J. Chem. Eng. Data* 24, 175-178.

Mailhot, H., Peters, R.H. (1988) Empirical relationships between the 1-octanol/water partition coefficient and nine physicochemical properties. *Environ. Sci. Technol.* 22, 1479-1488.

May, W.E., Wasik, S.P., Freeman, D.H. (1978a) Determination of the aqueous solubility of polynuclear aromatic hydrocarbons by a coupled-column liquid chromatographic technique. *Anal. Chem.* 50, 175-179.

May, W.E., Wasik, S.P., Freeman, D.H. (1978b) Determination of the solubility behavior of some polycyclic aromatic hydrocarbons in water. *Anal. Chem.* 50, 997-1000.

McAuliffe, C. (1966) Solubility in water of paraffin, cycloparaffin, olefin, acetylene, cycloolefin and aromatic hydrocarbons. *J. Phys. Chem.* 76, 1267-1275.

McDuffie, B. (1981) Estimation of octanol/water partition coefficient for organic pollutants using reversed phase HPLC. *Chemosphere* 10, 73-83.

McGowan, J.C., Mellors, A. (1986) *Molecular Volumes in Chemistry and Biology-Applications including Partitioning and Toxicity*. Ellis Horwood Limited, Chichester, England.

The Merck Index (1989) *An Encyclopedia of Chemicals, Drugs amd Biologicals*. 11th ed., Budavari, S., Ed., Merck and Co., Inc., Rahway, N.J.

Meylan, W.M., Howard, P.H. (1991) Bond contributin method for estimating Henry's law constants. *Environ. Toxicol. Chem.* 10, 1283-1293.

Miller, M.M., Ghodbane, S., Wasik, S.P., Tewari, Y.B., Martire, D.E. (1984) Aqueous solubilities, octanol/water partition coefficients and entropies of melting of chlorinated benzenes and biphenyls. *J. Chem. Eng. Data* 29, 184-190.

Miller, M.M., Wasik, S.P., Huang, G.L., Shiu, W.Y., Mackay, D. (1985) Relationships between octanol-water partition coefficient and aqueous solublity. *Environ. Sci. Technol.* 19, 522-529.

Neely, W.B., Branson, D.R., Blau, G.E. (1974) Partition coefficient to measure bioconcentration potential of organic chemicals in fish. *Environ. Sci. Technol.* 8, 1113-1115.

Nirmalakhandan, N.N., Speece, R.E. (1988a) Prediction of aqueous solubility of organic chemicals based on molecular structure. *Environ. Sci. Technol.* 22, 328-338.

Nirmalakhandan, N.N., Speece, R.E. (1988b) QSAR model for predicting Henry's law constant. *Environ. Sci. Technol.* 22, 1349-1357.

Nirmalakhandan, N.N., Speece, R.E. (1989) Prediction of aqueous solubility of organic chemicals based on molecular structure. 2. Application to PNAs, PCBs, PCDDs, etc. *Environ. Sci. Technol.* 23, 708-713.

Oliver, B.G. (1984) The relationship between bioconcentration factor in rainbow trout and physical-chemical properties for some halogenated compounds in: *QSAR in Environmental Toxicology*. Kaiser, K.L.E., Ed., pp. 300-317, D. Reidel Publishing Co., Dordrecht, Holland.

Oliver, B.G., Niimi, A.J. (1988) Trophodynamic analysis of polychlorinated biphenyl congeners and other chlorinated hydrocarbons in the Lake Ontario ecosystem. *Environ. Sci. Technol.* 22, 388-397.

Osborn, A.G., Douslin, D.R. (1975) Vapor pressures and derived enthalpies of vaporization of some condensed-ring hydrocarbons. *J. Chem. Eng. Data* 20, 229-231.

Paterson, S., Mackay, D. (1985) The fugacity concept in environmental modelling, In: *The Handbook of Environmental Chemistry.* Vol. 2/Part C, Hutzinger, O., Ed., pp. 121-140. Springer-Verlag, Heidelberg, Germany.

Pearlman, R.S. (1980) Molecular surface areas and volumes and their use in structure/activity relationships. In: *Physical Chemical Properties of Drugs.* Yalkowsky, S.H., Sinkula, A.A., Valvani, S.C., Eds., Medicinal Research Series, Vol. 10., pp. 321-317, Marcel Dekker, Inc., New York.

Pearlman, R.S. (1986) Molecular surface area and volume: Their calculation and use in predicting solubilities and free energies of desolvation. In: *Partition coefficient, Determination and Estimation.* Dunn III, W.J., Block, J.H., Pearlman R.S., Eds., pp. 3-20, Pergamon Press, New York.

Pearlman, R.S., Yalkowsky, S.H., Banerjee, S. (1984) Water solubilities of polynuclear aromatic and heteroaromatic compounds. *J. Phys. Chem. Ref. Data* 13, 555-562.

Pierotti, C., Deal, C., Derr, E. (1959) Activity coefficient and molecular structure. *Ind. Eng. Chem. Fundam.* 51, 95-101.

Rapaport, R.A., Eisenreich, S.J. (1984) Chromatographic determination of octanol-water partition coefficients (K_{ow}'s) for 58 polychlorinated biphenyl congeners. *Environ. Sci. Technol.* 18, 163-170.

Reid, R.C., Prausnitz, J.M., Polling, B.E. (1987) *The Properties of Gases and Liquids.* 4th ed., McGraw-Hill, New York, N.Y.

Rekker, R.F. (1977) *The Hydrophobic Fragmental Constant.* Elsevier, Amsterdam/New York N.Y.

Rordorf, B.F. (1985a) Thermodymanic and thermal properties of polyclorinated compounds: the vapor pressures and flow tube dinetic of ten dibenzo-para-diosins. *Chemosphere* 14, 885-892.

Rordorf, B.F. (1985b) Thermodynamic properties of polychlorinated ompounds: the vapor pressures and enthalpies of sublimation of ten dibenzo-p-dioxins. *Thermochimica Acta,* 85, 435-438.

Rordorf, B.F. (1986) Thermal properties of dioxins, furans and related compounds, *Chemosphere* 15, 1325-1332.

Sabljic, A. (1984) Predictions of the nature and strength of soil sorption of organic pollutants by molecular topology. *J. Agric. Food Chem.* 32, 243-246.

Sabljic, A. (1987) On the prediction of soil sorption coefficients of organic pollutants from molecular structure: Application of molecular topology model. *Environ. Sci. Technol.* 21, 358-366.

Sabljic, A., Lara, R., Ernst, W. (1989) Modelling association of highly chlorinated biphenyls with marine humic substances. *Chemosphere* 19, 1665-1676.

Sabljic, A., Güsten, H. (1989) Predicting Henry's law constants for polychlorinated biphenyls, *Chemosphere* 19, 1503-1511.

Sears, G.W., Hopke, E.R. (1947) Vapor pressures of naphthalene, anthracene and hexachlorobenzene in a low pressure region. *J. Am. Chem. Soc.* 71, 1632-1634.

Schwarzenbach, R.P., Westall, J. (1981) Transport of nonpolar compounds from surface water to groundwater. Laboratory sorption studies. *Environ. Sci. Technol.* 11, 1360-1367.

Shiu, W.Y., Mackay, D. (1986) A critical review of aqueous solubilities, vapor pressures, Henry's law constants, and octanol-water partition coefficients of the polychlorinated biphenyls. *J. Phys. Chem. Ref. Data* 15, 911-929.

Shiu, W.Y., Gobas, F.A.P.C., Mackay, D. (1987) Physical-chemical properties of three congeneric series of chlorinated aromatic hydrocarbons. In: *QSAR in Environmental Toxicology -II*. Kaiser, K.L.E., Ed., pp. 347-362. D. Reidel Publishing Co., Dordrecht, Holland.

Shiu, W.Y., Doucette, W., Gobas, F.A.P.C., Mackay, D., Andren, A.W. (1988) Physical-chemical properties of chlorinated dibenzo-p-dioxins. *Environ. Sci. Technol.* 22, 651-658.

Sinke, G.C. (1974) A method for measurement of vapor pressures of organic compounds below 0.1 torr. Naphthalene as reference substance. *J. Chem. Thermodyn.* 6, 311-316.

Sonnefeld, W.J., Zoller, W.H., May, W.E. (1983) Dynamic coupled-column liquid chromatographic determination of ambient temperature vapor pressures of polynuclear aromatic hydrocarbons. *Anal. Chem.* 55, 275-280.

Spencer, W.F., Cliath, M.M. (1969) Vapor density of dieldrin. *Environ. Sci. Technol.* 3, 670-674.

Spencer, W.F., Cliath, M.M. (1970) Vapor density and apparent vapor pressure of lindane (γ-BHC). *J. Agric. Food Chem.* 18, 529-530.

Spencer, W.F., Cliath, M.M. (1972) Volatility of DDT and related compounds. *J. Agric. Food Chem.* 20, 645-649.

Stephenson, R.M., Malanowski, A. (1987) *Handbook of the Thermodynamics of Organic Compounds*. Elsevier, New York.

Stiver, W., Mackay, D. (1989) The linear additivity principle in environmental modelling: Application to chemical behaviour in soil. *Chemosphere* 19, 1187-1198.

Suntio, L.R., Shiu, W.Y., Mackay, D. (1988) Critical review of Henry's law constants for pesticides. *Rev. Environ. Contam. Toxicol.* 103, 1-59.

Swann, R.L., Laskowski, D.A., McCall, P.J., Vander Kuy, K., Dishburger, H.J. (1983) A rapid method for the estimation of the environmental parameters octanol/water partition coefficient, soil sorption constant, water to air ratio, and water solubility. *Residue Rev.* 85, 17-28.

Szabo, G., Prosser, S., Bulman, R.A. (1990) Determination of the adsorption coefficient (K_{OC}) of some aromatics for soil by RP-HPLC on two immobilized humic acid phases. *Chemosphere* 21, 777-788.

Tomlinson, E., Hafkenscheid, T.L. (1986) Aqueous solution and partition coefficient estimation from HPLC data In: *Partition Coefficient, Determination and Estimation.* Dunn III, W.J., Block, J.H., Pearlman, R.S., Eds., pp. 101-141, Pergamon Press, New York.

Tsonopoulos, C., Prausnitz, J.M. (1971) Activity coefficients of aromatic solutes in dilute aqueous solutions. *Ind. Eng. Chem. Fundam.* 10, 593-600.

Tulp, M.T.M., Hutzinger, O. (1978) Some thoughts on the aqueous solubilities and partition coefficients of PCB, and the mathematical correlation between bioaccumulation and physico-chemical properties. *Chemosphere* 7, 849-760.

Veith, G.D., Austin, N.M., Morris, R.T. (1979) A rapid method for estimating log P for organic chemicals. *Water Res.* 13, 43-47.

Veith, G.D., Macek, K.J., Petrocelli, S.R., Caroll, J. (1980) An evaluation of using partition coefficients and water solubilities to estimate bioconcentration factors for organic chemicals in fish. In: *Aquatic Toxicology.* Eaton, J.G., Parrish, P.R., Hendrick, A.C., Eds, pp. 116-129, ASTM ATP 707, Am. Soc. for Testing and Materials, Philadelphia, Pa.

Verschueren, K. (1977) *Handbook of Environmental Data on Organic Chemicals.* Van Nostrand Reinhold, New York, N.Y.

Verschueren, K. (1983) *Handbook of Environmental Data on Organic Chemicals.* Van Nostrand Reinhold, New York, N.Y.

Warne, M., St. J., Connell, D.W., Hawker, D.W. (1990) Prediction of aqueous solubility and the octanol-water partition coefficient for lipophilic organic compounds using molecular descriptors and physicochemical properties. *Chemosphere* 16, 109-116.

Wasik, S.P., Miller, M.M., Tewari, Y.B., May, W.E., Sonnefeld, W.J., DeVoe, H., Zoller, W.H. (1983) Determination of the vapor pressure, aqueous solubility, and octanol/water partition coefficient of hydrophobic substances by coupled generator column/liquid chromatographic methods. *Res. Rev.* 85, 29-42.

Weast, R. (1972-73) *Handbook of Chemistry and Physics.* 53th ed., CRC Press, Cleveland, OH.

Weast, R. (1984) *Handbook of Chemistry and Physics.* 64th ed., CRC Press, Boca Raton, FL.

Weil, L., Dure, G., Quentin, K.L. (1974) Solubility in water of insecticide, chlorinated hydrocarbons and polychlorinated biphenyls in view of water pollution. *Z. Wasser Abwasser Forsch.* 7, 169-175.

Westcott, J.W., Bidleman, T.F. (1982) Determination of polychlorinated biphenyl vapor pressures by capillary gas chromatography. *J. Chromatogr.* 210, 331-336.

Westcott, J.W., Simon, J.J., Bidleman, T.F. (1981) Determination of polychlorinated biphenyl vapor pressures by a semimicro gas saturation method. *Environ. Sci. Technol.* 15, 1375-1378.

Whitehouse, B.G., Cooke, R.C. (1982) Estimating the aqueous solubility of aromatic hydrocarbons by high performance liquid chromatography. *Chemosphere* 11, 689-699.

Winholz, M. Ed. (1983) *The Merck Index, An Encyclopedia of Chemicals, Drugs and Biologicals.* 10th ed., Merck & Co. Inc. Rahway, New Jersey.

Woodburn, K.B., Doucette, W.J., Andren, A.W. (1984) Generator column determination of octanol/water partition coefficients for selected polychlorinated biphenyl congeners. *Environ. Sci. Technol.* 18, 457-459.

Yalkowsky, S.H. (1979) Estimation of entropies of fusion of organic compounds. *Ind. Eng. Chem. Fundam.* 18, 108-111.

Yalkowsky, S.H., Valvani, S.C. (1976) Partition coefficients and surface areas of some alkylbenzenes. *J. Med. Chem.* 19, 727-728.

Yalkowsky, S.H., Valvani, S.C. (1979) Solubility and partitioning. I: Solubility of nonelectrolytes in water. *J. Pharm. Sci.* 69, 912-922.

Yalkowsky, S.H., Orr, R.J., Valvani, S.C. (1979) Solubility and partitioning. 3. The solubility of halobenzenes in water. *I&EC Fundam.* 18, 351-353.

Yalkowsky, S.H., Valvani, S.S., Mackay, D. (1983) Estimation of the aqueous solubility of some aromatic compounds. *Res. Rev.* 85, 43-55.

Yoshida, K., Shigeoka, T., Yamauchi, F. (1983) Relationship between molar refraction and n-octanol/water partition coefficient. *Ecotox. Environ. Saf.* 7, 558-565.

Zwolinski, B.J. Wilhoit, R.C. (1971) *Handbook of Vapor Pressures and Heats of Vaporization of Hydrocarbons and Related Compounds.* API-44, TRC Publication No. 101, Texas A&M University, College Station, TX.

2. Polynuclear Aromatic Hydrocarbons (PAHs) and Related Aromatic Hydrocarbons

2.1 List of Chemicals and Data Compilations:

2.1   List of Chemicals and Data Compilations

Common Name: Indan
Synonym: hydrindene, 2,3-dihydroindene, 2,3-dihydro-1H-indene, indane
Chemical Name: indan
CAS Registry No: 496-11-7
Molecular Formula: C_9H_{10}
Molecular Weight: 118.18

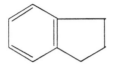

Melting Point (°C):
      <25   (Lande & Banerjee 1981)
      -51.4   (Weast 1982-83; Dean 1985; Mailhot & Peters 1988; Budavari 1989)
      -51.0   (Bjørseth 1983)
Boiling Point (°C):
      178.0   (Weast 1982-83; Bjørseth 1983; Pearlman et al. 1984; Mailhot & Peters 1988)
      176.5   (Dean 1985; Budavari 1989)
Density (g/cm³ at 20°C):
      0.9639 (Weast 1982-83; Dean 1985)
Molar Volume (cm³/mol):
      123.0   (from density, Lande & Banerjee 1981)
      122.57 (Mailhot & Peters 1988)
      143.7   (LeBas method)
Molecular Volume (Å³):
      121.6   (Pearlman et al. 1984)
Total Surface Area, TSA (Å²):
      151.1   (Yalkowsky & Valvani 1976)
      151.5   (Yalkowsky & Valvani 1979)
      156.7   (Pearlman et al. 1984)
      151.50 (quoted, Mailhot & Peters 1988)
Heat of Fusion, ΔH_{fus}, kcal/mol:
Entropy of Fusion, ΔS_{fus}, cal/mol K (e.u.):
Fugacity Ratio at 25 °C (assuming ΔS = 13.5 e.u.), F: 1.0

Water Solubility (g/m³ or mg/L at 25°C):
      88.9   (shake flask-GC, Price 1976)
      109.1 (shake flask-fluorescence, Mackay & Shiu 1977; quoted, Yalkowsky & Valvani
             1979; Mackay et al. 1980; Pearlman et al. 1984; Mailhot & Peters 1988)
      110.3 (quoted, Lande & Banerjee 1981)

110   (lit. mean, Yalkowsky et al. 1983)

110, 46.0 (quoted, calculated-K_{OW}, Valvani & Yalkowsky 1980)

88.9, 109.1; 100 (quoted lit. values; recommended, IUPAC 1989)

Vapor Pressure (Pa at 25°C):

     204.0   (extrpolated, comparative ebulliometry, Ambrose & Sprake 1976)

     196.91   (extrapolated-Antoine eqn., Stephenson & Malanowski 1987)

Henry's Law Constant (Pa m³/mol):

Octanol/Water Partition Coefficient, log K_{OW}:

     3.33   (Hansch & Leo 1979)

     3.33, 3.30 (quoted, calculated-TSA, Yalkowsky & Valvani 1976)

     3.57   (calculated-f const., Valvani & Yalkowsky 1980)

     3.33, 3.29 (quoted, calculated-solubility, Mackay et al. 1980)

     3.31   (calculated-f const., Yalkowsky et al. 1983)

     3.33, 3.57 (quoted, Mailhot & Peters 1988)

Bioconcentration Factor, log BCF:

Sorption Partition Coefficient, log K_{OC}:

Half-Lives in the Environment:

     Air:

     Surface water:

     Groundwater:

     Sediment:

     Soil:

     Biota:

Environmental Fate Rate Constants or Half-Lives:

     Volatilization:

     Photolysis:

     Oxidation:

     Hydrolysis:

     Biodegradation:

     Biotransformation:

     Bioconcentration, Uptake (k_1) and Elimination (k_2) Rate Constants:

Common Name: Naphthalene
Synonym: naphthene, tar camphor, moth balls
Chemical Name: naphthalene
CAS Registry No: 91-20-3
Molecular Formula: $C_{10}H_8$
Molecular Weight: 128.18

Melting Point (°C):
      80.0   (Parks & Huffman 1931)
      82-83 (Sahyun 1966)
      80.50 (Weast 1972-73, Cleland & Kingsbury 1977; quoted, Callahan et al. 1979;
            Mackay et al. 1980; Mabey et al. 1982; Arbuckle 1983; Weast 1984; Fu &
            Luthy 1985; Lande et al. 1985; Vadas et al. 1991)
      80.2 (Dean 1985; Yalkowsky & Valvani 1979, 1980; Karickhoff 1981; Mackay & Shiu
            1981; Yalkowsky 1981; Briggs 1981; Bruggeman et al. 1982; Windholz et al.
            1983; Patton et al. 1984; Miller et al. 1985; Chin et al. 1986; Gobas et al.
            1988; Banerjee et al. 1990; Pinal et al. 1991)
      81.0 (Bidleman 1984; Banerjee & Baughman 1991)
Boiling Point (°C):
      218   (Weast 1972-73; quoted, Mackay & Shiu 1981; Mabey et al. 1982; Banerjee et
            al. 1990; Pussemier et al. 1990)
      217.7 (Dean 1985)
Density (g/cm³ at 20°C):
      1.0253 (Weast 1983-84)
      1.162 (Dean 1985)
      1.152 (quoted, Verschueren 1977)
Molar Volume (cm³/mol):
      125.0 (from density, Aquan-Yuen et al. 1979, Lande & Banerjee 1981)
      133.2 (from density, Bohon & Claussen 1951)
      148.0 (LeBas method, Miller et al. 1985; quoted, Abernethy et al. 1988; Eastcott et al.
            1988; Gobas et al. 1988)
      0.753 (intrinsic volume: $V_I/100$, Kamlet et al. 1988; quoted, Hawker 1989)
      154.0 (selected, Valsaraj 1988)
      147.6 (LeBas method, Warne et al. 1991)

Molecular Volume (A^3):

     126.9 (Pearlman et al. 1984)

     127.69 (Pearlman 1986)

Total Surface Area, TSA (A^2):

     323.0 (Amidon et al. 1975, 1979)

     155.8 (Yalkowsky & Valvani 1979, 1980; quoted, Mackay et al. 1982; Lande et al. 1985)

     323.3 (calculated-1.5 Å solvent radius, Amidon & Anik 1980; quoted, Lande et al. 1985)

     155.8 (Pearlman et al. 1984; Pearlman 1986)

     154.2 (Sabljic 1987)

     161.0 (Pinal 1988; quoted, Pinal et al. 1990)

     155.8 (Valsaraj 1988; Valsaraj & Thibodeaux 1989)

     156.0 (quoted as HSA, Woodburn et al. 1989)

     156.5 (Warne et al. 1990)

Heat of Fusion, ΔH_{fus}, kcal/mol:

     4.61 (Parks & Huffman 1931)

     4.54 (Tsonopoulo & Prausnitz 1971; Fu & Luthy 1985)

     4.56 (Wauchope & Getzen 1972)

     4.491 (Weast 1972-73)

     4.536 (Dean 1985)

     4.56 (Podoll et al. 1989)

Entropy of Fusion, ΔS_{fus}, cal/mol K (e.u.):

     12.84 (Tsonopoulo & Prausnitz 1971)

     24.7 (Wauchope & Getzen 1972; quoted, Yalkowsky 1981)

     13.0 (Casellato et al. 1973; quoted, Yalkowsky 1981)

     13.1 (Ubbelohde 1978; quoted, Yalkowsky 1981)

     12.72, 13.50 (observed, estimated, Yalkowsky & Valvani 1980)

     12.59 (quoted, Hinckley et al. 1990)

Fugacity Ratio at 25 °C (assuming ΔS = 13.5 e.u.), F:

     0.28 (Mackay et al. 1980)

Water Solubility (g/m^3 or mg/L at 25°C):

     30.0 (shake flask-gravimetric, Hilpert 1916; quoted, Martin 1961; Gunther et al. 1968; IUPAC 1989)

     31.5 (shake flask-UV, Andrews & Keefer 1949; quoted, Pearlman et al. 1984; IUPAC 1989; Shiu et al. 1990)

     12.5 (shake flask-UV, Klevens 1950; quoted, Pearlman et al. 1984)

     34.4 (shake flask-UV, Bohon & Claussen 1951; quoted, Paul 1951; Vesala 1974; Mackay & Shiu 1981; Abernethy et al. 1986; IUPAC 1989; Shiu et al. 1990)

     30.6 (Stephen & Stephen 1963; quoted, Fu & Luthy 1985)

     20.4 (shake flask, Sahyun 1966; quoted, Shiu et al. 1990)

     33.47 (Gordon & Thorne 1967; quoted, Eganhouse & Calder 1976; Mackay & Shiu 1981; Pearlman et al. 1984)

38.4 (20°C, shake flask-UV, Eisenbrand & Baumann 1970)

31.2 (shake flask-UV, Wauchope & Getzen 1972; quoted, Vesala 1974; Mackay & Shiu 1981; Wasik et al. 1983; Pearlman et al. 1984; Walters & Luthy 1984; Billington et al. 1988; IUPAC 1989; Shiu et al. 1990)

33.0 (quoted, Mackay & Wolkoff 1973; quoted, Mackay & Leinonen 1975)

32.17 (shake flask-UV, Vesala 1974)

34.5 (quoted, Hine & Mookerjee 1975)

31.3 (shake flask-GC, Eganhouse & Calder, 1976; quoted, Mackay & Shiu 1981; Pearlman et al. 1984; Walters & Luthy 1984; IUPAC 1989; Shiu et al. 1990; Vadas et al. 1991)

22.0 (fluorescence, Schwarz & Wasik 1976; quoted, Mackay & Shiu 1981)

30.0 (quoted, Chiou et al. 1977; Freed et al. 1977)

31.7 (shake flask-fluorescence, Mackay & Shiu, 1977; quoted, Karickhoff et al. 1979; Yalkowsky & Valvani 1979, 1980; Mackay et al. 1979,1980; Hutchinson et al. 1980; Kenaga & Goring 1980; Chiou 1981; Chiou & Schmedding 1981; Karickhoff 1981; Lande & Banerjee 1981; Mackay & Shiu 1981; Amidon & Williams 1982; Chiou et al. 1982; Wasik et al. 1983; Pearlman et al. 1984; Walters & Luthy 1984; Lande et al. 1985; Miller et al. 1985; Chin et al. 1986; Billington et al. 1988; IUPAC 1989; Kayal & Connell 1990; Shiu et al. 1990; Vadas et al. 1991; Mackay 1991; Mackay & Shiu 1992)

30.0 (shake flask-fluorescence, Schwarz & Wasik 1977; quoted, Verschueren 1977; Mackay & Shiu 1981; IUPAC 1989)

30.3 (shake flask-fluorescence, Schwarz 1977; quoted, Mackay & Shiu 1981; Wasik et al. 1983; Billington et al. 1988; IUPAC 1989; Shiu et al. 1990)

31.69 (gen. col.-HPLC/UV, May et al. 1978; quoted, Callahan et al. 1979; Briggs 1981; Mackay & Shiu 1981; Mabey et al. 1982; May et al. 1983; Walters & Luthy 1984; IUPAC 1989)

34.4 (quoted, Callahan et al. 1979; quoted, Walters & Luthy 1984; Neuhauser et al. 1985)

21.3 (calculated-K_{OW}, Valvani & Yalkowsky 1980)

25.0 (calculated-K_{OW}, Yalkowsky & Valvani 1980)

32.4 (calculated-TSA, Amidon & Anik 1981)

28.4 (calculated-parachor, Briggs 1981)

125.3 (calculated-K_{OW}, Amidon & Williams 1982)

31.5 (RP-TLC, Bruggeman et al. 1982)

37.7 (quoted, Geyer et al. 1982)

32.0 (selected, Mills et al. 1982)

31.41, 14.30 (quoted, calculated-UNIFAC, Arbuckle 1983)

30.64 (gen. col.-HPLC/UV, Wasik et al. 1983; quoted, Miller et al. 1985, Eastcott et al. 1988)

32.2 (average lit. value, Pearlman et al. 1984; quoted, Banerjee 1985)

32.90 (gen. col.-HPLC/fluorescence, Walters & Luthy 1984)

31.1 (shake flask-UV, Bennet & Canady 1984)

33.56 (calculated, Lande et al. 1985)

30.6, 16.1 (shake flask-HPLC/UV, calculated-UNIFAC, Fu & Luthy 1985)

31.7 (selected, Miller et al. 1985; quoted, Vowles & Mantoura 1987; Gobas et al. 1988; Hawker 1989; Shorten et al. 1990)

137.35 (calculated-K_{ow} & M.P., Chin et al. 1986)

31.7 (Riddick et al. 1986; quoted, Howard 1989)

31.8 (quoted, Tomlinson & Hafkenshied 1986)

31.12 (vapor saturation-GC, Akiyoshi et al. 1987)

29.5 (calculated-QSAR of EPA, Passino & Smith 1987)

31.3, 31.9 (gen. col.-HPLC/UV, Billington et al. 1988)

30.2 (selected, Isnard & Lambert 1988, 1989)

34.0 (quoted, Valsaraj 1988; Valsaraj & Thibodeaux 1989)

32.0 (quoted, Podoll et al. 1990)

33.7 (shake flask-UV, Perez-Tejeda et al. 1990)

30.0 (shake flask-UV/fluorescence, Pinal et al. 1990)

32.0 (quoted, Edwards et al. 1991)

30.0 (quoted exptl., Pinal et al. 1991)

30.6 (gen. col.-HPLC, Vadas et al. 1991)

Vapor Pressure (Pa at 25°C):

14.26 (manometry, Sears & Hopke 1949; quoted, Bohon & Claussen 1951; Bidleman 1984)

10.8 (effusion method, Bradley & Cleasby 1953; quoted, Mackay & Shiu 1981; Wasik et al. 1983; Bidleman 1984)

10.98, 32.95 (manometry, extrapolated solid, subcooled liquid, Fowler et al. 1968)

30.66 (Antoine eqn., Zwolinski & Wilhoit 1971; quoted, Mackay & Wolkoff 1973; Mackay & Leinonen 1975)

11.60 (Weast 1972-73, quoted, Mackay et al. 1979; Mackay & Shiu 1981)

6.56 (calculate-evaporation rate, Gückel et al. 1973)

12.26 (effusion method, Radchenko & Kitiagorodskii 1974; quoted, Bidleman 1984)

10.9 (gas saturation, Sinke 1974; quoted, Mackay & Shiu 1981; Mackay et al. 1982; Shorten et al. 1990)

11.20 (quoted, Hine & Mookerjee 1975)

10.80 (lit. average-interpolated, API 1979; quoted, Wasik et al. 1983)

12.0 (calculated-HLC, Mackay et al. 1979; quoted, Bidleman 1984)

10.64 (gas saturation, Macknick & Prausnitz 1979; quoted, Bidleman 1984)

10.40 (extrapolated, Macknick & Prausnitz 1979; quoted, Mackay & Shiu 1981; Mackay 1991)

11.30 (effusion method, De Kruif 1980)

10.42 (effusion method, De Kruif et al. 1981)

11.41 (extrapolated-Antoine equation, Colomina et al. 1982)

6.558 (selected, Mills et al. 1982)

11.33 (gas saturation, Grayson & Fosbraey 1982; quoted, Bidleman 1984)

11.6 (calculated, Mabey et al. 1982)
11.75 (quoted, Arbuckle 1983)
6.53 (20°C, Mackay et al. 1983)
10.40 (gen. col.-HPLC, Wasik et al. 1983)
7.91 (GC-RT, Bidleman 1984)
10.4 (gas saturation-HPLC/UV, Sonnefeld et al. 1984)
66.7 (quoted, Neuhauser et al. 1985)
11.33 (selected, Howard et al. 1986)
10.9 (quoted, Riddick et al. 1986, Howard 1989)
11.27 (Antoine eqn., Stephenson & Malanowski 1987)
111 (subcooled liquid value, quoted, Eastcott et al. 1988)
11.14 (quoted, Valsaraj 1988)
11.33, 4.68 (quoted, calculated-UNIFAC, Banerjee et al. 1990)
41.88, 38.02 (subcooled liquid value, Hinckley et al. 1990)

Henry's Law Constant (Pa m^3/mol):
119.54 (Mackay & Wolkoff 1973; Mackay & Leinonen 1975)
42.10 (reported as exptl., calculated from log $(1/K_{AW})$ = log(C_g/C_w), Hine & Mookerjee 1975)
35.83, 31.93 (calculated-group contribution, bond contribution, Hine & Mookerjee 1975)
47.1 (calculated-P/C, Schwarz & Wasik 1977; quoted, Southworth 1979)
56.0 (batch stripping, Southworth, 1979; quoted, Howard 1989)
48.9 (batch stripping, Mackay et al. 1979; Mackay & Shiu 1981)
47.4 (calculated-P/C, Mackay et al. 1979)
43.0 (calculated-P/C, Mackay & Shiu 1981; quoted, Fendinger & Glotfelty 1990)
43.0 (recommended, Mackay & Shiu 1981)
44.6, 44.3 (batch stripping, calculated-P/C, Mackay et al. 1982)
46.6 (calculated-P/C, Mabey et al. 1982; quoted, Pankow & Rosen 1988)
48.94, 29.35 (quoted, calculated-UNIFAC, Arbuckle 1983)
38.0 (Mackay et al. 1983)
49.0 (calculated-P/C, Smith et al. 1983)
123.95 (Jury et al. 1984)
36.5 (20 °C, EPICS method, Yurteri et al. 1987)
29.2, 102 (20 °C, calculated-UNIFAC, P and S, Yurteri et al. 1987)
42.10, 31.21 (calculated-P/C, UNIFAC, Nirmalakhandan & Speece 1988)
50.08 (WERL Treatability database, quoted, Ryan et al. 1988)
56.2 (calculated-P/C, Eastcott et al. 1988)
48.63 (selected, Valsaraj 1988)
49.6, 74.37 (quoted, wetted-wall column-GC, Fendinger & Glotfelty 1990)
46.61 (Pankow 1990)
48.9 (Shorten et al. 1990)

Octanol/Water Partition Coefficient, log K_{OW}:

3.37   (shake flask, Fujita et al. 1964; quoted, Chiou et al. 1977; Freed et al. 1977; Mackay et al. 1980)

3.37, 3.42 (quoted, calculated, Kier et al. 1971)

3.37, 3.01, 3.45 (Leo et al. 1971; quoted, Kringstad et al. 1984)

3.37   (Hansch et al. 1973; quoted, Rekker 1977)

3.37   (Radding et al. 1976; quoted, Callahan et al. 1979)

3.37   (calculated-f const., Rekker 1977)

3.30   (calculated from Leo 1975, Southworth et al. 1978; quoted, Hawker & Connell 1986)

3.395 (fluorometry, Krishnamurthy & Wasik 1978)

3.30, 3.01, 3.37, 3.45, 3.59 (Hansch & Leo 1979)

3.41, 3.17 (quoted, HPLC-RT, Veith et al. 1979a,b)

3.36   (shake flask-UV, concn. ratio, Karickhoff et al. 1979; quoted, Chiou 1981; Chiou & Schmedding 1981; Karickhoff 1981; Govers et al. 1984; Walters & Luthy 1984; Ruepert et al. 1985; Kayal & Connell 1990; Pussemier et al. 1990; Szabo et al. 1990a,b; Shorten et al. 1990; Edwards et al. 1991)

3.35   (calculated-f const., Yalkowsky & Valvani 1979, 1980; quoted, Hutchinson et al. 1980)

3.30   (calculated-S & M.P., Mackay et al. 1980; quoted, Geyer et al. 1982,1984; Wasik et al. 1983)

3.21   (HPLC-k', Hanai et al. 1981)

3.18   (HPLC-k', D'Amboise & Hanai, 1981)

3.35   (gen. col.-HPLC/UV, Wasik et al. 1981, 1983)

3.30   (quoted, Amidon & Williams 1982)

3.35   (RP-TLC, Bruggeman et al. 1982; quoted, Kringstad et al. 1984)

3.36   (quoted, Chiou et al. 1982)

3.45   (HPLC-RT, Hammers et al. 1982)

3.29   (calculated-K_{OW}, Mabey et al. 1982)

3.59   (quoted, Mackay 1982; Schüürmann & Klein 1988; Banerjee & Baughman 1991)

3.36   (quoted, Mills et al. 1982)

3.01, 3.17, 3.30, 3.36, 3.37, 3.45, 3.59, 4.70 (compiled data, Trabalka & Garten 1982; quoted, Garten & Trabalka 1983)

3.36   (quoted, Casserly et al. 1983)

3.31, 3.35 (shake flask, HPLC, Eadsforth & Moser 1983)

3.36   (HPLC-k', Hafkenscheid & Tomlinson 1983)

3.35   (Leyder & Boulanger 1983)

4.52   (Mackay et al. 1983)

3.40   (Voice et al. 1983)

3.37, 3.57 (quoted, calculated-MR, Yoshida et al. 1983)

3.35, 3.42 (shake flask, ALPM, Garst & Wilson 1984)

3.57   (HPLC-retention volume, Garst & Wilson 1984)

3.45   (selected best lit. value, Garst & Wilson 1984)

3.30   (quoted, Patton et al. 1984)
3.30   (calculated-UNIFAC, Campbell & Luthy 1985)
3.35   (quoted, Fu & Luthy 1985)
3.35   (selected, Miller et al. 1985; quoted, Vowles & Mantoura 1987; Kamlet et al.
           1988; Eastcott et al. 1988, Hawker 1989; Mackay 1991)
3.30   (Hansch & Leo 1985; quoted, Howard 1989)
3.27   (quoted, Neuhauser et al. 1985)
3.30   (calculated- χ , as per Rekker & De Kort 1979, Ruepert et al. 1985)
3.36   (quoted, Brooke et al. 1986)
3.38   (RP-HPLC, Chin et al. 1986)
3.30   (THOR 1986; quoted, Schüürmann & Klein 1988)
3.32   (Leo 1986; quoted, Schüürmann & Klein 1988)
3.44, 3.29 (lit. average, HPLC-RT, Kock & Lord 1987)
3.31   (quoted, Pavlou 1987)
4.70   (quoted, Isnard & Lambert 1988, 1989)
3.40   (calculated-solvatochromic p. & V_I, Kamlet et al. 1988)
3.37   (quoted Ryan et al. 1988)
3.30   (quoted, Thomann 1989)
3.59   (Travis & Arms 1988)
3.45   (quoted, Pinal et al. 1990, 1991)
3.39   (quoted, Warne et al. 1990)
3.30   (quoted, Warne et al. 1991)

Bioconcentration Factor, log BCF:
1.64   (mussel *mytilus edulis*, Lee et al. 1972; quoted, Geyer et al. 1982)
1.49   (mussel *mytilus edulis*, Hansen et al. 1978; quoted, Geyer et al. 1982)
4.11   (bile of rainbow trout, Melancon & Lech 1978; quoted, Verschueren 1983)
2.12   (*daphnia pulex*, Southworth et al. 1978; quoted, Govers et al. 1984; Hawker &
           Connell 1986)
2.07   (*daphnia pulex,* by kinetic estimation, Southworth et al. 1978)
2.63   (fish, Veith et al. 1979)
2.63   (fathead minnow, Veith et al. 1980; quoted, Bysshe 1982)
2.62   (microorganisms-water, calculated from K_{ow}, Mabey et al. 1982)
2.63   (fish, Mackay 1982; quoted, Sabljic 1987; Schüürmann & Klein 1988)
4.10, 3.84, 4.25 (average, *selenastrum capricornutum*-dosed singly, dosed
           simultaneously, Casserly et al. 1983)
2.11   (*chlorella fusca*, Geyer et al. 1984)
2.43   (calculated-K_{ow}, Geyer et al. 1984)
1.48, 2.10, 3.0 (fish, algae, activated sludge, Freitag et al. 1985)
2.50   (bluegill sunfish, McCarthy & Jimenez 1985)
2.48   (bluegill sunfish with dissolved humic material, McCarthy & Jimenez 1985)
1.97   (calculated- χ , Sabljic 1987)

2.63 (selected, Isnard & Lambert 1988; quoted, Banerjee & Baughman 1991)

-3.70 (milk, reported as biotransfer factor, Travis & Arms 1988)

2.54 (calculated-K_{OW} , $S_{OCTANOL}$ & M.P., Banerjee & Baughman 1991)

Sorption Partition Coefficient, log K_{OC}:

    3.11 (average sorption isotherms on sediments, Karickhoff et al. 1979; Kenaga & Goring 1980; Schwarzenbach & Westall 1981; Govers et al. 1984; Sabljic 1984; Bahnick & Doucette 1988; Szabo et al. 1990a,b)

    3.15 (sediment, calculated-K_{OW}, Karickhoff et al. 1979)

    2.38 (22°C, suspended particulates, Herbes et al. 1980)

    2.94, 2.97, 3.00 (measured, calculated-K_{OW}, S, Karickhoff 1981)

    2.91 (calculated-K_{OW}, Schwarzenbach & Westall 1981)

    2.97 (calculated-K_{OW}, Mabey et al. 1982)

    3.18 (calculated, Sabljic 1984)

    3.0 (McCarty & Jamenez 1985; McCarty et al. 1985)

    2.95 (calculated, Pavlou 1987)

    2.93 (sediment, HPLC-k', Vowles & Mantoura 1987)

    3.27 (calculated- χ , Bahnick & Doucette 1988)

    2.73-3.91 (aquifer materials, Stauffer et al. 1989)

    3.15, 2.76 (Menlo Park soil, Eustis sand isotherm, Podoll et al. 1989)

    5.00 (sediments average, Kayal & Connell 1990)

    2.66 (soils average, Kishi et al. 1990)

    3.11 (soil, RP-HPLC, Szabo et al. 1990a; quoted, Pussemier et al. 1990)

    3.29 (sandy surface soil, Wood et al. 1990)

    2.97, 3.17 (dissolved organic matter, calculated-K_{OW}, Kan & Tomson 1990)

Half-Lives in the Environment:

    Air: volatility of 2.28×10^4 sec (experimental), 7.7×10^3 sec (calculated) for depth of water body of 22.5 m (23°C, Klöpffer et al. 1982); half-life less than one day for its reaction with photochemically produced hydroxyl radicals (Atkinson et al. 1984,1987; quoted, Howard 1989).

    Surface Water: 16 hours (calculated for river water 1 m deep, water velocity 0.5 m/s, wind velocity 1 m/s from air-water partition coefficients (Henry's law constants) (Southworth 1979; quoted, Hallett & Brecher 1984); 71 hours of photolysis half-life in near surface water, but 550 days for a depth of 5 meter (calculated from surface water in midsummer at 40°N latitude, Zepp & Schlotzhauer 1979; quoted, Halett & Brecher 1984; Mill & Mabey 1985; Howard 1989); calculated half-life of 7.15 hours, based on evaporative loss for a water depth of 1 meter at 25°C (Mackay & Leinonen 1975; quoted Verschueren 1977, 1983); an overall half-life in Rhine river of 2.3 days based on monitoring data (Zoeteman et al. 1980; quoted, Howard 1989); half-lives in coastal seawater from VOC experiments: 11.3 hours with $HgCl_2$ as poison and 0.8 hour without poison (Wakham et al. 1983).

Groundwater: estimated half-life is 0.6 years in the Netherlands (Zoeteman et al. 1981).

Sediment: half-lives: 4.9 hours in oil contaminated sediment and >88 days in uncontaminated sediments (Herbes & Schwall 1978; quoted, Howard 1989).

Soil: an overall half-life of 3.6 months in a solid waste site (Zoeteman et al. 1981; quoted, Howard 1989); 0.12-125 days (Sims & Overcash 1983; quoted, Bulman et al. 1987); 12 days for both 5 mg/kg and 50 mg/kg added (Bulman et al. 1987); > 50 days (Ryan et al. 1988); >80 days (Howard 1989); biodegradation rate constant 0.337 day^{-1} with a half-life of 2.1 days for Kidman sandy loam soil and 0.308 day^{-1} with 2.2 days for McLaurin sandy loam soil (Park et al. 1990); 0.02-46 weeks, <2.1 years (quoted, Ludington soil, Wild et al. 1991).

Environmental Fate Rate Constants or Half-Lives:

Volatilization/Evaporation: rate of evaporation estimated to be 1.675×10^{-9} mol cm^{-2} hour^{-1} at 20°C and air flow rate of 50 liter hour^{-1} (Gückel et al. 1973); calculated half-life is 7.15 hours from 1m depth of water (Mackay & Leinonen 1975; quoted, Haque et al. 1980); half-life of 16 hours for surface waters (Southworth 1979; quoted, Herbes et al. 1980; Hallett & Brecher 1984); when considering current velocity and wind speed in combined with typical reaeration rates for natural bodies of water which give a half-life for evaporation of 50 and 200 hours in a river and lake (Mills et al. 1982; quoted, Howard 1989); half-lives from solution: 28, 32 minutes (exptl., calcd., Mackay et al. 1983).

Photolysis: calculated half-life for direct sunlight photolysis of 50% conversion at 40°N at midday of midsummer, 71 hours (near surface water; quoted, Harris 1982) and 550 days (5 meter deep inland water) (Zepp & Schlotzhauer 1979; quoted, Herbes et al. 1980; Hallett & Brecher 1984; Mill & Mabey 1985; Howard 1989); rate constant of 0.028 hour^{-1} in distilled water with half-life of 25 hours (Fukuda et al. 1988); photodegradation in methanol-water (3:7, v/v) solution with initial concentration of 50.0 ppm by high pressure mercury lamp or sunlight with rate constant of 6.0×10^{-4} minute^{-1} and half-life of 19.18 hours (Wang et al. 1991).

Oxidation: calculated rate constants of < 360 M^{-1} h^{-1} for singlet oxygen and < 1 M^{-1} h^{-1} for peroxy radical (Mabey et al. 1982); rate constant for the reaction with OH radicals using relative rate technique was determined to be 2.42×10^{-11} cm^3 molecule^{-1} sec^{-1} at 294 ± 1K and an upper limit of 2×10^{-9} cm^3 molecule^{-1} sec^{-1} was obtained for the reaction with O$_3$ (Atkinson et al. 1984); rate constant for the gas phase reaction with OH radicals was 2.35×10^{-11} cm^3 molecule^{-1} sec^{-1} at 298 ± 1K based on the relative rate technique from propene (Biermann et al. 1985); rate constant for the gas-phase reaction with OH radical at 298°K and atmospheric pressure was 2.16×10^{-11} cm^3 molecule^{-1} sec^{-1} (Atkinson 1990).

Hydrolysis: not hydrolyzable (Mabey et al. 1982).

Biodegradation: ultimate loss process, 4×10^{-6} g/L/day (Lee & Ryan, 1976); 0.04 to 3.3×10^{-6} g/L/day (Lee & Anderson 1977); complete degradation in 8 days in gas-oil contaminated groundwater which was circulated through sand that had

70

been inoculated with groundwater under aerobic conditions (Kappeler & Wuhrmann 1978; quoted, Howard 1989); rate constant for microorganisms of 0.04-3 μg/L/day (Callahan et al. 1979); in deeper and slowly moving contaminated water with a half-life of 1-9 days (Herbes 1981; Wakeham et al. 1983; quoted, Howard 1989); half-lives: of 7, 24, 63 and 1700 days in an oil polluted estuarine stream, clean estuarine stream, coastal waters, and in the Gulf stream respectively (Lee 1977; quoted, Howard 1989); Aerobic half-life: 12 hours, based on dieaway test data for an oil polluted creek (Walker & Colwell 1976); 480 hours, for an estuarine river (Lee & Ryan 1976). Anaerobic half-life: 25 days at pH 8 and 258 days at pH 5 (Hambrick et al. 1980). Half-life in deeper and slower moving contaminated water is about 1-9 days (Herbes 1981; Wakeham et al. 1983); 0.23 hour^{-1} (microbial degradation rate constant, Herbes et al. 1980, Hallett & Brecher 1984); 100% degradation within 7 days for an average of three static-flask screening test (Tabak et al. 1981); nonautoclaved samples of 0.04 mg/litre in groundwater from hazardous waste site are completely degraded by microbes within one week (Lee et al. 1984); rate constant in groundwater with nutrients and acclimated microbes is of 0.024 day^{-1} with half-life of 28 days, 0.013 day^{-1} and 53 days in river water with acclimated microbes, 0.018 day^{-1} & 39 days in river water with nutrients and acclimated microbes (Vaishnav & Babeu 1987); degradation rate constants of 0.337 day^{-1} with half-life of 2.1 days for Kidman sandy loam and 0.308 day^{-1} with half-life of 2.2 days for McLarin sandy loam all at -0.33 bar soil moisture (Park et al. 1990).

Biotransformation: estimated rate constant for bacteria, 1×10^{-7} mL cell^{-1} h^{-1} (Mabey et al. 1982).

Bioconcentration, Uptake (k_1) or Elimination (k_2) Rate Constants:

k_1: 197 hour^{-1} (*daphnia pulex*, Southworth et al. 1978; quoted, Hawker & Connell 1986)

k_2: 1.667 hour^{-1} (*daphnia pulex*, Southworth et al. 1978; quoted, Hawker & Connell 1986)

log k_2: -0.70, -1.70 day^{-1} (fish, calculated-K_{ow}, Thomann 1989)

Common Name: 1-Methylnaphthalene
Synonym: α-methylnaphthalene
Chemical Name: 1-methylnaphthalene
CAS Registry No: 90-12-0
Molecular Formula: $C_{11}H_{10}$
Molecular Weight: 142.2

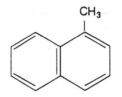

Melting Point (°C):
  -22   (Weast 1972-73, Mackay & Shiu 1981; Bildleman 1984; Miller et al. 1985)
  25   (Yalkowsky & Valvani 1979,1980; Lande & Banerjee 1981)
  -30.5 (Dean 1985)
Boiling Point (°C):
  244.6 (Weast, 1972-73; Dean 1985; Mackay & Shiu 1981)
  240-243 (Verschueren 1977, 1983)
  245   (Pearlman et al. 1984)
Density (g/cm³ at 20°C):
  1.022 (Weast 1982-83)
  1.6125 (Dean 1985)
Molar Volume (cm³/mol):
  139.0 (calculated, Lande & Banerjee 1981)
  0.851 (intrinsic volume: $V_I/100$, Kamlet et al. 1988; Hawker 1989)
  169.8 (LeBas method, Miller et al. 1985; Abernethy & Mackay 1987; Warne et al.
         1991)
Molecular Volume (A³):
  142.56 (Pearlman et al. 1984)
Total Surface Area, TSA (A²):
  172.5 (Yalkowsky & Valvani 1979)
  174.95 (Pearlman et al. 1984)
  172.5 (Valsaraj & Thibodeaux 1989)
Heat of Fusion, ΔH_{fus}, kcal/mol:
  1.160 (Dean 1985)
Entropy of Fusion, ΔS_{fus}, cal/mol K (e.u.):
Fugacity Ratio at 25 °C, F: 1.0

Water Solubility (g/m³ or mg/L at 25°C):

    25.8   (shake flask-GC, Eganhouse & Calder 1976; quoted, API 1978; Mackay & Shiu 1981; Pearlman et al. 1984; IUPAC 1989)

    28.5   (shake flask-fluorescence, Mackay & Shiu 1977; quoted, API 1978; Yalkowsky & Valvani 1979, 1980; Hutchinson et al. 1980; Mackay et al. 1980; Valvani & Yalkowsky 1980; Mackay & Shiu 1981; Wasik et al. 1981; Mackay et al. 1982; Pearlman et al. 1984; Miller et al. 1985; IUPAC 1989)

    29.9   (shake flask-fluorescence, Schwarz & Wasik 1977; quoted, Mackay & Shiu 1981)

    29.5   (shake flask-fluorescence, Schwarz 1977; quoted, Mackay & Shiu 1981; Pearlman et al. 1984; IUPAC 1989)

    30.4   (calculated-K_{OW}, Yalkowsky & Valvani 1980)

    26.9   (calculated-TSA, Amidon & Anik 1981)

    27.7   (quoted, Lande & Banerjee 1981)

    31.7   (gen. col.-HPLC, Wasik et al. 1981)

    27.02   (average lit. value, Pearlman et al. 1984)

    28.4   (selected, Miller et al. 1985; Vowles & Mantoura 1987; Eastcott et al. 1988, Hawker 1989)

    28.2   (quoted, Isnard & Lambert 1989)

    28.0   (recommended, IUPAC 1989)

    28.44   (quoted, Valsaraj & Thibodeaux 1989)

    28.0   (selected, Ma et al. 1990)

    28.0   (quoted, Warne et al. 1990)

Vapor Pressure (Pa at 25°C):

    7.165   (extrapolated, Camin & Rossini 1955)

    7.19   (interpolated, Zwolinski & Whilhoit 1971)

    8.82   (gas saturation, Macknick & Prausnitz 1979; quoted, Bidleman 1984)

    8.84   (extrapolated, Macknick & Prausnitz 1979; quoted, Mackay & Shiu 1981)

    8.13   (calculated, Mackay et al. 1982)

    7.90   (Mackay et al. 1982; quoted, Eastcott et al. 1988)

    8.84   (interpolated, Boublik et al. 1984)

    7.816   (interpolated, Antoine eqn., Dean 1985)

    0.895   (Antoine eqn., Stephenson & Malanowski 1987)

    6.31   (GC-RT, Bidleman 1984)

    7.00   (selected, Ma et al. 1990)

    8.83   (quoted, Hinckley et al. 1990)

Henry's Law Constant (Pa m³/mol):

    26.3   (batch stripping, Mackay et al. 1979,1982; quoted, Findinger & Glotfelty 1990)

    44.62   (calculated-P/C, Mackay et al. 1979; Mackay & Shiu 1981; quoted, Pankow et al. 1984)

    45.0   (recommended, Mackay & Shiu 1981)

    26.3, 35.9   (batch stripping, calculated-P/C, Mackay et al. 1982)

16.0   (calculated-P/C, Eastcott et al. 1988)

27.18, 44.08 (calculated-P/C, UNIFAC, Nirmalakhandan & Speece 1988)

62.0   (wetted-wall col., Fendinger & Glotfelty 1990)

Octanol/Water Partition Coefficient, log K_{OW}:

3.87   (fluorometry, Krishnamurthy & Wasik 1978)

3.87   (Hansch & Leo 1979; quoted, Mackay et al. 1980)

3.95   (calculated-S & M.P., Mackay et al. 1980)

3.86   (calculated-f const., Yalkowsky & Valvani 1979, 1980; quoted, Hutchinson et al. 1980)

3.87   (Chiou 1981; quoted, Valsaraj & Thibodeaux 1989)

3.87, 3.91 (quoted, calculated-MR, Yoshida et al. 1983)

3.87   (selected, Miller et al. 1985; quoted, Abernethy & Mackay 1987; Vowles & Mantoura 1987; Eastcott et al. 1988; Hawker 1989)

3.87, 3.92 (quoted, calculated-solvatochromic p. & V_I, Kamlet et al. 1988)

3.86   (quoted, Isnard & Lambert 1989)

5.08   (recommended, Sangster 1989)

3.87   (quoted, Warne et al. 1990, 1991)

Bioconcentration Factor, log BCF:

Sorption Partition Coefficient, log K_{OC}:

3.36   (sediment, HPLC-k', Vowles & Mantoura 1987)

2.96-3.83 (aquifer materials, Stauffer et al. 1989)

Half-Lives in the Environment:

Air:

Water: computed near-surface half-life of 22 hours and for direct photochemical transformation at latitude 40°N, midday, midsummer, and half-lives of 180 days (no sediment-water partitioning) and 190 days (with sediment-water partitioning) for direct photolysis in 5 m deep inland water body integrated over full summer day at latitude 40°N (Zeep & Schlotzhauer 1979); 180 days under summer sunlight (Mill & Mabey 1985).

Soil: biodagradation rate constant 0.415 day^{-1} with a half-life of 1.7 day for Kidman sandy loam soil and 0.321 day^{-1} with half-life of 2.2 days for McLaurin sandy loam soil (Park et al. 1990).

Sediment:

Biota:

Environmental Rate Constants or Half-Lives:

    Volatilization:

    Photolysis: calculated half-lives for direct sunlight photolysis of 50% conversion at 40°N latitude of midday in midsummer: 22 hours (near surface water), 180 days (5 meter deep inland water) and 190 days (inland water with a suspended sediment concentration of 20 mg/liter partitioning) (Zepp & Schlotzhauer 1979); 180 days under summer sunlight in surface water (Mill & Mabey 1985).

    Oxidation: room temperature rate constant for the gas-phase reaction with OH radical was 5.30×10^{-11} cm^3 $molecule^{-1}$ sec^{-1} (Atkinson & Aschmann 1987).

    Hydrolysis:

    Biodegradation: rate constants of 0.415 day^{-1} with half-life 1.7 days for Kidman sandy loam soil and 0.321 day^{-1} with half-life of 2.2 days for McLaurin sandy loam soil, all at -3.3 bar soil moisture (Park et al. 1990).

    Biotransformation:

    Bioconcentration, Uptake (k_1) and Elimination(k_2) Rate Constants:

Common Name: 2-Methylnaphthalene
Synonym: β-methylnaphthalene
Chemical Name: 2-methylnaphthalene
CAS Registry No: 91-57-6
Molecular Formula: $C_{11}H_{10}$
Molecular Weight: 142.19

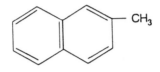

Melting Point (°C):
      34.6   (Weast 1972-73; quoted, Mackay et al. 1980; Lande & Banerjee 1981; Mackay
             & Shiu 1981; Mailhot & Peters 1988; Vadas et al. 1991)
      34.0   (Verschueren 1977,1983; Chiou et al. 1982; Miller et al. 1985; Mailhot &
             Peters 1988)
      35.0   (Yalkowsky & Valvani 1979, 1980)
Boiling Point (°C):
      241.1   (Weast 1977; Mackay & Shiu 1981)
      241-242 (Verschueren 1977, 1983; Mailhot & Peters 1988)
      241.0   (Vorňáková et al. 1978; Pearlman et al. 1984)
Density (g/cm³ at 20°C):
      1.0058 (Weast 1982-83)
      1.6026 (Dean 1985)
      1.0290 (Mailhot & Peters 1988)
Molar Volume (cm³/mol):
      141.0   (calculated, Lande & Banerjee 1981)
      170   (LeBas method, Miller et al. 1985; Abernethy & Mackay 1987)
      0.851   (intrinsic volume: $V_I/100$, Kamlet et al. 1988; Hawker 1989)
      138.19 (Mailhot & Peters 1988)
Molecular Volume (Å³):
      142.8   (Pearlman et al. 1984)
Total Surface Area, TSA (Å²):
      176.3   (Yalkowsky & Valvani 1979; Mailhot & Peters 1988)
      177.68 (Pearlman et al. 1984)
      175.7   (Sabljic 1987)
Heat of Fusion, ΔH_{fus}, kcal/mol:
      2.85   (Parks & Huffman 1931)
      2.89   (Tsonopoulos & Prausnitz 1971)

2.808   (Dean 1985)

Entropy of Fusion, ΔS_{fus}, cal/mol K (e.u.):
    9.38   (Tsonopoulos & Prausnitz 1971)

Fugacity Ratio at 25 °C (assuming $\Delta S = 13.5$ e.u.), F:
    0.803   (at M.P. = 34.6°C)

Water Solubility (g/m^3 or mg/L at 25°C):
    24.6   (shake flask-GC, Eganhouse & Calder 1976; quoted, Mackay & Shiu 1981; Pearlman et al. 1984; IUPAC 1989; Vadas et al. 1991)
    25.4   (shake flask-fluorescence, Mackay & Shiu 1977,1981; quoted, Karickhoff et al. 1979; Yalkowsky & Valvani 1979,1980; Hutchinson et al. 1980; Kenaga & Goring 1980; Mackay et al. 1980; Valvani & Yalkowsky 1980; Chiou 1981; Chiou & Schmedding 1981; Lande & Banerjee 1981; Mackay & Shiu 1981; Chiou et al. 1982; Pearlman et al. 1984; Miller et al. 1985; Eastcott et al. 1988; Mailhot & Peters 1988; IUPAC 1989; Vadas et al. 1991)
    20.0   (Vozňáková et al. 1978)
    24.1   (calculated-K_{OW}, Yalkowsky & Valvani 1980)
    25.0   (calculated-TSA, Amidon & Anik 1981)
    25.6   (average lit. value, Pearlman et al. 1984)
    25.1   (quoted, Isnard & Lambert 1988,1989)
    32.6   (quoted, Hawker 1989)
    25.0   (selected, Ma et al. 1990)
    27.3   (gen. col.-HPLC, Vadas et al. 1991)

Vapor Pressure (Pa at 25°C):
    9.07   (Camin & Rossini 1955)
    9.07   (extrapolated-Antoine eqn., Zwolinski & Wilhoit 1971; quoted, Mackay & Shiu 1981; Eastcott et al. 1988)
    9.07   (extrapolated, Boublik et al. 1973, 1984)
    5.60   (20°C, Vozňáková et al. 1978)
    9.07   (extrapolated from liq. state, Macknick & Prausnitz 1979; quote, Mackay & Shiu 1981)
    7.24   (calculation from extrapolated value with fugacity correction, Mackay & Shiu 1981)
    7.24, 8.13, 9.24 (Mackay et al. 1982)
    6.31   (GC-RT, Bidleman 1984)
    9.033   (extrapolated, Antoine eqn., Dean 1985)
    9.33   (extrapolated from Antoine eqn., Stephenson & Malanowski 1987)
    9.0   (selected, Ma et al. 1990)

Henry's Law Constant (Pa m^3/mol):
    40.52   (calculated-P/C, Mackay & Shiu 1981; quoted, Pankow et al. 1984)
    50.32   (calculated-P/C, Eastcott et al. 1988)

32.23   (wetted-wall col., Fendinger & Glotfelty 1990)

41.02, 58.75   (quoted, calculated-bond contribution method, Meylan & Howard 1991)

Octanol/Water Partition Coefficient, log K_{OW}:

3.864   (fluorometry, Krishnamurthy & Wasik 1978)

3.86   (Hansch & Leo 1979; quoted, Mackay et al. 1980)

4.11   (shake flask-UV, concn. ratio, Karickhoff et al. 1979; quoted, Chiou 1981)

3.86   (calculated-f const., Yalkowsky & Valvani 1979,1980; quoted, Hutchinson et al.
        1980)

4.11   (quoted, Kenaga & Goring 1980)

3.90   (calculated-S & M.P., Mackay 1980)

4.11   (quoted; Chiou & Schmedding 1981, Chiou et al. 1982)

3.70   (HPLC-k', Hanai et al. 1981)

3.86, 3.91   (quoted, calculated-MR, Yoshida et al. 1983)

3.86   (selected, Miller et al. 1985; quoted, Vowles & Mantoura 1987; Kamlet et al.
        1988; Hawker 1989)

3.87   (Abernethy & Mackay 1987)

3.92   (calculated-solvatochromic parameter, Kamlet et al. 1988)

3.86   (quoted, Isnard & Lambert 1988, 1989)

3.86, 4.11   (quoted, Mailhot & Peters 1988)

4.0, 5.08   (recommended, Sangster 1989)

4.0   (selected, Ma et al. 1990)

Bioconcentration Factor, log BCF:

2.61   (quoted from Davies & Dobbs 1984, Sabljic 1987)

2.65   (calculated- χ , Sabljic 1987)

2.61   (calculated-K_{OW} and S, Isnard & Lambert 1988)

Sorption Partition Coefficient, log K_{OC}:

3.93   (average sorption isotherms on sediments, Karickhoff et al. 1979; quoted, Kenaga
        & Goring 1980; Sabljic 1984)

3.40   (calculated- χ , Sabljic 1984)

3.40   (sediment, HPLC-k', Vowles & Mantoura 1987)

3.40   (Hodson & Williams 1988)

Half-Lives in the Environment:

Air:

Surface water: computed near-surface half-life of 54 hours and for direct photochemical
        transformation at latitude 40°N, midday, midsummer; half-lives of 410 days (no
        sediment-water partitioning) and 440 days (with sediment-water partitioning) by
        direct photolysis in a 5-m deep Inland Water Body (Zepp & Schlotzhauer 1979);
        410 days under summer sunlight (Mill & Mabey 1985); half-lives in coastal

seawater for VOC experiments: 15.1 hours with $HgCl_2$ as poison and 0.7 hour without poison (Wakham et al. 1983).

Groundwater:

Sediment:

Soil:

Biota:

Environmental Fate Rate Constants or Half-Lives:

Volatilization:

Photolysis: calculated half-lives for direct sunlight photolysis of 50% conversion at 40°N latitude of midday in midsummer: 54 hours (near surface water), 410 days (inland water) and 440 days (inland water with sediment partitioning)(Zepp & Schlotzhauer 1979); half-life of 410 days under summer sunlight in surface water (Mill & Mabey 1985); rate constant of 0.042 h^{-1} with half-life of 16.4 hours in distilled water (Fukuda et al. 1988).

Hydrolysis:

Oxidation: room temperature rate constant for the reaction with OH radicals was 5.23×10^{-11} cm^{-3} molecule^{-1} sec^{-1} (Atkinson & Aschmann 1986).

Biodegradation:

Biotransformation:

Bioconcentration, Uptake (k_1) and Elimination (k_2) Rate Constants:

Common Name: 1,3-Dimethylnaphthalene
Synonym:
Chemical Name: 1,3-dimethylnaphthalene
CAS Registry No: 575-41-7
Molecular Formula: $C_{12}H_{12}$
Molecular Weight: 156.23

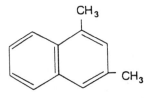

Melting Point (°C):
    -6   (Weast 1982-83, Dean 1985; quoted, Mailhot & Peters 1988)
    25   (Yalkowsky & Valvani 1979, 1980)
    < 25 (Mackay et al. 1980)

Boiling Point (°C):
    263 (Weast 1982-83; Dean 1985; Pearlman et al. 1984; Mailhot & Peters 1988)

Density (g/cm³ at 20°C):
    1.0144 (Weast 1982-83; Dean 1985; quoted, Mailhot & Peters 1988)

Molar Volume (cm³/mol):
    199.0 (LeBas method, Miller et al. 1985)
    0.949 (intrinsic volume: $V_I/100$, Kamlet et al. 1988; Hawker 1989)
    154.01 (Mailhot & Peters 1988)

Molecular Volume (Å³):
    158.45 (Pearlman et al. 1984)

Total Surface Area, TSA (Å²):
    192.9 (Yalkowsky & Valvani 1979; quoted, Mailhot & Peters 1988)
    196.79 (Pearlman et al. 1984)

Heat of Fusion, ΔH_{fus}, kcal/mol:

Entropy of Fusion, ΔS_{fus}, cal/mol K (e.u.):

Fugacity Ratio at 25 °C (assuming ΔS = 13.5 e.u.), F: 1.0

Water Solubility (g/m³ or mg/L at 25°C):
    8.00 (shake flask-fluorescence, Mackay & Shiu 1977; quoted, Yalkowsky & Valvani 1979, 1980; Mackay et al. 1980; Valvani & Yalkowsky 1980; Pearlman et al. 1984; Miller et al. 1985; Mailhot & Peters 1988; Hawker 1989; IUPAC 1989)
    11.90 (calculated-K_{ow}, Valvani & Yalkowsky 1980)
    10.80 (calculated-K_{ow}, Yalkowsky & Valvani 1980)
    8.01 (average lit. value, Yalkowsky et al. 1983)
    7.81 (average lit. value, Pearlman et al. 1984)

3.58   (calculated-QSAR of EPA, Passino & Smith 1987)
7.943   (quoted, Isnard & Lambert 1989)
8.0   (selected, Ma et al. 1990)

Vapor Pressure (Pa at 25°C):
6.950   (extrapolated, Antoine eqn., Dean 1985)

Henry's Law Constant (Pa m^3/mol):

Octanol/Water Partition Coefficient, log K_{OW}:
4.421   (fluorometry, Krishnamurthy & Wasik 1978)
4.42   (Hansch & Leo 1979; quoted, Mackay et al. 1980)
4.55   (calculated-S, Mackay et al. 1980)
4.38   (calculated-f const., Yalkowsky & Valvani 1979,1980)
4.36   (average lit. value, Yalkowsky et al. 1983)
4.42   (Miller et al. 1985; quoted, Kamlet et al. 1988; Hawker 1989)
4.42   (selected, Isnard & Lambert 1989)
4.44   (calculated-solvatochromic p., Kamlet et al. 1988)
4.38, 4.42   (quoted, Mailhot & Peters 1988)
4.50   (selected, Ma et al. 1990)

Bioconcentration Factor, log BCF:

Sorption Partition Coefficient, log K_{OC}:

Half-Lives in the Environment:
    Air:
    Surface water:
    Groundwater:
    Sediment:
    Soil:
    Biota:

Environmental Fate Rate Constants or Half-Lives:
    Volatilization:
    Photolysis:
    Oxidation:
    Hydrolysis:
    Biodegradation:
    Biotransformation:
    Bioconcentration, Uptake (k_1) and Elimination (k_2) Rate Constants:

Common Name: 1,4-Dimethylnaphthalene
Synonym:
Chemical Name: 1,4-dimethylnaphthalene
CAS Registry No: 571-58-4
Molecular Formula: $C_{12}H_{12}$
Molecular Weight: 156.23

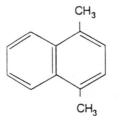

Melting Point (°C):
      7.66   (Weast 1982-83; Mackay et al. 1980; Miller et al. 1985; Mailhot & Peters 1988)
      -18   (Dean 1985)
      25.0   (Yalkowsky & Valvani 1979, 1980)
      < 25.0 (Lande & Banerjee 1981)
Boiling Point (°C):
      268   (Weast 1982-83; Pearlman et al. 1984; Mailhot & Peters 1987)
      264   (Dean 1985)
Density (g/cm³ at 20°C):
      1.0166   (Weast 1982-83)
      1.0157   (Dean 1985; Mailhot & Peters 1988)
Molar Volume (cm³/mol):
      154.0   (calculated, Lande & Banerjee 1981)
      199.0   (LeBas method, Miller et al. 1985)
      0.949   (intrinsic volume: $V_I/100$, Kamlet et al. 1988; Hawker 1989)
      153.82 (Mailhot & Peters 1988)
Molecular Volume (Å³):
      158.21 (Pearlman et al. 1984)
Total Surface Area, TSA (Å²):
      189.2   (Yalkowsky & Valvani 1979; quoted, Mailhot & Peters 1988)
      194.1   (Pearlman et al. 1984)
Heat of Fusion, ΔH_{fus}, kcal/mol:
Entropy of Fusion, ΔS_{fus}, cal/mol K (e.u.):
Fugacity Ratio at 25 °C (assuming ΔS = 13.5 e.u.), F: 1.0

Water Solubility (g/m^3 or mg/L at 25°C):

11.4   (shake flask-fluorescence, Mackay & Shiu 1977; quoted, Yalkowsky & Valvani 1979, 1980; Mackay et al. 1980; Valvani & Yalkowsky 1980; Lande & Banerjee 1981; Pearlman et al. 1984; Miller et al. 1985; Mailhot & Peters 1988; IUPAC 1989)

11.90   (calculated-K_{OW}, Valvani & Yalkowsky 1980)

10.80   (calculated-K_{OW}, Yalkowsky & Valvani 1980)

1.21   (calculated-TSA, Amidon & Anik 1981)

12.41   (average lit. value, Yalkowsky et al. 1983)

11.40   (average lit. value, Pearlman et al. 1984)

11.32   (quoted, Hawker 1989)

11.48   (quoted, Isnard & Lambert 1989)

11.4   (selected, Ma et al. 1990)

Vapor Pressure (Pa at 25°C):

2.27   (calculated-TSA, Amidon & Anik 1981)

4.50   (extrapolated, Antoine eqn., Dean 1985)

2.27   (selected, Ma et al. 1990)

Henry's Law Constant (Pa m^3/mol):

Octanol/Water Partition Coefficient, log K_{OW}:

4.372   (fluorometry, Krishnamurthy & Wasik 1978)

4.37   (Hansch & Leo 1979)

4.38   (calculated-TSA, Yalkowsky & Valvani 1979, 1980)

4.37, 4.39   (quoted, calculated-S & M.P., Mackay et al. 1980)

4.36   (calculated-f const., Yalkowsky et al. 1983)

4.37   (Miller et al. 1985; quoted, Hawker 1989)

4.37   (quoted, Isnard & Lambert 1989)

4.37, 4.44   (quoted, calculated-solvatochromic p. & V_I, Kamlet et al. 1988)

4.37, 4.38   (quoted, Mailhot & Peters 1988)

4.38   (selected, Ma et al. 1990)

Bioconcentration Factor, log BCF:

Sorption Partition Coefficient, log K_{OC}:

Half-Lives in the Environment:
        Air:
        Surface water:
        Groundwater:
        Sediment:
        Soil:
        Biota:

Environmental Fate Rate Constants or Half-Lives:
        Volatilization:
        Photolysis:
        Oxidation:
        Hydrolysis:
        Biodegradation:
        Biotransformation:
        Bioconcentration, Uptake (k_1) and Elimination (k_2) Rate Constants:

Common Name: 1,5-Dimethylnaphthalene
Synonym:
Chemical Name: 1,5-dimethylnaphthalene
CAS Registry No: 571-61-9
Molecular Formula: $C_{12}H_{12}$
Molecular Weight: 156.23

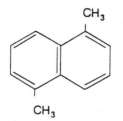

Melting Point (°C):
    81   (Yalkowsky & Valvani 1979, 1980; Mackay et al. 1980; Miller et al. 1985)
    82   (Weast 1982-83; Pearlman et al. 1984)
Boiling Point (°C):
    265   (Weast 1982-83; Pearlman et al. 1984)
Density (g/cm³ at 20°C):
Molar Volume (cm³/mol):
    199   (LeBas method, Miller et al. 1985)
    0.949   (intrinsic volume: $V_I/100$, Kamlet et al. 1988; Hawker 1989)
Molecular Volume (Å³):
    158.2   (Pearlman et al. 1984)
Total Surface Area, TSA (Å²):
    189.2   (Yalkowsky & Valvani 1979)
    194.1   (Pearlman et al. 1984)
Heat of Fusion, ΔH_{fus}, kcal/mol:
Entropy of Fusion, ΔS_{fus}, cal/mol K (e.u.):
Fugacity Ratio at 25 °C (assuming ΔS = 13.5 e.u.), F:
    0.274

Water Solubility (g/m³ or mg/L at 25°C):
    2.74   (shake flask-GC, Eganhouse & Calder 1976; quoted, Pearlman et al. 1984;
            IUPAC 1989)
    3.38   (shake flask-fluorescence, Mackay & Shiu 1977; Yalkowsky & Valvani 1979;
            Mackay et al. 1980,1982; Pearlman et al. 1984; Miller et al. 1985; Hawker
            1989; IUPAC 1989)
    3.26, 3.19   (quoted, Valvani & Yalkowsky 1980)
    3.19   (calculated-K_{OW}, Yalkowsky & Valvani 1980)

85

1.66   (calculated-TSA, Amidon & Anik 1981)
3.19   (average lit. value, Yalkowsky et al. 1983)
3.12   (average lit. value, Pearlman et al. 1984)
3.37   (selected, Miller et al. 1985; quoted, Hawker 1989)
3.31   (selected, Isnard & Lambert 1989)
3.38   (selected, Ma et al. 1990)

Vapor Pressure (Pa at 25°C):

Henry's Law Constant (Pa m^3/mol):
        35.5   (batch stripping-fluorescence, Mackay et al. 1982)

Octanol/Water Partition Coefficient, log K_{OW}:
        4.38   (fluorometry, Krishnamurthy & Wasik 1978)
        4.38   (Hansch & Leo 1979; selected, Miller et al. 1985; Eastcott et al. 1988)
        4.38   (calculated-f const., Yalkowsky & Valvani 1980)
        4.35   (calculated-TSA, Mackay et al. 1980)
        4.36   (calculated-f const., Yalkowsky et al. 1983)
        4.38   (quoted, Isnard & Lambert 1989)
        4.38, 4.44   (quoted, calculated-solvatochromic p. & V_I, Kamlet et al. 1988)

Bioconcentration Factor, log BCF:

Sorption Partition Coefficient, log K_{OC}:

Half-Lives in the Environment:
        Air:
        Surface water:
        Groundwater:
        Sediment:
        Soil:
        Biota:

Environmental Fate Rate Constants or Half-Lives:
        Volatilization:
        Photolysis:
        Hydrolysis:
        Oxidation:
        Biodegradation:
        Biotransformation:
        Bioconcentration, Uptake (k_1) and Elimination (k_2) Rate Constants:

Common Name: 2,3-Dimethylnaphthalene
Synonym: guaiene
Chemical Name: 2,3-dimethylnaphthalene
CAS Registry No: 581-40-8
Molecular Formula: $C_{12}H_{12}$
Molecular Weight: 156.23

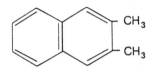

Melting Point (°C):
    105   (Weast 1972-73; Mackay et al. 1980; Miller et al. 1985; Mailhot & Peters 1988)
    102   (Yalkowsky & Valvani 1979, 1980; Lande & Banerjee 1981; Mailhot & Peters
            1988)
    103   (Pearlman et al. 1984)
    102-104 (Dean 1985)
Boiling Point (°C):
    268   (Weast 1982-83; Pearlman et al. 1984)
Density (g/cm³ at 20°C):
    1.003 (Weast 1982-83; Mailhot & Peters 1988)
    1.008 (Dean 1985)
Molar Volume (cm³/mol):
    156.0 (Lande & Banerjee 1981)
    199.0 (LeBas method, Miller et al. 1985)
    0.949 (intrinsic volume: $V_I/100$, Kamlet et al. 1988; Hawker 1989)
    155.76 (Mailhot & Peters 1988)
Molecular Volume (Å³):
    158.33 (Pearlman et al. 1984)
Total Surface Area, TSA (Å²):
    193.1 (Yalkowsky & Valvani 1979; quoted, Mailhot & Peters 1988)
    196.2 (Pearlman et al. 1984)
Heat of Fusion, ΔH_{fus}, kcal/mol:
Entropy of Fusion, ΔS_{fus}, cal/mol K (e.u.):
Fugacity Ratio at 25 °C (assuming ΔS = 13.5 e.u.), F:
    0.162 (at M.P. = 105°C)

Water Solubility (g/m³ or mg/L at 25°C):

    1.99   (shake flask-GC, Eganhouse & Calder 1976; quoted, IUPAC 1989)

    3.0   (shake flask-fluorescence, Mackay & Shiu 1977; quoted, Yalkowsky & Valvani 1979,1980; Mackay et al. 1980; Valvani & Yalkowsky 1980; Lande & Banerjee 1981; Pearlman et al. 1984; Miller et al. 1985; Mailhot & Peters 1988; IUPAC 1989)

    1.92   (calculated-K_{ow}, Valvani & Yalkowsky 1980)

    2.01   (calculated-K_{ow}, Yalkowsky & Valvani 1980)

    2.50   (calculated-TSA, Amidon & Anik 1981)

    2.65   (average lit. value, Yalkowsy et al. 1983)

    2.50   (average lit. value, Pearlman et al. 1984)

    2.26   (quoted, Hawker 1989)

    3.02   (quoted, Isnard & Lambert 1989)

    3.0   (selected, Ma et al. 1990)

Vapor Pressure (Pa at 25°C):

    0.627 (calculated-TSA, Amidon & Anik 1981)

    1.86   (extrapolated-Antoine eqn., Chao et al. 1983)

    0.91   (extrapolated-Antoine eqn., Boublik et al. 1984)

    1.543 (extrapolated, Antoine eqn., Dean 1985)

    0.437 (Antoine eqn., Stephenson & Malanowski 1987)

    1.0   (selected, Ma et al. 1990)

Henry's Law Constant (Pa m³/mol):

    92.16, 64.9 (quoted, calculated-bond contribution method, Meylan & Howard 1991)

    38.92 (calculated-P/C, Eastcott et al. 1988)

Octanol/Water Partition Coefficient, log K_{ow}:

    4.396 (fluorometry, Krishnamurthy & Wasik 1978)

    4.40   (Hansch & Leo 1979; quoted, Mackay et al. 1980)

    4.38   (calculated-f const., Yalkowsky & Valvani 1979,1980)

    4.18   (calculated-S & M.P., Mackay et al. 1980)

    4.36   (average lit. value, Yalkowsky et al. 1983)

    4.40   (selected, Miller et al. 1985; quoted, Kamlet et al. 1988; Hawker 1989)

    4.38, 4.40 (quoted, Mailhot & Peters 1988)

    4.40   (quoted, Isnard & Lambert 1989)

    4.44   (calculated-solvatochromic p., Kamlet et al. 1988)

Bioconcentration Factor, log BCF:

Sorption Partition Coefficient, log K_{oc}:

Half-Lives in the Environment:
        Air:
        Surface water:
        Groundwater:
        Sediment:
        Soil:
        Biota:

Environmental Fate Rate Constants or Half-Lives:
        Volatilization:
        Photolysis:
        Hydrolysis:
        Oxidation:
        Biodegradation:
        Biotransformation:
        Bioconcentration, Uptake (k_1) and Elimination (k_2) Rate Constants:

Common Name: 2,6-Dimethylnaphthalene
Synonym:
Chemical Name: 2,6-dimethylnaphthalene
CAS Registry No: 581-40-2
Molecular Formula: $C_{12}H_{12}$
Molecular Weight: 156.23

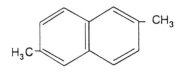

Melting Point (°C):
      110   (Weast 1983-84)
      110.2 (Dean 1985)
      108   (Yalkowsky & Valvani 1979,1980; Mackay et al. 1980; Mailhot & Peters 1988)
      111   (API 1981; quoted, Vadas et al. 1991)
      109   (Pearlman et al. 1984; Mailhot & Peters 1988)
Boiling Point (°C):
      262   (Dean 1985; Pearlman et al. 1984; Mailhot & Peters 1988)
Density (g/cm³ at 20°C):
      1.142 (Dean 1985)
Molar Volume (cm³/mol):
      199.0 (LeBas method, Miller et al. 1985)
      0.949 (intrinsic volume: $V_I/100$, Kamlet et al. 1988; Hawker 1989)
Molecular Volume (Å³):
      158.7 (Pearlman et al. 1984)
Total Surface Area, TSA (Å²):
      196.7 (Yalkowsky & Valvani 1979; quoted, Mailhot & Peters 1988)
      199.5 (Pearlman et al. 1984)
Heat of Fusion, ΔH_{fus}, kcal/mol:
      5.8   (Tsonopoulos & Prausnitz 1971)
      5.99 (calorimetry, Osborn & Douslin 1975)
Entropy of Fusion, ΔS_{fus}, cal/mol K (e.u.): 15.1 (Tsonopoulos & Prausnitz 1971)
Fugacity Ratio at 25 °C (assuming $\Delta S = 13.5$ e.u.), F:
      0.151 (at M.P. = 108 °C)

Water Solubility (g/m^3 or mg/L at 25°C):
- 1.30   (shake flask-GC, Eganhouse & Calder 1976; quoted, Pearlman et al. 1984; IUPAC 1989; Vadas et al. 1991)
- 2.0   (shake flask-fluorescence, Mackay & Shiu 1977; quoted, Yalkowsky & Valvani 1979, 1980; Mackay et al. 1980; Valvani & Yalkowski 1980; Pearlman et al. 1984; Miller et al. 1985; Eastcott et al. 1988, Mailhot & Peters 1988; IUPAC 1989; Vadas et al. 1991)
- 1.67   (calculated-K_{OW}, Valvani & Yalkowsky 1980)
- 1.79   (calculated-K_{OW}, Yalkowsky & Valvani 1980)
- 1.66   (calculated-TSA, Amidon & Anik 1981)
- 1.71   (average lit. value, Yalkowsky et al. 1983)
- 1.72   (average lit. value, Pearlman et al. 1984)
- 0.637   (calculated-QSAR of EPA, Passino & Smith 1987)
- 1.64   (quoted, Hawker 1989)
- 1.995   (quoted, Isnard & Lambert 1989)
- 2.0   (selected, Ma et al. 1990)
- 0.997   (gen. col.-HPLC, Vadas et al. 1991)

Vapor Pressure (Pa at 25°C):
- 0.75   (calculated-TSA, Amidon & Anik 1981)
- 1.38   (extrapolated, Antoine eqn., Chao et al. 1983)
- 1.41   (extrapolated, Antoine eqn., Boublik et al. 1984)
- 2.036   (extrapolated, Antoine eqn., Dean 1985)
- 0.378   (extrapolated, Antoine eqn., Stephenson & Malanowski 1987)
- 1.38   (quoted, Eastcott et al. 1988)
- 1.40   (selected, Ma et al. 1990)

Henry's Law Constant (Pa m^3/mol):
- 6.53   (calculated-P/C, Eastcott et al. 1988)

Octanol/Water Partition Coefficient, log K_{OW}:
- 4.313   (fluorometry, Krishnamurthy & Wasik 1978)
- 4.31   (Hansch & Leo 1979)
- 4.38   (calculated-f const., Yalkowsky & Valvani 1979, 1980)
- 4.31, 4.32   (quoted, calculated-S & M.P., Mackay et al. 1980)
- 4.36   (average lit. value, Yalkowsky et al. 1983)
- 4.31   (selected, Miller et al. 1985; quoted, Eastcott et al. 1988, Kamlet et al. 1988; Hawker 1989)
- 4.31, 4.38   (quoted, Mailhot & Peters 1988)
- 4.31   (quoted, Isnard & Lambert 1989)
- 4.31, 4.44   (quoted, calculated-solvatochromic p. & V_I, Kamlet et al. 1988)
- 4.31   (selected, Ma et al. 1990)

Bioconcentration Factor, log BCF:

Sorption Partition Coefficient, log K_{oc}:

Half-Lives in the Environment:
      Air:
      Surface water:
      Groundwater:
      Sediment:
      Soil:
      Biota:

Environmental Fate Rate Constants or Half-Lives:
      Volatilization:
      Photolysis: rate constant in distilled water of 0.045 hour^{-1} with half-life of 15.5 hours
           (Fukuda et al. 1988).
      Hydrolysis:
      Oxidation:
      Biodegradation:
      Biotransformation:
      Bioconcentration, Uptake (k_1) and Elimination (k_2) Rate Constants:

Common Name: 1-Ethylnaphthalene
Synonym: α-ethylnaphthalene
Chemical Name: 1-ethylnaphthalene
CAS Registry No:
Molecular Formula: $C_{12}H_{12}$
Molecular Weight: 156.23

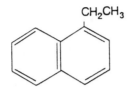

Melting Point (°C):
>   -13.88   (Weast 1972-73; Mackay et al. 1980; Mackay & Shiu 1981; Mailhot & Peters
>   1988)
>   25.0   (Yalkowsky & Valvani 1979,1980)
>   < 25.0 (Lande & Banerjee 1981)

Boiling Point (°C):
>   259.0   (Pearlman et al. 1984)
>   258.67 (Weast 1982-83; Mackay & Shiu 1981; Mailhot & Peters 1988)

Density (g/cm³ at 20°C):
>   1.0082 (Weast 1982-83; Mailhot & Peters 1988)

Molar Volume (cm³/mol):
>   155.0   (Lande & Banerjee 1981; Stephenson & Malanowski 1987)
>   199.0   (LeBas method, Miller et al. 1985)
>   0.949   (intrinsic volume: $V_I/100$, Kamlet et al. 1988; Hawker 1989)
>   154.96 (Mailhot & Peters 1988)
>   196.0   (Valsaraj 1988)

Molecular Volume (Å³):
>   158.58 (Pearlman et al. 1984)

Total Surface Area, TSA (Å²):
>   187.4   (Yalkowsky & Valvani 1979; quoted, Mailhot & Peters 1988)
>   192.28 (Pearlman et al. 1984)
>   187.4   (Vasalraj 1988)

Heat of Fusion, ΔH_{fus}, kcal/mol:
Entropy of Fusion, ΔS_{fus}, cal/mol K (e.u.):
Fugacity Ratio at 25°C (assuming $\Delta S = 13.5$ e.u.), F: 1.0

Water Solubility (g/m^3 or mg/L at 25°C):

10.7   (shake flask-fluorescence, Mackay & Shiu 1977; quoted, Yalkowsky & Valvani 1979; Mackay et al. 1980; Mackay & Shiu 1981; Wasik et al. 1981; Pearlman et al. 1984; Miller et al. 1985; Vowles & Mantoura 1987; Eastcott et al. 1988; Mailhot & Peters 1988; IUPAC 1989)

10.0   (shake flask-fluorescence, Schwarz & Wasik 1977)

10.0   (shake flask-fluorescence, Schwarz 1977; quoted, Mackay & Shiu 1981; Pearlman et al. 1984; IUPAC 1989)

10.80   (quoted, Valvani & Yalkowsky 1980; Yalkowsky & Valvani 1979, 1980; quoted, Lande & Banerjee 1981; Mailhot & Peters 1988)

11.60   (calculated-K_{ow}, Valvani & Yalkowsky 1980)

10.56   (calculated-K_{ow}, Yalkowsky & Valvani 1980)

10.36   (calculated-TSA, Amidon & Anik 1981)

11.58   (gen. col.-HPLC, Wasik et al. 1981)

9.86   (average lit. value, Yalkowsky et al. 1983)

10.31   (average lit. value, Pearlman et al. 1984)

10.90   (quoted, Valsaraj 1988)

10.81   (quoted, Hawker 1989)

10.715   (quoted, Isnard & Lambert 1989)

11.0   (quoted lit. average, Warne et al. 1990)

Vapor Pressure (Pa at 25°C):

2.51   (extrapolated from liq. state, Zwolinski & Wilhoit 1971; quoted, Mackay & Shiu 1981; Eastcott et al. 1988)

2.51   (extrapolated, Antoine eqn., Stephenson & Malanowski 1987)

2.53   (quoted, Valsaraj 1988)

Henry's Law Constant (Pa m^3/mol):

36.47   (selected, Valsaraj 1988)

14.8   (calculated-P/C, Eastcott et al. 1988)

Octanol/Water Partition Coefficient, log K_{ow}:

4.39   (calculated-f const., Yalkowsky & Valvani 1979, 1980; quoted, Mackay et al. 1980)

4.39, 4.42   (quoted, calculated-S & M.P., Mackay et al. 1980)

4.38   (calculated-f const., Yalkowsky et al. 1983)

4.39   (Miller et al. 1985; quoted, Vowles & Mantoura 1987; Kamlet et al. 1988; Hawker 1989)

3.87   (quoted, Eastcott et al. 1988)

4.39   (quoted, Mailhot & Peters 1988)

4.39   (quoted, Isnard & Lambert 1989)

4.39, 4.42   (quoted, calculated-solvatochromic p. & V_1, Kamlet et al. 1988)

4.39   (quoted, Warne et al. 1990)

Bioconcentration Factor, log BCF:

Sorption Partition Coefficient, log K_{OC}:
      3.77   (sediment, HPLC-k', Vowles & Mantoura 1987)

Half-Lives in the Environment:
      Air:
      Surface water:
      Groundwater:
      Sediment:
      Soil:
      Biota:

Environmental Fate Rate Constants or Half-Lives:
      Volatilization:
      Photolysis:
      Oxidation:
      Hydrolysis:
      Biodegradation:
      Biotransformation:
      Bioconcentration, Uptake (k_1) and Elimination (k_2) Rate Constants:

Common Name: 2-Ethylnaphthalene
Synonym: β-ethylnaphthalene
Chemical Name: 2-ethylnaphthalene
CAS Registry No: 939-27-5
Molecular Formula: $C_{12}H_{12}$
Molecular Weight: 156.23

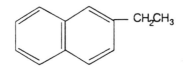

Melting Point (°C):
    -70   (Weast 1983-84)
    25.0   (Yalkowsky et al. 1983)
Boiling Point (°C):
    251-252 (Weast 1983-84)
    251   (Pearlman et al. 1984)
Density (g/cm³ at 20°C): 0.992 (Weast 1982-83)
Molar Volume (cm³/mol):
    199.0   (LeBas method, Miller et al. 1985)
    157.15   (calculated from density, Stephenson & Malanowski 1987)
Molecular Volume (Å³):
    158.9   (Pearlman et al. 1984)
Total Surface Area, TSA (Å²):
    199.91   (Pearlman et al. 1984)
Heat of Fusion, ΔH_{fus}, kcal/mol:
Entropy of Fusion, ΔS_{fus}, cal/mol K (e.u.):
Fugacity Ratio at 25°C (assuming ΔS = 13.5 e.u.), F: 1.0

Water Solubility (g/m³ or mg/L at 25°C):
    7.97   (shake flask-GC, Eganhouse & Calder 1976; quoted, Mackay & Shiu 1981;
               Pearlman et al. 1984; IUPAC 1989)
    7.99   (calculated-TSA, Amidon & Anik 1981)
    8.01   (average lit. value, Yalkowsky et al. 1983)
    7.97   (average lit. value, Pearlman et al. 1984)
    7.48   (Mantoura 1984; quoted, Vowles & Mantoura 1987)

Vapor Pressure (Pa at 25°C):
  3.24   (extrapolated from liq. state, Zwolinski & Wilhoit 1971; quoted, Mackay & Shiu 1981)
  4.21   (extrapolated, Macknick & Prausnitz 1979; quoted, Mackay & Shiu 1981)
  4.21   (Antoine eqn., Stephenson & Malanowski 1987)

Henry's Law Constant (Pa m³/mol):
  82.2   (calculated-P/C, Mackay & Shiu 1981)

Octanol/Water Partition Coefficient, log K_{OW}:
  4.377   (fluorometry, Krishnamurthy & Wasik 1978)
  4.38   (calculated-f const., Yalkowsky et al. 1983)
  4.43   (Vowles & Mantoura 1987)
  4.38   (quoted, Warne et al. 1990)

Bioconcentration Factor, log BCF:

Sorption Partition Coefficient, log K_{OC}:
  3.76   (sediment, HPLC-k', Vowles & Mantoura 1987)

Half-Lives in the Environment:
  Air:
  Surface water: half-life of 18.4 hours in distilled water (Fukuda et al. 1988).
  Groundwater:
  Sediment:
  Soil:
  Biota:

Environmental Fate Rate Constants or Half-Lives:
  Volatilization:
  Photolysis: rate constant in distilled water of 0.038 hour[-1] with half-life of 18.4 hours (Fukuda et al. 1988).
  Hydrolysis:
  Oxidation:
  Biodegradation:
  Biotransformation:
  Bioconcentration, Uptake (k_1) and Elimination (k_2) Rate Constants:

Common Name: 1,4,5-Trimethylnaphthalene
Synonym:
Chemical Name: 1,4,5-trimethylnaphthalene
CAS Registry No: 213-41-1
Molecular Formula: $C_{13}H_{14}$
Molecular Weight: 170.2

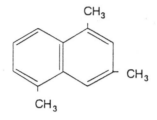

Melting Point (°C):
      64   (Weast 1983-84, Mackay et al. 1980; Mailhot & Peters 1988)
      25   (Yalkowsky & Valvani 1979,1980)
      <25   (Lande & Banerjee 1981)
Boiling Point (°C):
      285   (Zwolinski & Wilhoit 1971)
Density (g/cm³ at 20°C):
Molar Volume (cm³/mol):
      214   (LeBas method, Miller et al. 1985)
      169.0 (calculated, Lande & Banerjee 1981)
      1.047 (intrinsic volume: $V_I/100$, Kamlet et al. 1988; Hawker 1989)
Molecular Volume (Å³):
      170.72 (Pearlman et al. 1984)
Total Surface Area, TSA (Å²):
      200.9 (Yalkowsky & Valvani 1979; quoted, Mailhot & Peters 1988)
      204.4 (Pearlman et al. 1984)
Heat of Fusion, ΔH_{fus}, kcal/mol:
Entropy of Fusion, ΔS_{fus}, cal/mol K (e.u.):
Fugacity Ratio at 25 °C (assuming ΔS = 13.5 e.u.), F:
      0.411 (at M.P. = 64°C)

Water Solubility (g/m³ or mg/L at 25°C):
      2.1   (shake flask-fluorescence, Mackay & Shiu 1977; quoted, Yalkowsky & Valvani
             1979; Mackay et al. 1980; Pearlman et al. 1984; Miller et al. 1985; Mailhot &
             Peters 1988; IUPAC 1989)
      2.05 (quoted, Valvani & Yalkowsky 1980; Yalkowsky & Valvani 1979, 1980; quoted,
             Lande & Banerjee 1981)

4.48 (calculated-K_{OW}, Valvani & Yalkowsky 1980)
3.81 (calculated-K_{OW}, Yalkowsky & Valvani 1980)
2.09 (average lit. value, Yalkowsky et al. 1983)
2.04 (average lit. value, Pearlman et al. 1984)
5.77 (quoted, Hawker 1989)
2.08 (quoted, Isnard & Lambert 1989)

Vapor Pressure (Pa at 25°C):
    0.681 (subcooled liquid value, Chao et al. 1983; quoted, Eastcott et al. 1988)

Henry's Law Constant (Pa m^3/mol):
    23.50 (calculated-P/C, Eastcott et al. 1988)

Octanol/Water Partition Coefficient, log K_{OW}:
    4.90 (calculated-f const., Yakowsky & Valvani 1979, 1980)
    4.90, 4.79 (quoted, calculated-S & M.P., Mackay et al. 1980)
    4.87 (average lit. value, Yalkowsky et al. 1983)
    4.90 (selected, Miller et al. 1985; quoted, Eastcott et al. 1988; Hawker 1989)
    4.90, 4.94 (quoted, calculated-solvatochromic p. & V_1, Kamlet et al. 1988)
    4.90 (quoted, Mailhot & Peters 1988)

Bioconcentration Factor, log BCF:

Sorption Partition Coefficient, log K_{OC}:

Half-Lives in the Environment:
        Air:
        Surface water:
        Groundwater:
        Sediment:
        Soil:
        Biota:

Environmental Fate Rate Constants or Half-Lives:
        Volatilization:
        Photolysis:
        Oxidation:
        Hydrolysis:
        Biodegradation:
        Biotransformation:
        Bioconcentration, Uptake (k_1) and Elimination (k_2) Rate Constants:

99

Common Name: Biphenyl
Synonym: diphenyl
Chemical Name: biphenyl
CAS Registry No: 92-52-4
Molecular Formula: $C_{12}H_{10}$
Molecular Weight: 154.21

Melting Point (°C):
  68.6 (Parks & Huffman 1931)
  71   (Weast 1972-73; Mackay et al. 1980, Bruggeman et al. 1982; Burkhard et al.
          1985a; Shiu et al. 1987; Opperhuizen et al. 1988)
  87   (Banerjee et al. 1990)
Boiling Point (°C):
  255.9 (Weast 1972,1973)
  246.0 (Shiu & Mackay 1986)
Density (g/cm³ at 20°C): 0.866
Molar Volume (cm³/mol):
  184.6 (LeBas method, Miller et al. 1985; Shiu & Mackay 1986; Abernethy & Mackay
          1987; Shiu et al. 1987)
  0.920 (intrinsic volume, $V_I/100$, Kamlet et al. 1988; Hawker 1989b,1990; De Bruijn &
          Hermens 1990)
Molecular Volume (A³):
  207.0 (Opperhuizen et al. 1988)
  181.87, 163.7, 207.0 (De Bruijn & Hermens 1990)
Total Surface Area, TSA (A²):
  182.0 (Yalkowsky & Valvani 1979)
  192.2 (Mackay et al. 1980; Shiu et al. 1987)
  189.53 (Burkhard 1984)
  184.77 (planar, Doucette 1985)
  192.3 (nonplanar, Doucette 1985; Doucette & Andren 1988)
  195.2 (Shiu & Mackay 1986)
  224.1 (planar, shorthand, Opperhuizen et al. 1988)
  184.43 (planar, Hawker & Connell 1988a)
  182.0 (quoted as HSA, Woodburn et al. 1989)
  216.44, 189.44, 224.1 (De Bruijn & Hermens 1990)

Heat of Fusion, kcal/mol:
  4.44   (Parks & Huffman 1931)
  4.18   (Miller et al. 1984; Shiu & Mackay 1986)
Entropy of Fusion, cal/mol K (e.u.):
  12.20 (Miller et al. 1984; Shiu & Mackay 1986; Hinckley et al. 1990)
Fugacity Ratio, F (calculated, assuming $\Delta S_{fusion} = 13.5$ e.u.):
  0.35   (Mackay et al. 1980,1983; Shiu & Mackay 1986)
  0.352 (Shiu et al. 1987)

Water Solubility (g/m^3 or mg/L at 25°C):
  5.94   (shake flask-UV, Andrew & Keefer 1949,1950; quoted, Shiu et al. 1990)
  7.48   (shake flask-UV, Bohon & Claussen 1951; selected, Mackay & Wolkoff 1973;
           Mackay & Leinonen 1975; Shiu et al. 1990)
  3.87   (shake flask-UV, Sahyun 1966; quoted, Shiu et al. 1990)
  7.08   (shake flask-UV, Wauchope & Getzen 1972; quoted, Mackay et al. 1982; Shiu et
           al. 1990)
  7.0   (Hutzinger et al. 1974)
  7.45   (shake flask-GC, Eganhouse & Calder 1976; selected, Sh   t al. 1990)
  7.0   (shake flask-fluorescence, Mackay & Shiu 1977; selected   h   et al. 1981; Shiu
           et al. 1990)
  7.5   (quoted, Verschueren 1977, 1983; Kenaga 1980; Kenaga & Goring 1980; Neely
           1980; Shiu et al. 1990)
  8.5   (shake flask-nephelometric, Hollifield 1979)
  7.48   (quoted, Kilzer et al. 1979; selected, Sklarew & Girvin 1987)
  7.45   (quoted, Mackay et al. 1980; 1983a)
  7.05, 8.28 (quoted, estimated, Yalkowsky & Valvani 1980)
  8.09   (TLC-RT, Bruggeman et al. 1982)
  20.3   (quoted, subcooled liquid, Chiou et al. 1982; Chiou 1985)
  21.4   (quoted, subcooled liquid, Mackay et al. 1983a)
  7.21   (quoted lit. average, Yalkowsky et al. 1983; Erickson 1986)
  6.80   (quoted, Neely et al. 1983)
  2.82   (calculated-UNIFAC, Arbuckle 1983)
  23.6   (calculated-HPLC-k', converted from reported γ_w, Hafkenscheid & Tomlinson
           1983)
  6.71   (gen. col.-GC/ECD, Miller et al. 1984,1985; selected, Hawker 1989b)
  7.09   (quoted lit. mean, Pearlman et al. 1984; Shiu et al. 1990)
  22.82 (calculated-TSA, subcooled liquid value, Burkhard et al. 1985b)
  7.10, 8.08 (quoted, calculated-UNIFAC, Banerjee 1985)
  7.12, 20.3 (quoted, subcooled liquid, Chiou & Block 1986)
  7.0   (selected, Shiu & Mackay 1986; Shiu et al. 1987)
  7.50   (quoted, Chou & Griffin 1987)
  20.0   (selected, subcooled liquid, Shiu et al. 1987)
  8.09   (calculated-UNIFAC, converted from log γ, Arbuckle 1986)

19.2, 8.0 (exptl., calculated-UNIFAC, converted from log γ, Burkhard et al. 1986)

7.05   (vapor saturation-UV, Akiyoshi et al. 1987)

7.1, 6.5 (quoted; 29°C, shake flask-GC/FID; Stucki & Alexander 1987)

8.37   (calculated-UNIFAC, Banerjee & Howard 1988)

7.2   (gen. col.-HPLC/UV, Billington et al. 1988)

7.0   (quoted, Formica et al. 1988)

9.30   (quoted, Metcalfe et al. 1988)

16.9   (quoted, subcooled liq., Hawker 1989b)

7.05   (quoted, Nirmalakhandan & Speece 1989)

1.90   (calculated- χ , Nirmalakhandan & Speece 1989)

Vapor Pressure (Pa at 25°C):

1.30   (effusion method, Bright 1951)

0.031 (manometery, Augood et al. 1953; selected, Bidleman 1984)

1.273 (effusion method, Bradley & Cleasby 1953; selected, Bidleman 1984)

0.97   (Seki & Suzuki 1953)

1.03   (Aihara 1959)

7.60   (selected, Mackay & Wolkoff 1973; Mackay & Leinonen 1975; Mackay et al. 1982; Bopp 1983)

1.41   (effusion method, Radchenko & Kitiagorodskii 1974; selected, Bidleman 1984)

7.705 (subcooled liquid, Weast 1976-77)

1.866 (calculated-S & HLC, Mackay et al. 1979; selected, Bidleman 1984)

1.293 (quoted, Neely 1980)

1.33   (quoted, Neely 1981)

1.40   (HPLC-RT, Swann et al. 1983)

1.1-5.6 (quoted, Mackay et al. 1983a)

3.8   (quoted, subcooled liquid, Mackay et al. 1983a)

7.7   (quoted, subcooled liquid, Bopp 1983)

1.27   (quoted, Neely 1983; Erickson 1986)

5.608 (quoted, subcooled liquid, Bidleman 1984)

7.04; 6.22 (subcooled liquid value; GC-RT, Bildleman 1984)

5.61; 6.62 (subcooled liq. value, quoted; GC-RT, Foreman & Bidleman 1985)

1.19   (gas saturation, Burkhard 1984; selected, Sklarew & Girvin 1987)

1.01, 1.80 (quoted exptl., calculated-P/C, Burkhard et al. 1985a)

0.423, 0.703, 0.594 (calculated-MW, GC-RI, χ, Burkhard et al. 1985a)

2.03   (subcooled liq., calculated, GC-RI, Burkhard et al. 1985b)

2.43   (selected, Shiu & Mackay 1986; Shiu et al. 1987; Sklarew & Girvin 1987)

6.9   (selected, subcooled liquid, Shiu et al. 1987)

4.90   (quoted, Metcalfe et al. 1988)

1.186, 0.639 (quoted, calculated-UNIFAC, Banerjee et al. 1990)

5.610, 5.00 (quoted, Hinckley et al. 1990)

Henry's Law Constant (Pa m³/mol):

   157   (calculated-P/C, Mackay & Leinonen 1975)
   41.34   (batch stripping, Mackay et al. 1979)
   30.4   (batch stripping, Mackay et al. 1980; Mackay & Shiu 1981; Mackay et al. 1982)
   28.0   (calculated-P/C, Mackay & Shiu 1981)
   66.27   (quoted, Neely 1982)
   30-66   (calculated-P/C, Mackay et al. 1983a)
   27.0   (selected, Mackay et al. 1983)
   69.37   (calculated-P/C, Bopp 1983)
   34.65   (calculated-UNIFAC, Arbuckle 1983)
   13.68   (calculated-P/C, Burkhard et al. 1985b)
   53.5   (calculated-P/C, Shiu & Mackay 1986; Shiu et al. 1987)
   124   (calculated- χ , Nirmalakhandan & Speece 1988)
   19.57   (wetted-wall column, Fendinger & Glofelty 1990)

Octanol/Water Partition Coefficient, log K_{OW}:

   3.16   (shake flask-UV, Rogers & Cammarata 1969; quoted, Sangster 1989)
   3.19   (calculated-molecular orbital indices, Rogers & Cammarata 1969)
   4.09   (shake flask, Leo et al. 1971; Hansch & Leo 1979; quoted, Chiou et al. 1982)
   4.04   (shake flask, Hansch et al. 1973)
   4.17, 4.09, 3.16, 4.04   (Neely et al. 1974; Hansch & Leo 1979)
   3.95   (HPLC-k', Rekker & De Kort 1979; quoted, Ruepert et al. 1985)
   3.75   (HPLC-RT, Veith et al. 1979)
   4.09   (quoted, Veith et al. 1979; Veith & Kosian 1983)
   4.03   (calculated-f const., Yalkowsky & Valvani 1979, 1980; quoted, Arbuckle 1983)
   4.04   (shake flask-HPLC, Banerjee et al. 1980; quoted, Sangster 1989)
   3.88   (quoted, Kenaga & Goring 1980; Chou & Griffin 1987)
   4.14   (calculated-S & M.P., Mackay et al. 1980)
   4.10   (TLC, Bruggeman et al. 1982; quoted, Erickson 1986)
   4.06, 4.08 (quoted, HPLC-k', Hammers et al. 1982)
   3.70   (HPLC-RT, Woodburn 1982)
   3.16-4.09, 3.91 (shake flask, range, average, Eadsforth & Moser, 1983; selected, Sangster 1989)
   3.91-4.15, 4.05 (HPLC, range, average, Eadsforth & Moser 1983)
   4.03   (calculated-HPLC-k', Hafkenscheid & Tomlinson 1983)
   4.10   (quoted, Mackay et al. 1983; Kaiser 1983)
   3.79   (calculated-f const., Yalkowsky et al. 1983)
   4.09, 4.23 (quoted, calculated-molar fraction, Yoshida et al. 1983)
   4.15   (calculated-TSA, Burkhard 1984)
   3.76   (generator column-GC/ECD, Miller et al. 1984,1985; selected, Sklarew & Girvin 1987; Hawker 1989b; Sangster 1989)
   3.89   (generator column-HPLC, Woodburn et al. 1984; selected, Burkhard & Kuehl 1986; Sklarew & Girvin 1987; Sangster 1989)

3.70   (HPLC-RT, Woodburn et al. 1984)

3.79   (HPLC-RT, Rapaport & Eisenreich 1984; selected, Sklarew & Girvin 1987)

4.11-4.13 (HPLC-RV, Garst 1984)

4.09   (quoted exptl., Garst 1984)

4.10   (HPLC-RV, Garst & Wilson 1984)

3.88   (quoted, Freitag et al. 1985)

3.97   (calculated- χ as per Rekker & De Kort 1979, Ruepert et al. 1985)

3.90   (selected, Shiu & Mackay 1986; Abernethy & Mackay 1987; Shiu et al. 1987;
        Sklarew & Girvin 1987)

4.06   (quoted, Tomlinson & Hafkenscheid 1986)

3.63, 4.00 (HPLC-k', calculated, De Kock & Lord 1987)

3.89   (generator column-GC, Doucette & Andren 1987,1988)

4.03, 4.42 (calculated-π, TSA, Doucette & Andren 1987)

4.04, 3.62 (quoted, calculated-UNIFAC, Banerjee & Howard 1988)

4.03, 3.79, 3.69, 3.88, 4.35, 4.52 (calculated-π, f-const., HPLC-RT, MW, χ , TSA,
        Doucette & Andren 1988)

4.33   (quoted, Formica et al. 1988)

3.90, 3.99, 3.98 (calculated-f const., π, solvatochromic p., Kamlet et al. 1988)

4.09   (calculated-TSA, Hawker & Connell 1988a & b)

4.30   (Metcalfe et al. 1988)

3.90   (quoted, Hawker 1989b)

3.88   (quoted, Isnard & Lambert 1989)

4.008   (slow stirring, De Bruijn et al. 1989; De Bruijn & Hermens 1990)

4.10   (calculated-π, De Bruijn et al. 1989)

3.98   (recommended, Sangster 1989)

4.09   (quoted, Hawker 1990)

Bioconcentration Factor, log BCF:

2.64   (trout, calculated-k_1/k_2, Neely et al. 1974; quoted, Connell & Hawker 1988;
        Hawker 1990)

3.12   (rainbow trout, Veith et al. 1979; Veith & Kosian 1983)

2.53   (fish, flowing water, Kenaga & Goring 1980; Kenaga 1980)

2.30   (calculated-S, Kenaga 1980)

2.10   (calculated- χ , Koch 1983)

2.73, 2.45, 3.41 (algae, fish, activated sludge, Freitag et al. 1985; selected, Halfon &
        Reggiani 1986)

3.0   (fish, selected, Metcalfe et al. 1988)

Sorption Partition Coefficient, log K_{oc}:

3.15   (calculated, Kenaga 1980)

3.95   (suspended particulate-matter, calculated-K_{ow}, Burkhard 1984)

3.57, 3.77 (Lake Erie at 9.6 mg/L DOC, Landrum et al. 1984)

5.58, 4.04 (Huron River at 7.8 mg/L DOC, Landrum et al. 1984)

3.40   (calculated, soil, Chou & Griffin 1987)
3.23   (soils, batch equilibration adsorption coefficient, Kishi et al. 1990)

Sorption Partition Coefficient, log K_P:
    2.146   (lake sediment, calculated-K_{OW}, f_{OC}, Formica et al. 1988)

Half-Lives in the Environment:
    Air:
    Surface water: half-life in river water estimated to be 1.5 days (Bailey et al. 1983).
    Groundwater:
    Sediment:
    Soil:
    Biota: estimated half-life from fish in simulated ecosystem to be 29 hours (Neely 1980).

Environmental Fate Rate Constants and Half-Lives:
    Volatilization/Evaporation: half-life of evaporation from water depth of 1 meter estimated
        to be 7.52 hours (Mackay & Leinonen 1975) and rate of volatilization, 0.92
        g/m^2h (Mackay 1986; Metcalfe et al. 1988).
    Photolysis: photodegradation in methanol-water (3:7, v/v) with initial concentration of
        16.2 ppm by high pressure mercury lamp or sunlight with rate constant of
        5.1×10^{-4} minute^{-1} and half-life of 22.61 hours (Wang et al. 1991).
    Hydrolysis:
    Oxidation: rate constant for reaction with OH radicals in the gas phase using relative rate
        technique comparing with cyclohexane determined to be 8.5×10^{-12} cm^3 molecule^{-1}
        sec^{-1} at 295K for a 24-hour average OH radical concentration of 5×10^5 cm^{-3}
        (Atkinson & Aschmann 1985); room temperature rate constants for reaction with
        OH radicals in the gas phase calculated to be 7.9×10^{-12} cm^3 molecule^{-1} sec^{-1} and
        5.8 to 8.2×10^{-12} cm^3 molecule^{-1} sec^{-1} with a calculated tropospheric lifetime of
        3 days (Atkinson 1987).
    Biodegradation: 100% degraded by activated sludge in 47 hour cycle (Monsanto Co.
        1972); microbial degradation with pseudo first-order rate constant of 109 year^{-1}
        in the water column and 1090 year^{-1} in the sediment (Wong & Kaiser 1975;
        selected, Neely 1981); rate of biodegradation in water from Port Valdez
        estimated to be 9.3-9.8 nmol/L/day with an initial biphenyl concentration of
        4.4-4.7 μmol/L (data of Aug. 1977, Reichardt et al. 1981) and 3.2 nmol/L/day
        with initial concentration of 2.9 μmol/L (data of Aug. 1978, Reichardt et al.
        1981); half life of biodegradation estimated to be 1.5 days by using water die-
        away test (Bailey et al. 1983).
    Biotransformation:

Bioconcentration, Uptake (k_1) and Elimination (k_2) Rate Constants:

k_1:	6.79 hour^{-1}	(trout mussele, Neely et al. 1974)
k_2:	0.0155 hour^{-1}	(trout mussele, Neely et al. 1974)
k_1:	6.8 hour^{-1}	(trout, selected, Hawker & Connell 1985)
$1/k_2$:	65 hour^{-1}	(trout, selected, Hawker & Connell 1985)
log k_1:	2.21 day^{-1}	(fish, Connell & Hawker 1988)
log $1/k_2$:	0.43 day^{-1}	(fish, Connell & Hawker 1988)
$1/k_2$:	65 hour	(trout, Hawker & Connell 1988b)
log k_2:	0.43 day^{-1}	(fish, selected, Thomann 1989)

Common Name: 4-Methylbiphenyl
Synonym:
Chemical Name: 4-methylbiphenyl
CAS Registry No: 644-08-6
Molecular Formula: $C_{13}H_{12}$
Molecular Weight: 168.24

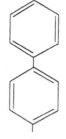

CH_3

Melting Point (°C): 49-50
Boiling Point (°C): 267-268
Density (g/cm³ at 27°C): 1.105
Molar Volume (cm³/mol):
Molecular Volume (Å³):
Total Surface Area, TSA (Å²):
     212.9 (Doucette & Andren 1987)
Heat of Fusion, ΔH_{fus}, kcal/mol:
Entrophy of Fusion, ΔS_{fus}, cal/mol K, or e.u.):
Fugacity Ratio at 25 °C (assuming ΔS = 13.5 e.u.), F:
    0.579 (at M.P. = 49 °C)

Water Solubility (g/m³ or mg/L at 25°C):
    4.05 (gen. col.-HPLC/GC, Doucette & Andren 1988a)

Vapor Pressure (Pa at 25°C):

Henry's Law Constant (Pa m³/mol):

Octanol/Water Partition Coefficient, log K_{OW}:
    4.63, 4.68, 4.91 (gen. col.-HPLC/GC, calculated-group contribution, TSA, Doucette & Andren 1987)

Bioconcentration Factor, log BCF:

Sorption Partition Coefficient, log K_{oc}:

Half-Lives in the Environment:
  Air:
  Surface water:
  Groundwater:
  Sediment:
  Soil:
  Biota:

Environmental Fate Rate Constants or Half-Lives:
  Volatilization:
  Photolysis:
  Oxidation:
  Hydrolysis:
  Biodegradation:
  Biotransformation:
  Bioconcentration, Uptake (k_1) and Elimination (k_2) Rate Constants:

Common Name: 4,4'-Dimethylbiphenyl
Synonym:
Chemical Name: 4,4'-dimethylbiphenyl
CAS Registry No: 613-33-2
Molecular Formula: $C_{14}H_{14}$
Molecular Weight: 182.27

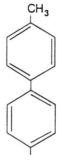

Melting Point (°C): 125
Boiling Point (°C): 295
Density (g/cm³ at 125°C): 0.917
Molar Volume (cm³/mol):
Molecular Volume (Å³):
Total Surface Area, TSA (Å²):
      233.4   (Doucette & Andren 1987)
Heat of Fusion, ΔH_{fus}, kcal/mol:
Entropy of Fusion, ΔS_{fus}, cal/mol K, or e.u.):
Fugacity Ratio at 25 °C (assuming ΔS = 13.5 e.u.), F: 0.103

Water Solubility (g/m³ or mg/L at 25°C):
      0.175   (gen. col.-GC, Doucette & Andren 1988a)

Vapor Pressure (Pa at 25°C):

Henry's Law Constant (Pa m³/mol):

Octanol/Water Partition Coefficient, log K_{OW}:
      5.09, 5.33, 5.40   (gen. col.-GC/ECD, calculated-group contribution, TSA, Doucette &
            Andren 1987)

Bioconcentration Factor, log BCF:

Sorption Partition Coefficient, log K_{OC}:

Half-Lives in the Environment:
        Air:
        Surface water:
        Groundwater:
        Sediment:
        Soil:
        Biota:

Environmental Fate Rate Constants or Half-Lives:
        Volatilization:
        Photolysis:
        Oxidation:
        Hydrolysis:
        Biodegradation:
        Biotransformation:
        Bioconcentration, Uptake (k_1) and Elimination (k_2) Rate Constants:

Common Name: Diphenylmethane
Synonym: diphenyl methane
Chemical Name: diphenylmethane
CAS Registry No: 101-85-5
Molecular Formula: $C_{13}H_{12}$
Molecular Weight: 188.24

Melting Point (°C):
      25.2 (Parks & Huffman 1931)
      25.3 (Weast 1982-83)
      26-27 (Stephenson & Malanowski 1987)
      25.0 (Bright 1951)
Boiling Point (°C):
      264.3 (Weast 1982-83)
      261-266 (Stephenson & Malanowski 1987)
Density (g/cm^3 at 20°C):
      1.006 (Weast 1982-83)
Molar Volume (cm^3/mol):
      187.12 (from density)
      190.2 (Stephenson & Malanowski 1987)
Molecular Volume (Å^3):
      171.7 (Pearlman et al. 1984)
Total Surface Area, TSA (Å^2):
      202.1 (Pearlman et al. 1984)
Heat of Fusion, ΔH_{fus}, kcal/mol:
      4.44 (Parks & Huffman 1931)
      4.40 (Stephenson & Malanowski 1987)
Entrophy of Fusion, ΔS_{fus}, cal/mol K, or e.u.):
      14.8 (Stephenson & Malanowski 1987)
Fugacity Ratio at 25°C (assuming $\Delta S = 13.5$ e.u.), F: 1.0

Water Solubility (g/m^3 or mg/L at 25°C):
      14.10 (shake flask/UV, Andrews and Keefer 1949; quoted, IUPAC 1989)
      16.40 (Deno & Berkheimer 1960)
      3.76 (Lu & Metcalf 1978)

3.00 (shake flask-nephelometry, Hollfield 1979)
16.02 (lit. mean, Yalkowsky et al. 1983)
16.19 (lit. mean, Pearlman et al. 1984; quoted, Tomlinson & Halkenscheid 1986)
14.12 (quoted, Isnard & Lambert 1989)

Vapor Pressure (Pa at 25°C):
1.09 (effusion method, Bright 1951)
0.0520 (extrapolated, Antoine eqn., Dean 1985)
0.0885 (interpolated, Antoine eqn., Stephenson & Malanowski 1987)

Henry's Law Constant (Pa m^3/mol):

Octanol/Water Partition Coefficient, log K_{ow}:
4.14 (Hansch & Leo 1979; quoted, Tomlinson & Hafkensheid 1986)
4.32 (calculated-f const., Yalkowsky et al. 1983)
4.14, 4.22, 4.36 (quoted, HPLC-RT, calculated-f const., Burkhard et al. 1985)
4.14 (quoted, Isnard & Lambert 1989)

Bioconcentration Factor, log BCF:

Sorption Partition Coefficient, log K_{oc}:

Half-Lives in the Environment:
     Air:
     Surface water:
     Groundwater:
     Sediment:
     Soil:
     Biota:

Environmental Fate Rate Constants or Half-Lives:
     Volatilization:
     Photolysis:
     Oxidation:
     Hydrolysis:
     Biodegradation:
     Biotransformation:
     Bioconcentration, Uptake (k_1) and Elimination (k_2) Rate Constants:

Common Name: Bibenzyl
Synonym: 1,2-Diphenylethane, dibenzyl
Chemical Name: 1,2-diphenylethane
CAS Registry No: 103-29-7
Molecular Formula: $C_{14}H_{14}$
Molecular Weight: 182.27

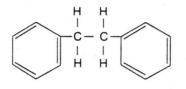

Melting Point (°C):
       51.4 (Parks & Huffman 1931)
       52.2 (Weast 1982-83)
       51.6-51.8 (Bright 1951)
Boiling Point (°C):
       285 (Weast 1982-83; Pearlman et al. 1984)
Density (g/cm³ at 60°C): 0.9583 (Weast 1982-83)
Molar Volume (cm³/mol):
       190.2 (Stephenson & Malanowski 1987)
       206.8 (LeBas method)
Molecular Volume (Å³):
       188   (Pearlman et al. 1984)
Total Surface Area, TSA (Å²):
       237   (Pearlman et al. 1984)
Heat of Fusion, ΔH_{fus}, kcal/mol:
       5.60 (Parks & Huffman 1931)
       7.30 (Stephenson & Malanowski 1987)
Entrophy of Fusion, ΔS_{fus}, cal/mol K, or e.u.):
       22.5 (Stephenson & Malanowski 1987)
Fugacity Ratio at 25°C (assuming ΔS = 13.5 e.u.), F:
       0.538 (at M.P. = 52.2°C)

Water Solubility (g/m³ or mg/L at 25°C):
       4.37 (shake flask-UV, Andrews & Keefer 1950b, quoted, Tsonopoulos & Prausnitz
               1971)
       4.37 (quoted, Pearlman et al. 1984)
       1.89, 0.44 (gen. col.-HPLC/UV, HPLC-RT, Swann et al. 1983)

113

Vapor Pressure (Pa at 25°C):
   0.198 (effusion method, Bright 1951)
   0.406 (interpolated, Antoine eqn., Stephenson & Malanowski 1987)

Henry's Law Constant (Pa m³/mol):

Octanol/Water Partition Coefficient, log K_{OW}:
   4.79, 4.82 (Hansch & Leo 1979)
   4.81 (quoted, Veith et al. 1979)
   4.79, 4.76 (quoted, HPLC-k', Hammers et al. 1982)
   3.76, 3.67 (quoted, HPLC-RT, Swann et al. 1983)

Bioconcentration Factor, log BCF:

Sorption Partition Coefficient, log K_{OC}:

Half-Lives in the Environment:
       Air:
       Surface water:
       Groundwater:
       Sediment:
       Soil:
       Biota:

Environmental Fate Rate Constants or Half-Lives:
       Volatilization:
       Photolysis:
       Oxidation:
       Hydrolysis:
       Biodegradation:
       Biotransformation:
       Bioconcentration, Uptake (k_1) and Elimination (k_2) Rate Constants:

114

Common Name: *trans*-1,2-Diphenylethene
Synonym: *trans*-stilbene, *trans*-diphenylethylene, E-stilbene
Chemical Name: *trans*-1,2-diphenylethene
CAS Registry No: 103-30-0
Molecular Formula: $C_{14}H_{12}$
Molecular Weight: 180.25

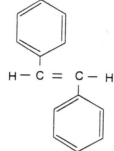

Melting Point (°C):
      124-125 (Weast 1982-82)
      124   (Stephenson & Malanowksi 1987)
Boiling Point (°C):
      305   (Weast 1982-83)
      306-307 (Stephenson & Malanowski 1987)
Density (g/cm³ at 20°C):
      0.9707 (Weast 1982-83)
Molar Volume (cm³/mol):
Molecular Volume (Å³):
Total Surface Area, TSA (Å²):
Heat of Fusion, ΔH_{fus}, kcal/mol:
      7.20 (Stephenson & Malanowski 1987)
Entrophy of Fusion, ΔS_{fus}, cal/mol K, or e.u.):
      18.1 (Stephenson & Malanowski 1987)
Fugacity Ratio at 25 °C (assuming ΔS = 13.5 e.u.), F: 0.105

Water Solubility (g/m³ or mg/L at 25°C):
      0.29   (shake flask-UV, Andrews & Keefer 1950, quoted, Tsonopoulos & Prausnitz
              1971; IUPAC 1989)
      0.292 (lit. mean, Yalkowsky et al. 1983)

Vapor Pressure (Pa at 25°C):
      0.0693 (24.4°C, manometer-spinning rotor gauge, Van Ekeren et al. 1983)
      0.0647 (interpolated, Stephenson & Malanowski 1987)

Henry's Law Constant (Pa m^3/mol):

Octanol/Water Partition Coefficient, log K_{ow}:
    4.81   (Hansch & Leo 1979)
    4.53   (calculated-f const., Yalkowsky et al. 1983)
    4.81   (recommended, Sangster 1989)

Bioconcentration Factor, log BCF:

Sorption Partition Coefficient, log K_{oc}:

Half-Lives in the Environment:
        Air:
        Surface water:
        Groundwater:
        Sediment:
        Soil:
        Biota:

Environmental Fate Rate Constants or Half-Lives:
        Volatilization:
        Photolysis:
        Oxidation:
        Hydrolysis:
        Biodegradation:
        Biotransformation:
        Bioconcentration, Uptake (k_1) and Elimination (k_2) Rate Constants:

Common Name: Acenaphthene
Synonym: 1,8-hydroacenaphthylene, ethylenenaphthalene, periethylenenaphthalene
Chemical Name: 1,8-hydroacenaphthylene
CAS Registry No: 83-32-9
Molecular Formula: $C_{12}H_{10}$
Molecular Weight: 154.20

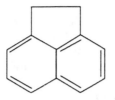

Melting Point (°C):
  96.2 (Weast 1982-83; quoted, Callahan et al. 1979; Banerjee et al. 1980; Mackay et
        al. 1980; Yalkowsky & Valvani 1979, 1980; Lande & Banerjee 1981; Mackay
        & Shiu 1981; Mabey et al. 1982; Arbuckle 1983; Miller et al. 1985; Gobas et
        al. 1988)
  90-95 (quoted, Verschueren 1983)
  93.0 (Pearlman et al. 1984; Banerjee & Baughman 1991)
  94.0 (Yalkowsky 1981; Banerjee et al. 1990)
Boiling Point (°C):
  277.5 (quoted, Mackay & Shiu 1981)
  279   (Weast 1982-82; quoted, Verschueren 1983; Pearlman et al. 1984; Banerjee et al.
        1990; Pussemier et al. 1990)
Density (g/cm³ at 20°C):
  1.069 (95°C, Dean 1985; quoted, Verschueren 1983)
  1.042 (95°C, Pussemier et al. 1990)
Molar Volume (cm³/mol):
  151.0 (calculated, Lande & Banerjee 1981)
  173.0 (LeBas method, Miller et al. 1985; Gobas et al. 1988)
  0.916 (intrinsic volume: $V_I/100$, Kamlet et al. 1988; Hawker 1989)
Molecular Volume (Å³):
  148.82 (Pearlman et al. 1984)
Total Surface Area, TSA (Å²):
  175.0 (Yalkowsky & Valvani 1979)
  180.77 (Pearlman et al. 1984)
Heat of Fusion, ΔH_{fus}, kcal/mol:
  4.95 (Tsonopoulos & Prausnitz 1971)
  5.23 (differential calorimetry, Wauchope & Getzen 1972)
  5.13 (calorimetry, Osborn & Douslin 1975)

117

Entropy of Fusion, ΔS_{fus}, cal/mol K (e.u.):
    13.5   (Tsonopoulos & Prausnitz 1971)
    14.4   (Wauchope & Getzen 1972; quoted, Yalkowsky 1981)
    14.3   (Casellato et al. 1973; quoted, Yalkowsky 1981)
    13.6   (Ubbelohde 1978; quoted, Yalkowsky 1981)
Fugacity Ratio at 25°C (assuming ΔS = 13.5 e.u.), F:
    0.197   (at M.P. = 96.2°C)

Water Solubility (g/m³ or mg/L at 25°C):
    6.14   (Deno & Berkheimer 1960; quoted, Pearlman et al. 1984)
    3.88   (shake flask-UV, Wauchope & Getzen 1972; quoted, Vesala 1974; Mackay &
            Shiu 1981; Mackay et al. 1982; Pearlman et al. 1984; Walters & Luthy 1984;
            IUPAC 1989)
    3.59   (shake flask-UV, Vesala 1974)
    3.47   (shake flask-GC, Eganhouse & Calder 1976; quoted, Mackay & Shiu 1981;
            Pearlman et al. 1984; Walters & Luthy 1984; IUPAC 1989)
    3.42   (quoted, Callahan et al. 1979; Mabey et al. 1982; Neuhauser et al. 1985)
    3.93   (shake flask-fluorescence, Mackay & Shiu 1977; Yalkowsky & Valvani 1979;
            Mackay et al. 1980; Mackay & Shiu 1981; Pearlman et al. 1984; Walters &
            Luthy 1984; Miller et al. 1985; IUPAC 1989)
    7.37   (shake flask-LSC, Banerjee et al. 1980; quoted, Pearlman et al. 1984; IUPAC
            1989)
    3.96   (shake flask-fluorescence, Valvani & Yalkowsky 1980; Yalkowsky & Valvani
            1979,1980; quoted, Lande & Banerjee 1981)
    4.45   (calculated-K_{OW}, Valvani & Yalkowsky 1980)
    4.88   (calculated-K_{OW}, Yalkowsky & Valvani 1980)
    3.96   (Lande & Banerjee 1981)
    2.42   (shake flask-GC, Rossi & Thomas 1981; quoted, Pearlman et al. 1984; IUPAC)
    3.40   (quoted, Mills et al. 1982)
    4.66, 0.090 (quoted, calculated-UNIFAC, Arbuckle 1983)
    4.25   (average lit. value, Yalkowsky et al. 1983)
    4.47   (average lit. value, Pearlman et al. 1984; quoted, Banerjee 1985; Farrington 1991)
    4.16   (HPLC/fluorescence, Walters & Luthy 1984)
    3.93   (selected, Miller et al. 1985; quoted, Gobas et al. 1988; Hawker 1989)
    3.98   (quoted, Brooke et al. 1986)
    3.76   (quoted, Shorten et al. 1990)

Vapor Pressure (Pa at 25°C):
    0.207   (Hoyer & Peperle 1958; quoted, Mabey et al. 1982)
    4.02   (extrapolated from liquid state, Weast 1972-73; quoted, Mackay & Shiu 1981;
            Mackay et al. 1982)
    3.07   (extrapolated from liquid state, Boublik et al. 1973; quoted, Mackay & Shiu 1981)
    3.07   (extrapolated from liquid state, Dean 1985)

0.373 (manometry-extrapolated, Osborn & Douslin 1975; quoted, Wasik et al. 1983)

0.60 (solid, fugacity ratio correlation, Mackay & Shiu 1981)

0.378 (calculated as per Mackay et al. 1979, Arbuckle 1983)

0.287 (gas saturation, Sonnefeld et al. 1983)

0.287 (gen. col.-HPLC, Wasik et al. 1983)

0.133 (quoted of Callahan et al. 1979, Neuhauser et al. 1985)

0.307 (selected, Howard et al. 1986; quoted, Banerjee et al. 1990)

0.3109 (Antoine eqn., Stephenson & Malanowski 1987)

0.122 (calculated-UNIFAC, Banerjee et al. 1990)

3.07 (quoted, Shorten et al. 1990)

Henry's Law Constant (Pa m^3/mol):

14.79 (batch stripping, Mackay et al. 1979; Mackay & Shiu 1981; quoted, Arbuckle 1983; Fendinger & Glotfelty 1990)

15.7 (batch stripping, Mackay & Shiu 1981; Mackay et al. 1982; quoted, Fendinger & Glotfelty 1990)

24.53 (calculated-P/C, Mackay & Shiu 1981; quoted, Pankow et al. 1984)

24.0 (recommended, Mackay & Shiu 1981)

9.22 (calculated-P/C, Mabey et al. 1982)

12.64 (calculated-UNIFAC, Arbuckle 1983)

24.42 (batch stripping, Warner et al. 1987)

24.79 (WERL Treatability database, quoted, Ryan et al. 1988)

6.45 (wetted-wall col., Fendinger & Glotfelty 1990)

14.8 (quoted, Shorten et al. 1990)

Octanol/Water Partition Coefficient, log K_{OW}:

4.33 (calculated as per Leo et al. 1971; Callahan et al. 1979; quoted, Neuhauser et al. 1985)

3.92 (shake flask-LSC, Veith et al. 1979, 1980)

4.03 (calculated-f const., Yalkowsky & Valvani 1979,1980; quoted, Mackay et al. 1980; Walters & Luthy 1984)

3.92 (Banerjee et al. 1980; quoted, Arbuckle 1983; Kayal & Connell 1990)

4.15 (calculated-S & M.P., Mackay et al. 1980)

4.49 (estimated, RP-HPLC-RT, Veith et al. 1980)

4.45 (calculated as per Leo et al. 1971, Veith et al. 1980)

3.98 (calculated-f const., Mabey et al. 1982)

3.92 (quoted, Mackay 1982; Schüürmann & Klein 1988)

3.32 (quoted, Mills et al. 1982)

3.70 (average lit. value, Yalkowsky et al. 1983)

3.92 (selected, Miller et al. 1985; quoted, Chin et al. 1986; Kamlet et al. 1988; Hawker 1989)

3.97 (calculated- χ as per Rekker & De Kort 1979, Ruepert et al. 1985)

4.33 (calculated-S & M.P., Chin et al. 1986)

3.92 (THOR 1986; quoted, Schüürmann & Klein 1988)
4.07 (Leo 1986; quoted, Schüürmann & Klein 1988)
4.22 (calculated-solvatochromic p., Kamlet et al. 1988)
4.17 (quoted, Pavlou 1987)
3.97 (quoted, Isnard & Lambert 1988; Banerjee & Baughman 1991)
4.13 (quoted, Ryan et al. 1988)
3.92 (recommended, Sangster 1989)
4.03 (quoted, Banerjee et al. 1990)
4.03 (quoted, Shorten et al. 1990)

Bioconcentration Factor, log BCF:
2.59 (bluegill sunfish, Veith et al. 1979,1980)
3.08 (calculated-K_{OW}, Veith et al. 1980)
2.59 (bluegill sunfish, Barrows et al. 1980)
2.59, 2.60 (quoted exptl., calculated-K_{OW}, Mackay 1982; quoted, Schüürmann & Klein 1988)
2.59 (bluegill sunfish, Davies & Dobbs, 1984)
2.60, 3.30 (quoted exptl., calculated- χ , Sabljic 1987)
2.58 (Isnard & Lambert 1988; quoted, Banerjee & Baughman 1991)
2.45 (calculated-$S_{OCTANOL}$ & M.P., Banerjee & Baughman 1991)

Sorption Partition Coefficient, log K_{OC}:
3.66 (calculated-K_{OW}, Mabey et al. 1982)
3.67 (calculated, Pavlou 1987)
5.38 (sediments average, Kayal & Connell 1990)
3.79 (RP-HPLC on CIHAC, Szabo 1990b)
3.59 (RP-HPLC on PIHAC, Szabo 1990b; quoted, Pussemier et al. 1990)

Half-Lives in the Environment:
Air: 0.879-8.79 hours, based on estimated photooxidation half-life in air (Atkinson 1987; quoted, Howard et al. 1991).
Surface water: 3-300 hours, based on photolysis half-life in water (Howard et al. 1991).
Groundwater: 590-4896 hours, based on estimated unacclimated aqueous aerobic biodegradation half-life (Howard et al. 1991).
Sediment:
Soil: 295-2448 hours, based on aerobic soil column test data (Kincannon & Lin 1985; quoted, Howard et al. 1991); > 50 days (Ryan et al. 1988).
Biota: < 1.0 day in the tissue of bluegill sunfish (Veith et al. 1980).

Environmental Fate Rate Constants or Half-Lives:

Volatilization:

Photolysis: half-lives on different atmospheric substrates determined in the rotary photoreactor (appr. 25 μg/gm on substrate): 2.0 hours on silica gel, 2.2 hours on alumina and 44 hours on fly ash (Behymer & Hites 1985); rate constant in distilled water of 0.23 hour^{-1} with half-life of 3 hours (Fukuda et al. 1988).

Hydrolysis: not hydrolyzable (Mabey et al. 1982).

Oxidation: rate constant of <3600 M^{-1} h^{-1} for singlet oxygen and 8000 M^{-1} h^{-1} for peroxy radical at 25°C (calculated, Mabey et al. 1982); photooxidation half-life in air: 0.879-8.79 hours, based on estimated rate constant for reaction with hydroxy radical in air (Atkinson 1987; quoted, Howard et al. 1991).

Biodegradation: significant degradation within 7 days for a domestic sewer test (Tabak et al. 1981); aerobic half-life: 295-2448 hours, based on aerobic soil column test data (Kincannon & Lin 1985; quoted, Howard et al. 1991); anaerobic half-life: 1180-9792 hours, based on estimated unacclimated aqueous aerobic biodegradation half-life (Howard et al. 1991).

Biotransformation: 3x10^{-9} ml cell^{-1} h^{-1} , estimated bacterial transformation rate constant (Mabey et al. 1982).

Bioconcentration, Uptake (k_1) and Elimination (k_2) Rate Constants:

Common Name: Acenaphthylene
Synonym:
Chemical Name: acenaphthylene
CAS Registry No: 208-96-8
Molecular Formula: $C_{12}H_8$
Molecular Weight: 152.20

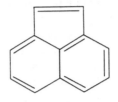

Melting Point (°C):
  92   (Weast 1977; quoted, Callahan et al. 1979; Mabey et al. 1982)
  80-83   (Verschueren 1983)
  92-93   (Dean 1985)
  96.2   (Lande et al. 1985)
Boiling Point (°C):
  265-275 (Weast 1975; Dean 1985; quoted, Mabey et al. 1982)
  280   (Verschueren 1983)
Density (g/cm³ at 20°C):
  0.899   (Dean 1985)
Molar Volume (cm³/mol):
Molecular Volume (Å³):
Total Surface Area, TSA (Å²):
  182.0   (Valvani et al. 1976; Yalkowsky & Valvani 1979; quoted, Lande et al. 1985)
Heat of Fusion, ΔH_{fus}, kcal/mol:
Entropy of Fusion, ΔS_{fus}, cal/mol K (e.u.):
Fugacity Ratio at 25 °C (assuming ΔS = 13.5 e.u.), F: 0.217

Water Solubility (g/m³ or mg/L at 25°C):
  3.93   (misquoted from Mackay & Shiu 1977; quoted, API 1978; Callahan et al. 1979;
        Mabey et al. 1982; Verschueren 1983; Walters & Luthy 1984; Warner et al.
        1987)
  3.93   (selected, Mills et al. 1982)
  16.1   (HPLC/fluorescence, Walters & Luthy 1984; quoted, Shorten et al. 1990)
  3.88, 2.94 (quoted, predicted, Lande et al. 1985)

Vapor Pressure (Pa at 25°C):
    3.87   (20°C, estimated, Callahan et al. 1979; quoted, Mabey et al. 1982)

    0.893   (effusion method, Sonnefeld et al. 1983)

    0.893   (gen. col.-HPLC, Wasik et al. 1983)

    1.105   (interpolated, Antoine eqn., Stephenson & Malanowski 1987)

Henry's Law Constant (Pa m^3/mol):
    146.9   (calculated-P/C, Mabey et al. 1982)

    11.55   (batch stripping, Warner et al. 1987)

    11.90   (WERL Treatability database, quoted, Ryan et al. 1988)

    11.40   (wetted-wall col., Fendinger & Glotfelty 1990)

Octanol/Water Partition Coefficient, log K_{OW}:
    4.07   (calculated as per Leo et al. 1971, Callahan et al. 1979; quoted, Chin et al. 1986)

    3.94   (Yalkowsky & Valvani 1979; quoted, Walters & Luthy 1984; Shorten et al. 1990)

    3.72   (calculated-f const., Mabey et al. 1982)

    4.08   (selected, Mills et al. 1982)

    4.06, 4.06   (quoted, calculated-MR, Yoshida et al. 1983)

    3.90   (calculated- χ as per Rekker & De Kort 1979, Ruepert et al. 1985)

    3.55   (RP-HPLC, calculated-S & M.P., Chin et al. 1986)

    4.07   (quoted, Pavlou 1987)

    4.07   (quoted, Ryan et al. 1988)

Bioconcentration Factor, log BCF:
    3.0   (microorganisms-water, calculated-K_{OW}, Mabey et al. 1982)

    2.58   (Isnard & Lambert 1988)

Sorption Partition Coefficient, log K_{OC}:
    3.4   (calculated-K_{OW}, Mabey et al. 1982)

    3.59   (soil, calculated, Pavlou 1987)

    3.83, 3.75   (soil, RP-HPLC on CIHAC, on PIHAC, Szabo et al. 1990b)

Half-Lives in the Environment:
    Air: 0.191-1.27 hours, based on photooxidation half-life in air (Atkinson 1987; quoted, Howard et al. 1991).

    Surface water: 1020-1440 hours, based on estimated unacclimated aqueous aerobic biodegradation half-life (Kincannon & Lin 1985; quoted, Howard et al. 1991).

    Groundwater: 2040-2880 hours, based on estimated unacclimated aqueous aerobic biodegradation half-life (Howard et al. 1991).

    Sediment:

    Soil: 1020-1440 hours, based on soil column study data (Kincannon & Lin 1985; quoted, Howard et al. 1991); > 50 days (Ryan et al. 1988).

    Biota:

Environmental Fate Rate Constants or Half-Lives:

Volatilization:

Photolysis: not environmentally significant (Mabey et al. 1982); half-lives on different atmospheric particulate substrates determined in the rotary photoreactor (appr. 25 μg/gm on substrate): 0.7 hour on silica gel, 2.2 hours on alumina and 44 hours on fly ash (Behymer & Hites 1985).

Hydrolysis: not hydrolyzable (Mabey et al. 1982; Howard et al. 1991).

Oxidation: rate constants: 4×10^7 M^{-1} hr^{-1} for singlet oxygen and 5×10^3 M^{-1} hr^{-1} for peroxy radical (calculated, Mabey et al. 1982); photooxidation half-life of 0.191-1.27 hours, based on estimated rate constants for reaction in air with hydroxy radical (Atkinson 1987; quoted, Howard 1991) and ozone (Atkinson & Carter 1984; quoted, Howard et al. 1991).

Biodegradation: >98% degradation within 7 days, based on domestic sewer for an average of three static-flask screening test (Tabak et al. 1981); aerobic half-life of 1020-1440 hours, based on soil column study data (Kincannon & Lin 1985; quoted, Howard et al. 1991); anaerobic half-life of 4080-5760 hours, based on estimated unacclimated aqueous aerobic biodegradation half-life (Howard et al. 1991).

Biotransformation: 3×10^{-9} ml cell^{-1} hr^{-1}, estimated rate constant for bacteria (Mabey et al. 1982).

Bioconcentration, Uptake (k_1) and Elimination (k_2) Rate Constants:

Common Name: Fluorene
Synonym: 2,3-benzindene, diphenylenemethane
Chemical Name: diphenylenemethane
CAS Registry No: 86-73-7
Molecular Formula: $C_{13}H_{10}$
Molecular Weight: 166.23

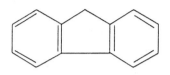

Melting Point (°C):
      116-118 (Sahyun 1966)
      117.0   (Windholz 1976; Yalkowsky et al. 1983b)
      116-117 (Weast 1977; quoted, Callahan et al. 1979; Mabey et al. 1982)
      116.0   (API 1978; Yalkowsky & Valvani 1979,1980; Mackay et al. 1980; Briggs 1981;
                 Lande & Banerjee 1981; Mackay & Shiu 1981; Bruggeman et al. 1982;
                 Bidleman 1984; Weast et al. 1984; Miller et al. 1985; Chin et al. 1986; Gobas
                 et al. 1988; Mailhot & Peters 1988; Vadas et al. 1991)
      114.8   (Dean 1985)
      113.0   (Yalkowsky 1981)
      114.0   (Pearlman et al. 1984; Banerjee & Baughman 1991)
Boiling Point (°C):
      293-295 (Weast 1975; quoted, Mabey et al. 1982; Pussemier et al. 1990)
      293   (API 1978)
      298.0   (Pearlman et al. 1984)
      295   (Dean 1985; Mackay & Shiu 1981; Mailhot & Peters 1988)
Density (g/cm³ at 20°C):
      1.203   (Mailhot & Peters 1988; Pussemier et al. 1990)
Molar Volume (cm³/mol):
      138.0   (calculated, Lande & Banerjee 1981)
      188.0   (LeBas method, Miller et al. 1985; Gobas et al. 1988)
      0.960   (intrinsic volume: V_I, Kamlet et al. 1988; Hawker 1989)
      138.18 (Mailhot & Peters 1988)
Molecular Volume (Å³):
      160.4   (Pearlman et al. 1984)

Total Surface Area, TSA (Å^2):

    193.6 (Yalkowsky & Valvani 1979; quoted, Mailhot & Peters 1988)

    194.0 (Pearlman et al. 1984)

Heat of Fusion, ΔH_{fus}, kcal/mol:

    4.67 (Wauchope & Getzen 1972)

    4.68 (Osborn & Douslin 1975)

Entropy of Fusion, ΔS_{fus}, cal/mol K (e.u.):

    12.1 (Wauchope & Getzen 1972; quoted, Yalkowsky 1981)

    11.6 (Casellato et al. 1973; quoted, Yalkowsky 1981)

    12.2 (Mackay & Shiu 1977; quoted, Hinckley 1989)

Fugacity Ratio at 25 °C (assuming ΔS = 13.5 e.u.), F:

    0.13 (Mackay et al. 1980)

    0.126 (at M.P. = 116°C)

Water Solubility (g/m^3 or mg/L at 25°C):

    1.90 (Pierotti et al. 1959; quoted, Mackay & Shiu 1977)

    1.66 (shake flask, binding to bovine serum albumin, Sahyun 1966)

    1.90 (shake flask-UV, Wauchope & Getzen 1972; quoted, Futoma et al. 1981; Mackay & Shiu 1981; Wasik et al. 1983; Pearlman et al. 1984; Walters & Luthy 1984; Miller et al. 1985; Billington et al. 1988; Mailhot & Peters 1988; IUPAC 1989; Vadas et al. 1991)

    1.98 (shake flask-fluorescence, Mackay & Shiu 1977; quoted, API 1978; Callahan et al. 1979; Yalkowsky & Valvani 1979; Mackay et al. 1980; Futoma et al. 1981; Lande & Banerjee 1981; Mackay & Shiu 1981; Mackay et al. 1982; Wasik et al. 1983; Pearlman et al. 1984; Walters & Luthy 1984; Warner et al. 1987; Billington et al. 1988; Mailhot & Peters 1988; IUPAC 1989; Shorten et al. 1990; Vadas et al. 1991)

    4.64 (Lu et al. 1978; quoted, Pearlman et al. 1984)

    1.68 (gen. col.-HPLC, May et al. 1978; quoted, Briggs 1981; Wise et al. 1981; Futoma et al. 1981; Pearlman et al. 1984; Walters & Luthy 1984; Mailhot & Peters 1988; IUPAC 1989)

    1.69 (gen. col.-HPLC/UV, May et al. 1978; quoted, Callahan et al. 1979; Mabey et al. 1982; Billington et al. 1988)

    2.0 (quoted, Yalkowsky & Valvani 1979; quoted, Lande & Banerjee 1981; Chin et al. 1986)

    2.0, 1.23 (quoted, calculated-K_{OW}, Valvani & Yalkowsky 1980)

    2.0, 1.32 (quoted, calculated-K_{OW}, Yalkowsky & Valvani 1980)

    1.50 (calculated-parachor, Briggs 1981)

    1.95 (RP-TLC, Bruggeman et al. 1982)

    1.90 (selected, Mills et al. 1982; quoted, Gobas et al. 1988; Hawker 1989; Capel et al. 1991)

    1.62 (24°C, gen. col.-HPLC, May et al. 1983)

    1.71 (average lit. value, Yalkowsky et al. 1983a)

2.05   (quoted, Yalkowsky et al. 1983b)
1.68   (gen. col.-HPLC, Wasik et al. 1983)
1.83   (average lit. value, Pearlman et al. 1984; quoted, Billington et al. 1988)
1.90   (HPLC/fluorescence, Walters & Luthy 1984)
1.90   (selected, Miller et al. 1985)
1.80   (quoted, Neuhauser et al. 1985)
10.98   (calculated-K_{ow} & M.P., Chin et al. 1986)
1.62   (calculated-QSAR of EPA, Passino & Smith 1987)
1.96   (gen. col.-HPLC/UV, Billington et al. 1988)
1.995   (quoted, Isnard & Lambert 1988, 1989)
2.23   (gen. col.-HPLC, Vadas et al. 1991)

Vapor Pressure (Pa at 25°C):
0.087   (effusion method, Bradley & Cleasby 1953; quoted, Mackay & Shiu 1981; Wasik et al. 1983; Bidleman 1984)
1.66   (extrapolated from liquid state, Weast 1972-73; quoted, Mackay & Shiu 1981; Mackay et al. 1982)
1.13   (extrapolated from liquid state, Boublik et al. 1973; quoted, Mackay & Shiu 1981)
1.133   (extrapolated, Antoine eqn., Dean 1985)
0.127   (manometry, Osborn & Douslin 1975; quoted, Wasik et al. 1983; Bidleman 1984)
0.0946 (Irwin 1982; quoted, Mabey et al. 1982)
0.079   (gas saturation-HPLC/UV, Sonnefeld et al. 1983; quoted, Bidleman 1984)
0.080   (gen. col.-HPLC, Wasik et al. 1983)
0.403   (Yamasaki et al. 1984; quoted, Capel et al. 1991)
0.133   (quoted, Neuhauser et al. 1985)
0.088   (extrapolated from Antoine eqn., Stephenson & Malanowski 1987)
0.793, 0.652 (subcooled liq. values, quoted, GC-RT, Hinckley et al. 1990)
0.0886 (quoted, Shorten et al. 1990)

Henry's Law Constant (Pa m^3/mol):
7.75   (batch stripping, Mackay & Shiu 1981; quoted, Fendinger & Glotfelty 1990)
7.74   (calculated-P/C, Mackay & Shiu 1981; quoted, Fendinger & Glotfelty 1990)
8.50   (recommended, Mackay & Shiu 1981)
6.48   (calculated-P/C, Mabey et al. 1982)
10.13   (batch stripping, Mackay et al. 1982)
11.85   (batch stripping, Warner et al. 1987)
10.57   (calculated-P/C, Nirmalakhandan & Speece 1988)
5.06   (calculated-UNIFAC, Nirmalakhandan & Speece 1988)
11.90   (WERL Treatability Database, quoted, Ryan et al. 1988)
6.45   (wetted-wall col., Fendinger & Glotfelty 1990)
33.4   (calculated-P/C, Capel et al. 1991)
10.1   (Shorten et al. 1990)

Octanol/Water Partition Coefficient, log K_{OW}:
- 4.18 (Hansch & Leo 1979; quoted, Mackay et al. 1980; Brooke et al. 1986)
- 4.12 (Chou & Jurs 1979; quoted, Kayal & Connell 1990)
- 4.18 (calculated-π const., Callahan et al. 1979; quoted, Neuhauser et al. 1985)
- 4.47 (calculated-f const., Yalkowsky & Valvani 1979,1980; quoted, Walters & Luthy 1984)
- 4.18 (HPLC-k', Rekker & De Kort 1979; quoted, Ruepert et al. 1985)
- 4.12, 4.03 (quoted, calculated-f const., Chou & Jurs 1979)
- 4.27 (calculated-S & M.P., Mackay 1980)
- 3.91 (HPLC-k', Hanai et al. 1981)
- 4.18 (RP-TLC, Bruggeman et al. 1982)
- 4.38 (Mackay 1982; quoted, Schüürmann & Klein 1988)
- 4.18 (Pomona 1982; quoted, Mabey et al. 1982)
- 4.18 (selected, Mills et al. 1982)
- 4.18 (shake flask-UV, Yalkowsky et al. 1983b)
- 4.18, 4.42 (quoted, calculated-MR, Yoshida et al. 1983)
- 4.38, 4.23 (quoted, HPLC-RT, Rapaport et al. 1984)
- 4.18 (selected, Miller et al. 1985; quoted, Chin et al. 1986; Kamlet et al. 1988; Hawker 1989; Banerjee & Baughman 1991; Capel et al;. 1991; Landrum et al. 1991)
- 4.12 (calculated-χ as per Rekker & De Kort 1979, Ruepert et al. 1985)
- 4.10 (RP-HPLC, calculated-S & M.P., Chin et al. 1986)
- 4.18 (quoted, Pavlou 1987)
- 4.18 (THOR 1986; quoted, Schüürmann & Klein 1988)
- 4.23 (Leo 1986; quoted, Schüürmann & Klein 1988)
- 4.38 (selected, Isnard & Lambert 1988,1989)
- 4.11 (calculated-solvatochromic p., Kamlet et al. 1988)
- 4.18, 4.38, 4.47 (quoted, Mailhot & Peters 1988)
- 4.18 (quoted, Ryan et al. 1988)
- 4.18 (quoted, Pussemier et al. 1990)
- 4.47 (quoted, Shorten et al. 1990)

Bioconcentration Factor, log BCF:
- 3.67 (microorganisms-water, calculated-K_{OW}, Mabey et al. 1982)
- 3.11 (calculated-K_{OW}, Mackay 1982; quoted, Schüürmann & Klein 1988; Banerjee & Baughman 1991)
- 2.70 (*daphnia magna*, Newsted & Giesy 1987)
- 3.11, 3.39 (quoted exptl., calculated-χ , Sabljic 1987b)
- 3.11 (calcd., Isnard & Lambert 1988)
- 2.62 (calculated-K_{OW}, $S_{OCTANOL}$ & M.P., Banerjee & Baughman 1991)

Sorption Partition Coefficient, log K_{OC}:
    4.15   (calculated-K_{OW}, Mabey et al. 1982)
    5.47   (sediments average, Kayal & Connell 1990)
    3.76   (RP-HPLC, Pussemier et al. 1990)
    4.15, 4.21   (RP-HPLC on CIHAC, on PIHAC, Szabo 1990b)

Half-Lives in the Environment:
    Air: 6.81-68.1 hours, based on estimated photooxidation half-life in air (Atkinson 1987; quoted, Howard et al. 1991).
    Surface water: 768-1440 hours, based on aerobic soil dieaway test data (Sims 1990; quoted, Howard et al. 1991).
    Groundwater: 1536-2880 hours, based on estimated unacclimated aqueous aerobic biodegradation half-life (Howard et al. 1991).
    Sediment:
    Soil: 768-1440 hours, based on aerobic soil dieaway test data (Coover & Sims 1987; quoted, Howard et al. 1991); >50 days (Ryan et al. 1988).
    Biota:

Environmental Fate Rate Constants or Half-Lives:
    Volatilization:
    Photolysis: half-lives on different atmospheric particulate substrates determined in rotary photoreactor (appr. 25 μg/gm on substrate): 110 hours on silica gel, 62 hours on alumina and 37 hours on fly ash (Behymer & Hites 1985).
    Hydrolysis: no hydrolyzable groups (Howard et al. 1991).
    Oxidation: calculated rate constants of <360 M^{-1} hr^{-1} for singlet oxygen and 3×10^3 M^{-1} hr^{-1} for peroxy radical (Mabey et al. 1982); photooxidation half-life in air of 6.81-68.1 hours, based on estimated rate constant for reaction with hydroxyl radical in air (Atkinson 1987; quoted, Howard et al. 1991).
    Biodegradation: significant degradation with gradual adaptation within 7 days for an average of three static-flask screening test (Tabak et al. 1981); nonautoclaved groundwater samples of approx. 0.06 mg/litre are degraded at rates of about 30% per week by microbes (Lee et al. 1984); aerobic half-life of 768-1440 hours, based on aerobic soil dieaway test data (Coover & Sims 1987; quoted, Howard et al. 1991); anaerobic half-life of 3072-5760 hours, based on estimated unacclimated aqueous aerobic biodegradation half-life (Howard et al. 1991).
    Biotransformation: estimated rate constant for bacteria, 3×10^{-9} ml cell^{-1} hr^{-1} (Mabey et al. 1982).
    Bioconcentration, Uptake (k_1) and Elimination (k_2) Rate Constants:

Common Name: 1-Methylfluorene
Synonym:
Chemical Name: 1-methylfluorene
CAS Registry No: 1730-37-6
Molecular Formula: $C_{14}H_{12}$
Molecular Weight: 180.25

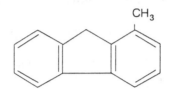

Melting Point (°C): 85
Boiling Point (°C): 318
Density (g/cm³ at 20°C):
Molar Volume (cm³/mol):
  210 (LeBas method, Miller et al. 1985)
  1.058 (intrinsic volume: $V_I/100$, Kamlet et al. 1988; quoted, Hawker 1989)
Molecular Volume (Å³):
Total Surface Area, TSA (Å²):
Heat of Fusion, ΔH_{fus}, kcal/mol:
Entropy of Fusion, ΔS_{fus}, cal/mol K (e.u.):
Fugacity Ratio at 25 °C (assuming $\Delta S = 13.5$ e.u.), F: 0.255

Water Solubility (g/m³ or mg/L at 25°C):
  1.092, 4.867 (measured, subcooled liquid value, Miller et al. 1985)
  1.092 (quoted, Hawker 1989)
  1.096 (quoted, Isnard & Lambert 1989)

Vapor Pressure (Pa at 25°C):

Henry's Law Constant (Pa m³/mol):

Octanol/Water Partition Coefficient, log K_{OW}:
  4.97 (calculated, Miller et al. 1985; quoted, Eastcott et al. 1988; Hawker 1989)
  4.97, 4.63 (quoted, calculated-solvatochromic p. & V_I, Kamlet et al. 1988)
  4.97 (quoted, Isnard & Lambert 1989)

Bioconcentration Factor, log BCF:

Sorption Partition Coefficient, log K_{OC}:

Half-Lives in the Environment:
        Air:
        Surface water:
        Groundwater:
        Sediment:
        Soil:
        Biota:
Environmental Fate Rate Constants or Half-Lives:
        Volatilization:
        Photolysis:
        Hydrolysis:
        Oxidation:
        Biodegradation:
        Biotransformation:
        Bioconcentration, Uptake (k_1) and Elimination (k_2) Rate Constants:

Common Name: Phenanthrene
Synonym: *o*-diphenyleneethylene, phenanthren
Chemical Name: phenanthrene
CAS Registry No: 85-01-8
Molecular Formula: $C_{14}H_{10}$
Molecular Weight: 178.24

Melting Point (°C):
  96.3   (Parks & Huffman 1931)
  101-102 (Sahyun 1966)
  101   (Weast 1977; quoted, API 1978; Callahan et al. 1979; Yalkowsky & Valvani
            1979,1980; Mackay et al. 1980; Briggs 1981; Karickhoff 1981; Steen &
            Karickhoff 1981; Lande & Banerjee 1981; MacKay & Shiu 1981; Bruggeman
            et al. 1982; Mabey et al. 1982; Whitehouse & Cooke 1982; Arbuckle 1983;
            Bidleman 1984; Patton et al. 1984; Weast et al. 1984; Lande et al. 1985; Miller
            et al. 1985; Chin et al. 1986; Gobas et al. 1988; Vadas et al. 1991)
  100   (Verschueren 1977; Yalkowsky et al. 1983; Banerjee et al. 1990)
  98.0   (Yalkowsky 1981)
  97.0   (Dean 1985; Pearlman et al. 1984; Banerjee & Baughman 1991)
Boiling Point (°C):
  339.0   (Mackay & Shiu 1981)
  338   (Bjørseth 1983)
  340   (Dean 1985; Verschueren 1977; API 1978; Pearlman et al. 1984; Banerjee et al.
            1990)
Density (g/cm³ at 20°C):
  1.174   (Dean 1985)
  0.9800   (4°C, Weast 1982-83)
Molar Volume (cm³/mol):
  182.0   (calculated, Lande & Banerjee 1981)
  199.0   (LeBas method, Miller et al. 1985; Abernethy & Mackay 1987; Gobas et al.
            1988)
  1.015   (intrinsic volume: $V_I/100$, Kamlet et al. 1987,1988; Hawker 1989)

132

Molecular Volume (Å^3):

    169.5   (Pearlman et al. 1984)

    1170.65 (Pearlman 1986)

Total Surface Area, TSA (Å^2):

    384.0   (supercooled liq., Amidon et al. 1974; quoted, Amidon et al. 1979)

    198.0   (Yalkowsky & Valvani 1979; quoted, Whitehouse & Cooke 1982; Pearlman et al. 1984; Lande et al. 1985)

    384.3   (calculated-1.5 Å solvent radius, Amidon & Anik 1980)

    284.3   (Amidon & Anik 1981; quoted, Lande et al. 1985)

    199.38 (Pearlman 1986)

    249.0   (calculated as per Gavezzotti 1985; Sabljic 1987)

    198.0   (quoted as HSA, Woodburn et al. 1989)

Heat of Fusion, ΔH_{fus}, kcal/mol:

    4.45   (Parks & Huffman 1931)

    4.45   (Tsonopoulos & Prausnitz 1971; Fu & Luthy 1985)

    3.89   (differential calorimetry, Wauchope & Getzen 1972)

    3.93   (calorimetry, Osborn & Douslin 1975)

Entropy of Fusion, ΔS_{fus}, cal/mol K (e.u.):

    11.9   (quoted, Tsonopoulos & Prausnitz 1971)

    10.5   (Wauchope & Getzen 1972; quoted, Yalkowsky 1981)

    10.8   (Casellato et al. 1973; quoted, Yalkowsky 1981)

    11.9   (Mackay & Shiu 1977; quoted, Hinckley 1989)

    12.1   (Ubbelohde 1978; quoted, Yalkowsky 1981)

    11.4   (De Kruif 1980; quoted, Hinckley 1989)

    12.07, 13.5   (observed, estimated, Yalkowsky & Valvani 1980)

    11.6   (selected, Hinckley 1989)

Fugacity Ratio at 25 °C (assuming ΔS = 13.5 e.u.), F:

    0.177   (at M.P. = 101°C)

    0.18   (Mackay et al. 1980)

Water Solubility (g/m^3 or mg/L at 25°C):

    1.65   (nephelometry, Davis & Parker 1942)

    1.600   (27°C, nephelometry, Davis et al. 1942; quoted, Barone et al. 1967; Futoma et al. 1981; Pearlman et al. 1984; Billington et al. 1988; IUPAC 1989)

    0.994   (shake flask-UV, Andrews & Keefer 1949; quoted, IUPAC 1989)

    1.600   (shake flask-UV, Klevens 1950; quoted, Barone et al. 1967; Futoma et al. 1981; Mackay & Shiu 1981; Pearlman et al. 1984)

    1.18   (Pierotti et al. 1959; quoted, Mackay & Shiu 1977)

    0.71   (shake flask, binding to bovine serum albumin, Sahyun 1966)

    2.67   (20°C, shake flask-UV, Eisenbrand & Baumann 1970)

    1.180   (shake flask-UV, Wauchope & Getzen 1972; quoted, Vesala 1974; API 1978; Futoma et al. 1981; Mackay & Shiu 1981; Wasik et al. 1983; Pearlman et al. 1984; Walters & Luthy 1984; Billington et al. 1988; IUPAC 1989)

1.21   (shake flask-UV, Vesala 1974)

1.07   (shake flask-GC, Eganhouse & Calder 1976; quoted, API 1978; Mackay & Shiu 1981; Whitehouse & Cooke 1982; Walters & Luthy 1984; IUPAC 1989)

1.290   (shake flask-fluorescence, Mackay & Shiu 1977; quoted, Callahan et al. 1979; Karickhoff et al. 1979; Mackay et al. 1979; May et al. 1979; Yalkowsky & Valvani 1979,1980; Hutchinson et al. 1980; Kenaga 1980a; Kenaga & Goring 1980; Mackay et al. 1980; Valvani & Yalkowsky 1980; Chiou 1981; Chiou & Schmedding 1981; Futoma et al. 1981; Karickhoff 1981; Lande & Banerjee 1981; Mackay & Shiu 1981; Steen & Karickhoff 1981; Chiou et al. 1982; Mackay et al. 1982; Whitehouse & Cooke 1982; Arbuckle 1983; Wasik et al. 1983; Pearlman et al. 1984; Walters & Luthy 1984; Fu & Luthy 1985; Lande et al. 1985; Chin et al. 1986; Kamlet et al. 1987; Landrum et al. 1987; Billington et al. 1988; IUPAC 1989; Shorten et al. 1990; Vadas et al. 1991; Mackay 1991)

1.002   (Rossi 1977; Neff 1979; quoted, Eadie et al. 1982; Whitehouse & Cooke 1982)

1.151   (shake flask-UV, Schwarz 1977; quoted, May et al. 1979; Futoma et al. 1981; Mackay & Shiu 1981; Wasik et al. 1983; Pearlman et al. 1984; Billington et al. 1988; IUPAC 1989)

1.002   (gen. col.-HPLC/UV, May et al. 1978; quoted, Callahan et al. 1979; May et al. 1979; Briggs 1981; Wise et al. 1981; Futoma et al. 1981; Mackay & Shiu 1981; Mabey et al. 1982; Pearlman et al. 1984; Walters & Luthy 1984; Billington et al. 1988; IUPAC 1989; Vadas et al. 1991)

1.29   (quoted, Kenaga 1980)

1.38   (calculated-K_{ow}, Valvani & Yalkowsky 1980)

0.99   (calculated-parachor, Briggs 1981)

1.260   (Lande & Banerjee 1981)

1.38; 3.99 (quoted, calculated-K_{ow}, Amidon & Williams 1982)

1.26   (RP-HPLC, Bruggeman et al. 1982)

1.0-1.3 (selected, Mills et al. 1982)

0.234   (calculated-UNIFAC, Arbuckle 1983)

0.956   (24.3°C, gen. col.-HPLC, May et al. 1983)

0.816   (quoted, Verschueren 1983)

1.00   (gen. col.-HPLC, Wasik et al. 1983)

1.10   (average lit. value, Yalkowsky et al. 1983)

1.28   (average lit. value, Pearlman et al. 1984; quoted, Banerjee 1985; Billington et al. 1988; Tomlinson & Halkenscheid 1986; Stucki & Alexander 1987; Farrington 1991)

1.000   (gen. col.-HPLC/UV, Wasik et al. 1983)

1.29   (HPLC/fluorescence, Walters & Luthy 1984)

0.31   (calculated-UNIFAC, Fu & Luthy 1985)

1.26, 1.36 (quoted, predicted, Lande et al. 1985)

1.180   (selected, Miller et al. 1985; quoted, Vowles & Mantoura 1987; Gobas et al. 1988; Hawker 1989; Capel et al. 1991)

134

11.25 (RP-HPLC, calculated-K_{ow} & M.P., Chin et al. 1986)

1.17 (calculated-QSAR of EPA, Passino & Smith 1987)

1.69 (29°C, shake flask-GC/FID, Stucki & Alexander 1987)

0.0446 (vapor saturation-UV, Akiyoshi et al. 1987)

1.26 (calculated-V_1, M.P. & hydrogen bond, Kamlet et al. 1987)

1.08 (gen. col.-HPLC, Billington et al. 1988)

1.290 (quoted, Isnard & Lambert 1988,1989)

1.200 (selected, Ma et al. 1990)

1.282 (quoted of Mackay & Shiu 1981, Edwards et al. 1991)

1.00 (gen. col.-HPLC, Vadas et al. 1991)

Vapor Pressure (Pa at 25°C):

0.0997 (effusion method, Inokuchi et al. 1952; quoted, Bidleman 1984)

0.0227 (effusion method, Bradley & Cleasby 1953; quoted, Mackay & Shiu 1981; Wasik et al. 1983; Bidleman 1984)

0.464 (extrapolated from liquid state, Weast 1972-73; quoted, Mackay & Shiu 1981)

0.159 (extrapolated from liquid state, Boublik et al. 1973; quoted, Mackay & Shiu 1981)

0.0187 (lit. average-interpolated, API 1979; quoted, Wasik et al. 1983)

0.0288 (calculated-HLC, Mackay et al. 1979; quoted, Arbuckle 1983; Bidleman 1984)

0.0263 (gas saturation, Macknick & Prausnitz 1979; quoted, Bidleman 1984)

0.0267 (extrapolated, Macknick & Prausnitz 1979; quoted, Mackay & Shiu 1981)

0.018 (effusion, De Kruif 1980)

0.0227 (calculated-TSA, Amidon & Anik 1981)

0.128 (calculated-Critical Tables, Mabey et al. 1982)

0.091 (20°C, selected, Mills et al. 1982)

0.0144 (20°C, Mackay et al. 1983)

0.0159 (gas saturation-HPLC/UV, Sonnefeld et al. 1983; quoted, Bidleman 1984)

0.016 (gen. col.-HPLC/UV, Wasik et al. 1983)

0.0688 (subcooled liq., Bidleman 1984)

0.070 (Yamasaki et al. 1984; quoted, Capel et al. 1991)

0.0149 (selected, Howard et al. 1986; quoted, Banerjee et al. 1990)

0.025 (Antoine eqn., Stephenson & Malanowski 1987)

0.0127, 0.0827 (lit. mean, subcooled liq. value, Bidleman & Foreman 1987)

0.106, 0.100 (quoted, GC-RT, Hinckley 1989)

0.0181 (calculated-UNIFAC, Banerjee et al. 1990)

0.018 (selected, Ma et al. 1990)

0.134, 0.10 (subcooled liq. values, Hinckley et al. 1990)

0.0267 (quoted, Shorten et al. 1990)

0.0161 (selected, Mackay 1991)

Henry's Law Constant (Pa m³/mol):

5.55 (batch stripping, Southworth 1979)

4.96   (calculated-P/C, Southworth 1979)

22.896 (calculated-P/C, Mabey et al. 1982)

3.981 (batch stripping, Mackay et al. 1979; Mackay & Shiu 1981; Mackay et al. 1982; quoted, Arbuckle 1983; Fendinger & Glotfelty 1990)

3.65   (batch stripping, Mackay & Shiu 1981; Mackay et al. 1982; quoted, Fendinger & Glotfelty 1990)

4.0   (recommended, Mackay & Shiu 1981)

3.85   (calculated-UNIFAC, Arbuckle 1983)

4.0   (calculated-P/C, Smith et al. 1983)

3.97   (Jury et al. 1984)

2.60, 2.06 (calculated-P/C, UNIFAC, Nirmalakhandan & Speece 1988)

3.97   (WERL Treatability database, quoted, Ryan et al. 1988)

0.691 (15°C, calculated, Baker & Eisenreich 1990)

2.38   (wetted-wall col., Fendinger & Glotfelty 1990)

10.13 (calculated-P/C, Capel et al. 1991)

3.98   (Shorten et al. 1990)

Octanol/Water Partition Coefficient, log K_{OW}:

    4.46   (Hansch & Fujita 1964; quoted, Mackay et al. 1980)

    4.46, 4.66 (quoted, calculated, Kier et al. 1971)

    4.46   (Leo et al. 1971; quoted, Hansch et al. 1973; Rekker 1977)

    4.46   (Radding et al. 1976; quoted, Callahan et al. 1979)

    4.45   (calculated from Leo 1975, Southworth et al. 1978)

    4.67   (calculated-f const., Rekker 1977)

    4.46   (Hansch & Leo 1979; quoted, Amidon & Williams 1982; Haky & Young 1984; Fu & Luthy 1985; Tomlinson & Halkenscheid 1986; De Bruijn et al. 1989)

    4.46   (Veith et al. 1979)

    4.63   (calculated-f const., Yalkowsky & Valvani 1979,1980; quoted, Hutchinson et al. 1980)

    4.57   (shake flask-UV, concn ratio, Karickhoff et al. 1979; quoted, Chiou 1981; Chiou & Schmedding 1981; Karickhoff 1981; Chiou et al. 1982; Geyer et al. 1984; Govers et al. 1984; Walters & Luthy 1984; Ruepert et al. 1985; Kayal & Connell 1990; Szabo et al. 1990a,b; Shorten et al. 1990; Edwards et al. 1991; Mackay 1991)

    4.52   (quoted, Kenaga & Goring 1980)

    4.64   (calculated-S & M.P., Mackay et al. 1980)

    4.57   (quoted, Steen & Karickhoff 1981)

    4.45   (HPLC-k', McDuffie 1981)

    4.63   (RP-TLC, Bruggeman et al. 1982)

    4.53   (HPLC, Hammers et al. 1982; quoted, De Bruijn et al. 1989)

    4.45   (calculated-f const., Mabey et al. 1982)

    4.46   (Mackay 1982; quoted, Chin et al. 1986; Schüürmann & Klein 1988)

    4.46   (selected, Mills et al. 1982)

4.63   (quoted, Whitehouse & Cooke 1982)
4.57   (quoted, Casserly et al. 1983)
4.52, 4.31 (shake flask, HPLC, Eadsforth & Moser 1983)
4.46   (HPLC-k', Hafkenscheid & Tomlinson 1983)
3.60, 5.92 (Mackay et al. 1983)
4.63   (average lit. value, Yalkowsky et al. 1983)
4.52, 4.72 (quoted, calculated-MR, Yoshida et al. 1983)
4.46   (quoted, Geyer et al. 1984)
4.46, 4.28 (quoted, HPLC-k', Haky & Young 1984)
4.45   (Mackay & Hughes 1984; quoted, Hawker & Connell 1986)
4.46   (quoted, Patton et al. 1984)
4.57   (Miller et al. 1985; quoted, Vowles & Mantoura 1987; Kamlet et al. 1988; De Bruijn et al. 1989; Hawker 1989; Banerjee & Baughman 1991; Capel et al. 1991; Landrum et al. 1991)
4.57   (calculated- χ as per Rekker & De Kort 1979, Ruepert et al. 1985)
4.46   (quoted, Abernethy & Mackay 1986,1987)
4.52   (quoted, Brooke et al. 1986)
4.39   (RP-HPLC, calculated-S & M.P., Chin et al. 1986)
4.45   (quoted, Hawker & Connell 1986)
4.49   (Leo 1986; quoted, Schüürmann & Klein 1988)
4.46   (THOR 1986; quoted, Schüürmann & Klein 1988)
4.52   (quoted, Pavlou 1987)
4.46   (selected, Isnard & Lambert 1988,1989)
4.44   (calculated-solvatochromatic p., Kamlet et al. 1988)
4.16   (quoted, Landrum 1988)
4.46   (quoted Ryan et al. 1988)
4.56   (De Bruijn et al. 1989)
4.50   (selected, Ma et al. 1990)
4.60   (Bayona et al. 1991)
4.16   (Landrum & Stubblefield 1991)

Bioconcentration Factor, log BCF:
3.42   (*pimephales promelas*, Carlson et al. 1978)
2.51   (*daphnia pulex*, Southworth et al. 1978; quoted, Govers et al. 1984)
2.57   (kinetic estimation, Southworth et al. 1978)
3.42   (fathead minnow, Veith et al. 1979, 1980; quoted, Bysshe 1982)
2.73, 3.3 (calculated-S, K_{oc}, Kenaga 1980)
3.80   (mixed microbial population, Steen & Karickhoff 1981)
2.79   (calculated as per Kenaga & Goring 1979, Eadie et al. 1982)
4.28   (*p. hoyi*, Eadie et al. 1982)
3.67   (microorganisms-water, Mabey et al. 1982)
3.14   (Mackay 1982)

137

4.38, 4.03, 4.57 (average, *selenastum capricornutu*m-dosed singly, dosed silmutaneously, Casserly et al. 1983)

3.25 (*chlorella fusca*, Geyer et al. 1984)

3.20 (calculated-K_{OW}, Geyer et al. 1984)

2.51 (Govers et al. 1984)

2.51 (*daphnia pulex*, Mackay & Hughes 1984; quoted, Hawker & Connell 1986)

2.97, 3.25, 3.25 (activated sludge, algae, fish, Freitag et al. 1985)

4.18 (*p. hoyi* of Lake Michigan interstitial waters, Landrum et al. 1985)

4.28 (*p. hoyi* of high sediment study site, Landrum et al. 1985)

2.51 (*daphnia magna*, Newsted & Giesy 1987)

3.42, 3.40 (quoted exptl., calculated- χ , Sabljic 1987b)

3.42 (calcd., Isnard & Lambert 1988; Banerjee & Baughman 1991)

3.21 (10-20°C, *h. limbata*, Landrum & Poore 1988)

4.45 (4°C, *p. hoyi*, quoted, Landrum & Poore 1988)

3.77 (4°C, *s. heringianus*, quoted, Landrum & Poore 1988)

3.43 (4°C, *mysis relicta*, quoted, Landrum & Poore 1988)

3.22 (calculated-K_{OW}, $S_{OCTANOL}$ & M.P., Banerjee & Baughman 1991)

0.756, 1.487 (*polychaete sp, capitella capitata*, Bayona et al. 1991)

Sorption Partition Coefficient, log K_{OC}:

4.36 (average of sorption isotherms on sediments, Karickhoff et al. 1979; Kenaga & Goring 1980; Govers et al. 1984)

4.36, 3.58 (quoted, calculated as per Kenaga & Goring 1980, Kenaga 1980a)

4.08 (quoted, Karickhoff 1981)

4.18, 4.22, 3.90 (calculated-K_{OW}, S & M.P., S, Karickhoff 1981)

4.15 (calculated-K_{OW}, Mabey et al. 1982)

4.36, 4.27 (quoted, calculated- χ , Sabljic 1984)

4.60 (fluorescence quenching interaction with AB humic acid, Gauthier et al. 1986)

4.28 (sediment, HPLC-k', Vowles & Mantoura 1987)

3.97 (calculated, Pavlou 1987)

4.28 (Hodson & Williams 1988)

6.12 (sediments average, Kayal & Connell 1990)

4.28 (calculated-lit. K_P, Szabo et al. 1990a,b)

4.22, 4.28 (RP-HPLC on CIHAC, on PIHAC, Szabo 1990b)

4.42 (sandy surface soil, Wood et al. 1990)

Half-Lives in the Environment:

Air: 2.01-20.1 hours, based on photooxidation half-life in air (Atkinson 1987; quoted, Howard et al. 1991).

Surface water: computed near-surface of a water body, half-life of 8.4 hours and for direct photochemical transformation at latitude 40°N, midday, midsummer with half-lives of 59 days (no sediment-water partitioning), 69 days (with sediment-

138

water partitioning) on direct photolysis in a 5-m deep inland water body (Zepp & Schlotzhauer 1979; quoted, Hallett & Brecher 1984); 3-25 hours, based on aqueous photolysis half-life (Lyman et al. 1982; quoted, Howard et al. 1991); calculated half-life of 59 days under sunlight for summer at 40°N latitude (Mill & Mabey 1985).

Groundwater: 768-9600 hours, based on estimated unacclimated aqueous aerobic biodegradation half-life (Howard et al. 1991).

Sediment:

Soil: 2.5-26 days (Sims & Overcash 1983; quoted, Bulman et al. 1987); half-life of 9.7 days for 5 mg/kg treatment and 14 days for 50 mg/kg (Bulman et al. 1987); biodegradation rate constant of 0.0447 day^{-1} with a half-life of 16 days in Kidman sandy loam soil and 0.0196 day^{-1} with half-life of 35 days in McLaurin sandy loam soil (Park et al. 1990); biodegradation half-life of 11 days in Kendaia soil (Manilal & Alexander 1991); 384-4800 hours, based on aerobic soil dieaway test data (Coover & Sims 1987; Sims 1990; quoted, Howard et al. 1991); > 50 days (Ryan et al. 1988); 0.4-26 weeks, 5.7 years (quoted, Luddington soil, Wild et al. 1991).

Biota: with depuration half-life of 40.9 hours in *s. heringianus* (Frank et al. 1986).

Environmental Fate Rate Constants or Half-Lives:

Volatilization: half-lives from solution: 97, 108 minutes (exptl., calculated, Mackay et al. 1983).

Photolysis: calculated half-lives of direct sunlight photolysis of 50% conversion at 40°N latitude of midday in midsummer: 8.4 hours (near surface water; quoted by Herbes et al. 1980; Harris 1982), 59 days (inland water) and 69 days (inland water with sediment partitioning) in a 5-m deep inland water body (Zepp & Schlotzhauer 1979); atmospheric and aqueous half-life of 3 hours, based on measured aqueous photolysis quantum yields and calculated for midday summer sunlight at 40°N latitude (Zepp & Stoltzhauer 1979; quoted, Howard et al. 1991) and of 25 hours after adjusting for approximate winter sunlight intensity (Lyman et al. 1982; quoted, Howard et al. 1991); half-lives on different atmospheric particulate substrates (appr. 25 μg/gm on substrate): 150 hours on silica gel, 45 hours on alumina and 49 hours on fly ash (Behymer & Hites 1985); half-life of 59 days under sunlight (Mill & Mabey 1985); rate constant in distilled water of 0.11 hour^{-1} with half-life of 6.3 hours (Fukuda et al. 1988); photodegradation in methanol-water (2:3, v/v) solution with an initial concentration of 5.0 ppm under high pressure mercury lamp or sunlight having rate constant of 6.53x10^3 minute and half-life of 1.78 hour (Wang et al. 1991).

Hydrolysis: not hydrolyzable (Mabey et al. 1982); no hydrolyzable groups (Howard et al. 1991).

Oxidation: aquatic fate rate of 0.01 liter M^{-1} sec^{-1} with half-life of $8x10^6$ days (Callahan et al. 1979); rate constant of <360 M^{-1} hr^{-1} for singlet oxygen and <36 M^{-1} hr^{-1} for peroxy radical (Mabey et al. 1982); rate constant for the gas phase reaction with OH radicals was $3.4x10^{-11}$ cm^3 $molecule^{-1}$ sec^{-1} at 298 ± 1 K , based on relative rate technique for propene (Biermann et al. 1985); photooxidation half-life of 2.01-20.1 hours, based on measured rate constant for reaction with hydroxyl radical in air (Atkinson 1987; quoted, Howard et al. 1991).

Biodegradation: 100% degradation within seven days for a domestic sewage of an average of three static-flask screening test (Tabak et al. 1981);aerobic half-life of 384-4800 hours, based on aerobic soil dieaway test data (Coover & Sims 1987; quoted, Howard et al. 1991); rate constants of 0.0447 day^{-1} with half-life of 16 days for Kidman sandy loam and 0.0196 day^{-1} with half-life of 35 days for McLarin sandy loam all at -0.33 bar soil moisture (Park et al. 1990); anaerobic half-life of 1536-19200 hours, based on estimated unacclimated aqueous aerobic biodegradation half-life (Howard et al. 1991); half-life in inorganic solution was found to be 4 days and in Kendaia soil was 11 days (Manilal & Alexander 1991).

Biotransformation: for bacteria, $1.6x10^7$ ml $cell^{-1}$ hr^{-1} (Paris et al. 1980; quoted, Mabey et al. 1982).

Bioconcentration, Uptake (k_1) and Elimination (k_2) Rate Constants:

k_1: 203 $hour^{-1}$ (*daphnia pulex*, Southworth et al. 1978)

k_2: 0.543 $hour^{-1}$ (*daphnia pulex*, Southworth et al. 1978)

log k_1: 2.31 $hour^{-1}$ (*daphnia pulex*, as per the correlation of Mackay & Hughes 1984, Hawker & Connell 1986)

log k_2: -0.27 $hour^{-1}$ (*daphnia pulex*, as per the correlation of Mackay & Hughes 1984, Hawker & Connell 1986)

k_1: 129.0 mL/g/h (4°C, *p. hoyi*, Landrum 1988; quoted, Landrum & Poore 1988)

k_2: 0.0046 $hour^{-1}$ (4°C, *p. hoyi*, Landrum 1988; quoted, Landrum & Poore 1988)

k_1: 52.5 $hour^{-1}$ (10-20°C, *h. limbata*, Landrum & Poore 1988)

k_2: 0.032 $hour^{-1}$ (10-20°C, *h. limbata*, Landrum & Poore 1988)

k_1: 94.0 $hour^{-1}$ (4°C, *s. Heringianus*, quoted, Landrum & Poore 1988)

k_2: 0.016 $hour^{-1}$ (4°C, *s. heringianus*, quoted, Landrum & Poore 1988)

k_1: 32.0 $hour^{-1}$ (4°C, *mysis relicta*, quoted, Landrum & Poore 1988)

k_2: 0.012 $hour^{-1}$ (4°C, *mysis relicta*, quoted, Landrum & Poore 1988)

Common Name: 1-Methylphenanthrene
Synonym:
Chemical Name: 1-methylphenanthrene
CAS Registry No: 832-69-6
Molecular Formula: $C_{15}H_{12}$
Molecular Weight: 192.26

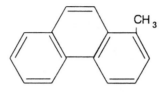

Melting Point (°C):
  123 (Weast 1982-83; Bjorseth 1983)
  119 (Pearlman et al. 1984)
Boiling Point (°C):
  359 (Weast 1982-83; Bjorseth 1983)
  390 (Pearlman et al. 1984)
Density (g/cm³ at 20°C):
Molar Volume (cm³/mol):
Molecular Volume (Å³):
  185.135 (Pearlman et al. 1984)
Total Surface Area, TSA (Å²):
  217.004 (Pearlman et al. 1984)
Heat of Fusion, ΔH_{fus}, kcal/mol:
Entropy of Fusion, ΔS_{fus}, cal/mol K (e.u.):
Fugacity Ratio at 25 °C (assuming ΔS = 13.5 e.u.), F:
  0.107 (at M.P. = 123°C)

Water Solubility (g/m³ or mg/L at 25°C):
  0.255 (24.1°C, gen. col.-HPLC/UV, May et al. 1978a,1983)
  0.269 (gen. col.-HPLC, May et al. 1978b; quoted, Pearlman et al. 1984; IUPAC 1989)
  0.27   (lit. mean, Yalkowsky et al. 1983)

Vapor Pressure (Pa at 25°C):

Henry's Law Constant (Pa m³/mol):

Octanol/Water Partition Coefficient, log K_{OW}:
    5.14 (calculated-f const., Yalkowsky et al. 1983)

Bioconcentration Factor, log BCF:

Sorption Partition Coefficient, log K_{OC}:

Half-Lives in the Environment:
    Air:
    Surface water:
    Groundwater:
    Sediment:
    Soil:
    Biota:

Environmental Fate Rate Constants or Half-Lives:
    Volatilization:
    Photolysis: photodegradation of 5 ppm initial concentration in methanol-water (3:7, v/v) by high pressure mercury lamp or sunlight with a rate constant of 1.84×10^3 minute^{-1} and half-life of 6.27 hours (Wang et al. 1991).
    Hydrolysis:
    Oxidation:
    Biodegradation:
    Biotransformation:
    Bioconcentration, Uptake (k_1) and Elimination (k_2) Rate Constants:

Common Name: Anthracene
Synonym: paranaphthalene, green oil, tetra olive NZG
Chemical Name: anthracene
CAS Registry No: 120-12-7
Molecular Formula: $C_{14}H_{10}$
Molecular Weight: 178.24

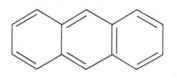

Melting Point (°C):
- 216.5 (Parks & Huffman 1931)
- 215.8 (Goursot et al. 1970; quoted, Rordorf 1986)
- 218   (Sahyun 1966; Yalkowsky 1981)
- 216   (API 1978; Yalkowsky & Valvani 1979,1980; Briggs 1981; Karickhoff 1981; Bruggeman et al. 1982; Chiou et al. 1982; Hashimoto et al. 1982; Whitehouse & Cooke 1982; Bidleman 1984; Pearlman et al. 1984; Miller et al. 1985; Chin et al. 1986; Mailhot 1987; Gobas et al. 1988; Banerjee et al. 1990; Murphy et al. 1990)
- 219   (Verschueren 1983; quoted, Pinal et al. 1990,1991)
- 167   (Bidleman 1984)
- 216.2 (Weast 1972-73; Dean 1985; Mackay et al. 1980; Lande & Banerjee 1981; Mackay & Shiu 1981; Weast et al. 1984; Lande et al. 1985; Vadas et al. 1991)

Boiling Point (°C):
- 340   (Weast 1982-83; Dean 1985; API 1978; Mackay & Shiu 1981; Pearlman et al. 1984; Rordorf 1986; Mailhot 1987; Banerjee et al. 1990; Pussemier et al. 1990)
- 340   (calculated, Rordorf 1986)

Density (g/cm³ at 20°C):
- 1.283 (Weast 1982-83; Pussemier et al. 1990)
- 1.250 (27°C, Dean 1985; quoted, Mailhot 1987)

Molar Volume (cm³/mol):
- 139.0 (calculated, Lande & Banerjee 1981)
- 197.0 (LeBas method, Miller et al. 1985; Gobas et al. 1988)
- 1.015 (intrinsic volume: $V_I/100$, Kamlet et al. 1987,1988; Hawker 1989)
- 143.0 (calculated-density, Mailhot 1987)

Molecular Volume ($\mathring{A}^3$):
  170.29 (Pearlman et al. 1984)
  171.42 (Pearlman 1986)
Total Surface Area, TSA ($\mathring{A}^2$):
  391.0 (calculated-with hydration cell, Amidon et al. 1974, 1975)
  202.2 (Yalkowsky & Valvani 1979; quoted, Whitehouse & Cooke 1982; Lande et al. 1985)
  390.9 (calculated-with hydration cell, Amidon & Anik 1980; quoted, Lande et al. 1985)
  200.0 (Mackay et al. 1982; quoted, Valsaraj & Thibodeaux 1989)
  200.16 (Pearlman et al. 1984)
  202.0 (Nkedi-Kizza et al. 1985)
  202.0, 381.0 (calculated-1.5 $\mathring{A}$ radius, Lande et al. 1985)
  203.51 (Pearlman 1986)
  201.6 (calculated as per Gavezzotti 1985, Sabljic 1987)
  203.0 (Pinal 1988; quoted, Pinal et al. 1990)
  202.0 (quoted as HSA, Woodburn et al. 1989)
  202.3 (Warne et al. 1990)
Heat of Fusion, ΔH_{fus}, kcal/mol:
  6.89 (Parks & Huffman 1931)
  6.9 (Tsonopoulos & Prausnitz 1971)
Entropy of Fusion, ΔS_{fus}, cal/mol K (e.u.):
  14.1 (quoted, Tsonopoulos & Prausnitz 1971)
  14.1 (Wauchope & Getzen 1972; quoted, Yalkowsky 1981)
  14.0 (Casellato et al. 1973; quoted, Yalkowsky 1981)
  14.4 (Mackay & Shiu 1977; quoted, Hinckley 1989)
  14.1 (Ubbelohde 1978; quoted, Yalkowsky 1981)
  13.3 (De Kruif 1980; quoted, Hinckley 1989)
  14.09, 13.5 (observed, estimated, Yalkowsky & Valvani 1980)
  14.1 (Rordorf 1986)
  13.8 (selected, Hinckley 1989)
Fugacity Ratio at 25°C (assuming $\Delta S = 13.5$ e.u.), F:
  0.01 (Mackay et al. 1980)

Water Solubility (g/m³ or mg/L at 25°C):
  0.075 (27°C, shake flask-nephelometry, Davis et al. 1942; quoted, Barone et al. 1967; Futoma et al. 1981; Billington et al. 1988; IUPAC 1989)
  0.075 (shake flask-UV, Klevens 1950; quoted, Barone et al. 1967; Futoma et al. 1981; Pearlman et al. 1984; IUPAC 1989)
  0.075 (Pierotti et al. 1959; Weimer & Prausnitz 1965; quoted, Mackay & Shiu 1977)
  0.112 (shake flask, binding to bovine serum albumin-UV, Sahyun 1966)
  0.043 (20°C, shake flask-UV, Eisenbrand & Baumann 1970; quoted, Billington et al. 1988)
  0.079 (HPLC, Locke 1974)

0.075 (shake flask-UV, Wauchope & Getzen 1972; quoted, Mackay & Shiu 1977; Futoma et al. 1981; Mackay & Shiu 1981; Pearlman et al. 1984; Walters & Luthy 1984; Nkedi-Kizza 1985; Tomlinson & Halkenscheid 1986; Billington et al. 1988; IUPAC 1989; Vadas et al. 1991)

0.030 (fluorescence/UV, Schwarz & Wasik 1976; quoted, Mackay & Shiu 1981)

0.075 (selected, Herbes 1977)

0.073 (shake flask-fluorescence, Mackay & Shiu 1977; quoted, API 1978; Callahan et al. 1979; Karickhoff et al. 1979; May et al. 1979; Yalkowsky & Valvani 1979, 1980; Hutchinson et al. 1980; Kenaga 1980a; Kenaga & Goring 1980; Mackay et al. 1980; Valvani & Yalkowsky 1980; Chiou 1981; Chiou & Schmedding 1981; Futoma et al. 1981; Karickhoff 1981; Lande & Banerjee 1981; Mackay & Shiu 1981; Mackay et al. 1982; Whitehouse & Cooke 1982; Wasik et al. 1983; Pearlman et al. 1984; Walters & Luthy 1984; Lande et al. 1985; Miller et al. 1985; Chin et al. 1986; Billington et al. 1988; IUPAC 1989; Valsaraj & Thibodeaux 1989; Kayal & Connell 1990; Shorten et al. 1990; Vadas et al. 1991)

0.0446 (Rossi 1977; Neff 1979; quoted, API 1978; Eadie et al. 1982; Whitehouse & Cooke 1982)

0.0410 (shake flask-UV, Schwarz 1977; quoted, May et al. 1979; Futoma et al. 1981; Mackay & Shiu 1981; Chiou et al. 1982; Whitehouse & Cooke 1982; Wasik et al. 1983; Pearlman et al. 1984; Walters & Luthy 1984; Billington et al. 1988; IUPAC 1989; Mackay 1991)

0.074 (Lu et al. 1978; quoted, Pearlman et al. 1984)

0.0446 (gen. col.-HPLC/UV, May et al. 1978; quoted, Callahan et al. 1979; May et al. 1979; Wise et al. 1981; Briggs 1981; Futoma et al. 1981; Mackay & Shiu 1981; Mabey et al. 1982; Walters & Luthy 1984; Billington et al. 1988; IUPAC 1989)

0.094 (calculated-K_{OW}, Valvani & Yalkowsky 1980)

0.115 (calculated-K_{OW}, Yalkowsky & Valvani 1980; quoted, Amidon & Williams 1982; Kamlet et al. 1987)

0.070 (calculated-parachor, Briggs 1981)

0.053 (selected, Geyer et al. 1981)

0.399 (calculated-K_{OW}, Amidon & Williams 1982)

0.074 (RP-TLC, Bruggeman et al. 1982)

0.05-0.07 (selected, Mills et al. 1982)

0.033 (20°C, gen. col.-fluorescence, Hashimoto et al. 1982)

0.057 (quoted average, Whitehouse & Cooke 1982)

0.0434 (24.6°C, gen. col.-HPLC, May et al. 1983)

0.030, 0.051 (gen. col.-HPLC/UV, Swann et al. 1983)

0.0446 (gen. col.-HPLC/UV, Wasik et al. 1983)

0.049 (average lit. value, Yalkowsky et al. 1983)

0.066 (average lit. value, Pearlman et al. 1984; quoted, Kamlet et al. 1987; Billington et al. 1988)

0.0698 (gen. col.-HPLC/fluorescence, Walters & Luthy 1984)

0.075, 0.198 (quoted, predicted, Lande et al. 1985)

0.075, 0.046 (selected, Miller et al. 1985; quoted, Gobas et al. 1988; Hawker 1989; Capel et al. 1991)

0.551 (calculated-K_{OW} & M.P., Chin et al. 1986)

0.0446 (vapor saturation-UV, Akiyoshi et al. 1987)

0.071, 0.068 (quoted average, calculated-V_1, M.P. & hydrogen bond, Kamlet et al. 1987)

0.063 (quoted, Mailhot 1987)

0.108 (calculated-QSAR of EPA, Passino & Smith 1987)

0.0443, 0.034 (gen. col.-HPLC/UV, Billington et al. 1988)

0.076 (selected, Isnard & Lambert 1988,1989)

0.041 (20°C, shake flask/UV, ring test, Kishi & Hashimoto 1989)

0.073 (quoted 1982, Valsaraj & Thibodeaux 1989)

0.075 (selected, Murphy et al. 1990)

0.070 (23°C, shake flask-RPLC/UV/fluorescence, Pinal et al. 1991)

0.058 (gen. col.-HPLC, Vadas et al. 1991)

Vapor Pressure (Pa at 25°C):

0.001014 (manometry, Sears & Hopke 1949; quoted, Bidleman 1984)

3.60×10^{-3} (effusion method, Inokuchi et al. 1952; quoted, Bidleman 1984)

8.31×10^{-4} (effusion method, Bradley & Cleasby 1953; quoted, Mackay & Shiu 1981; Wasik et al. 1983; Bidleman 1984)

1.04×10^{-3} (fluorescence spectroscopy, Stevens 1953; quoted, Bidleman 1984)

8.62×10^{-4} (effusion method, Kelley & Rice 1964; quoted, Wasik et al. 1983; Bidleman 1984)

3.87×10^{-7} (Wakayama & Inokuchi 1967; quoted, Bidleman 1984)

0.0024 (effusion method-interpolated, Widerman & Vaughn 1969; quoted, Wasik et al. 1983)

0.001113 (Knudsen effusion weight-loss method, Malaspina et al. 1973; quoted, Wasik et al. 1983; Bidleman 1984)

1.47×10^{-5} (effusion method, Murray et al. 1974; quoted, Bidleman 1984)

0.026 (20°C, Radding et al. 1976; quoted, Callahan et al. 1979)

9.04×10^{-4} (effusion method, Taylor & Crooks 1976; quoted, Bidleman 1984)

5.59×10^{-3} (gas saturation, Power et al. 1977; quoted, Bidleman 1984)

4.93×10^{-4} (lit. average-interpolated, API 1979; quoted, Wasik et al. 1983)

1.41×10^{-3} (gas saturation, Macknick & Prausnitz 1979; quoted, Mackay & Shiu 1981; Bidleman 1984)

7.50×10^{-4} (effusion method, De Kruif 1980)

0.00049 (calculated-TSA, Amidon & Anik 1981)

1.83×10^{-3} (gas saturation, Grayson & Fosbraey 1982; quoted, Bidleman 1984)

0.00227 (Jaber 1982; quoted, Mabey et al. 1982)

0.0266 (20°C, selected, Mills et al. 1982)

1.44×10^{-3} (extrapolated, gas saturation, Bender et al. 1983)

7.89x10^{-4} (gas saturation-UV, Sonnefeld et al. 1983; quoted, Bidleman 1984)

8.00x10^{-4} (gen. col.-HPLC, Wasik et al. 1983)

0.065   (Yamasaki et al. 1984; quoted, Capel et al. 1991)

0.040   (quoted, Nkedi & Kizza 1985)

3.60x10^{-4} (selected, Howard et al. 1986; quoted, Banerjee et al. 1990)

8.05x10^{-4} (extrapolated, gas saturation, Hansen & Eckert 1986)

1.14x10^{-3} (extrapolated-Antoine eqn., Stephenson & Malanowski 1987)

0.000573, 0.0056 (lit. mean, subcooled liq. value, Bidleman & Foreman 1987)

0.095, 0.0926 (quoted, GC-RT, Hinckley 1989)

0.086, 0.0940 (quoted, subcooled liq. values, Hinckley et al. 1990)

1.25x10^{-3} (calculated-UNIFAC, Banerjee et al. 1990)

0.00144   (quoted, Shorten et al. 1990)

Henry's Law Constant (Pa m^3/mol):

6.59   (25°C, batch stripping, Southworth 1977)

29.75   (calculated-P/C, Southworth 1979)

7.19, 59.50 (batch stripping, calculated-P/C, Mackay & Shiu 1981; quoted, Fendinger
            & Glotfelty 1990)

6.0   (recommended, Mackay & Shiu 1981)

8.712   (calculated-P/C, Mabey et al. 1982)

73.0, 676 (batch stripping, quoted, Mackay et al. 1982)

10.0   (calculated-P/C, Smith et al. 1983)

1.80, 110 (calculated-P/C, UNIFAC, Nirmalakhandan & Speece 1988)

27.27   (WERL Treatability database, quoted, Ryan et al. 1988)

1.96   (wetted-wall col./GC, Fendinger & Glotfelty 1990)

73   (Shorten et al. 1990)

16.21   (calculated-P/C, Capel et al. 1991)

Octanol/Water Partition Coefficient, log K_{OW}:

4.45   (Hansch & Fujita 1964)

4.45   (Leo et al. 1971; quoted, Kringstad et al. 1984; Ogata et al. 1984)

4.45   (Hansch et al. 1973; quoted, Rekker 1977)

4.45   (Radding et al. 1976; quoted, Callahan et al. 1979)

4.67   (calculated-f const., Rekker 1977)

4.45   (Herbes & Risi 1978)

4.45   (calculated from Leo 1975, Southworth et al. 1978)

4.45   (Hansch & Leo 1979; quoted, Tomlinsons & Hafkenscheid 1986)

4.54   (shake flask-UV, concn. ratio, Karickhoff et al. 1979; quoted, Chiou 1981;
        Chiou & Schmedding 1981; Karickhoff 1981; Chiou et al. 1982; Geyer et al.
        1984; Govers et al. 1984; Walters & Luthy 1984; Ruepert et al. 1985; Kayal &
        Connell 1990; Szabo et al. 1990a,b; Shorten et al. 1990)

4.45, 3.45 (quoted, HPLC-RT, Veith 1979)

4.45, 4.45 (quoted, calcd.-f constant, Chou & Jurs 1979)

147

4.63   (calculated-f const., Yalkowsky & Valvani 1979, 1980; quoted, Hutchinson et al. 1980; Whitehouse & Cooke 1982)

4.34   (Kenaga & Goring 1980)

4.45, 4.73 (quoted, calculated-S & M.P., Mackay et al.1980)

4.54   (Chiou 1981; quoted, Valsaraj & Thibodeaux 1989)

4.49   (HPLC-k', McDuffie 1981)

4.38   (HPLC-k', Hanai et al. 1981)

4.63   (RP-TLC, Bruggeman et al. 1982; quoted, Kringstad et al. 1984)

4.54   (quoted, Chiou et al. 1982; Brooke et al. 1986)

4.20   (HPLC-k', D'Amboise 1982)

4.45   (calculated-f const., Mabey et al. 1982)

4.34   (quoted, Mackay 1982; Schüürmann & Klein 1988)

4.45   (selected, Mills et al. 1982)

4.45   (HPLC-k', Hafkenscheid & Tomlinson 1983)

4.63   (average lit. value, Yalkowsky et al. 1983)

4.45   (Verschueren 1983; quoted, Pinal et al. 1990, 1991)

4.45, 4.72 (quoted, calculated-MR, Yoshida et al. 1983)

4.54, 4.45 (quoted, HPLC-RT, Rapaport et al. 1984)

4.45   (quoted, Mackay & Hughes 1984; Hawker & Connell 1986)

4.45   (Hansch & Leo 1985; quoted, Banerjee & Howard 1988)

4.54   (selected, Miller et al. 1985; quoted, Chin et al. 1986; Kamlet et al. 1988;Hawker 1989; Capel et al. 1991; Landrum et al. 1991)

4.57   (calculated- χ as per Rekker & De Kort 1979, Ruepert et al. 1985)

4.49   (quoted, Mailhot 1987)

4.40   (quoted, Pavlou 1987)

4.45   (THOR 1986; quoted, Schüürmann & Klein 1988)

4.49   (Leo 1986; quoted, Schüürmann & Klein 1988)

4.15   (calculated-UNIFAC, Banerjee & Howard 1988)

4.54   (selected, Isnard & Lambert 1988,1989)

4.44   (calculated-solvatochromic p., Kamlet et al. 1988)

4.45   (quoted, Ryan et al. 1988)

4.30   (quoted, Thomann 1989)

4.45   (quoted, Murphy et al. 1990)

4.54   (Pussemier et al. 1990)

4.47   (quoted, Warne et al. 1990)

4.50   (Bayona et al. 1991)

4.63   (selected, Mackay 1991)

Bioconcentration Factor, log BCF:

3.08, 2.68   (*daphnia, pimephales*, Southworth 1977)

2.88   (*daphnia pulex*, Herbes & Risi 1978; quoted, Hawker & Connell 1986)

2.96   (*daphnia pulex*, Southworth et al. 1978; quoted, Govers et al. 1984)

3.08   (kinetic estimation, Southworth et al. 1978)
3.55   (calculated as per Kenaga & Goring 1979, Eadie et al. 1982)
3.89, 3.76 (algae, calculated-S, Geyer et al. 1981)
4.22   (*p. hoyi*, Eadie et al. 1982)
3.67   (microorganisms-water, calculated-K_{OW}, Mabey et al. 1982)
3.02, 2.96 (quoted, calculated-K_{OW}, Mackay 1982; quoted, Schüürmann & Klein 1988)
2.83   (bluegill sunfish, Spacie et al. 1983; quoted, Linder et al. 1985)
2.95   (bluegill sunfish, estimated, Spacie et al. 1983)
3.08   (bluegill sunfish, calculated-K_{OW}, Spacie et al. 1983; quoted, Linder et al. 1985)
3.83   (activated, sludge, Freitag et al. 1984)
3.89, 3.26 (algae, calculated-K_{OW}, Geyer et al. 1984)
2.21   (goldfish, Ogata et al. 1984)
2.96   (*daphnia pulex*, Mackay & Hughes 1984; quoted, Hawker & Connell 1986)
3.96   (rainbow trout, calculated, Linder et al. 1985)
2.96, 3.89, 3.83 (fish, algae, activated sludge, Freitag et al. 1985)
2.99   (*daphnia magna*, Newsted & Giesy 1987)
2.96, 3.57 (quoted exptl., calculated-χ , Sabljic 1987b)
2.96   (calculated, Isnard & Lambert 1988)
0.820, 1.373 (*polychaete sp, capitella capitata*, Bayona et al. 1991)

Sorption Partition Coefficient, log K_{OC}:
4.42   (average of sorption isotherms on sediments, Karickhoff et al. 1979; quoted, Kenaga & Goring 1980; Kenaga 1980a; Bahnick & Doucette 1988)
3.74   (22°C, suspended particulates, Herbes et al. 1980)
4.27   (calculated-K_{OW}, Kenaga 1980a)
4.20; 4.15, 4.25, 4.63 (soil, shake flask-UV; calculated-K_{OW}, S & M.P., S, Karickhoff 1981)
4.15   (calculated-K_{OW}, Mabey et al. 1982)
2.96   (quoted, Govers et al. 1984)
3.95, 4.73 (Lake Erie with 9.6 mg C/liter, Landrum et al. 1984)
4.87, 5.70 (Huron River with 7.8 mg C/liter, Landrum et al. 1984)
4.42, 4.26 (soil: quoted, calculated- χ , Sabljic 1984)
4.20   (soil, shake flask-LSC, Nkedi-Kizza 1985)
4.93   (fluorescence quenching interaction with AB humic acid, Gauthier et al. 1986)
3.87   (soil, calculated, Pavlou 1987)
4.31   (soil, calculated- χ , Bahnick & Doucette 1988)
4.38   (HPLC, Hodson & Williams 1988)
5.76   (sediments average, Kayal & Connell 1990)
4.24, 4.25 (calculated-K_{OW}, S, Murphy et al. 1990)
4.41   (RP-HPLC, Pussemier et al. 1990)
4.41   (quoted, Szabo et al. 1990a,b)
4.53, 4.42 (RP-HPLC on CIHAC, on PIHAC, Szabo et al. 1990b)

Half-Lives in the Environment:

    Air: 0.58-1.7 hours, based on photolysis half-life in water (Southworth 1979 and Lyman et al. 1982; quoted, Howard et al. 1991).

    Surface water: half-lives for removal from water column at 25°C in midsummer sunlight were, 10.5 hours for deep, slow, somewhat turbid water; 21.6 hours for deep, slow, muddy water; 8.5 hours for deep, slow, clear water; 3.5 hours for shallow, fast, clear water; and 1.4 hour for very shallow, fast, clear water (Southworth 1977); computed near-surface half-life of a water body, 0.75 hour and for direct photochemical transformation at latitude 40°N, midday, midsummer and half-lives of 4.5 days (no sediment-water partitioning), 5.2 days (with sediment-water partitioning) for direct photolysis in a 5-m deep inland water body (Zepp & Schlotzhauer 1979); 0.58-1.7 hours, based on photolysis half-life in water (Lyman et al. 1982; quoted, Howard et al. 1991); 4.5 days at 40°N under summer sunlight (Mill & Mabey 1985).

    Groundwater: 2400-22080 hours, based on estimated unacclimated aqueous aerobic biodegradation half-life ((Howard et al. 1991).

    Sediment:

    Soil: 3.3 to 175 days (Sims & Overcash 1983; quoted, Bulman et al. 1987); 17 days for 5 mg/kg treatment and 45 days for 50 mg/kg treatment (Bulman et al. 1987); degradation rate constant of 0.0052 day^{-1} with a half-life of 134 days for Kidman sandy loam soil and 0.138 day^{-1} with half-life of 50 days for McLauren sandy loam soil (Park et al. 1990); half-lives were: 1200-11040 hours, based on aerobic soil dieaway test data (Coover & Sims 1987; Sims 1990; quoted, Howard et al. 1991); 0.5-26 weeks, 7.9 years (quoted, Luddington soil, Wild et al. 1991).

    Biota: 17 hours in bluegill sunfish (Spacie et al. 1983); with depuration half-life of 37.75 hours in *s. heringianus* (Frank et al. 1986).

Environmental Fate Rate Constants or Half-Lives:

    Volatilization: removal rate constants from the water column at 25°C in midsummer sunlight were: 0.002 hour^{-1} in deep, slow, somewhat turbid water; 0.001 hour^{-1} in deep, slow muddy water; 0.002 hour^{-1} in deep slow, clear water; 0.042 hour^{-1} in shallow, fast, clear water; and 0.179 hour^{-1} in very shallow, fast, clear water (Southworth 1977); aquatic half-life of 18-300 hours (Callahan et al. 1979); calculated half-life of 62 hours for a river 1 meter deep with water velocity of 0.5 m/sec and wind velocity of 1 m/sec (Southworth 1979; quoted, Herbes et al. 1980; Hallett & Brecher 1984).

    Photolysis: removal rate constants from the water at 25°C in midsummer sunlight were: 0.004 hour^{-1} in deep, slow somewhat turbid water; <0.001 hour^{-1} in deep, slow, muddy water; 0.018 hour^{-1} in deep, slow, clear water; 0.086 hour^{-1} in shallow, fast, clear water; and 0.238 hour^{-1} in very shallow, fast, clear water (Southworth 1977); 24-hour photolytic half-lives of about 1.6 hours in summer and 4.8 hours in winter at 35°N latitude (Southworth 1977); direct sunlight photolysis rate constant in winter at 35°N latitude of 0.15 hour^{-1} (Callahan et al. 1979; quoted, Mabey et al. 1982); calculated half-lives for direct sunlight photolysis of 50% conversion at 40°N latitude of midday in midsummer: 0.75 hours (near surface water; quoted, Herbes et al. 1980; Harris 1982), 4.5 days (inland water) and 5.2 days (inland water with sediment partitioning) and 0.75 hour for direct photochemical transformation near water surface (Zepp & Schlotzhauer 1979); atmospheric and aqueous photolysis half-life of 0.58 hour, based on measured aqueous photolysis rate constant for midday summer sunlight at 35°N latitude (Southworth 1979; quoted, Howard et al. 1991) and adjusted for approximate winter sunlight intensity (Lyman et al. 1982; quoted, Howard et al. 1991); half-lives on different atmospheric particulate substrates (appr. 25 μg/gm on substrate): 2.9 hours on silica gel, 0.5 hour on alumina and 48 hours on flyash (Behymer & Hites 1985); 4.5 days for summer at 40°N latitude under sunlight in surface water (Mill & Mabey 1985); rate constant in distilled water of 0.66 hour^{-1} with half-life of 1.0 hour (Fukuda et al. 1988); photodegradation for initial concentration of 5 ppm in methanol-water (1:1, v/v) solution by high pressure mercury lamp or sunlight with a rate constant of 0.023 minute^{-1} and half-life of 0.50 hour (Wang et al. 1991).

Hydrolysis: <0.001 hour^{-1} at 25°C (Southworth 1977); not hydrolyzable (Mabey et al. 1982); no hydrolyzable groups (Howard et al. 1991).

Oxidation: aquatic fate rate of 50 liter mol^{-1} sec^{-1} with half-life of 1600 days (Callahan et al. 1979); calculated rate constants of 5×10^8 M^{-1} hr^{-1} for singlet oxygen and 2.2×10^5 M^{-1} hr^{-1} for peroxy radical (Mabey et al. 1982); photooxidation half-life in water of 1111-38500 hours, based on measured rate constant for reaction with hydroxyl radical in water (Radding et al. 1976; quoted, Howard et al. 1991); rate constant for the gas phase reaction with OH radicals was 1.1×10^{-12} cm^3 molecule^{-1} sec^{-1} at 298 $\pm$ 1K, based on the relative rate technique for propene (Biermann et al. 1985); photooxidation half-life in air of 0.501-5.01 hours, based on estimated rate constant for reaction with hydroxyl radical in air (Atkinson 1987).

Biodegradation: the rate constant of microbial degradation in Third Creek water incubated 18 hours at 25°C was 0.061 hour^{-1}; removal rate constants from water column at 25°C in midsummer sunlight were: 0.060 hour^{-1} in deep, slow, somewhat turbid water; 0.030 hour^{-1} in deep, slow, muddy water; 0.061 hour^{-1} in deep, slow, clear water; 0.061 hour^{-1} in shallow, fast, clear water; and 0.061 hour^{-1} in very shallow, fast, clear water (Southworth 1977); microbial degradation rate constant was reported to be 0.035 hour^{-1} (Herbes et al. 1980; quoted, Hallett & Brecher 1984); significant degradation in 7 days with rapid adaptation for an average of three static-flask screening test (Tabak et al. 1981); aerobic half-life of 1200-11040 hours, based on aerobic soil dieaway test data (Coover & Sims 1987; Sims 1990; quoted, Howard et al. 1991); rate constants of 0.0052 day^{-1} with half-life of 134 days for Kidman sandy loam and 0.0138 day^{-1} with half-life of 50 days for McLarin sandy loam all at -0.33 bar soil moisture (Park et al. 1990); anaerobic half-life of 4800-44160 hours, based on estimated unacclimated aqueous aerobic biodegradation half-life (Howard et al. 1991).

Biotransformation: aquatic fate rate of < 0.0612 hour^{-1} with half-life of > 11.3 hours (Callahan et al. 1979); estimated rate constant for bacteria of 3×10^{-9} ml cell^{-1} hr^{-1} (Mabey et al. 1982).

Bioconcentration, Uptake (k_1) and Elimination (k_2) Rate Constants:

log k_1: 2.89 hour^{-1} (*daphnia pulex*, Herbes & Risi 1978; quoted, Hawker & Connell 1986)

log k_2: 0.0043 hour^{-1} (*daphnia pulex*, Herbes & Risi 1978; quoted, Hawker & Connell 1986)

k_1: 702 hour^{-1} (*daphnia pulex*, Southworth et al. 1978)

k_2: 0.589 hour^{-1} (*daphnia pulex*, Southworth et al. 1978)

k_1: average 1.73×10^{-3} to 36 hour^{-1} (bluegill sunfish, Spacie et al. 1983; quoted, Linder et al. 1985)

k_2: 0.040 hour^{-1} (bluegill sunfish, Spacie et al. 1983; quoted, Linder et al. 1985)

log k_1: 2.85 hour^{-1} (*daphnia pulex*, correlated to Mackay & Hughes 1984, Hawker & Connell 1986)

log k_2: -0.23 hour^{-1} (*daphnia pulex*, correlated to Mackay & Hughes 1984, Hawker & Connell 1986)

k_1: 1.46, 16.9 hour^{-1} (rainbow trout, Linder et al. 1985)

k_2: $1.58-1.88 \times 10^{-3}$ hour^{-1} (rainbow trout, Linder et al. 1985)

k_1: 87.2 hour^{-1} (4°C, *s. heringianus*, Frank et al. 1986)

k_2: 0.019 hour^{-1} (*s. heringianus*, Frank et al. 1986)

k_1: 131.1 ml/g/h (4°C, *p. hoyi*, Landrum 1988)

k_2: 0.0033 hour^{-1} (4°C, *p. hoyi*, Landrum 1988)

log k_2: 0.2, -0.01 day^{-1} (fish, calculated-K_{OW}, Thomann 1989)

log k_2: -0.96 day^{-1} (oyster, calculated-K_{OW}, Thomann 1989)

Common Name: 2-Methylanthracene
Synonym:
Chemical Name: 2-methylanthracene
CAS Registry No: 613-12-7
Molecular Formula: $C_{15}H_{12}$
Molecular Weight: 192.26

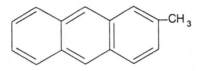

Melting Point (°C):
    209   (Weast 1982-83; Yalkowsky & Valvani 1979, 1980; Lande & Banerjee 1981;
            Whitehouse & Cooke 1982; Bjørseth 1983; Mailhot & Peters 1988)
    207   (Briggs 1981; Mailhot & Peters 1988)
    199   (Yalkowsky et al. 1983)
    205   (Pearlman et al. 1984)
Boiling Point (°C):
    359   (sublimation, Bjørseth 1983)
Density (g/cm³ at 20°C):
    1.8100 (Mailhot & Peters 1988)
Molar Volume (cm³/mol):
    219   (LeBas method)
    106.0   (calculated, Lande & Banerjee 1981)
    1.113   (intrinsic volume: $V_I/100$, Kamlet et al. 1987)
    106.22 (Mailhot & Peters 1988)
Molecular Volume (Å³):
    186.18 (Pearlman et al. 1984)
Total Surface Area, TSA (Å²):
    222.6   (Yalkowsky & Valvani 1979; quoted, Whitehouse & Cooke 1982; Mailhot &
            Peters 1988)
    224.01 (Pearlman et al. 1984)
Heat of Fusion, ΔH_{fus}, kcal/mol:
Entropy of Fusion, ΔS_{fus}, cal/mol K (e.u.):
Fugacity Ratio at 25 °C (assuming ΔS = 13.5 e.u.), F:
    0.0151 (at m.p. = 209°C)

Water Solubility (g/m^3 or mg/L at 25°C):

 0.039 (shake flask-fluorescence, Mackay & Shiu 1977; quoted, May et al. 1979; Yalkowsky & Valvani 1979; Lande & Banerjee 1981; Whitehouse & Cooke 1982; Pearlman et al. 1984; Mailhot & Peters 1988; IUPAC 1989)

 0.0219 (gen. col.-HPLC/UV, May et al. 1978; quoted, May et al. 1979; Briggs 1981; Whitehouse & Cooke 1982; Pearlman et al. 1984; Mailhot & Peters 1988; IUPAC 1989)

 0.039 (shake flask-fluorescence, Valvani & Yalkowsky 1980; Yalkowsky & Valvani 1979,1980; quoted, Lande & Banerjee 1981; Kamlet et al. 1987)

 0.041 (calculated-K_{OW}, Valvani & Yalkowsky 1980)

 0.046 (calculated-K_{OW}, Yalkowsky & Valvani 1980)

 0.027 (calculated-parachor, Briggs 1981)

 0.029 (quoted average, Whitehouse & Cooke 1982)

 0.0191 (23.1°C, gen. col.-HPLC, May et al. 1983)

 0.023 (average lit. value, Yalkowsky et al. 1983)

 0.031 (average lit. value, Pearlman et al. 1984; quoted, Kamlet et al. 1987)

 0.034, 0.027 (quoted average, calculated-V_1, M.P., hydrogen bond, Kamlet et al. 1987)

 0.281 (calculated-QSAR of EPA, Passino & Smith 1987)

 0.0389 (selected, Isnard & Lambert 1989)

Vapor Pressure (Pa at 25°C):

Henry's Law Constant (Pa m^3/mol):

Octanol/Water Partition Coefficient, log K_{OW}:

 5.15 (calculated-f const., Yalkowsky & Valvani 1979,1980; quoted, Whitehouse & Cooke 1982)

 5.14 (average lit. value, Yalkowsky et al. 1983)

 5.15 (quoted, Mailhot & Peters 1988)

 5.15 (selected, Isnard & Lambert 1989)

Bioconcentration Factor, log BCF:

Sorption Partition Coefficient, log K_{OC}:

Half-Lives in the Environment:
 Air:
 Surface water:
 Groundwater:

Sediment:
Soil:
Biota:

Environmental Fate Rate Constants or Half-Lives:
Volatilization:
Photolysis:
Hydrolysis:
Oxidation:
Biodegradation:
Biotransformation:
Bioconcentration, Uptake (k_1) and Elimination (k_2) Rate Constants:

Common Name: 9-Methylanthracene
Synonym:
Chemical Name: 9-methylanthracene
CAS Registry No: 779-02-2
Molecular Formula: $C_{15}H_{12}$
Molecular Weight: 192.26

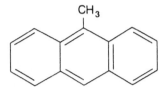

Melting Point (°C):
  81.5 (Weast 1982-83; Yalkowsky & Valvani 1979, 1980; Mackay et al. 1980; Lande
        & Banerjee 1981; Whitehouse & Cooke 1982; Weast et al. 1984; Mailhot &
        Peters 1988; Vadas et al. 1991)
  82.0 (Karickhoff 1981)
  80   (Pearlman et al. 1984)
  81.3 (Miller et al. 1985)
Boiling Point (°C):
  345   (Pearlman et al. 1984)
  196.0 (Weast 1982-83; Mailhot & Peters 1988)
Density (g/cm³ at 20°C):
  1.065 (Mailhot & Peters 1988)
Molar Volume (cm³/mol):
  181.0 (calculated, Lande & Banerjee 1981)
  219.0 (LeBas method, Miller et al. 1985; Abernethy & Mackay 1987)
  1.113 (intrinsic volume: V_I/100, Kamlet et al. 1987,1988; Hawker 1989)
  184.32 (Mailhot & Peters 1988)
Molecular Volume (Å³):
  184.89 (Pearlman et al. 1984)
Total Surface Area, TSA (Å²):
  215.1 (Yalkowsky & Valvani 1979; quoted, Whitehouse & Cooke 1982; Mailhot &
        Peters 1988)
  216.14 (Pearlman et al. 1984)
  223.1 (calculated as per Gavezzotti 1985, Sabljic 1987)
Heat of Fusion, ΔH_{fus}, kcal/mol:
Entropy of Fusion, ΔS_{fus}, cal/mol K (e.u.):
Fugacity Ratio at 25 °C (assuming ΔS = 13.5 e.u.), F:
  0.276 (at M.P. = 81.5°C)

156

Water Solubility (g/m³ or mg/L at 25°C):

    0.261 (shake flask-fluorescence, Mackay & Shiu 1977; quoted, Karickhoff et al. 1979; Yalkowsky & Valvani 1979,1980; Kenaga & Goring 1980; Mackay et al. 1980; Valvani & Yalkowsky 1980; Karickhoff 1981; Lande & Banerjee 1981; Whitehouse & Cooke 1982; Pearlman et al. 1984; Miller et al. 1985; Kamlet et al. 1987; Mailhot & Peters 1988; IUPAC 1989; Vadas et al. 1991)

    0.261 (quoted, Kenaga & Goring 1980)

    0.801 (calculated-K_{OW}, Valvani & Yalkowsky 1980)

    0.731 (calculated-K_{OW}, Yalkowsky & Valvani 1980)

    0.206 (average lit. value, Yalkowsky et al. 1983)

    0.269 (average lit. value, Pearlman et al. 1984)

    0.261 (selected, Miller et al. 1985; quoted, Hawker 1989)

    0.682 (calculated-V_I, M.P. & hydrogen bond, Kamlet et al. 1987)

    0.257 (quoted, Isnard & Lambert 1988, 1989)

    0.530 (gen. col.-HPLC, Vadas et al. 1991)

Vapor Pressure (Pa at 25°C):

    0.00224 (extrapolated-Antoine eqn., Stephenson & Malanowski 1987)

Henry's Law Constant (Pa m³/mol):

Octanol/Water Partition Coefficient, log K_{OW}:

    5.12 (calculated-π const., Southworth et al. 1978)

    5.07 (shake flask-UV, concn. ratio, Karickhoff et al. 1979; quoted, Karickhoff 1981)

    5.65 (calculated-S & M.P., Mackay et al. 1980)

    5.15 (calculated-f const., Valvani & Yalkowsky 1980; Yalkowsky & Valvani 1979,1980; quoted, Mackay et al. 1980; Whitehouse & Cooke 1982)

    5.07 (quoted, Mackay 1982; Schüürmann & Klein 1988)

    5.14 (average lit. value, Yalkowsky et al. 1983)

    5.56 (quoted, Mackay & Hughes 1984; Hawker & Connell 1986)

    5.07, 5.61, 5.14 (quoted, HPLC-RT, calculated-f const., Burkhard et al. 1985)

    5.07 (Hansch & Leo 1985; quoted, Banerjee & Howard 1988)

    5.07 (calculated, Miller et al. 1985; quoted, Kamlet et al. 1988; Hawker 1989)

    5.14 (Leo 1986; quoted, Schüürmann & Klein 1988)

    5.07 (THOR 1986; quoted, Schüürmann & Klein 1988)

    5.12 (quoted, Abernethy & Mackay 1987)

    4.96 (calculated-solvatochromic p., Kamlet et al. 1988)

    4.61 (calculated-UNIFAC, Banerjee & Howard 1988)

    5.07 (selected, Isnard & Lambert 1988, 1989)

    5.07, 5.15 (quoted, Mailhot & Peters 1988)

Bioconcentration Factor, log BCF:

  3.66 (*daphnia pulex*, Southworth et al. 1978)

  3.59 (kinetic estimation, Southworth et al. 1978)

  3.66, 3.75 (quoted, calculated-K_{ow}, Mackay 1982; quoted, as correlation to Mackay & Hughes 1984, Hawker & Connell 1986; Schüürmann & Klein 1988)

  3.66, 3.94 (quoted exptl., calculated- χ , Sabljic 1987b)

  3.66 (calculated, Isnard & Lambert 1988)

Sorption Partition Coefficient, log K_{OC}:

  4.81 (average of isotherms on sediments, Karickhoff et al. 1979; quoted, Kenaga & Goring 1980; Karickhoff 1981; Sabljic 1984)

  4.68, 5.06, 4.32 (calculated-K_{ow}, S & M.P., S, Karickhoff 1981)

  4.50 (calculated- χ , Sabljic 1984)

Half-Lives in the Environment:

  Air:

  Surface water: computed half-life near-surface of a water body, for direct photochemical transformation was 0.13 hour, and half-lives for direct photolysis in a 5-m deep inland water body were, 0.79 day (no sediment-water partitioning) and 1.2 day (with sediment-water partitioning) (Zepp & Schlotzhauer 1979); 0.79 day for summer at 40°N latitude under sunlight (Mill & Mabey 1985).

  Groundwater:

  Sediment:

  Soil:

  Biota:

Environmental Fate Rate Constants or Half-Lives:

  Volatilization:

  Photolysis: half-life of 0.13 hour for direct photochemical transformation near water surface and 0.78 day for no sediment-water partitioning; and 1.2 day with sediment-water partitioning (Zepp & Scholtzhauer 1979); 0.79 day for summer at 40°N latitude under sunlight in surface water (Mill & Mabey 1985); photodegradation in methanol-water (2:3, v/v) solution for initial concentration of 5 ppm by high pressure mercury lamp or sunlight with rate constant of 0.163 minute^{-1} and half-life of 0.07 hour (Wang et al. 1991).

  Hydrolysis:

  Oxidation:

  Biodegradation:

  Biotransformation:

  Bioconcentration, Uptake (k_1) and Elimination (k_2) Rate Constants:

    k_1: 561 hour^{-1} (*daphnia pulex*, Southworth et al. 1978)

    k_2: 0.144 hour^{-1} (*daphnia pulex*, Southworth et al. 1978) log k_1: 2.75 hour^{-1} (*daphnia pulex*, correlated as per Mackay & Highes 1984, Hawker & Connell 1986)

log k_2: -0.84 hour^{-1} (*daphnia pulex*, correlated as per Mackay & Hughes 1984, Hawker & Connell 1986)

Common Name: 9,10-Dimethylanthracene
Synonym:
Chemical Name: 9,10-dimethylanthracene
CAS Registry No: 781-43-1
Molecular Formula: $C_{16}H_{14}$
Molecular Weight: 206.29

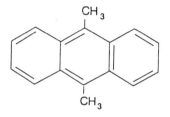

Melting Point (°C):
      182 (Yalkowsky & Valvani 1979,1980; Mackay et al. 1980; Miller et al. 1985; Mailhot
           & Peters 1988)
      183 (Bjørseth 1983; Pearlman et al. 1984)
      181 (Yalkowsky et al. 1983)
Boiling Point (°C):
Density (g/cm³ at 20 °C):
Molar Volume (cm³/mol):
      241.0 (LeBas method, Miller et al. 1985)
      1.211 (intrinsic volume: $V_I/100$, Kamlet et al. 1987,1988; Hawker 1989)
Molecular Volume (Å³):
      199.51 (Pearlman et al. 1984)
Total Surface Area, TSA (Å²):
      228.0 (Yalkowsky & Valvani 1979; quoted, Mailhot & Peters 1988)
      230.12 (Pearlman et al. 1984)
Heat of Fusion, ΔH_{fus}, kcal/mol:
Entropy of Fusion, ΔS_{fus}, cal/mol K (e.u.):
Fugacity Ratio at 25 °C (assuming $\Delta S = 13.5$ e.u.), F:
      0.028 (at M.P. = 182°C)
      0.030 (Mackay et al. 1980)

Water Solubility (g/m³ or mg/L at 25°C):
      0.056 (shake flask-fluorescence, Mackay & Shiu 1977; quoted, Yalkowsky & Valvani
           1979,1980; Mackay et al. 1980; Valvani & Yalkowsky 1980; Pearlman et al.
           1984; Miller et al. 1985; Kamlet et al. 1987; Mailhot & Peters 1988; IUPAC
           1989)
      0.029 (calculated-K_{ow}, Valvani & Yalkowsky 1980; Yalkowsky & Valvani 1980)

0.039 (average lit. value, Yalkowsky et al. 1983)
0.056 (average lit. value, Pearlman et al. 1984)
0.056 (selected, Miller et al. 1985; quoted, Hawker 1989)
0.0176 (calculated-V_I, M.P. & hydrogen bond, Kamlet et al. 1987)

Vapor Pressure (Pa at 25°C):
    1.53×10^{-4} (extrapolated-Antoine eqn., Stephenson & Malanowski 1987)

Henry's Law Constant (Pa m^3/mol):

Octanol/Water Partition Coefficient, log K_{OW}:
    5.67 (calculated-f const., Yalkowsky & Valvani 1979,1980; quoted, Mackay et al. 1980;
        Mailhot & Peters 1988)
    5.25 (calculated-S & M.P., Mackay et al. 1980)
    5.65 (average lit. value, Yalkowsky et al. 1983)
    5.67, 5.42 (quoted, calculated-MR, Yoshida et al. 1983)
    5.25 (calculated, Miller et al. 1985; quoted, Kamlet et al. 1988; Hawker 1989)
    5.48 (calculated-solvatochromic p., Kamlet et al. 1988)

Bioconcentration Factor, log BCF:

Sorption Partition Coefficient, log K_{OC}:

Half-Lives in the Environment:
        Air:
        Surface water:
        Groundwater:
        Sediment:
        Soil:
        Biota:

Environmental Fate Rate Constants or Half-Lives:
        Volatilization:
        Photolysis: calculated half-life of direct sunlight photolysis of 50% conversion of midday
            in midsummer at 40°N latitude for near-surface water, 0.35 hour (Zepp &
            Schlotzhauer 1979; quoted, Harris 1982); photodegradation in methanol-water
            (2:3, v/v) solution for initial concentration of 5 ppm by high pressure mercury
            lamp or sunlight with a rate constant of 0.0633 minute^{-1} and half-life of 0.18
            hour (Wang et al. 1991).
        Hydrolysis:

161

Oxidation:

Biodegradation:

Biotransformation:

Bioconcentration, Uptake (k_1) and Elimination (k_2) Rate Constants:

Common Name: Pyrene
Synonym: benzo[def]phenanthrene
Chemical Name: pyrene
CAS Registry No: 129-00-0
Molecular Formula: $C_{16}H_{10}$
Molecular Weight: 202.26

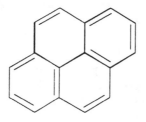

Melting Point (°C):
　　150　　(Cleland & Kingsbury 1977; quoted, API 1978; Callahan et al. 1979; Briggs
　　　　　　1981; Yalkowsky 1981; Mabey et al. 1982; Pearlman et al. 1984; Banerjee et
　　　　　　al. 1990; Banerjee & Baughman 1991)
　　156　　(Weast 1972-73; Yalkowsky & Valvani 1979, 1980; Mackay et al. 1980;
　　　　　　Karickhoff 1981; Steen & Karickhoff 1981; Lande & Banerjee 1981; Mackay
　　　　　　& Shiu 1981; Bruggeman et al. 1982; Chiou et al. 1982; Whitehouse & Cooke
　　　　　　1982; Bjørseth 1983; Windholz et al. 1983; Bidleman 1984; Weast et al. 1984;
　　　　　　Miller et al. 1985; Chin et al. 1986; Gobas et al. 1988; Vadas et al. 1991)
Boiling Point (°C):
　　393　　(Weast 1975; quoted, Mabey et al. 1982; Bjørseth 1983; Pearlman et al. 1984;
　　　　　　Pussemier 1990)
　　>360　(API 1978)
　　360　　(Mackay & Shiu 1981)
　　388　　(Pearlamn et al. 1984; quoted, Tsai et al. 1991)
　　404　　(Banerjee et al. 1990)
Density (g/cm³ at 20°C):
　　1.271　(23°C, Weast 1982-83; Pussemier et al. 1990)
Molar Volume (cm³/mol):
　　213.8　(Reid et al. 1979; quoted, Tsai et al. 1991)
　　159.0　(calculated, Lande & Banerjee 1981)
　　214.0　(LeBas method, Miller et al. 1985; Abernethy & Mackay 1987; Gobas et al. 1988)
　　1.156　(intrinsic volume: $V_I/100$, Kamlet et al. 1988; Hawker 1989)
Molecular Volume (Å³):
　　186.0　(Pearlman et al. 1984)

163

Total Surface Area, TSA ($\mathring{A}^2$):

    403.0 (calculated with hydration cell, Amidon et al. 1974,1975,1979)

    216.0 (Valvani et al. 1976; quoted, Pinal et al. 1990)

    213.0 (Yalkowsky & Valvani 1976,1979; quoted, Whitehouse & Cooke 1982)

    401.5 (calculated-1.5 $\mathring{A}$ solvent radius, Amidon & Anik 1980)

    218.6 (Mackay et al. 1982; quoted, Valsaraj & Thibodeaux 1989)

    213.47 (Pearlman et al. 1984)

    213.4 (calculated as per Gavezzotti 1985, Sabljic 1987)

    213.0 (quoted as HSA, Woodburn et al. 1989)

Heat of Fusion, ΔH_{fus}, kcal/mol:

Entropy of Fusion, ΔS_{fus}, cal/mol K (e.u.):

    8.6   (Wauchope & Getzen 1972; quoted, Yalkowsky 1981)

    9.6   (Casellato et al. 1973; quoted, Yalkowsky 1981)

    13.1 (Mackay & Shiu 1977; quoted, Hinckley 1989)

    13.1 (selected, Hinckley 1989)

Fugacity Ratio at 25 °C (assuming ΔS = 13.5 e.u.), F:

    0.0505 (at M.P. = 156 °C)

    0.050 (Mackay et al. 1980)

Water Solubility (g/m^3 or mg/L at 25°C):

    0.165 (27°C, nephelometry, Davis et al. 1942; quoted, Hollifield 1979; Futoma et al. 1981; Billington et al. 1988; IUPAC 1989)

    0.175 (shake flask-UV, Klevens 1950; quoted, Futoma et al. 1981; quoted, Mackay & Shiu 1981; Pearlman et al. 1984)

    0.148 (Pierotti et al. 1959; quoted, Mackay & Shiu 1977)

    1.56 (shake flask-UV/fluorescence, quoted, Barone et al. 1967)

    0.105 (20°C, shake flask-UV, Eisenbrand & Baumann 1970; quoted, Billington et al. 1988)

    0.148 (shake flask-UV, Wauchope & Getzen 1972; quoted, Futoma et al. 1981; Mackay & Shiu 1981; Pearlman et al. 1984; Walters & Luthy 1984; Billington et al. 1988)

    0.171 (shake flask-fluorescence/UV, Schwarz & Wasik 1976; quoted, Mackay & Shiu 1981)

    0.135 (shake flask-fluorescence, Mackay & Shiu 1977; quoted, API 1978; Callahan et al. 1979; Karickhoff et al. 1979; May et al. 1979; Yalkowsky & Valvani 1979,1980; Hutchinson et al. 1980; Kenaga & Goring 1980; Mackay et al. 1980; Valvani & Yalkowsky 1980; Chiou 1981; Chiou & Schmedding 1981; Futoma et al. 1981; Karickhoff 1981; Lande & Banerjee 1981; Mackay & Shiu 1981; Steen & Karickhoff 1981; Chiou et al. 1982; Mackay et al. 1982; Whitehouse & Cooke 1982; Pearlman et al. 1984; Walters & Luthy 1984; Miller et al. 1985; Landrum et al. 1987; Billington et al. 1988; IUPAC 1989; Kayal & Connell 1990; Shorten et al. 1990; Vadas et al. 1991; Mackay 1991)

0.132   (Rossi 1977; Neff 1979; quoted, Eadie et al. 1982; Whitehouse & Cooke 1982)
0.1295 (shake flask-fluorescence, Schwarz 1977; quoted, API 1978; May et al. 1979;
        Futoma et al. 1981; Whitehouse & Cooke 1982; Pearlman et al. 1984; Walters
        & Luthy 1984; Billington et al. 1988; IUPAC 1989; Vadas et al. 1991)
0.132   (gen. col.-HPLC/UV, May et al. 1978; quoted, May et al. 1979; Briggs 1981;
        Futoma et al. 1981; Mackay & Shiu 1981; Mabey et al. 1982; Pearlman et al.
        1984; Walters & Luthy 1984; Billington et al. 1988; IUPAC 1989; Vadas et al.
        1991)
0.032   (shake flask-nephelometry, Hollifield 1979)
0.135   (shake flask-LSC, Means et al. 1979; 1980)
0.134   (calculated-K_{ow}, Valvani & Yalkowsky 1980; Yalkowsky & Valvani 1979,1980;
        quoted, Lande & Banerjee 1981; Chin et al. 1986)
0.131   (calculated-K_{ow}, Valvani & Yalkowsky 1980)
0.120   (calculated-parachor, Briggs 1981)
0.134   (Lande & Banerjee 1981)
0.130   (shake flask-GC/UV, Rossi & Thomas 1981; quoted, IUPAC 1989)
0.134   (RP-TLC, Bruggeman et al. 1982)
0.140   (selected, Mills et al. 1982)
0.132   (quoted average, Whitehouse & Cooke 1982)
0.136   (25.5°C, gen. col.-HPLC, May et al. 1983)
0.131   (average lit. value, Yalkowsky et al. 1983)
0.160   (Verschueren 1983; quoted, Pinal et al. 1990)
0.129   (average lit. value, Pearlman et al. 1984; quoted, Billington et al. 1988)
0.133   (gen. col.-HPLC/fluorescence, Walters & Luthy 1984)
0.135   (selected, Miller et al. 1985; quoted, Vowles & Mantoura 1987; Gobas et al.
        1988; Hawker 1989; Capel et al. 1991)
0.135; 0.557   (quoted; RP-HPLC, calculated-K_{ow} & M.P., Chin et al. 1986)
0.145   (quoted, Tomlinson & Hafkenscheid 1986)
0.118   (gen. col.-HPLC/UV, Billington et al. 1988)
0.142   (quoted, Isnard & Lambert 1988, 1989)
0.132   (recommended, IUPAC 1989)
0.135   (quoted, Valsaraj & Thibodeaux 1989)
0.150   (LSC, Eadie et al. 1990)
0.132   (selected, Ma et al. 1990)
0.131   (quoted, Edwards et al. 1991)
0.107   (gen. col.-HPLC, Vadas et al. 1991)

Vapor Pressure (Pa at 25°C):
        0.00339 (effusion method, Inokuchi et al. 1952)
        0.000882 (extrapolated from Antoine eqn., Bradley & Cleasby 1953; quoted, Mackay &
                 Shiu 1981; Wasik et al. 1983; Bidleman 1984)
        0.00033 (Hoyer & Peperle 1958; quoted, Mabey et al. 1982; Tsai et al. 1991)

0.00091 (effusion method, Pupp et al. 1974; quoted, Bidleman 1984)
0.00027 (lit. average-interpolated, API 1979; quoted, Wasik et al. 1983)
0.00088 (extrapolated from Antoine eqn., Amidon & Anik 1981)
0.00089 (Mackay & Shiu 1981)
0.00059 (gas saturation, Sonnefeld et al. 1983)
0.0006 (gen. col.-HPLC, Wasik et al. 1983)
0.0049 (subcooled liquid, Bidleman 1984)
0.00442 (Yamasaki et al. 1984; quoted, Capel et al. 1991)
0.00033 (selected, Howard et al. 1986; quoted, Banerjee et al. 1990)
0.00055 (Antoine eqn., Stephenson & Malanowski 1987)
0.000413, 0.00973 (lit. mean, subcooled liq. value, Bidleman & Foreman 1987)
0.010, 0.014 (quoted, GC-RT, Hinckley 1989)
0.00017 (calculated-UNIFAC, Banerjee et al. 1990)
0.0005 (selected, Ma et al. 1990)
0.000886 (quoted, Shorten et al. 1990)
0.0006 (selected, Mackay 1991)

Henry's Law Constant (Pa m^3/mol):
1.89, 1.07 (batch stripping, calculated-P/C, Southworth 1979)
1.10,1.20 (batch stripping, recommended, Mackay & Shiu 1981)
0.5167 (calculated-P/C, Mabey et al. 1982)
1.16, 1.84 (calculated-P/C, UNIFAC, Nirmalakhandan & Speece 1988)
694.1 (WERL Treatability database, quoted, Ryan et al. 1988)
0.95 (calculated-P/C, Eastcott et al. 1988)
0.125 (15°C, calculated, Baker & Eisenreich 1990; quoted, Tsai et al. 1991)
12.4 (calculated-C_A/C_W, Jury et al. 1990)
6.59 (calculated-P/C, Capel et al. 1991)
1.10 (Shorten et al. 1990)

Octanol/Water Partition Coefficient, log K_{OW}:
4.90 (calculated-π conat., Southworth et al. 1978)
5.32 (calculated-f const., 1971, Callahan et al. 1979; quoted, Ogata et al. 1984)
4.88 (Hansch & Leo 1979; quoted, Mackay et al. 1980; Landrum et al. 1987; Tomlinson & Hafkensheid 1986)
5.18 (shake flask-UV, concn. ratio, Karickhoff et al. 1979; quoted, Kenaga & Goring 1980; Chiou 1981; Chiou & Schmedding 1981; Karickhoff 1981; Schwarzenbach & Westall 1981; Chiou et al. 1982; Govers et al. 1984; Walters & Luthy 1984; Ruepert et al. 1985; Kayal & Connell 1990; Edwards et al. 1991)
4.88 (Veith et al. 1979)
5.22 (calculated-f const., Yalkowsky & Valvani 1980; quoted, Hutchinson et al. 1980)

5.12   (calculated-S & M.P., Mackay 1980)
5.09   (shake flask-LSC, Means et al. 1980)
4.88, 4.90   (quoted, calculated- χ , Chou & Jurs 1980)
5.18   (Chiou 1981; quoted, Valsaraj & Thibodeaux 1989)
5.03   (HPLC-k', Hanai et al. 1981)
5.18   (quoted, Steen & Karickhoff 1981)
5.05   (HPLC-k', McDuffie 1981)
5.22   (RP-TLC, Brugeman et al. 1982)
4.50   (HPLC-k', D'Amboise & Hanai 1982)
4.88   (HPLC-k', Hammers et al. 1982)
4.90   (calculated-f const., Mabey et al. 1982)
4.88   (quoted, Mackay 1982; Schüürmann & Klein 1988)
5.30   (selected, Mills et al. 1982)
5.18   (quoted, Casserly et al. 1983)
4.88   (HPLC-k', Hafkenscheid & Tomlinson 1983)
5.18, 4.96   (quoted, HPLC-RT, Rapaport 1984)
4.90   (quoted, Mackay & Hughes 1984; quoted, Hawker & Connell 1986)
5.22   (average lit. value, Yalkowsky et al. 1983)
5.32, 5.21   (quoted, calculated-MR, Yoshida et al. 1983)
5.02, 5.52, 4.95   (quoted lit. average, HPLC-RT/MS, calculated-f. const., Burkhard et al. 1985)
4.80   (Hansch & Leo 1985; quoted, Banerjee & Howard 1988)
5.18   (selected, Miller et al. 1985; quoted, Chin et al. 1986; Vowles & Mantoura 1987; Eastcott et al. 1988; Kamlet et al. 1988; Hawker 1989; Pussemier et al. 1990; Szabo et al. 1990a,b; Banerjee & Baughman 1991; Capel et al. 1991; Landrum et al. 1991)
5.17   (calculated- χ as per Rekker & De Kort 1979, Ruepert et al. 1985)
4.45   (Yalkowsky 1985; quoted, Pinal et al. 1990)
4.90   (quoted, Abernethy & Mackay 1986,1987)
5.03   (quoted, Brooke et al. 1986)
4.97   (RP-HPLC, calculated-S & M.P., Chin et al. 1986)
4.95   (Leo 1986; quoted; Schüürmann & Klein 1988)
4.88   (THOR 1986; quoted; Schüürmann & Klein 1988)
5.18   (quoted, Pavlou 1987)
4.68   (calculated-UNIFAC, Banerjee & Howard 1988)
4.88   (selected, Isnard & Lambert 1988,1989)
4.85   (calculated-solvatochromic p., Kamlet et al. 1988)
5.32   (quoted, Ryan et al. 1988)
5.00   (recommended, Sangster 1989)
5.10   (selected, Ma et al. 1990)
5.18   (quoted, Shorten et al. 1990)
5.20   (Bayona et al. 1991)
6.70   (calculated, Broman et al. 1991)

Bioconcentration Factor, log BCF:

    3.43   (*daphnia pulex*, Southworth et al. 1978; quoted, Govers et al. 1984)

    3.52   (kinetic estimation, Southworth et al. 1978)

    3.43   (Veith et al. 1979)

    4.38   (mixed microbial population, Steen & Karickhoff 1981)

    3.29   (calculated as per Kenaga & Goring 1979, Eadie et al. 1982)

    4.65   (*p. hoyi*, Eadie et al. 1982)

    4.08   (microorganisms-water, calculated from K_{OW}, Mabey et al. 1982)

    3.43, 3.56 (quoted exptl., calculated-K_{OW}, Mackay 1982; quoted, Banerjee & Baughman 1991)

    4.56, 4.22, 4.75   (average, *selenustrum capricornutum*-dosed singly, dosed simultaneously, Casserly et al. 1983)

    3.43   (Govers et al. 1984)

    2.66   (goldfish, GC, concn. ratio, Orgata et al. 1984)

    3.43   (*daphnia pulex*, Mackay & Hughes 1984; quoted, Hawker & Connell 1986)

    3.43   (*daphnia magna*, Newsted & Giesy 1987)

    3.43, 4.12 (quoted exptl., calculated-χ , Sabljic 1987)

    3.43   (calculated, Isnard & Lambert 1988)

    3.43   (quoted, Schüürmann & Klein 1988)

    3.65, 3.81, 2.35 (mussel, clam, shrimp, Gobas & Mackay 1989)

    2.85, 2.70 (*polychaete, shrimo-hepatopancreas*, Gobas & Mackay 1989)

    3.49   (calculated-K_{OW},$S_{OCTANOL}$ & M.P., Banerjee & Baughman 1991)

    0.716, 1.124 (*polychaete sp, capitella capitata*, Bayona et al. 1991)

Sorption Partition Coefficient, log K_{OC}:

    4.92   (average of isotherms on sediments, Karickhoff et al. 1979; Kenaga & Goring 1980; Schwarzenbach & Westall 1981; Govers et al. 1984)

    4.92   (sediment, shake flask-sorption isotherm, Karickhoff et al. 1979)

    4.81   (average value of soil and sediment, shake flask-LSC, sorption isotherms, Means et al. 1979)

    4.92   (Kenaga & Goring 1980; quoted, Bahnick & Doucette 1988)

    4.80   (average value of 12 soil/sediment samples, shake flask-LSC, sorption isotherms, Means et al. 1980)

    4.78, 4.80 (soil/sediment: calculated-K_{OW}, regress of k_p versus substrate properties, Means et al. 1980)

    4.83, 4.79, 4.64, 4.51 (sorption isotherms, calculated-K_{OW}, S & M.P., S, Karickhoff 1981; quoted, Szabo et al. 1990a,b)

    4.22   (calculated-K_{OW}, Schwarzenbach & Westall 1981)

    4.58   (calculated-K_{OW}, Mabey et al. 1982)

    4.92, 4.81 (quoted, calculated-χ , Sabljic 1984)

    3.11, 3.46 (sediment suspensions, Karickhoff & Morris 1985)

    5.23   (fluorescence quenching interaction with AB humic acid, Gauthier et al. 1986)

5.08 (fluorescence quenching interaction with AB fulvic acid, Gauthier et al. 1986)

4.46-4.81 (marine humic acids, fluorescence quenching technique, Gauthier et al. 1987)

4.94-5.51 (soil humic acids, fluorescence quenching technique, Gauthier et al. 1987)

4.73, 5.02 (soil fulvic acids, fluorescence quenching technique, Gauthier et al. 1987)

5.02 (dissolved humic materials, Aldrich humic acid, fluorescence quenching technique, Gauthier et al. 1987)

4.52 (calculated, Pavlou 1987)

5.13 (sediment, HPLC-k', Vowles & Mantoura 1987)

4.84 (calculated- χ , Bahnick & Doucette 1988)

4.80, 5.13 (Hodson & Williams 1988)

5.65 (LSC, Eadie et al. 1990)

5.29 (soil, Jury et al. 1990)

6.51 (sediments average, Kayal & Connell 1990)

4.83 (RP-HPLC, Pussemier et al. 1990)

4.82, 4.77 (RP-HPLC on CIHAC, on PIHAC, Szabo et al. 1990b)

6.50 (specified particulate log K_{OC}, Broman et al. 1991)

4.0 (predicted particulate log K_{OC}, Broman et al. 1991)

Half-Lives in the Environment:

Air: 0.68-2.04 hours, based on estimated sunlight photolysis half-life in water (Zepp & Stotzhauer 1979; Lyman et al. 1982; quoted, Howard et al. 1991).

Surface water: computed near-surface half-life for direct photochemical transformation was, 0.58 hour at latitude 40°N, midday, midsummer and direct photolysis in a 5-m deep inland water body were 4.2 days with no sediment-water partitioning and 5.9 days with sediment-water partitioning (Zepp & Schlotzhauer 1979); 0.68-2.04 hours, based on estimated sunlight photolysis half-life in water (Lyman et al. 1982; quoted, Howard et al. 1991); 4.2 days for summer at 40°N latitude under sunlight (Mill & Mabey 1985); half-life of 0.68 hour, based on direct photolysis in sunlight at midday, mid-summer and 40°N latitude (quoted, Zepp 1991).

Groundwater: 10080-91200 hours, based on estimated unacclimated aqueous aerobic biodegradation half-life (Howard et al. 1991).

Sediment:

Soil: 3-35 hours (Sims & Overcash 1983; quoted, Bulman et al. 1987); half-life of 58 days for 5 mg/kg treatment and 48 days for 50 mg/kg treatment (Bulman et al. 1987); 5040-45600 hours, based on aerobic soil dieaway test data at 10-30°C (Coover & Sims 1987; Sims 1990; quoted, Howard et al. 1991); >50 days (Ryan et al. 1988); degradation rate constant, 0.0027 day⁻¹ with half-life of 260 days for Kidman sandy loam soil and 0.0035 day⁻¹ with half-life of 199 days for McLaurin sandy loam soil (Park et al. 1990); 500 days in soil (Jury et al. 1990); 0.4 to more than 90 weeks, 8.5 years (quoted, Luddington soil, Wild et al. 1991).

Biota: with a depuration half-life of 40.8 hours in *s. heringianus* (Frank et al. 1986).

Environmental Fate Rate Constants or Half-Lives:

Volatilization: sublimation rate constant of 1.1×10^{-4} sec^{-1} was measured as loss from glass surface at 24°C at an air flow rate of 3 liter/min (Cope & Kalkwarf 1987).

Photolysis: calculated half-lives of direct sunlight photolysis of 50% conversion in midday of midsummer at 40°N: 4.2 days (inland water) and 5.9 days (inland water with sediment partitioning) (Zepp & Schlotzhauer 1979); rate constant of 1.014 hour^{-1} (Zepp 1980); atmospheric and aqueous photolysis half-life of 0.68 hour, based on measured aqueous photolysis quantum yields calculated for midday summer sunlight at 40°N latitude (Zepp & Schlotzhauer 1979; quoted, Harris 1982; Howard et al. 1991) and 2.04 hours after adjusting for approximate winter sunlight intensity (Lyman et al. 1982; quoted, Howard et al. 1991); half-lives on different atmospheric particulate substrates (appr. 25 μg/gm on substrate): 21 hours on silica gel, 31 hours on alumina and 46 hours on fly ash (Behymer & Hites 1985); half-life of 4.2 days for summer sunlight photolysis in surface water (Mill & Mabey 1995); ozonation rate constant on glass surface of $<1.05 \times 10^{-4}$ m/s was measured at 24°C with $[O_3] = 0.16$ ppm and light intensity of 1.3 kW/m^2 (Cope & Kalkwarf 1987); photodegradation half-life for adsorption on airborne particulates was found ranging from 1 hour in summer to days in winter by sunlight (Valerio et al. 1991); photolysis half-life in water of 0.68 hour, based on direct photolysis in sunlight at midday, midsummer, latitude 40°N (Zepp 1991).

Hydrolysis: not hydrolyzable (Mabey et al. 1982); no hydrolyzable groups (Howard et al. 1991).

Oxidation: rate constant of 5×10^8 M^{-1} hr^{-1} for singlet oxygen and 2.2×10^4 M^{-1} hr^{-1} for peroxy radical (Mabey et al. 1982); half-life of 1000 days (Callahan et al. 1979); photooxidation half-life in air of 0.802-8.02 hours, based on estimated rate constant for reaction with hydroxyl radical in air (Atkinson 1987; quoted, Howard et al. 1991).

Biodegradation: significant degradation within seven days for a domestic sewage 28-days test for an average of three static-flask screenig (Tabak et al. 1981); aerobic half-life of 5040-45600 hours, based on aerobic soil dieaway test data at 10-30°C (Coover & Sims 1987; Sims 1990; quoted, Howard et al. 1991); rate constant in atmosphere of 0.29 hour^{-1} (Dragoscu & Friedlander 1989; quoted, Tsai et al. 1991); rate constants of 0.0027 day^{-1} with half-life of 260 days for Kidman sandy loam and 0.0035 day^{-1} with half-life of 199 days for McLarin sandy loam all at -0.33 bar soil moisture (Park et al. 1990); anaerobic half-life of 20160-182400 hours, based on estimated unacclimated aqueous aerobic biodegradation half-life (Howard et al. 1991).

Biotransformation: estimated to be 1×10^{-10} ml cell^{-1} hr^{-1} for bacteria (Mabey et al. 1982).

Bioconcentration, Uptake (k_1) and Elimination (k_2) Rate Constants:

k_1:   1126 hour^{-1} (*daphnia pulex*, Southworth et al. 1978)

k_2:   0.343 hour^{-1} (*daphnia pulex*, Southworth et al. 1978)

log k_1: 3.05 hour^{-1} (*daphnia pulex*, correlated as per Mackay & Hughes 1984, Hawker & Connell 1986)

log k_2: -0.46 hour^{-1} (*daphnia pulex*, correlated as per Mackay & Hughes 1984, Hawker & Connell 1986)

k_1:   113.0 hour^{-1} (4°C, *s. heringianus*, Frank et al. 1986)

k_2:   0.017 hour^{-1} (*s. heringianus*, Frank et al. 1986)

k_1:   199.2 mL/g/h   (4°C, *p. hoyi*, Landrum 1988)

k_2:   0.0012 hour^{-1} (4°C, *p. hoyi*, Landrum 1988)

Common Name: Fluoranthene
Synonym: idryl, 1,2-benzacenaphthene, benzo[j,k]fluorene, benz[a]acenaphthylene, fluoranthrene
Chemical Name: 1,2-benzacenaphthene
CAS Registry No: 206-44-0
Molecular Formula: $C_{16}H_{10}$
Molecular Weight: 202.26

Melting Point (°C):
   107-110 (Dean 1985)
   111   (Cieland & Kingsbury 1977; quoted, Callahan et al. 1979; Yalkowsky & Valvani
            1979, 1980; Mackay et al. 1980; Briggs 1981; Lande & Banerjee 1981; Mackay
            & Shiu 1981; Bruggeman et al. 1982; Mabey et al. 1982; Bidleman 1984;
            Patton et al. 1984; Weast et al. 1984; Miller et al. 1985; Chin et al. 1986;
            Gobas et al. 1988; Vadas et al. 1991)
   107   (Verschueren 1977, 1983; quoted, Pinal et al. 1990,1991)
   110.3 (Hashimoto et al. 1982)
   110   (API 1978; Yalkowsky 1981; Pearlman et al. 1984)
Boiling Point (°C):
   217   (at 50 mmHg, Weast 1975; quoted, Mabey et al. 1982)
   250   (Verschueren 1977,1983)
   255   (Dean 1985)
   393   (API 1978)
   375   (Weast 1985; Mackay & Shiu 1981; Pearlman et al. 1984; Tsai et al. 1991)
Density (g/cm³ at 0°C):
   1.252 (Weast 1982-83, Dean 1985)
Molar Volume (cm³/mol):
   217.3 (Reid et al. 1979; quoted, Tsai et al. 1991)
   162.0 (calculated, Lande & Banerjee 1981)
   217.0 (LeBas method, Miller et al. 1985; Gobas et al. 1988)
Molecular Volume (Å³):
   187.74 (Pearlman et al. 1984)
Total Surface Area, TSA (Å²):
   218.0 (Yalkowsky & Valvani 1979; Lande et al. 1985)
   218.63 (Pearlman et al. 1984)
   218.0 (quoted as HSA, Woodburn et al. 1989)

172

Heat of Fusion, ΔH_{fus}, kcal/mol:
Entropy of Fusion, ΔS_{fus}, cal/mol K (e.u.):

    11.8   (Casellato et al. 1973; quoted, Yalkowsky 1981)

    11.4   (Hinckley et al. 1989)

Fugacity Ratio at 25 °C (assuming $\Delta S = 13.5$ e.u.), F:

    0.141 (at M.P. = 111°C)

    0.14   (Mackay et al. 1980)

Water Solubility (g/m³ or mg/L at 25°C):

    0.240 (27°C, nephelometry, Davis et al. 1942; quoted, Hollifield 1979; Futoma et al. 1981; IUPAC 1989)

    0.265 (shake flask-UV, Klevens 1950; quoted, Futoma et al. 1981; Mackay & Shiu 1981; Pearlman et al. 1984; IUPAC 1989)

    0.240 (20°C, shake flask-UV, Eisenbrand & Baumann 1970)

    0.265 (shake flask-UV, Wauchope & Getzen 1972; quoted, Mackay & Shiu 1977; Futoma et al. 1981; Walters & Luthy 1984)

    0.236 (fluorescence/UV, Schwarz & Wasik 1976; quoted, Mackay & Shiu 1981)

    0.260 (shake flask-fluorescence, Mackay & Shiu 1977; quoted, API 1978; Callahan et al. 1979; May et al. 1979; Yalkowsky & Valvani 1979,1980; Mackay et al. 1980; Valvani & Yalkowsky 1980; Futoma et al. 1981; Lande & Banerjee 1981; Mackay & Shiu 1981; Mabey et al. 1982; Pearlman et al. 1984; Walters & Luthy 1984; Lande et al. 1985; Miller et al. 1985; Chin et al. 1986; IUPAC 1989; Kayal & Connell 1990; Shorten et al. 1990; Vadas et al. 1991)

    0.206 (Rossi 1977; Neff 1979; quoted, Eadie et al. 1982)

    0.206 (gen. col.-HPLC/UV, May et al. 1978ab; quoted, Wise et al. 1981; Briggs 1981; Futoma et al. 1981; Pearlman et al. 1984; Walters & Luthy 1984; IUPAC 1989; Vadas et al. 1991)

    0.240, 0.120 (quoted, shake flask-nephelometry, Hollifield 1979)

    0.368 (calculated-K_{OW}, Valvani & Yalkowsky 1980)

    0.351 (calculated-K_{OW}, Yalkowsky & Valvani 1980)

    0.255 (Lande & Banerjee 1981)

    0.200 (20°C, quoted, Schmidt-Bleek et al. 1982)

    0.255 (RP-TLC, Bruggeman et al. 1982)

    0.26 (selected, Mills et al. 1982)

    0.190 (20°C, gen. col.-fluorescence, Hashimoto et al. 1982)

    0.203 (24.6°C, gen. col.-HPLC, May et al. 1983)

    0.265 (quoted, Verschueren 1983; Pinal et al. 1990)

    0.243 (average lit. value, Yalkowsky et al. 1983)

    0.243 (average lit. value, Pearlman et al. 1984)

    0.199 (gen. col.-HPLC/fluorescence, Walters & Luthy 1984)

    0.256, 0.119 (quoted, predicted, Lande et al. 1985)

0.260 (selected, Miller et al. 1985; quoted, Gobas et al. 1988; Shorten et al. 1990; Capel et al. 1991)

0.200 (quoted, Neuhauser et al. 1985)

1.43   (RP-HPLC, calculated-K_{OW} & M.P., Chin et al. 1986)

0.283 (vapor saturation-UV, Akiyoshi et al. 1987)

0.263 (quoted, Isnard & Lambert 1989)

0.240 (recommended, IUPAC 1989)

0.222, 0.373 (gen. col.-HPLC/fluorescence, average value of Japan, OECD tests, Kishi & Hashimoto 1989)

0.166 (shake flask-fluorescence, Kishi & Hashimoto 1989)

0.260 (selected, Ma et al. 1990)

0.265 (shake flask-HPLC/UV/fluorescence, Pinal et al. 1991)

0.177 (gen. col.-HPLC, Vadas et al. 1991)

Vapor Pressure (Pa at 25°C):

6.67×10^{-4} (Hoyer & Peperle 1958; quoted, Mabey et al. 1982; Tsai et al. 1991)

0.254   (corrected by fugacity ratio to solid state, Mackay & Shiu 1981)

0.0007   (20°C, quoted, Schmidt-Bleek et al. 1982)

0.00121 (gas saturation, Sonnefeld et al. 1983; quoted, Bidleman 1984)

0.00124 (gen. col.-HPLC, Wasik et al. 1983)

1.79   (subcooled liq., extrapolated from Antoine eqn., Boublik et al. 1984; quoted, Shorten et al. 1990)

0.0067   (subcooled liq., Bidleman 1984)

0.0070   (Yamasaki et al. 1984; quoted, Capel et al. 1991)

1.65×10^{-4} (extrapolated, Antoine eqn., Dean 1985)

0.00133 (quoted, Neuhauser et al. 1985)

0.00105 (extrapolated, Antoine eqn., Stephenson & Malanowski 1987)

0.00068, 0.0056 (lit., subcooled liq. value, Bidleman & Foreman 1987)

0.992   (WERL Treatability database, quoted, Ryan et al. 1988)

0.00642, 0.00636 (quoted, GC-RT, Hinckley 1989)

0.00861, 0.00635 (quoted, subcooled liq. values, Hinckley et al. 1990)

Henry's Law Constant (Pa m^3/mol):

220 (calculated-P/C, Mackay & Shiu 1981; quoted, Shroten et al. 1990)

0.659 (calculated-P/C, Mabey et al. 1982)

0.87 (calculated-P/C, Eastcott et al. 1988)

0.134 (15°C, calculated, Baker & Eisenreich 1990; quoted, Tsai et al. 1991)

5.47 (calculated-P/C, Capel et al. 1991)

Octanol/Water Partition Coefficient, log K_{OW}:

    4.78 (calculated-S, Zepp & Schlotzhauer 1979)

    5.33 (calculated-f const, Callahan et al. 1979)

    5.22 (calculated-f const., Yalkowsky & Valvani 1979, 1980; quoted, Mackay et al. 1980; Walters & Luthy 1984)

    5.29 (calculated-S & M.P., Mackay et al. 1980)

    5.22 (RP-TLC, Bruggeman et al. 1982)

    4.90 (calculated-f const., Mabey et al. 1982; quoted, Kayal & Connell 1990)

    5.53 (selected, Mills et al. 1982)

    4.70 (quoted, Schmidt-Bleek et al. 1982)

    5.22 (average lit. value, Yalkowsky et al. 1983)

    5.20, 5.22 (quoted, calculated-MR, Yoshida et al. 1983)

    5.22 (quoted, Patton et al. 1984)

    5.22 (selected, Miller et al. 1985; quoted, Chin et al. 1986; De Bruijn et al. 1989; Capel et al. 1991; Landrum et al. 1991)

    5.17 (calculated- χ as per Rekker & De Kort 1979, Ruepeert et al. 1985)

    4.90 (Yalkowsky 1985; quoted, Pinal et al. 1990)

    4.84 (RP-HPLC, calculated-S & M.P., Chin et al. 1986)

    5.33 (quoted, Pavlou 1987; Ryan et al. 1988)

    5.22 (quoted, Isnard & Lambert 1989)

    5.16 (slow stirring-GC, De Bruijn et al. 1989)

    5.20 (recommended, Sangster 1989)

    5.30 (selected, Ma et al. 1990)

    5.20 (Bayona et al. 1991)

    4.90 (quoted, Pinal et al. 1991)

    6.50 (calculated, Broman et al. 1991)

Bioconcentration Factor, log BCF:

    3.18 (calculated as per Kenaga & Goring 1979, Eadie et al. 1982)

    4.90 (*p. hoyi*, Eadie et al. 1982)

    4.08 (microorganisms-water, calculated from K_{OW}, Mabey et al. 1982)

    3.24 (*daphnia magna*, Newsted & Giesy 1987)

    0.756, 1.079 (*polycaaete sp, capitella capitata*, Bayona et al. 1991)

Sorption Partition Coefficient, log K_{OC}:

    4.58 (calculated-K_{OW}, Mabey et al. 1982)

    4.65 (calculated, Pavlou 1987)

    6.38 (sediments average, Kayal & Connell 1990)

    4.74, 4.62 (RP-HPLC on CIHAC, on PIHAC, Szabo et al. 1990b)

    6.30 (specified particulate log K_{OC}, Broman et al. 1991)

    4.0 (predicted dissolved log K_{OC}, Broman et al. 1991)

Half-Lives in the Environment:

Air: 2.02-20.2 hours, based on estimated sunlight photolysis half-life in water (Howard et al. 1991).

Surface water: computed near-surface half-life for photochemical transformation of a water body, 21 hours (latitude 40°N, midday, midsummer) and direct photolysis in a 5-m deep inland water body were 160 days with no sediment-water partitioning and 200 days with sediment-water partitioning (Zepp & Schlotzhauer 1979); 21-63 hours, based on photolysis half-life in water (Lyman et al. 1982; quoted, Howard et al. 1991); 160 days for summer sunlight at 40°N latitude (Mill & Mabey 1985).

Groundwater: 6720-21120 hours, based on estimated unacclimated aqueous aerobic biodegradation half-life (Howard et al. 1991).

Sediment:

Soil: 44-182 days (Sims & Overcash 1983; quoted, Bulman et al. 1987); 39 days for 5 mg/kg treatment and 34 days for 50 mg/kg treatment (Bulman et al. 1987); bidegradation rate constant of 0.0018 day^{-1} with a half-life of 377 days for Kidman sandy loam soil, and 0.0026 day^{-1} with half-life of 268 days for McLaurin sandy loam soil (Park et al. 1990); half-lives were: 3360-10560 hours, based on aerobic soil dieaway test data at 10-30°C (Coover & Sims 1987; quoted, Howard et al. 1991); >50 days (Ryan et al. 1988); 17.961 weeks, 7.8 years (quoted, Luddington soil, Wild et al. 1991).

Biota: half-life of depuration by oysters, 5 days (Lee et al. 1978; quoted, Verschueren 1983).

Environmental Fate Rate Constants or Half-Lives:

Volatilization:

Photolysis: calculated half-lives of direct sunlight photolysis for 50% conversion at 40°N latitude of midday in midsummer: 160 days (5 meter deep inland water) and 200 days (inland water with sediment partitioning) at 40°N latitude (Zepp & Schlotzhauer 1979); atmospheric and aqueous photolysis half life of 21 hours, based on measured sunlight photolysis rate constant in water adjusted for midday summer sunlight at 40°N latitude (quoted, Howard et al. 1991) and of 63 hours after adjusting for approximate winter sunlight intensity (Lyman et al. 1982; quoted, Howard et al. 1991); half-life of 160 days under summer sunlight in surface water (Mill & Mabey 1985); half-lives on different atmospheric particulate substrates (appr. 25 μg/gm on substrate): 74 hours on silica gel, 23 hours on alumina and 44 hours on fly ash (Behymer & Hites 1985).

Hydrolysis: not hydrolyzable (Mabey et al. 1982); no hydrolyzable groups (Howard et al. 1991).

Oxidation: rate constant of <3600 M^{-1} hr^{-1} for singlet oxygen and <360 M^{-1} hr^{-1} for peroxy radical (Mabey et al. 1982); photooxidation half-life of 2.02-20.2 hours, based on estimated rate constant for reaction with hydroxyl radical in air (Atkinson 1987; quoted, Howard et al. 1991).

Biodegradation: aquatic fate rate of 2.2×10^{-3} μmol hr^{-1} mg^{-1} with bacterial protein (Barnsley 1975; quoted, Callahan et al. 1979); significant with gradual degradation for a domestic sewer test for an average three static-flask screening (Tabak et al. 1981); aerobic half-life of 3360-10560 hours, based on aerobic soil dieaway test data at 10-30°C (Coover & Sims 1987; quoted, Howard et al. 1991); rate constant in atmosphere of 0.19 hour^{-1} (Dragoescu & Friedlander 1989; quoted, Tsai et al. 1991); rate constants of 0.0018 day^{-1} with half-life of 377 days for Kidman sandy loam and 0.0026 day^{-1} with half-life of 268 days for McLarin sandy loam all at -0.33 bar soil moisture (Park et al. 1990); anaerobic half-life of 13440-42240 hours, based on estimated unacclimated aqueous aerobic biodegradation half-life (Howard et al. 1991).

Biotransformation: estimated rate constant for bacteria, 1×10^{-10} ml cell^{-1} hr^{-1} (Mabey et al. 1982).

Bioconcentration, Uptake (k_1) and Elimination (k_2) Rate Constants:

Common Name: Benzo[a]fluorene
Synonym: 1,2-benzofluorene, 11H-benzo[a]fluorene, chrysofluorene
Chemical Name: benzo[a]fluorene, 1,2-benzofluorene
CAS Registry No: 238-84-3
Molecular Formula: $C_{17}H_{12}$
Molecular Weight: 216.29

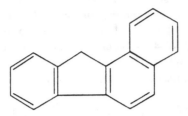

Melting Point (°C):
    187   (Yalkowsky & Valvani 1979,1980; Mackay et al. 1980; Yalkowsky et al. 1983;
           Miller et al. 1985; Mailhot & Peters 1988)
    190   (Bjørseth 1983)
    188   (Pearlman et al. 1984)
Boiling Point (°C):
    407   (Bjørseth 1983)
    403   (Pearlman et al. 1984)
    398   (Mailhot & Peters 1988)
Density (g/cm³ at 20°C):
Molar Volume (cm³/mol):
    240.0 (LeBas method, Miller et al. 1985)
Molecular Volume (Å³):
    203.78 (Pearlman et al. 1984)
Total Surface Area, TSA (Å²):
    237.40 (Yalkowsky & Valvani 1979; quoted, Mailhot & Peters 1988)
    240.33 (Pearlman et al. 1984)
Heat of Fusion, ΔH_{fus}, kcal/mol:
Entropy of Fusion, ΔS_{fus}, cal/mol K (e.u.):
Fugacity Ratio at 25°C (assuming ΔS = 13.5 e.u.), F:
    0.0249 (at M.P. = 187°C)
    0.02   (Mackay et al. 1980)

Water Solubility (g/m³ or mg/L at 25°C):
    0.045 (shake flask-fluorescence, Mackay & Shiu 1977; quoted, Yalkowsky & Valvani
           1979; Mackay et al. 1980; Wise et al. 1981; Pearlman et al. 1984; Miller et al.
           1985; Mailhot & Peters 1988; IUPAC 1989)

0.045 (quoted, Valvani & Yalkowsky 1980; Yalkowsky & Valvani 1979, 1980; Mailhot
    & Peters 1988)
0.023 (calculated-K_{OW}, Valvani & Yalkowsky 1980; Yalkowsky & Valvani 1980)
0.045 (average lit. value, Yalkowsky et al. 1983)
0.045 (average lit. value, Pearlman et al. 1984)
0.0454 (selected, Miller et al. 1985; quoted, Capel et al. 1991)

Vapor Pressure (Pa at 25°C):

Henry's Law Constant (Pa m^3/mol):

Octanol/Water Partition Coefficient, log K_{OW}:
    5.75 (calculated-f const., Valvani & Yalkowsky 1980; Yalkowsky & Valvani
        1979,1980; quoted, Mackay et al. 1980; Mailhot & Peters 1988)
    5.32 (calculated-S & M.P., Mackay et al. 1980)
    5.27 (average lit. value, Yalkowsky et al. 1983)
    5.75, 5.57 (quoted, calculated-MR, Yoshida et al. 1983)
    5.32 (calculated, Miller et al. 1985; quoted, Capel et al. 1991)
    5.69 (calculated- χ as per Rekker & De Kort 1979, Ruepert et al. 1985)

Bioconcentration Factor, log BCF:

Sorption Partition Coefficient, log K_{OC}:

Half-Lives in the Environment:
    Air:
    Surface water:
    Groundwater:
    Sediment:
    Soil:
    Biota:

Environmental Fate Rate Constants or Half-Lives:
    Volatilization:
    Photolysis:
    Hydrolysis:
    Oxidation:
    Biodegradation:
    Biotransformation:
    Bioconcentration, Uptake (k_1) and Elimination (k_2) Rate Constants:

Common Name: Benzo[b]fluorene
Synonym: 2,3-benzofluorene, 11H-benzo[b]fluorene, isonaphthofluorene
Chemical Name: benzo[b]fluorene
CAS Registry No: 243-17-4
Molecular Formula: $C_{17}H_{12}$
Molecular Weight: 216.29

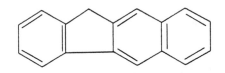

Melting Point (°C):
    209   (Yalkowsky & Valvani 1979,1980; Miller et al. 1985; Gobas et al. 1988)
    210   (Pearlman et al. 1984)
Boiling Point (°C):
    406   (Bjørseth 1983)
Density (g/cm³ at 20°C):
Molar Volume (cm³/mol):
    240.0 (LeBas method, Miller et al. 1985; Gobas et al. 1988)
Molecular Volume (Å³):
    203.78 (Pearlman et al. 1984)
Total Surface Area, TSA (Å²):
    239.90 (Yalkowsky & Valvani 1979)
    240.34 (Pearlman et al. 1984)
Heat of Fusion, ΔH_{fus}, kcal/mol:
Entropy of Fusion, ΔS_{fus}, cal/mol K (e.u.):
Fugacity Ratio at 25°C (assuming ΔS = 13.5 e.u.), F:
    0.151 (at M.P. = 209°C)

Water Solubility (g/m³ or mg/L at 25°C):
    0.0020 (shake flask-fluorescence, Mackay & Shiu 1977; quoted, Yalkowsky & Valvani
         1979; Wise et al. 1981; Pearlman et al. 1984; Miller et al. 1985; IUPAC 1989)
    0.0116 (quoted, Valvani & Yalkowsky 1980; Yalkowsky & Valvani 1980)
    0.0140 (calculated-K_{OW}, Valvani & Yalkowsky 1980)
    0.0143 (calculated-K_{OW}, Yalkowsky & Valvani 1980)
    0.0020 (average lit. value, Yalkowsky et al. 1983)
    0.0020 (average lit. value, Pearlman et al. 1984)

0.0020 (selected, Miller et al. 1985; quoted, Eastcott et al. 1988; Gobas et al. 1988; Capel et al. 1991)

Vapor Pressure (Pa at 25°C):

Henry's Law Constant (Pa m^3/mol):

Octanol/Water Partition Coefficient, log K_{OW}:
    5.75   (calculated-f const., Valvani & Yalkowsky 1980; Yalkowsky & Valvani 1979,1980)
    5.27   (average lit. value, Yalkowsky et al. 1983)
    5.75   (calculated, Miller et al. 1985; quoted, Eastcott et al. 1988; Capel et al. 1991)
    5.69   (calculated- χ as per Rekker & De Kort 1979, Ruepert et al. 1985)

Bioconcentration Factor, log BCF:

Sorption Partition Coefficient, log K_{OC}:

Half-Lives in the Environment:
    Air:
    Surface water:
    Groundwater:
    Sediment:
    Soil:
    Biota:

Environmental Fate Rate Constants or Half-Lives:
    Volatilization:
    Photolysis:
    Hydrolysis:
    Oxidation:
    Biodegradation:
    Biotransformation:
    Bioconcentration, Uptake (k_1) and Elimination (k_2) Rate Constants:

Common Name: Chrysene
Synonym: 1,2-benzophenanthrene, benzo(a)phenanthrene, 1,2,5,6-dibenzonaphthalene
Chemical Name: chrysene
CAS Registry No: 218-01-9
Molecular Formula: $C_{18}H_{12}$
Molecular Weight: 228.30

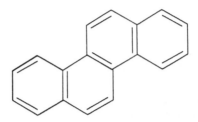

Melting Point (°C):
    256   (Weast 1982-83; Radding et al. 1976; quoted, Callahan et al. 1979; Mabey et al.
           1982; Whitehouse & Cooke 1982; Weast et al. 1984; Vadas et al. 1991)
    255   (Yalkowsky & Valvani 1979,1980; Mackay et al. 1980; Lande & Banerjee 1981;
           Bruggeman et al. 1982; Miller et al. 1985; Gobas et al. 1988; Mailhot & Peters
           1988)
    254   (API 1978; Briggs 1981; Yalkowsky 1981; Windholz 1983; Mailhot & Peters
           1988; Pinal et al. 1991)
    252   (Pearlman et al. 1984)
Boiling Point (°C):
    448   (Weast 1975; quoted, API 1978; Mabey et al. 1982; Pearlman et al. 1984; Mailhot
           & Peters)
    441   (Bjørseth 1983)
    488   (Mailhot & Peters 1988)
Density (g/cm³ at 20°C):
    1.274 (Weast 1982-83)
    1.2826 (Mailhot & Peters 1988)
Molar Volume (cm³/mol):
    179.0 (calculated, Lande & Banerjee 1981)
    251.0 (LeBas method, Miller et al. 1985; Gobas et al. 1988)
    178.0 (Mailhot & Peters 1988)
Molecular Volume (Å³):
    212.06 (Pearlman et al. 1984)
Total Surface Area, TSA (Å²):
    241.0 (Yalkowsky & Valvani 1979; quoted, Whitehouse & Cooke 1982; Mailhot &
           Peters 1988)
    240.15 (Pearlman et al. 1984)
    241.0 (quoted as HSA, Woodburn et al. 1989)

Heat of Fusion, ΔH_{fus}, cal/mol:
Entropy of Fusion, ΔS_{fus}, cal/mol K (e.u.):
  11.8 (Casellato et al. 1973; quoted, Yalkowsky 1981)
  14.9 (Ubbelohde 1978; quoted, Yalkowsky 1981)
Fugacity Ratio at 25 °C (assuming ΔS = 13.5 e.u.), F:
  0.005 (Mackay et al. 1980)
  0.00528 (at M.P. = 255°C)

Water Solubility (g/m³ or mg/L at 25°C):
  0.0015 (27°C, shake flask-nephelometry, Davis et al. 1942; quoted, Hollifield 1979; Futoma et al. 1981; Billington et al. 1988; IUPAC 1989)
  0.006 (shake flask-UV, Klevens 1950; quoted, Futoma et al. 1981; Pearlman et al. 1984; IUPAC 1989)
  0.0015 (Weimer & Prausnitz 1965; quoted, Mackay & Shiu 1977)
  0.006 (shake flask-UV, Wauchope & Getzen 1972; quoted, Mackay & Shiu 1977; Walters & Luthy 1984)
  0.002 (shake flask-fluorescence, Mackay & Shiu 1977; quoted, API 1978; Callahan et al. 1979; May et al. 1979; Yalkowsky & Valvani 1979, 1980; Mackay et al. 1980; Valvani & Yalkowsky 1980; Futoma et al. 1981; Lande & Banerjee 1981; Whitehouse & Cooke 1982; Pearlman et al. 1984; Walters & Luthy 1984; Lande et al. 1985; Miller et al. 1985; Billington et al. 1988; Mailhot & Peters 1988; IUPAC 1989; Kayal & Connell 1990; Vadas et al. 1991)
  0.0018 (Rossi 1977; Neff 1979; quoted, Eadie et al. 1982)
  0.0018 (gen. col.-HPLC/UV, May et al. 1978a,b; quoted, Callahan et al. 1979; Briggs 1981; Futoma et al. 1981; Wise et al. 1981; Mabey et al. 1982; Whitehouse & Cooke 1982; Pearlman et al. 1984; Walters & Luthy 1984; Billington et al. 1988; Mailhot & Peters 1988; IUPAC 1989; Vadas et al. 1991)
  0.017 (shake flask-nephelometry, Hollifield 1979)
  0.0036 (calculated-K_{OW}, Valvani & Yalkowsky 1980)
  0.0040 (calculated-K_{OW}, Yalkowsky & Valvani 1980)
  0.0015 (calculated-parachor, Briggs 1981)
  0.00199 (RP-TLC, Bruggeman et al. 1982)
  0.0019 (quoted average, Whitehouse & Cooke 1982)
  0.00189 (25.3°C, gen. col.-HPLC, May et al. 1983)
  0.0020 (average lit. value, Yalkowsky et al. 1983)
  0.0018 (average lit. value, Pearlman et al. 1984; quoted, Billington et al. 1988)
  0.00327 (gen. col.-HPLC/fluorescence, Walters & Luthy 1984)
  0.002 (Lande & Banerjee 1981)
  0.002 (selected, Mills et al. 1982)
  0.002, 0.0069 (quoted, predicted, Lande et al. 1985)
  0.0012 (selected, Miller et al. 1985; quoted, Gobas et al. 1988)
  0.00102, 0.0012 (gen. col.-HPLC/UV, Billington et al. 1988)

0.0019 (recommended, IUPAC 1989)
0.002 (quoted, Isnard & Lambert 1989)
0.002 (selected, Ma et al. 1990)
0.0021 (quoted exptl., Pinal et al. 1991)
0.0016 (gen. col.-HPLC, Vadas et al. 1991)

Vapor Pressure (Pa at 25°C):
5.70×10^{-7} (effusion method, De Kruif 1980)
8.4×10^{-7} (Hoyer & Peperle 1958; quoted, Mabey et al. 1982)
6.08×10^{-7} (extrapolated-Antoine eqn., Stephenson & Malanowski 1987)
4.0×10^{-6} (selected, Ma et al. 1990)

Henry's Law Constant (Pa m^3/mol):
0.1064 (calculated-P/C, Mabey et al. 1982)
218.15 (WERL Treatability database, quoted, Ryan et al. 1988)
0.45 (calculated-P/C, Eastcott et al. 1988)

Octanol/Water Partition Coefficient, log K_{OW}:
5.94 (calculated-π const., Zepp & Schlotzhauer 1979)
5.61 (Radding et al. 1976; quoted, Callahan et al. 1979; Kayal & Connell 1990)
5.91 (calculated-f const., Yalkowsky & Valvani 1979; Valvani & Yalkowsky 1980; quoted, Mackay et al. 1980; Walters & Luthy 1984)
5.01 (calculated-f const., Yalkowsky & Valvani 1980)
6.01 (calculated-S & M.P., Mackay 1980)
5.79 (HPLC-k', Hanai et al. 1981)
5.91 (RP-TLC, Bruggeman et al. 1982)
5.61 (calculated-f const., Mabey et al. 1982)
5.60 (selected, Mills et al. 1982)
5.91 (calculated-S, Yalkowsky et al. 1983)
5.91, 5.88 (quoted, calculated-MR, Yoshida et al. 1983)
5.79 (calculated, Miller et al. 1985; quoted, Landrum et al. 1991)
5.84 (calculated- χ as per Rekker & De Kort 1979, Ruepert et al. 1985)
5.61 (quoted, Pavlou 1987; Ryan et al. 1988)
5.91 (quoted, Mailhot & Peters 1988)
5.79 (quoted, Isnard & Lambert 1989)
5.61 (recommended, Sangster 1989)
5.70 (selected, Ma et al. 1990)
5.80 (Bayona et al. 1991)
5.61 (quoted, Pinal et al. 1991)
7.10 (calculated-K_{OC}, Broman et al. 1991)

Bioconcentration Factor, log BCF:

    4.34   (calculated as per Kenaga & Goring 1979, Eadie et al. 1982)

    4.31   (*p. hoyi*, Eadie et al. 1982)

    4.72   (microorganisms-water, calculated from K_{ow}, Mabey et al. 1982)

    3.785 (*Daphnia magna*, Newsted & Giesy 1987)

    1.17, 0.792 (*polychaete sp, capitella capitata*, Bayona et al. 1991)

Sorption Partition Coefficient, log K_{OC}:

    3.66   (calculated-K_{ow}, Mabey et al. 1982)

    4.89   (calculated, Pavlou 1987)

    6.27   (sediments average, Kayal & Connell 1990)

    6.90   (specified particulate log K_{OC}, Broman et al. 1991)

    4.0   (predicted dissolved log K_{OC}, Broman et al. 1991)

Half-Lives in the Environment:

    Air: 0.802-8.02 hours, based on estimated photooxidation half-life in air (Atkinson 1987; quoted, Howard et al. 1987); 1.3 hour for adsorption on wood soot particles in an outdoor teflon chamber with an estimated rate constant of 0.0092 minute^{-1} at 1 cal cm^{-2} min^{-1}, 10 g/m^3 H$_2$O and 20°C (Kamens et al. 1988).

    Surface water: the computed half-life of near-surface of a water body for direct photochemical transformation, 4.4 hours at latitude 40°N, midday, midsummer and the direct photolysis in 5-m deep inland water, 13 hours with no sediment-water partitioning and 68 days with sediment-water partitioning (Zepp & Schlotzhauer 1979); half-lives were estimated to be 4.4-13 hours, based on photolysis half-life in water (Lyman et al. 1982; quoted, Howard et al. 1991).

    Groundwater: 17808-48000 hours, based on estimated unacclimated aqueous aerobic biodegradation half-life (Howard et al. 1991).

    Sediment:

    Soil: more than 5.5 days (Sims & Overcash 1983; quoted, Bulman et al. 1987); 328 days for 5 mg/kg treatment and 224 days for 50 mg/kg treatment (Bulman et al. 1987); biodegradation rate constant of 0.0019 day^{-1} with a half-life of 371 days for Kidman sandy loam soil, and 0.0018 hour^{-1} with 387 days for McLaurin sandy loam soil (Park et al. 1990); half-lives were: 8904-24000 hours, based on aerobic soil dieaway test data (Coover & Sims 1987; quoted, Howard et al. 1991); >50 days (Ryan et al. 1988).

    Biota:

Environmental Fate Rate Constants or Half-Lives:

    Volatilization:

    Photolysis: calculated half-lives of direct sunlight photolysis for 50% conversion at 40°N latitude of midday in midsummer: 4.4 hours (near-surface water; quoted, Herbes et al. 1980), 13 days (5 meter deep inland water) and 68 days (inland water with sediment partitioning) (Zepp & Schlotzhauer 1979); atmospheric and aqueous photolysis half-life of 4.4 hours, based on measured aqueous photolysis quantum yields and calculated for midday summer sunlight at 40°N latitude (Zepp & Schlotzhauer 1979; quoted, Harris 1982; Howard et al. 1991) and 13 hours after adjusting for approximate winter sunlight intensity (Lyman et al. 1982; quoted, Howard et al. 1991); half-lives on different atmospheric particulate substrates (appr. 25 μg/gm on substrate): 100 hours on silica gel, 78 hours on alumina and 38 hours on fly ash (Behymer & Hites 1985); first order daytime decay constants: 0.0056 min^{-1} for soot particles loading of 1000-2000 ng/mg and 0.0090 min^{-1} with 30-350 ng/mg loading (Kamens et al. 1988); photodegradation in ethanol-water (1:1, v/v) solution for initial concentration of 5.0 ppm by high pressure mercury lamp or sunlight with a rate constant of 7.07x10^{-3} minute with half-life of 1.63 hour (Wang et al. 1991).

Hydrolysis: not hydrolyzable (Mabey et al. 1982); no hydrolyzable groups (Howard et al. 1991).

Oxidation: rate constant of >1x10^6 M^{-1} hr^{-1} for singlet oxygen and 1x10^3 M^{-1} hr^{-1} for peroxy radical (Mabey et al. 1982); photooxidation half-life in air of 0.802-8.02 hours, based on estimated rate constant for reaction with hydroxyl radical in air (Atkinson 1987; quoted, Howard et al. 1991).

Biodegradation: significant degradation with gradual adaptation within 7 days for a domestic sewer 28 days test for an average of three static-flask screening (Tabak et al. 1981); aerobic half-life of 8904-24000 hours, based on aerobic soil dieaway test data (Coover & Sims 1987; quoted, Howard et al. 1991); rate constants of 0.0019 day^{-1} with half-life of 371 days for Kidman sandy loam and 0.0018 day^{-1} with half-life of 387 days for McLarin sandy loam all at -0.33 bar soil moisture (Park et al. 1990); anaerobic half-life of 35616-96000 hours, based on estimated unacclimated aqueous aerobic biodegradation half-life (Howard et al. 1991).

Biotransformation: estimated to be 1x10^{-10} ml cell^{-1} h^{-1} for bacteria (Mabey et al. 1982).

Bioconcentration, Uptake (k_1) and Elimination (k_2) Rate Constants:

Common Name: Triphenylene
Synonym: 9,10-benzophenanthrene, isochrysene, 1,2,3,4-dibenznaphthalene
Chemical Name: triphenylene
CAS Registry No: 217-59-4
Molecular Formula: $C_{18}H_{12}$
Molecular Weight: 228.30

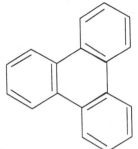

Melting Point (°C):
    199   (Weast 1982-83; Bjørseth 1983; Dean 1985; Budavari 1989)
    197   (Pearlman et al. 1984)
Boiling Point (°C):
    425   (Weast 1982-83; Dean 1985; Pearlman et al. 1984; Budavari 1989)
    439   (Bjørseth 1983)
Density (g/cm³ at 20°C):
    1.302 (Dean 1985; Budavari 1989)
Molar Volume (cm³/mol):
    251   (LeBas method)
    1.277 (V_I/100, intrinsic molar vol., Kamlet et al. 1988)
Molecular Volume (Å³):
    211.3 (Pearlman et al. 1984)
Total Surface Area, TSA (Å²):
    236.0 (Yalkowsky & Valvani 1979)
    235.99(Pearlman et al. 1984)
Heat of Fusion, ΔH_{fus}, kcal/mol:
Entropy of Fusion, ΔS_{fus}, cal/mol K (e.u.):
Fugacity Ratio at 25°C (assuming $\Delta S = 13.5$ e.u.), F:
    0.0190 (Mackay et al. 1980)

Water Solubility (g/m³ or mg/L at 25°C):
    0.0388 (27°C, nephelometry, Davis et al. 1942; quoted, Pearlman et al. 1984; IUPAC
        1989)
    0.043 (shake flask-UV, Klevens 1950; quoted, Pearlman et al. 1984; IUPAC 1989)
    0.043 (shake flask-UV, Wauchope & Getzen 1972; quoted, API 1978)

0.043   (shake flask-fluorescence, Mackay & Shiu 1977; quoted, API 1978; Yalkowsky & Valvani 1979; Mackay et al. 1980; Pearlman et al. 1984; IUPAC 1989)

0.066   (gen. col.-HPLC/UV, May et al. 1978)

0.0049 (20.5°C, gen. col.-HPLC, May et al. 1983)

0.0162 (lit. mean, Yalkowsky et al. 1983)

0.041   (lit. mean, Pearlman et al. 1984)

0.024   (quoted, Tomlinson & Hafkenscheid 1986)

0.041   (vapor saturation-UV, Akiyoski et al. 1987)

0.043, 0.0194 (quoted, calculated-χ , Nirmalakahandan & Speece 1989)

0.0425, 0.0425, 0.033 (quoted, calculated- χ , K_{ow}, Yalkowsky & Mishra 1990)

Vapor Pressure (Pa at 25°C):

2.30×10^{-6} (effusion, De Kruif 1980)

3.85×10^{-7} (extrapolated, Antoine eqn., Stephenson & Malanowski 1987)

1.17×10^{-8} (extrapolated, Antoine eqn.,subcooled liq., Stephenson & Malanowsky 1987)

Henry's Law Constant (Pa m^3/mol):

Octanol/Water Partition Coefficient, log K_{ow}:

5.45   (shake flask-UV, Karickhoff et al. 1979; quoted, Walters & Luthy 1984)

5.45   (calculated-TSA, Yalkowsky & Valvani 1979)

5.45, 5.24 (quoted, calculated-S, Mackay et al. 1980)

5.45, 5.88 (quoted, calculated-MR, Yoshida et al. 1983)

5.45   (estimated-f const., Valvani & Yalkowsky 1980)

5.20   (calculated-f const., Yalkowsky et al. 1983)

5.84   (calculated- χ as per Rekker & De Kort 1979, Ruepert et al. 1985)

5.66   (calculated-f const., Yalkowsky & Mishra 1990)

7.10   (calculated, Broman et al. 1991)

Bioconcentration Factor, log BCF:

3.96   (*daphnia magna*, Newsted & Giesy 1987)

Sorption Partition Coefficient, log K_{oc}:

6.90   (specified particulate log K_{oc}, Broman et al. 1991)

4.0   (predicted dissolved log K_{oc}, Broman et al. 1991)

Half-Lives in the Environment:

Air:

Surface water:

Groundwater:

Sediment:

Soil:
Biota:

Environmental Fate Rate Constants or Half-Lives:
Volatilization:
Photolysis:
Hydrolysis:
Oxidation:
Biodegradation:
Biotransformation:
Bioconcentration, Uptake (k_1) and Elimination (k_2) Rate Constants:

189

Common Name: *p*-Terphenyl
Synonym: 1,4-diphenylbenzene
Chemical Name: *p*-terphenyl
CAS Registry No: 92-94-4
Molecular Formula: $C_{18}H_{14}$
Molecular Weight: 230.31

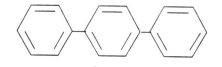

Melting Point (°C):
     213   (Weast 1982-83)
     195   (Stephenson & Malanowski 1987)
Boiling Point (°C):
     250   (sublime, Weast 1982-82)
     376   (Stephenson & Malanowski 1987)
Density (g/cm³ at 20 °C):
Molar Volume (cm³/mol):
     258.2 (LeBas method)
Molecular Volume (Å³):
Total Surface Area, TSA (Å²):
Heat of Fusion, ΔH_{fus} kcal/mol:
Entropy of Fusion, ΔS_{fus} cal/mol K (e.u.):
Fugacity Ratio at 25 °C (assuming ΔS = 13.5 e.u.), F:
     0.0138 (at M.P. 213 °C)

Water Solubility (g/m³ or mg/L at 25°C):
     0.0180   (vapor saturation-UV/fluo., Akiyoshi et al. 1987)

Vapor Pressure (Pa at 25°C):
     4.86×10^{-6} (extrapolated from solid, Stephenson & Malanowski 1987)
     1.78×10^{-5} (extrapolated from liq., Stephenson & Malanowski 1987)

Henry's Law Constant (Pa m³/mol):

Octanol/Water Partition Coefficient, log K_{OW}:
      6.03, 5.88 (HPLC-RV, Garst 1984)
      6.03   (recommended, Sangster 1989)

Bioconcentration Factor, log BCF:

Sorption Partition Coefficient, log K_{OC}:

Half-Lives in the Environment:
      Air:
      Surface water:
      Groundwater:
      Sediment:
      Soil:

Environmental Fate Rate Constants or Half-Lives:
      Volatilization:
      Photolysis:
      Hydrolysis:
      Oxidation:
      Biodegradation:
      Biotransformation:
      Bioconcentration, Uptake (k_1) and Elimination (k_2) Rate Constants:

Common Name: Naphthacene
Synonym: benz[b]anthracene, 2,3-benzanthracene, tetracene
Chemical Name: benz[b]anthracene
CAS Registry No: 92-24-0
Molecular Formula: $C_{18}H_{12}$
Molecular Weight: 228.30

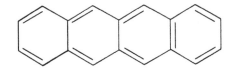

Melting Point (°C):
    357   (Yalkowsky & Valvani 1979)
    257   (Bjørseth 1983)
    335   (Karickhoff 1981; Lande et al. 1985)
Boiling Point (°C):
    450   (sublimation, Bjørseth 1983)
Density (g/cm³ at 20°C):
Molar Volume (cm³/mol):
    251   (LeBas method)
    1.227   ($V_I/100$, intrinsic molar vol., Kamlet et al. 1988)
Molecular Volume (Å³):
    213.7 (Pearlman et al. 1984)
Total Surface Area, TSA (Å²):
    248.0 (Yalkowsky & Valvani 1979)
    248.5 (Pearnman et al. 1984)
    248.0 (Lande et al. 1985)
    249.0 (Sabljic 1987)
Heat of Fusion, ΔH_{fus}, kcal/mol:
Entropy of Fusion, ΔS_{fus}, cal/mol K (e.u.):
Fugacity Ratio at 25°C (assuming ΔS = 13.5 e.u.), F:
    0.0005   (Mackay et al. 1980)
    0.000518 (at M.P. = 160°C)

Water Solubility (g/m³ or mg/L at 25°C):
    0.0010 (27°C, shake flask-nephelometry, Davis et al. 1942; quoted, IUPAC 1989)
    0.0015 (approximate, shake flask-UV, Klevens 1950; quoted, Pearlman et al. 1984;
        IUPAC 1989)

0.0010 (Weimer & Prausnitz 1965; quoted, Mackay & Shiu 1977)

0.0036 (shake flask-UV, Eisenbrand & Baumann 1970)

0.00057 (shake flask-fluorescence, Mackay & Shiu 1977; quoted, May et al. 1979; Yalkowsky & Valvani 1979; Karickhoff et al. 1979, Pearlman et al. 1984; IUPAC 1989)

0.00047 (quoted, Valvani & Yalkowsky 1980; Yalkowsky & Valvani 1980)

0.044 (shake flask-nephelometry, Hollifield 1979)

0.00033 (calculated-K_{OW}, Valvani & Yalkowsky 1980)

0.00044 (calculated-K_{OW}, Yalkowsky & Valvani 1980)

0.00138 (lit. mean, Yalkowsky et al. 1983)

0.00103 (lit. mean, Pearlman et al. 1984)

0.00138, 0.0003 (quoted, predicted, Lande et al. 1985)

0.00057, 0.0194, 0.000848 (quoted, subcooled value calculated-χ , K_{OW}, Yalkowsky & Mishara 1990)

Vapor Pressure (Pa at 25°C):

7.30×10^{-9} (effusion method, De Kruif 1980)

3.70×10^{-8} (extrapolated, Antoine eqn., Stephenson & Malanowski 1987)

Henry's Law Constant (Pa m^3/mol):

Octanol/Water Partition Coefficient, log K_{OW}:

5.90 (shake flask-UV, concn. ratio, Karickhoff et al., 1979; quoted, Karickhoff 1981; Schwarzenbach & Westall 1981; Ruepert et al. 1985)

5.91 (calculated-f const., Yalkowsky & Valvani 1979, 1980; Valvani & Yalkowsky 1980)

5.91, 5.54 (quoted, calculated-S, Mackay et al. 1980)

6.02 (HPLC-k', McDuffie 1981)

5.91 (calculated-f const., Yalkowsky et al. 1983)

5.91, 5.88 (quoted, calculated-MR, Yoshida et al. 1983)

5.90 (quoted, Govers et al. 1984)

5.90 (calculated, Miller et al. 1985, quoted, Eastcott et al. 1988)

5.84 (calculated- χ as per Rekker & De Kort 1979, Ruepert et al. 1985)

5.90, 5.48 (quoted, calculated-solvatochromic p., Kamlet et al. 1988)

5.90, 5.26 (quoted, calculated-UNIFAC, Banerjee & Howard 1988)

5.66 (quoted, Yalkowsky & Mishara 1990)

Bioconcentration Factor, log BCF:

Sorption Partition Coefficient, log K_{OC}:

    5.81   (sediment, sorption isotherms, shake flask-UV/GC, Karickhoff et al. 1979; quoted, Govers et al. 1984)

    5.81; 5.51, 5.25, 5.99 (quoted; calculated-K_{OW}, M.P. and S, S, Karickhoff 1981)

    5.81, 4.74 (quoted, calculated-K_{OW}, for natural sorbants, Schwarzenbach & Westall 1981)

    5.81, 5.74 (quoted, calculated-χ , Sabhjic 1984)

Half-Lives in the Environment:

    Air:

    Surface water:

    Groundwater:

    Sediment:

    Soil:

    Biota:

Environmental Fate Rate Constants or Half-Lives:

    Volatilization:

    Photolysis: photodegradation in ethanol-water (1:1, v/v) solution for initial concentration of 5.0 ppm by high pressure mercury lamp or sunlight with a rate constant of 0.051 minute^{-1} and half-life of 0.23 hour (Wang et al. 1991).

    Hydrolysis:

    Oxidation:

    Biodegradation:

    Biotransformation:

    Bioconcentration, Uptake (k_1) and Elimination (k_2) Rate Constants:

Common Name: Benz[a]anthracene
Synonym: 1,2-benzanthracene, 2,3-benzophenanthrene, naphthanthracene, BaA, B(a)A, tetraphene
Chemical Name: 1,2-benzanthracene
CAS Registry No: 56-55-3
Molecular Formula: $C_{18}H_{12}$
Molecular Weight: 228.3

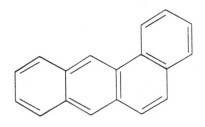

Melting Point (°C):
    156.9, 158-159 (measured, quoted, Murray et al. 1974)
    155-157, 160, 162 (quoted, Verschueren 1977, 1983)
    158   (API 1978; Pearlman et al. 1984)
    155-157, 167 (quoted, Smith et al. 1978; Callahan et al. 1979)
    160   (Yalkowsky & Valvani 1979, 1980; Mackay & Shiu 1981; Steen & Karickhoff
        1981; Miller et al. 1985; Mailhot & Peters 1988; Banerjee et al. 1990)
    162   (Weast 1982-83; Mackay et al. 1980; Briggs 1981; Whitehouse & Cooke 1982;
        Mailhot & Peters 1988)
    167   (Bidleman 1984)
Boiling Point (°C):
    435   (sublimation at 760 torr, Weast 1982-83; quoted, Verschueren 1983; Mailhot &
        Peters 1988)
Density (g/cm³ at 20°C):
    1.2544 (Mailhot & Peters 1988)
Molar Volume (cm³/mol):
    248.0 (LeBas method, Miller et al. 1985)
    1.277 (intrinsic volume: $V_I/100$, Kamlet et al. 1988; Hawker 1989)
    182.0 (Mailhot & Peters 1988)
Molecular Volume (Å³):
    212.9
    212.86 (Pearlman et al. 1984)
Total Surface Area, TSA (Å²):
    244.3 (Yalkowsky & Valvani 1979; quoted, Whitehouse & Cooke 1982; Lande et al.
        1985; Mailhot & Peters 1988)
    244.32 (Pearlman et al. 1984)
Heat of Fusion, ΔH_{fus}, kcal/mol:

Entropy of Fusion, ΔS_{fus}, cal/mol K (e.u.):
Fugacity Ratio at 25°C (assuming ΔS = 13.5 e.u.), F:
    0.0461 (at M.P. = 160°C)
    0.040 (Mackay et al. 1980)

Water Solubility (g/m^3 or mg/L at 25°C):
    0.011 (27°C, nephelometry, Davis & Parker 1942; quoted, Smith et al. 1978; Mill et
        al. 1981; Mackay & Shiu 1981; Verschueren 1983; IUPAC 1989)
    0.010 (shake flask-UV, Klevens 1950; quoted, Pearlman et al. 1984; IUPAC 1989)
    0.010 (shake flask-UV, Wauchope & Getzen 1972; quoted, Mackay & Shiu 1977; API
        1978; Walters & Luthy 1984)
    0.014 (shake flask-fluorescence, Mackay & Shiu 1977; quoted, API 1978; Callahan et
        al. 1979; May et al. 1979; Yalkowsky & Valvani 1979, 1980; Mackay et al.
        1980; Valvani & Yalkowsky 1980; Mackay & Shiu 1981; Steen & Karickhoff
        1981; Whitehouse & Cooke 1982; Pearlman et al. 1984; Walters & Luthy 1984;
        Lande et al. 1985; Miller et al. 1985; Mailhot & Peters 1988; IUPAC 1989;
        Kayal & Connell 1990; Shorten et al. 1990)
    0.0094 (gen. col.-HPLC/UV, May et al. 1978ab; quoted, Callahan et al. 1979; Briggs
        1981; Wise et al. 1981; Whitehouse & Cooke 1982; Pearlman et al. 1984;
        Walters & Luthy 1984; Mailhot & Peters 1988; IUPAC 1989)
    0.010 (Smith et al. 1978; quoted, Verschueren 1983)
    0.044 (nephelometry, Hollifield 1979)
    0.033 (calculated-K_{OW}, Valvani & Yalkowsky 1980)
    0.031 (calculated-K_{OW}, Yalkowsky & Valvani 1980)
    0.013 (calculated-parachor, Briggs 1981)
    0.0115 (quoted average, Whitehouse & Cooke 1982)
    0.0086 (gen. col.-HPLC, May et al. 1983)
    0.010 (average lit. value, Yalkowsky et al. 1983; Mailhot & Peters 1988)
    0.011 (average lit. value, Pearlman et al. 1984)
    0.010 (selected, Mills et al. 1982)
    0.0168 (HPLC/fluorescence, Walters & Luthy 1984)
    0.014, 0.0236 (quoted, predicted, Lande et al. 1985)
    0.014 (quoted, Isnard & Lambert 1988,1989)
    0.011 (quoted, Hawker 1989; Capel et al. 1991)
    0.011 (recommended, IUPAC 1989)

Vapor Pressure (Pa at 25°C):
    2.17×10^{-5} (solid, extrapolated from Antoine eqn., Kelly & Rice 1964; quoted, Bidleman
        1984)
    2.20×10^{-5} (effusion method, Kelly & Rice 1964; quoted, Bidleman 1984)
    3.87×10^{-7} (effusion method, Wakayama & Inokuchi 1967; quoted, Bidleman 1984)
    1.47×10^{-5} (solid, extrapolated from Antoine eqn., Murray et al. 1974)
    1.47×10^{-5} (effusion method, Murray et al. 1974; quoted, Bidleman 1984)

6.67x10^{-7} (20°C, Pupp et al. 1974; quoted, Smith et al. 1978; Mackay & Shiu 1981)

2.93x10^{-6} (20°C, Hoyer & Peperle 1958; quoted, Mabey et al. 1982)

7.30X10^{-6} (effusion method, De Kruif 1980)

2.71x10^{-5} (gas saturation-HPLC/UV, Sonnefeld et al. 1983; quoted, Bidleman 1984)

2.49x10^{-4} (Yamasaki et al. 1984; quoted, Capel et al. 1991)

4.10x10^{-6} (selected, Howard et al. 1986; quoted, Banerjee et al. 1990)

1.51x10^{-5}, 2.17x10^{-5} (extrapolated-Antoine eqn., Stephenson & Malanowski 1987)

5.43x10^{-4}, 4.06x10^{-6} (quoted, GC-RT, subcooled liq. values, Hinckley 1989)

5.43x10^{-4} (subcooled liquid value, quoted, Hinckley et al. 1990)

3.40x10^{-5} (calculated-UNIFAC, Banerjee et al. 1990)

Henry's Law Constant (Pa m^3/mol):

    0.813   (batch col., Southworth 1979)

    0.248   (calculated-P/C, Southworth 1979)

    0.1013   (20°C, calculated-P/C, Mabey et al. 1982)

    0.102   (WERL Treatability database, quoted, Ryan et al. 1988)

    0.092   (15°C, calculated, Baker & Eisenreich 1990)

    5.17   (calculated-P/C, Capel et al. 1991)

Octanol/Water Partition Coefficient, log K_{OW}:

    5.61   (Radding et al. 1976; quoted, Callahan et al. 1979; Kayal & Connell 1990)

    5.49   (calculated-S, Zepp & Schlotzhauer 1979)

    5.60   (calculated-π const., Southworth et al. 1978)

    5.61   (Veith et al. 1979)

    5.91   (calculated-f const., Yalkowsky & Valvani 1979, 1980; quoted, Whitehouse & Cooke 1982; Walters & Luthy 1984; Mailhot & Peters 1988; Shorten et al. 1990)

    5.91, 6.10 (quoted, calculated-S & M.P., Mackay et al. 1980)

    5.61   (Steen & Karickhoff 1981)

    5.61   (calculated-f const., Mabey et al. 1982)

    5.61   (quoted, Mackay 1982; Kamlet et al. 1988; Schüürmann & Klein 1988)

    5.70   (quoted, Mills et al. 1982)

    5.91   (average lit. value, Yalkowsky et al. 1983)

    5.91, 5.88 (quoted, calculated-MR, Yoshida et al. 1983)

    5.60   (quoted, Mackay & Hughes 1984; Hawker & Connell 1986)

    5.91   (calculated, Miller et al. 1985; quoted, Hawker 1989; Capel et al. 1991)

    5.84   (calculated-χ as per Rekker & De Kort 1979, Ruepert et al. 1985)

    5.66   (Leo 1986; quoted, Schüürmann & Klein 1988)

    5.61   (quoted, Pavlou 1987; Mailhot & Peters 1988; Ryan et al. 1988)

    5.91   (quoted, Isnard & Lambert 1988,1989)

    5.48   (calculated-solvatochromic p., Kamlet et al. 1988)

    5.90   (quoted, Landrum 1988)

    5.61   (quoted, Ryan et al. 1988)

5.91   (recommended, Sangster 1989)
7.50   (calculated-K_{OC}, Broman et al. 1991)

Bioconcentration Factor, log BCF:
4.56   (Smith et al. 1978; quoted, Steen & Karickhoff 1981)
4.0   (*daphnia pulex*, Southworth et al. 1978)
3.67   (kinetic estimation, Southworth et al. 1978)
4.0   (fathead minnow, Veith et al. 1979)
4.56, 5.0 (bacteria, Baughman & Paris 1981)
4.72   (microorganisms-water, calculated from K_{OW}, Mabey et al. 1982)
4.40, 4.29 (quoted, calculated-K_{OW}, Mackay 1982)
4.39   (activated sludge, Freitag et al. 1984)
4.0   (*daphnia pulex*, correlated as per Mackay & Hughes 1984, Howell & Connell 1986)
4.39, 3.50, 2.54 (activated sludge, algae, fish, Freitag et al. 1985)
4.01   (*daphnia magna*, Newsted & Giesy 1987)
4.00, 4.41 (quoted, calculated- χ , Sabljic 1987)
4.0   (quoted, Isnard & Lambert 1988)
4.0   (quoted, Schüürmann & Klein 1988)

Sorption Partition Coefficient, log K_{OC}:
4.52   (22°C, suspended particulates, Herbes et al. 1980)
5.30   (calculated-K_{OW}, Mabey et al. 1982)
4.57   (calculated, Pavlou 1987)
6.30   (sediments average, Kayal & Connell 1990)
7.30   (specified particulate log K_{OC}, Broman et al. 1991)
4.0   (predicted dissolved log K_{OC}, Broman et al. 1991)

Half-Lives in the Environment:
Air: 1-3 hours, based on estimated photolysis half-life in air (Smith et al. 1978; Lyman et al. 1982; quoted, Howard et al. 1991); 0.4 hour for adsorption on soot particles in an outdoor teflon chamber with an estimated rate constant of 0.0265 minute^{-1} at 1 cal cm^{-2} min^{-1}, 10 g/m^3 H$_2$O and 20°C (Kamens et al. 1988).
Surface water: 1-3 hours, based on estimated photolysis half-life in water (Smith et al. 1978; Lyman et al. 1982; quoted, Howard et al. 1991).
Groundwater: computed near-surface half-life for direct photochemical transformation of a natural water body, 0.034 hour at a latitude of 40°N, midday, midsummer sunlight and direct photolysis half-life of 0.20 day (no sediment-water partitioning) and 0.95 day (with sediment-water partitioning) in a 5-m deep inland water body (Zepp & Schlotzhauer (1979); 0.20 day under summer sunlight (Mill & Mabey 1985); half-lives were: 4896-32640 hours, based on

estimated unacclimated aqueous aerobic biodegradation half-life (Howard et al. 1991).

Sediment:

Soil: 4-6250 days (Sims & Overcash 1983; quoted, Bulman et al. 1987); half-life of 240 days for 5 mg/kg treatment and 130 days for 50 mg/kg treatment (Bulman et al. 1987); biodegradation rate constant of 0.0026 day^{-1} with a half-life of 261 days for Kidman sandy loam soil and 0.0043 day^{-1} with half-life of 162 days for McLaurin sandy loam soil (Park et al. 1990); half-lives were estimated to be: 2448-16320 hours, based on aerobic dieaway test data at 10-30°C (Coover & Sims 1987; Groenewegen & Stolp 1976; quoted, Howard et al. 1991); >50 days (Ryan et al. 1988).

Biota: half-life for depuration by oysters, 9 days (Lee et al. 1978; quoted, Verschueren 1983).

Environmental Fate Rate Constants or Half-Lives:

Volatilization: aquatic fate rate of 8×10^3 hour^{-1} with half-life about 90 hours (Callahan et al. 1979); half-lives predicted by one compartment model: >1000 hours in stream, eutrophic pond or lake and oligotrophic lake (Smith et al. 1978); calculated half-life of 500 hours for a river of 1 meter deep with water velocity of 0.5 m/sec and wind velocity of 1 m/sec (Southworth 1979; quoted, Herbes et al. 1980; Hallett & Brecher 1984).

Photolysis: aquatic fate rate of about 6×10^{-5} sec^{-1} with half-life of 10-50 hours (Callahan et al. 1979); half-lives predicted by one compartment model: 20 hours in stream, 50 hours in eutrophic pond or lake and 10 hours in oligotrophic lake (Smith et al. 1978); half-life in aquatics, 0.58 hours (quoted of EPA Report 600/7-78-074, Haque et al. 1980); half-life for early day in March, 0.2 day (Mill et al. 1981); rate constant of 1.93 hour^{-1} (Zepp 1980; quoted, Mill & Mabey 1985); calculated photolysis rate constant in pure water at 366 nm and in sunlight at 23-28°C was 13.4×10^{-5} sec^{-1} (early March) with a half-life of 5 hours and at 313 nm with 1% acetonitrile in filter-sterilized natural water was 2.28×10^{-5} sec^{-1} (Mill et al. 1981); rate constant for summer midday at 40°N latitude, 1.39 hour^{-1} (quoted, Mabey et al. 1982); atmospheric and aqueous photolysis half-life of 1-3 hours, based on measured photolysis rate constant for midday March sunlight on a cloudy day (Smith et al. 1978; quoted, Harris 1982; Howard et al. 1991) and adjusted for approximate summer and winter sunlight intensity (Lyman et al. 1982; quoted, Howard et al. 1991); half-lives on different atmospheric particulate substrates (appr. 25 μg/gm on substrate): 4.0 hours on silica gel, 2.0 hours on alumina and 38 hours on fly ash (Behymer & Hites 1985); first order daytime decay rate constants: 0.0125 minute^{-1} for soot particles loading of 1000-2000 ng/mg and 0.0250 minute^{-1} for soot particles loading of 30-350 ng/mg (Kamens et al. 1988); photodegradation in ethanol-water (2:3, v/v) solution for initial concentration of 12.5 ppm by high pressure mercury lamp or sunlight with a rate constant of 0.0251 minute^{-1} with half-life of 0.46 hour (Wang et al. 1991).

Hydrolysis: not hydrolyzable (Mabey et al. 1982); no hydrolyzable groups (Howard et al. 1991).

Oxidation: aquatic fate rate of 5x10^3 M^{-1} sec^{-1} with half-life of 38 hours (Callahan et al. 1979); photooxidation half-life in air of 0.801-8.01 hours, based on estimated rate constant for reaction with hydroxyl radical in air (Atkinson 1987; quoted, Howard et al. 1991); photooxidation half-life in water of 77-3850 hours, based on measured rate constant for reaction with hydroxyl radical in water (Radding et al. 1976; quoted, Howard et al. 1991); half-lives predicted by one compartment model: 38 hours in stream, eutrophic pond or lake and oligotrophic lake based on peroxy radical concentration of 10^{-9} M (Smith et al. 1978); rate constants of 5x10^8 M^{-1} hr^{-1} for singlet oxygen and 2x10^4 M^{-1} hr^{-1} for peroxy radical (Mabey et al. 1982).

Biodegradation: aerobic half-life of 2448-16320 hours, based on aerobic soil dieaway test data at 10-30°C (Groenewegen & Stolp 1976; Coover & Sims 1987; quoted, Howard et al. 1991); not observed during enrichment procedures (Smith et al. 1978; quoted, Verschueren 1983); no significant degradation in 7 days for an average of three static-flask screening test (Tabak et al. 1981); rate constants of 0.0026 day^{-1} with half-life of 261 days for Kidman sandy loam and 0.0043 day^{-1} with half-life of 162 days for McLarin sandy loam all at -0.33 bar soil moisture (Park et al. 1990); anaerobic half-life of 9792-65280 hours, based on estimated unacclimated aqueous aerobic biodegradation half-life (Howard et al. 1991).

Biotransformation: rate constant estimated to be 1x10^{-10} ml cell^{-1} hr^{-1} for bacteria (Mabey et al. 1982).

Bioconcentration, Uptake (k_1) and Elimination (k_2) Rate Constants:

k_1:   669 hour^{-1} (*daphnia pulex*, Southworth et al. 1978)
k_2:   0.144 hour^{-1} (*daphnia pulex*, Southworth et al. 1978)
log k_1: 2.83 hour^{-1} (*daphnia pulex*, correlated as per Mackay & Hughes 1984, Hawker & Connell 1986)
log k_2: -0.84 hour^{-1} (*daphnia pulex*, correlated as per Mackay & Hughes 1984, Hawker & Connell 1986)
k_1:   138.6 mL/g/h   (4°C, *p. hoyi*, Landrum 1988)
k_2:   0.0022 hour^{-1} (4°C, *p. hoyi*, Landrum 1988)

200

Common Name: Benzo[b]fluoranthene
Synonym: 2,3-benzofluoranthene, 3,4-benzofluoranthene, benz[e]acephenanthrylene, B[b]F
Chemical Name: 2,3-benzofluoranthene
CAS Registry No: 205-99-2
Molecular Formula: $C_{20}H_{12}$
Molecular Weight: 252.32

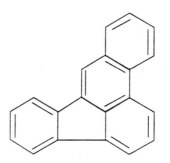

Melting Point (°C):
     167-168 (IARC 1973; quoted, Callahan et al. 1979; Mabey et al. 1982)
     168   (Bjørseth 1983; Pearlman et al. 1984)
Boiling Point (°C):
     481   (Bjørseth 1983)
Density (g/cm³ at 20°C):
Molar Volume (cm³/mol):
     268.9 (LeBas method)
Molecular Volume (Å³):
     230.32 (Pearlman et al. 1984)
Total Surface Area, TSA (Å²):
     260.78 (Pearlman et al. 1984)
Heat of Fusion, ΔH_{fus}, kcal/mol:
Entropy of Fusion, ΔS_{fus}, cal/mol K (e.u.):
Fugacity Ratio at 25°C (assuming ΔS = 13.5 e.u.), F:
     0.0385 (at M.P. = 168°C)

Water Solubility (g/m³ or mg/L at 25°C):
     0.0015 (gen. col.-HPLC/UV, Wise et al. 1981; quoted, Pearlman et al. 1984)
     0.014 (Mabey et al. 1982)
     0.0015 (average lit. value, Pearlman et al. 1984; quoted, Capel et al. 1991)
     0.014 (selected, Ma et al. 1990)

Vapor Pressure (Pa at 25°C):
     6.67×10^{-5} (20°C, estimated, Callahan et al. 1979; quoted, Mabey et al. 1982)
     2.12×10^{-5} (Yamasaki et al. 1984; quoted, Capel et al. 1991)
     5.00×10^{-7} (selected, Ma et al. 1990)

201

Henry's Law Constant (Pa m^3/mol):
    1.236   (calculated-P/C, Mabey et al. 1982)
    0.054   (15°C, calculated, Baker & Eisenreich 1990)
    3.55   (calculated-P/C, Capel et al. 1991)

Octanol/Water Partition Coefficient, log K_{OW}:
    6.57   (calculated as per Leo et al. 1971, Callahan et al. 1979)
    6.06   (Mabey et al. 1982)
    6.60   (selected, Mills et al. 1982)
    6.44   (calculated- χ as per Rekker & De Kort 1979, Ruepert et al. 1985)
    5.78   (HPLC, Wang et al. 1986)
    6.57   (quoted, Pavlou 1987; Ryan et al. 1988)
    5.78   (recommended, Sangster 1989)
    6.20   (selected, Ma et al. 1990)
    6.40   (Bayona et al. 1991)
    6.50   (calculated as per Mackay et al. 1980, Capel et al. 1991)

Bioconcentration Factor, log BCF:
    5.15   (microorganisms-water, calculated from K_{OW}, Mabey et al. 1982)
    4.00   (*daphnia magna*, Newsted & Giesy 1987)
    0.959, 0.230 (*polychaete sp, capitella capitata*, Bayona et al. 1991)

Sorption Partition Coefficient, log K_{OC}:
    5.74   (calculated-K_{OW}, Mabey et al. 1982)
    5.70   (calculated, Pavlou 1987)

Half-Lives in the Environment:
    Air: 1.43-14.3 hours, based on estimated photooxidation half-life in air (Atkinson 1987; quoted, Howard et al. 1991); 1.3 hour for adsorption on soot particles in an outdoor teflon chamber with an estimated rate constant of 0.0091 minute^{-1} at 1 cal cm^{-2} min^{-1}, 10 g/m^3 H$_2$O and 20°C (Kamens et al. 1988).
    Surface water: 8.7-720 hours, based on estimated aqueous photolysis half-life (Lane & Katz 1977; Muel & Sauem 1985; quoted, Howard et al. 1991).
    Groundwater: 17280-29280 hours, based on estimated unacclimated aqueous aerobic biodegradation half-life (Howard et al. 1991).
    Sediment:
    Soil: biodegradation rate constant of 0.0024 day^{-1} with a half-life of 294 days for Kidman sandy loam soil, and 0.0033 day^{-1} with half-life of 211 days for McLaurin sandy loam soil (Park et al. 1990); half-lives were: 8640-14640 hours, based on aerobic dieaway test data (Coover & Sims 1987; quoted, Howard et al. 1991); 42 weeks, 9.0 years (quoted, Luddington soil, Wild et al. 1991).
    Biota:

Environmental Fate Rate Constants or Half-Lives:

Volatilization:

Photolysis: atmospheric and aqueous half-life of 8.7-720 hours, based on measured rate of photolysis in heptane irradiated with light >290 nm (Lane & Katz 1977; Muel & Saguem 1985; quoted, Howard et al. 1991); first order daytime decay rate constnats: 0.0065 minute^{-1} for 1000-2000 ng/mg soot particles loading and 0.0090 minute^{-1} with 30-350 ng/mg loading (Kamens et al. 1988).

Hydrolysis: not hydrolyzable (Mabey et al. 1982; no hydrolyzable groups (Howard et al. 1991).

Oxidation: rate constant of 4×10^7 M^{-1} hr^{-1} for singlet oxygen and 5×10^3 M^{-1} hr^{-1} for peroxy radical (Mabey et al. 1982); photooxidation half-life in air of 1.43-14.3 hours, based on estimated rate constant for reaction with hydroxyl radical in air (Atkinson 1987; quoted, Howard et al. 1991).

Biodegradation: aerobic half-life of 8640-14640 hours, based on aerobic soil dieaway test data (Coover & Sims 1987; quoted, Howard et al. 1991); rate constants of 0.0024 day^{-1} with half-life of 294 days for Kidman sandy loam and 0.0033 day^{-1} with half-life of 211 days for McLarin sandy loam all at -0.33 bar soil moisture (Park et al. 1990); anaerobic half-life of 34560-58560 hours, based on estimated unacclimated aqueous aerobic degradation half-life (Howard et al. 1991).

Biotransformation: estimated to be 3×10^{-12} mL cell^{-1} h^{-1} for bacteria (Mabey et al. 1982).

Bioconcentration, Uptake (k_1) and Elimination (k_2) Rate Constants:

Common Name: Benzo[j]fluoranthene
Synonym: 7,8-benzofluoranthene, 10,11-fluoranthene
Chemical Name: benzo[j]fluoranthene
CAS Registry No: 205-82-2
Molecular Formula: $C_{20}H_{12}$
Molecular Weight: 252.32

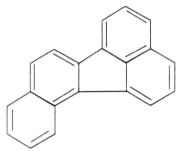

Melting Point (°C):
      166 (Bjørseth 1983; Pearlman et al. 1984)
Boiling Point (°C):
      480 (Bjørseth 1983)
Density (g/cm³ at 20°C):
Molar Volume (cm³/mol):
      268.9 (LeBas method)
Molecular Volume (Å³):
      230.41 (Pearlman et al. 1984)
Total Surface Area, TSA (Å²):
      259.71 (Pearlman et al. 1984)
Heat of Fusion, ΔH_{fus}, kcal/mol:
Entropy of Fusion, ΔS_{fus}, cal/mol K (e.u.):
Fugacity Ratio at 25°C (assuming ΔS = 13.5 e.u.), F:
      0.0403 (at M.P. = 166°C)

Water Solubility (g/m³ or mg/L at 25°C):
      0.0025 (gen. col.-HPLC/UV, Wise et al. 1981; quoted, Pearlman et al. 1984)
      0.0025 (average lit. value, Pearlman et al. 1984)

Vapor Pressure (Pa at 25°C):

Henry's Law Constant (Pa m³/mol):

Octanol/Water Partition Coefficient, log K_{OW}:
      6.44   (calculated- χ as per Rekker & De Kort 1979, Ruepert et al. 1985)
      6.40   (Bayona et al. 1991)

Bioconcentration Factor, log BCF:
      0.914, -0.222 (*polychaete sp, capitella capitata*, Bayona et al. 1991)

Sorption Partition Coefficient, log K_{OC}:

Half-Lives in the Environment:
      Air:
      Surface water:
      Groundwater:
      Sediment:
      Soil:
      Biota:

Environmental Fate Rate Constants or Half-Lives:
      Volatilization:
      Photolysis:
      Hydrolysis:
      Oxidation:
      Biodegradation:
      Biotransformation:
      Bioconcentration, Uptake (k_1) and Elimination (k_2) Rate Constants:

Common Name: Benzo[k]fluoranthene
Synonym: 8,9-benzofluoranthene, 11,12-benzofluoranthene, B[k]F
Chemical Name: 8,9-benzofluoranthene
CAS Registry No: 207-08-9
Molecular Formula: $C_{20}H_{12}$
Molecular Weight: 252.32

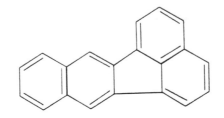

Melting Point (°C):
      198-217, 217 (measured, quoted, Murray et al. 1974)
      217   (Weast 1977; Quoted, Callahan et al. 1979; Mabey et al. 1982; Bjørseth 1983;
               Pearlman et al. 1984)
Boiling Point (°C):
      481 (Bjørseth 1983)
      480 (Pearlman et al. 1984)
Density (g/cm³ at 20°C):
Molar Volume (cm³/mol):
      268.9 (LeBas method)
Molecular Volume (Å³):
      231.12 (Pearlman et al. 1984)
Total Surface Area, TSA (Å²):
      264.95 (Pearlman et al. 1984)
Heat of Fusion, ΔH_{fus}, kcal/mol:
Entropy of Fusion, ΔS_{fus}, cal/mol K (e.u.):
Fugacity Ratio at 25 °C (assuming ΔS = 13.5 e.u.), F:
      0.0126 (at M.P. = 217°C)

Water Solubility (g/m³ or mg/L at 25°C):
      0.0008 (gen. col.-HPLC/UV, Wise et al. 1981)
      0.0043 (calculated-K_{OW}, Mabey et al. 1982)
      0.00076 (quoted, Pearlman et al. 1984; Kayal & Connell 1990)
      0.00081 (average lit. value, Pearlman et al. 1984)
      0.008 (selected, Ma et al. 1990)
      0.0007 (quoted; Capel et al. 1991)

Vapor Pressure (Pa at 25°C):
    1.28×10^{-8} (20°C, Radding et al. 1976; quoted, Callahan et al. 1979)
    6.70×10^{-5} (20°C, Mabey et al. 1982)
    2.07×10^{-5} (Yamasaki et al. 1984; quoted, Capel et al. 1991)
    5.20×10^{-8}, 4.93×10^{-6} (20°C, lit. mean, subcooled liquid value, Bidleman & Foreman 1987; quoted, Ma et al. 1990)
    1.29×10^{-7} (extrapolated, Antoine eqn., Stephenson & Malanowski 1987)

Henry's Law Constant (Pa m^3/mol):
    3.921 (calculated-P/C, Mabey et al. 1982)
    0.111 (15°C, calculated, Baker & Eisenreich 1990)
    7.50 (calculated-P/C, Capel et al. 1991)

Octanol/Water Partition Coefficient, log K_{ow}:
    6.84 (calculated-f const., Callahan et al. 1979; Pavlou 1987; Ryan et al. 1988; Kayal & Connell 1990)
    6.06 (calculated-f const., Mabey et al. 1982)
    6.85 (selected, Mills et al. 1982)
    6.44 (calculated- χ as per Rekker & De Kort 1979, Ruepert et al. 1985)
    6.84 (quoted, Pavlou 1987; Ryan et al. 1988)
    6.20 (selected, Ma et al. 1990)
    6.40 (Bayona et al. 1991)
    6.50 (calculated-S and M.P., Capel et al. 1991)
    7.20 (calculated-K_{oc}, Broman et al. 1991)

Bioconcentration Factor, log BCF:
    5.15 (microorganisms-water, calculated from K_{ow}, Mabey et al. 1982)
    4.12 (*daphnia magna*, Newsted & Giesy 1987)
    1.149, 0.255 (*polychaete sp, capitella capitata*, Bayona et al. 1991)

Sorption Partition Coefficient, log K_{oc}:
    5.74 (calculated-K_{ow}, Mabey et al. 1982).
    5.92 (calculated, Pavlou 1987)
    5.99 (sediments average, Kayal & Connell 1990)
    7.00 (specified particulate log K_{oc}, Broman et al. 1991)
    4.00 (predicted dissolved log K_{oc}, Broman et al. 1991)

Half-Lives in the Environment:
    Air: 1.1-11 hours, based on estimated photooxidation half-life in air (Atkinson 1987; quoted, Howard et al. 1991); 0.8 hour for adsorption on soot particles in an outdoor teflon chamber with an estimated rate constant of 0.0138 minute^{-1} at 1 cal cm^{-2} min^{-1} and 10 g/m^3 H$_2$O at 20°C (Kamens et al. 1988).

Surface water: 3.8-499 hours, based on photolysis half-life in water (Smith et al. 1978; Muel & Saguem 1985; quoted, Howard et al. 1991).

Groundwater: 42680-102720 hours, based on estimated unacclimated aqueous aerobic biodegradation half-life (Howard et al. 1991).

Sediment:

Soil: 21840-51360 hours, based on aerobic soil dieaway test data (Bossert et al. 1984; Coover & Sims 1987; quoted, Howard et al. 1991); >50 days (Ryan et al. 1988); mean half-life of 8.7 years for Luddington soil (Wild et al. 1991).

Biota:

Environmental Fate Rate Constants or Half-Lives:

Volatilization:

Photolysis: atmospheric and aqueous photolysis half-life of 3.8-499 hours, based on measured rate of photolysis in heptane under November sunlight and adjusted by ratio of sunlight photolysis half-lives in water vs. heptane (Smith et al. 1978; Muel & Saguem 1985; quoted, Howard et al. 1991); first order daytime decay constants: 0.0047 minute^{-1} for soot particles loading of 1000-2000 ng/mg and 0.0013 minute^{-1} with 30-350 ng/mg loading (Kamens et al. 1988).

Hydrolysis: not hydrolyzable (Mabey et al. 1982); no hydrolyzable groups (Howard et al. 1991).

Oxidation: rate constant of $4x10^7$ M^{-1} hr^{-1} for singlet oxygen and $5x10^3$ M^{-1} hr^{-1} for peroxy radical (Mabey et al. 1982); photooxidation half-life of 1.1-11 hours, based on estimated rate constant for reaction with hydroxyl radical in air (Atkinson 1987; quoted, Howard et al. 1991).

Biodegradation: aerobic half-life of 21840-51360 hours, based on aerobic soil dieaway test data (Coover & Sims 1987; quoted, Howard et al. 1991); anaerobic half-life of 87360-205440 hours, based on estimated unacclimated aqueous aerobic biodegradation half-life (Howard et al. 1991).

Biotransformation: estimated to be $3x10^{-12}$ ml cell^{-1} h^{-1} for bacteria (Mabey et al. 1982).

Bioconcentration, Uptake (k$_1$) and Elimination (k$_2$) Rate Constants:

Common Name: Benzo[a]pyrene
Synonym: BaP, B(a)P, 3,4-benzopyrene
Chemical Name: benzo[a]pyrene
CAS Registry No: 50-32-8
Molecular Formula: $C_{20}H_{12}$
Molecular Weight: 252.32

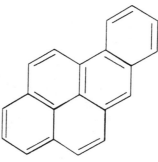

Melting Point (°C):

    176.4, 176.5-177.5 (measured, quoted, Murray et al. 1974)

    179 (Smith et al. 1978; quoted, Callahan et al. 1979; Mabey et al. 1982; Verschueren 1983; Pearlman et al. 1984; Mailhot & Peters 1988)

    175 (Yalkowsky & Valvani 1979,1980; Mackay & Shiu 1981; Steen & Karickhoff 1981; Bruggeman et al. 1982; Miller et al. 1985; Mailhot & Peters 1988; Banerjee et al. 1990)

    176.5 (Mackay et al. 1980; Patton et al. 1984; Mailhot & Peters 1988)

    176 (Yalkowsky et al. 1983)

    177 (Bidleman 1984)

Boiling Point (°C):

    311 (at 10 torr, quoted, Smith 1978; Verschueren 1983; Mailhot & Peters 1988)

    496 (Weast 1977)

    493 (Bjørseth 1983)

    495 (Pearlman et al. 1984)

Density (g/cm³ at 20°C):

Molar Volume (cm³/mol):

    263 (LeBas method, Miller et al. 1985)

    1.418 (intrinsic volume: $V_I/100$, Kamlet et al. 1988; Hawker 1989)

Molecular Volume (A³):

    228.6 (Pearlman et al. 1984)

Total Surface Area, TSA (A²):

    256.0 (Yalkowsky & Valvani 1979; quoted, Mailhot & Peters 1988)

    255.6 (Pearlman et al. 1984)

Heat of Fusion, ΔH_{fus}, kcal/mol:

Entropy of Fusion, ΔS_{fus}, cal/mol K (e.u.):

    9.27 (differential scanning calorimetry, Hinckley et al. 1990)

Fugacity Ratio, F:
      0.03   (Mackay et al. 1980)
      0.0328 (at M.P. = 175°C)

Water Solubility (g/m^3 or mg/L at 25 °C):
      0.004   (27°C, nephelometry, Davis et al. 1942; quoted, Barone et al. 1967; Mackay &
                Shiu 1981; Billington et al. 1988; IUPAC 1989)
      0.004   (nephelometry, Davis & Parker 1942)
      0.004   (Weimer & Prausnitz 1965; quoted, Mackay & Shiu 1977)
      0.0043   (shake flask-UV/fluorescence, Barone et al. 1967)
      0.0061   (average, Barone et al. 1967; quoted, Pearlman et al. 1984)
      0.0005   (20°C, shake flask-UV, Eisenbrand & Baumann 1970)
      0.00121 (Haque & Schmedding 1975; quoted, Smith et al. 1978; Mackay & Shiu 1981;
                Mill et al. 1981; Whitehouse & Cooke 1982)
      0.0038   (shake flask-fluorescence, Mackay & Shiu 1977; quoted, API 1978; Callahan et
                al. 1979; May et al. 1979; Yalkowsky & Valvani 1979,1980; Mackay et al.
                1980; Valvani & Yalkowsky 1980; Futoma et al. 1981; Mackay & Shiu 1981;
                Steen & Karickhoff 1981; Mabey et al. 1982; Whitehouse & Cooke 1982;
                Pearlman et al. 1984; Lande et al. 1985; Miller et al. 1985; Landrum et al.
                1987; Billington et al. 1988; Mailhot & Peters 1988; IUPAC 1989; Kayal &
                Connell 1990; Mackay 1991)
      0.000172 (quoted, Lu et al. 1977)
      0.0002   (Rossi 1977; Neff 1979; quoted, Eadie et al. 1982)
      0.0078   (calculated-K_{OW}, Valvani & Yalkowsky 1980)
      0.0069   (calculated-K_{OW}, Yalkowsky & Valvani 1980)
      0.0038   (RP-TLC, Brugeman et al. 1982)
      0.0038   (selected, Mills et al. 1982)
      0.00162 (gen. col.-HPLC, May et al. 1983)
      0.003   (quoted, Verschueren 1983; Mailhot & Peters 1988)
      0.0044   (average lit. value, Yalkowsky et al. 1983)
      0.0063   (average lit. value, Pearlman et al. 1984)
      0.0038   (selected value, Pearlman et al. 1984; Ma et al. 1990)
      0.0040   (quoted, Banerjee 1985; Billington et al. 1988; Farrington 1991)
      0.0035, 0.0080 (quoted, predicted, Lande et al. 1985)
      0.0038   (selected, Miller et al. 1985; quoted, Hawker 1989; Capel et al. 1991)
      0.0016   (gen. col.-HPLC/UV, Billington et al. 1988)
      0.00398 (quoted, Isnard & Lambert 1989)
      0.000504 (LSC, Eadie et al. 1990)
      0.002   (10°C, estimated, McLachlan et al. 1990)

Vapor Pressure (Pa at 25°C):
      7.32x10^{-7} (effusion method, Murray et al. 1974; quoted, Bidleman 1984)
      6.67x10^{-7} (quoted, Smith et al. 1978; Callahan et al. 1979; Mackay & Shiu 1981)

7.47×10^{-7} (Murray et al. 1974; quoted, Mabey et al. 1982)

6.67×10^{-7} (selected, Mills et al. 1982)

4.5×10^{-7}, 3.36×10^{-7} (GC-RT, Bidleman 1984)

1.5×10^{-5} (subcooled liq., Bidleman 1984)

1.22×10^{-5} (Yamasaki et al. 1984; quoted, Capel et al. 1991)

7.3×10^{-7} (selected, Howard et al. 1986; quoted, Banerjee et al. 1990)

8.53×10^{-10} (estimated, Ryan & Cohen 1986)

3.2×10^{-7}, 1.23×10^{-5} (lit. mean, subcooled liq. value, Bidleman & Foreman 1987)

7.51×10^{-7} (extrapolated, Antoine eqn., Stephenson & Malanowski 1987)

8.22×10^{-6}, 7.26×10^{-6} (quoted, GC-RT, Hinckley 1989)

2.53×10^{-5}, 1.02×10^{-5} (subcooled liquid values, Hinckley et al. 1990)

7.0×10^{-7} (calculated-UNIFAC, Banerjee et al. 1990)

6.0×10^{-8} (10°C, estimated, McLachlan et al. 1990)

7.05×10^{-7}, 2.18×10^{-7} (GC-RT, Hinckley et al. 1990)

3.0×10^{-7} (selected, Ma et al. 1990)

7.0×10^{-7} (selected, Mackay 1991)

Henry's Law Constant (Pa m^3/mol):

   <0.248, 0.052 (calculated-P/C, Southworth 1979)

   0.5   (calculated-P/C, Mabey et al. 1982)

   1214.7 (WERL Treatability database, quoted, Ryan et al. 1988)

   0.056 (calculated-P/C, Eastcott et al. 1988)

   0.009 (15°C, calculated, Baker & Eisenreich 1990)

   0.0079 (10°C, McLachlan et al. 1990)

   0.810 (calculated-P/C, Capel et al. 1991)

Octanol/Water Partition Coefficient, log K_{ow}:

   6.04   (Radding et al. 1976; quoted, Callahan et al. 1979; Kayal & Connell 1990)

   4.05   (quoted, Lu et al. 1977)

   6.31   (Smith et al. 1978)

   5.78   (Hansch & Leo 1979)

   5.81   (calculated-S, Zepp & Scholtzhauer 1979)

   6.50   (calculated-f const., Yalkowsy & Valvani 1979, 1980; quoted, Mackay et al. 1980; Mailhot & Peters 1988)

   6.32, 6.57 (quoted, calculated-S & M.P., Mackay et al. 1980; quoted, Landrum et al. 1987)

   6.20   (HPLC-k', Hanai et al. 1981)

   6.34   (Steen & Karickhoff 1981)

   5.93   (Briggs 1981)

   6.50   (RP-TLC, Bruggeman et al. 1982)

   6.20   (GC, Hanai et al. 1982)

   6.06   (calculated-f const., Mabey et al. 1982)

   6.0   (quoted, Mills et al. 1982)

6.50   (average lit. value, Yalkowsky et al. 1983)
6.0   (quoted, Mallon & Harris 1984)
5.95   (quoted, Landrum et al. 1984)
6.53   (quoted, Patton et al. 1984)
6.74, 7.77, 7.99 (HPLC-RT, Sarna et al. 1984)
6.50, 6.42   (quoted, HPLC-RT, Rapaport et al. 1984)
6.31   (quoted, Pavlou & Weston 1983, 1984; Pavlou 1987)
5.97   (Hansch & Leo 1985)
5.98   (selected, Miller et al. 1985; quoted, Kamlet et al. 1988; Hawker 1989; Capel
        et al. 1991; Landrum et al. 1991)
6.44   (calculated- χ as per Rekker & De Kort 1979, Ruepert et al. 1985)
6.42   (quoted, Hawker & Connell 1986)
6.06   (quoted, Gobas et al. 1987)
5.97, 6.12 (quoted, calculated-UNIFAC, Banerjee & Howard 1988)
6.00   (quoted, Isnard & Lambert 1989)
5.89   (calculated-solvatochromic p., Kamlet et al. 1988)
6.5   (quoted, Landrum 1988)
6.06, 6.34, 6.50 (quoted, Mailhot & Peters 1988)
6.04   (quoted, Ryan et al. 1988)
6.35   (recommended, Sangster 1989)
6.0   (estimated, McLachlan et al. 1990)
6.04   (selected, Ma et al. 1990)
6.20   (Bayona et al. 1991)
6.20   (quoted, Landrum & Stubblefield 1991)
6.04   (selected, Mackay 1991)
8.50   (calculated-K_{oc}, Broman et al. 1991)

Bioconcentration Factor, log BCF:
        1.10   (*macrochirus*, Leversee et al. 1981)
        3.45   (*daphnia magna*, Leversee et al. 1981)
        5.65   (mixed microbial population, quoted of Smith et al. 1978, Steen & Karickhoff
                1981)
        4.88   (calculated as per Kenaga & Goring 1979, Eadie et al. 1982)
        4.74   (*p. hoyi*, Eadie et al. 1982)
        5.15   (microorganisms-water, calculated from K_{ow}, Mabey et al. 1982)
        3.90   (*Daphnia magna*, McCarthy 1983)
        3.69   (*lepomis macrochirus*, Spacie et al., 1983)
        4.45   (bluegill sunfish, calculated-K_{ow}, Spacie et al. 1983)
        3.69   (bluegill sunfish, estimated, Spacie et al. 1983)
        2.69   (bluegill sunfish, Spacie et al. 1983)
        4.00   (activated sludge, Freitag et al. 1984)
        3.42   (bluegills, McCarthy & Jimenez 1985)

2.35, 2.45 (bluegills-with dissolved humic material, McCarthy & Jimenez 1985)

2.68, 3.52, 4.0 (fish, algae, activated sludge, Freitag et al. 1985)

3.51 (worms, Frank et al. 1986)

6.95, 6.51 (*p. hoyi* of Lake Michigan interstitial waters, Landrum et al. 1985)

3.34 (*p. hoyi* of Government Pond of Grand Haven in Michigan, Landrum et al. 1985)

2.69 (Gobas et al. 1987)

4.11 (*daphnia magna*, Newsted & Giesy 1987)

3.7, 4.8 (quoted exptl., calculated- χ , Sabljic 1987)

4.69, 3.93 (calculated for amphipods and mysids, Evans & Landrum 1989)

3.77 (10-20°C, *h. limbata*, Landrum & Poore 1988)

4.61 (4°C, *p. hoyi*, Landrum & Poore 1988)

3.86 (4°C, *s. heringianus*, Landrum & Poore 1988)

3.87 (4°C, *mysis relicta*, Landrum & Poore 1988)

1.140, -0.155 (*polychaete sp, capitella capitata*, Bayona et al. 1991)

Sorption Partition Coefficient, log K_{OC}:

6.74 (calculated-K_{OW}, Mabey et al. 1982)

5.48 (calculated, Pavlou & Weston 1983,1984; Pavlou 1987)

6.66 (LSC, Eadie et al. 1990)

6.26 (sediments average, Kayal & Connell 1990)

8.3 (specified particulate log K_{OC}, Broman et al. 1991)

4.0 (predicted dissolved log K_{OC}, Broman et al. 1991)

Half-Lives in the Environment:

Air: 0.37-1.1 hours, based on estimated photolysis half-life in air (Lyman et al. 1982; quoted, Howard et al. 1991); 0.5 hour for adsorption on soot particles in an outdoor teflon chamber with an estimated rate constant of 0.0234 minute^{-1} at 1 cal cm^{-2} min^{-1}, 10 g/m^3 H$_2$O and 20°C (Kamens et al. 1988).

Surface Water: 2 hours in methanol solution irradiated at 254 nm (Lu et al. 1977); computed near-surface half-life for direct photochemical transformation of a natural water body 0.54 hr at latitude 40°N, midday, midsummer, and direct photolysis, 3.2 days (no sediment-water partitioning) and 13 days (with sediment-water partitioning) in a 5-m deep inland water body (Zepp & Schlotzhauer 1979); half-lives were: 0.37-1.1 hours, based on photolysis half-life in water (Lyman et al. 1982; quoted, Howard et al. 1991); 0.045 day under mid-December sunlight (Mill & Mabey 1985).

Groundwater: 2736-25440 hours, based on estimated unacclimated aqueous aerobic biodegradation half-life (Howard et al. 1991).

Sediment:

Soil: more than 2 days (Sims & Overcash 1983; quoted, Bulman et al. 1987); 347 days for 5 mg/kg treatment and 218 days for 50 mg/kg treatment (Bulman et al. 1987); biodegradation rate constant of 0.002 day^{-1} with a half-life of 309 days for Kidman sandy loam soils and 0.0030 day^{-1} with half-life of 229 days for Mclaurin sandy loam soils (Park et al. 1990); half-lives were: 1368-12720 hours, based on aerobic soil dieaway test data at 10-30°C (Groenewegen & Stolp 1976; Coover & Sims 1987; quoted, Howard et al. 1991); >50 days (Ryan et al. 1988); 0.3->300 weeks, 8.2 years (literature, Luddington soil, Wild et al. 1991).

Biota: half-life of depuration by oysters, 18 days (Lee et al. 1978; quoted, Verschueren 1983); 67 hours in bluegill sunfish (Spacie et al. 1983); with depuration half-life of 52 hours in *s. heringianus* (Frank et al. 1986); calculated half-lives in different tissues of sea bass: 12.4 days for fat, 6.5 days for kidney, 5.1 days for kidney, 5.1 days for intestine, 4.8 days for gallblader, 4.5 days for spleen, 2.9 days for muscle, 2.4 days for whole body, 2,3 days for gonads, 2.3 days for gills, and 2.2 days for liver (Lemaire et al. 1990).

Environmental Fate Rate Constants or Half-Lives:

Volatilization: aquatic fate rate of 300 hour^{-1} with half-life of 22 hours (Callahan et al. 1979); half-lives predicted by one compartment model: 140 hours in river water, 350 hours in eutrophic pond, 700 hours in eutrophic lake and oligotrophic lake (Smith et al. 1978); calculated half-life of 1500 hours for a river of 1 meter deep with water velocity of 0.5 m/sec and wind velocity of 1 m/sec (Southworth 1979; quoted, Herbes et al. 1980; Hallett & Brecher 1984); sublimation rate constant from glass surface of $< 1 \times 10^{-5}$ sec^{-1} was measured at 24°C at an airflow rate of 3 L/min (Cope & Kalkwarf 1987).

Photolysis: photolysis half-life of 2 hours in methanol solution when irradiated at 254 nm (Lu et al. 1977); 0.58 hour^{-1} for winter at midday at 40°N latitude (Smith et al. 1978; quoted, Mabey et al. 1982); computed near-surface half-life for direct photochemical transformation of a natural water body, 0.54 hour at latitude 40°N, midday, midsummer, and direct photolysis half-life of 3.2 days (no sediment-water partitioning) and 13 days (with sediment-water partitioning) in a 5-m deep inland water body (Zepp & Schlotzhauer 1979; quoted, Hallett & Brecher 1984); aquatic fate rate of 2.8×10^{-4} sec^{-1} with half-life of 1-2 hours (Callahan et al. 1979); photolytic half-life in aquatics, 0.53 hour (quoted of EPA Report 600/7-78-074, Haque et al. 1980); rate constant of 1.30 hour^{-1} (Zepp 1980); half-lives predicted by one compartment model: 3.0 hours in river water based on the the photolysis rates estimated for summer sunlight, 7.5 hours in eutrophic pond or eutrophic lake, and 1.5 hours in oligotrophic lake (Smith et al. 1978; quoted, Harris 1982); 2.8×10^{-4} sec^{-1} with half-life of 1-2 hours (Callahan et al., 1979); calculated rate constant of direct photolysis in pure water at 366 nm and in sunlight at 23-28°C was 3.86×10^{-4} sec^{-1} (late January) with a half-life of 0.69 hours and at 313 nm with 1-20% acetonitrile as cosolvent in filter-sterilized natural water was 1.05×10^{-5} sec^{-1} (mid-December) with a half-life of 1.1 hour (Mill et al. 1981); 0.37-1.1 hours, based on estimated photolysis

214

half-life in air (Lyman et al. 1982; quoted, Howard et al. 1991); sunlight photolysis half-life of 0.045 day for mid-December (Mill & Mabey 1985); half-lives on different atmospheric particulate substrates (appr. 25 μg/gm on substrate): 4.7 hours on silica gel, 1.4 hours on alumina and 31 hours on fly ash (Behymer & Hites 1985); ozonation rate constant of $< 6.1 \times 10^{-4}$ m/s was measured at 24°C with $[O_3] = 0.16$ ppm and light intensity of 1.3 kW/m² (Cope & Kalkwarf 1987); first order daytime decay rate constants: 0.0090 minute^{-1} for soot particles loading of 1000-2000 ng/mg and 0.0211 minute^{-1} with 30-350 ng/mg loading (Kamens et al. 1988); photodegradation half-life was found ranging from 1 hour in summer to days in winter (Valerio et al. 1991); photodegradation in ethanol-water (3:7, v/v) solution for initial concentration of 2.5 ppm by high pressure mercury lamp or sunlight with a rate constant of 0.0322 minute^{-1} and half-life of 0.35 hour (Wang et al. 1991).

Oxidation: aquatic fate rate of 1680 M^{-1} sec^{-1} with half-life of 96 hours (Callahan et al. 1979); rate constant of 5×10^8 M^{-1} h^{-1} for singlet oxygen and 2×10^4 M^{-1} hr^{-1} for peroxy radical (Mabey et al. 1982); half-lives predicted by one comparment model: >340 hours in river water, eutrophic pond or lake and oligotrophic lake (Smith et al. 1978); very slow, not an important process (Callahan et al. 1979); photooxidation half-life in air: 0.428-4.28 hours, based on estimated rate constant for reaction with hydroxy radical in air (Atkinson 1987; quoted, Howard et al. 1991).

Hydrolysis: not hydrolyzable (Mabey et al. 1982; Howard et al. 1991).

Biodegradation: >10000 hours (quoted, Smith et al. 1978); 0.2-0.9 μ mol^{-1} mg^{-1} for bacterial protein (Callahan et al. 1979); rate constant in soil and water estimated to be 3.5×10^{-5} hour^{-1} (Ryan & Cohen 1986); aerobic half-life: 57 days to 1.45 years at 10-30°C, soil dieaway test (Coover & Sims 1987; quoted, Howard et al. 1991); rate constants of 0.0022 day^{-1} with half-life of 309 days for Kidman sandy loam and 0.0030 day^{-1} with half-life of 229 days for McLarin sandy loam all at -0.33 bar soil moisture (Park et al. 1990); anaerobic half-life: 228 days to 5.8 years, based on estimated unacclimated aqueous aerobic biodegradation half-life (Coover & Sims 1987; quoted, Howard et al. 1991).

Biotransformation: estimated to be 3×10^{-12} ml cell^{-1} hr^{-1} for bacteria (Mabey et al. 1982).

Bioconcentration, Uptake (k_1) and Elimination (k_2) Rate Constants:

$\quad$ k_1: $\quad$ 49 hour^{-1} (bluegill sunfish, Spacie et al. 1983)

$\quad$ k_2: 0.010 hour^{-1} (bluegill sunfish, Spacie et al. 1983)

$\quad$ k_1: 75.9 mL/g/h (*pontoporeia hoyi*, Evans & Landrum 1989)

$\quad$ k_1: 39.9 mL/g/h (*mysis relicta*, Evans & Landrum 1989)

$\quad$ k_2: 0.0017 hour^{-1} (*amphipods*, Evans & Landrum 1989)

$\quad$ k_2: 0.0047 hour^{-1} (*mysids*, Evans & Landrum 1989)

$\quad$ k_1: 131.1 mL/g/h (4°C, *p. hoyi*, Landrum 1988)

$\quad$ k_2: 0.0033 hour^{-1} (4°C, *p. hoyi*, Landrum 1988)

$\quad$ k_1: 81.3 hour^{-1} (10-20°C, *h. limbata*, Landrum & Poore 1988)

k_2: 0.014 hour^{-1} (10-20°C, *h. limbata*, Landrum & Poore 1988)

k_1: 116.8 hour^{-1} (4°C, *p. hoyi*, Landrum & Poore 1988)

k_2: 0.0016 hour^{-1} (4°C, *p. hoyi*, Landrum & Poore 1988)

k_1: 87.8 hour^{-1} (4°C, *s. heringianus*, quoted, Landrum & Poore 1988)

k_2: 0.012 hour^{-1} (4°C, *s. heringianus*, quoted, Landrum & Poore 1988)

k_1: 112.0 hour^{-1} (4°C, *mysis relicta*, quoted, Landrum & Poore 1988)

k_2: 0.013 hour^{-1} (4°C, *mysis relicta*, quoted, Landrum & Poore 1988)

Common Name: Benzo[e]pyrene
Synonym: B[e]P, 4,5-benzopyrene
Chemical Name: 4,5-benzopyrene
CAS Registry No: 192-97-20
Molecular Formula: $C_{20}H_{12}$
Molecular Weight: 252.3

Melting Point (°C):
      178, 178-179 (measured, quoted, Murray et al. 1974)
      179 (Bjorseth 1983)
      178 (Pearlman et al. 1984)
Boiling Point (°C):
      493 (Bjorseth 1983)
Density (g/cm³ at 20°C):
Molar Volume (cm³/mol):
      263 (LeBas method, Miller et al. 1985)
      1.418 ($V_I/100$, intrinsic molar vol., Kamlet et al. 1988)
Molecular Volume (Å³):
      251.457 (Pearlman et al. 1984)
Total Surface Area, TSA (Å²):
      227.781 (Pearlman et al. 1984)
Heat of Fusion, ΔH_{fus}, kcal/mol:
Entropy of Fusion, ΔS_{fus}, cal/mol K (e.u.):
      10.11   (differential scanning calorimetry, Hinckley et al. 1990)
Fugacity Ratio at 25°C (assuming ΔS = 13.5 e.u.), F:
      0.0307 (at M.P. = 178°C)

Water Solubility (g/m³ or mg/L at 25°C):
      0.0035 (27°C, nephelometry, Davis et al. 1942)
      0.00732 (shake flask-UV/fluorescence, Barone et al. 1967)
      0.0040 (shake flask-fluorescence, Schwarz 1977)
      0.00732, 0.004; 0.0063 (quoted values; lit. mean, Pearlman et al. 1984)
      0.004, 0.007 (quoted, calculated-χ , Nirmalakhandan & Speece 1989)

Vapor Pressure (Pa at 25°C):
  7.40x10^{-7} (effusion method, extrapolated, Murray et al. 1974)
  7.32x10^{-7}, 1.28x10^{-5} (lit. mean, subcooled liquid value, Bidleman & Foreman 1987)
  2.25x10^{-5}, 7.28x10^{-6} (subcooled liquid values, Hinckley et al. 1990)

Henry's Law Constant (Pa m^3/mol):

Octanol/Water Partition Coefficient, log K_{ow}:
  6.44 (calculated- χ as per Rekker & De Kort 1979, Ruepert et al. 1985)
  7.40 (calculated-K_{OC}, Broman et al. 1991)

Bioconcentration Factor, log BCF:

Sorption Partition Coefficient, log K_{OC}:
  7.20 (specified particulate log K_{OC}, Broman et al. 1991)
  4.00 (predicted dissolved log K_{OC}, Broman et al. 1991)

Half-Lives in the Environment:
  Air:
  Surface water:
  Groundwater:
  Sediment:
  Soil:
  Biota:

Environmental Fate Rate Constants or Half-Lives:
  Volatilization:
  Photolysis:
  Oxidation:
  Hydrolysis:
  Biodegradation:
  Biotransformation:
  Bioconcentration, Uptake (k_1) and Elimination (k_2) Rate Constants:

218

Common Name: Perylene
Synonym: peri-dinaphthalene
Chemical Name: perylene
CAS Registry No: 198-55-0
Molecular Formula: $C_{20}H_{12}$
Molecular Weight: 252.32

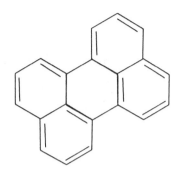

Melting Point (°C):
 273 (API 1978)
 277 (Yalkowsky & Valvani 1979,1980; Mackay et al. 1980; Lande & Banerjee 1981;
   Lande et al. 1985; Miller et al. 1985; Gobas et al. 1988)
 278 (Whitehouse & Cooke 1982; Yalkowsky et al. 1983; Pearlman et al. 1984)
 274 (Riederer 1990)
Boiling Point (°C):
 500 (API 1978)
 503 (Pearlman et al. 1984)
Density (g/cm³ at 20°C):
Molar Volume (cm³/mol):
 187.0 (calculated, Lande & Banerjee 1981)
 263.0 (LeBas method, Miller et al. 1985; Gobas et al. 1988)
Molecular Volume (Å³):
 227.78 (Pearlman et al. 1984)
Total Surface Area, TSA (Å²):
 251.5 (Yalkowsky & Valvani 1979; quoted, Whitehouse & Cooke 1982)
 251.46 (Pearlman et al. 1984)
Heat of Fusion, ΔH_{fus}, kcal/mol:
 5.62 (quoted, Tsonopoulos & Prausnitz 1971)
Entropy of Fusion, ΔS_{fus}, cal/mol K (e.u.):
 10.2 (quoted, Tsonopoulos & Prausnitz 1971)
Fugacity Ratio at 25 °C (assuming ΔS = 13.5 e.u.), F:
 0.00321 (at M.P. = 277°C)
 0.0030 (Mackay et al. 1980)

Water Solubility (g/m^3 or mg/L at 25°C):

    0.0005   (27°C, Davis et al. 1942; quoted, Futoma et al. 1981)

    <0.0005 (Weimer & Prausnitz 1965; quoted, Mackay & Shiu 1977)

    0.00011 (20°C, shake flask-UV, Eisenbrand & Baumann 1970)

    0.0004   (shake flask-fluorescence, Mackay & Shiu 1977; quoted, API 1978; Yalkowsky & Valvani 1979; Mackay et al. 1980; Futoma et al. 1981; Lande & Banerjee 1981; Whitehouse & Cooke 1982; Pearlman et al. 1984; Lande et al. 1985; Miller et al. 1985; IUPAC 1989)

    0.0004   (quoted, Valvani & Yalkowski 1980; Yalkowsky & Valvani 1979,1980; quoted, Lande & Banerjee 1981; Lande et al. 1985; Whitehouse & Cooke 1982)

    0.00073 (calculated-K_{OW}, Valvani & Yalkowsky 1980)

    0.00076 (calculated-K_{OW}, Yalkowsky & Valvani 1980)

    0.00021 (average lit. value, Yalkowsky et al. 1983)

    0.00030 (average lit. value, Pearlman et al. 1984)

    0.00040, 0.00078 (quoted, predicted, Lande et al. 1985)

    0.0004   (selected, Miller et al. 1985; quoted, Gobas et al. 1988; Capel et al. 1991)

    0.00041 (quoted, Isnard & Lambert 1989)

    0.00037 (quoted, Tomlinson & Hafkenscheid 1986)

    0.0004   (quoted, Riederer 1990)

Vapor Pressure (Pa at 25°C):

    5.31x10^{-9} (extrapolated-Antoine eqn., Stephenson & Malanowski 1987)

    7.00x10^{-7} (quoted, Riederer 1990)

Henry's Law Constant (Pa m^3/mol):

    0.440   (calculated-P/C, Riederer 1990)

Octanol/Water Partition Coefficient, log K_{OW}:

    6.06   (calculated-π const., Southworth et al. 1978)

    6.50   (calculated-f const., Yalkowsky & Valvani 1979,1980; quoted, Whitehouse & Cooke 1982)

    6.50, 6.53 (quoted, calculated-S & M.P., Mackay et al. 1980)

    5.91   (average lit. value, Yalkowsky et al. 1983)

    4.80, 6.37 (quoted, calculated-MR, Yoshida et al. 1983)

    6.44   (calculated- χ as per Rekker & De Kort 1979, Ruepert et al. 1985)

    6.30, 5.10 (HPLC-RV predicted, Brooke et al. 1986)

    5.30   (HPLC-RV measured, Brooke et al. 1986)

    6.06   (correlated as per Mackay & Hughes 1984, Hawker & Connell 1986)

    6.50   (calculated, Miller et al. 1985; quoted, Capel et al. 1991; Landrum et al. 1991)

    6.50   (quoted, Isnard & Lambert 1989)

    6.25   (recommended, Sangster 1989)

    6.50   (quoted, Riederer 1990)

    6.40   (Bayona et al. 1991)

Bioconcentration Factor, log BCF:
    3.86   (*daphnia pulex*, Southworth et al. 1978)
    3.73   (kinetic estimation, Southworth et al. 1978)
    4.36   (activated sludge, Freitag et al. 1984)
    3.30, 4.36, <1.0 (algae, activated sludge, fish, Klein et al. 1984)
    3.85   (*daphnia pulex*, correlated as per Mackay & Hughes 1984, Hawker & Connell
            1986)
    3.30, 4.36, <1.0 (algae, activated sludge, fish, Freitag et al. 1985)
    3.86   (*daphnia magna*, Newsted & Giesy 1987)
    1.196, -0.398 (*polychaete sp, capitella capitata*, Bayona et al. 1991)

Sorption Partition Coefficient, log K_{OC}:

Half-Lives in the Environment:
    Air:
    Surface water:
    Groundwater:
    Sediment:
    Soil:
    Biota:

Environmental Fate Rate Constants or Half-Lives:
    Volatilization: sublimation rate constant of $<1\times10^{-5}$ sec^{-1} from glass surface was
            measured at 24°C at an airflow rate of 3 liter/min (Cope & Kalkwarf 1987).
    Photolysis: half-lives on different atmospheric particulate substrates (appr. 25 μg/gm on
            substrate): 3.9 hours on silica gel, 1.2 hours on alumina and 35 hours on flyash
            (Behymer & Hites 1985); ozonation rate constant of $<4.7\times10^{-5}$ m/s was
            measured from glass surface at 24°C with $[O_3]$ = 0.16 ppm and light intensity
            of 1.3 kW/m^2 (Cope & Kalkwarf 1987); photodegradation in ethanol-water (2:3,
            v/v) solution for initial concentration of 5.0 ppm by high pressure mercury lamp
            or sunlight with a rate constant of 3.62×10^{-3} minute^{-1} and half-life of 3.18 hours
            (Wang et al. 1991).
    Hydrolysis:
    Oxidation:
    Biodegradation:
    Biotransformation:
    Bioconcentration, Uptake (k_1) and Elimination (k_2) Rate Constants:
        k_1:   752 hour^{-1} (*daphnia pulex*, Southworth et al. 1978)
        k_2:   0.139 hour^{-1} (*daphnia pulex*, Southworth et al. 1978)
        log k_1: 2.88 hour^{-1} (*daphnia pulex*, correlated as per Mackay & Hughes 1984,
                Hawker & Connell 1986)
        log k_2: -0.86 hour^{-1} (*daphnia pulex*, correlated as per Mackay & Hughes
                1984, Hawker & Connell 1986)

Common Name: 7,12-Dimethylbenz[a]anthracene
Synonym: 7,12-dimethylbenz[a]anthracene,   9,10-dimethyl-1,2-benzanthracene,   7,12-
    dimethylbenzanthracene
Chemical Name: 7,12-dimethylbenz[a]anthacene
CAS Registry No: 57-97-6
Molecular Formula: $C_{20}H_{16}$
Molecular Weight: 256.35

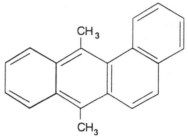

Melting Point (°C):
        122   (quoted, Mackay & Shiu 1977)
        122-123 (Budavari 1989)
        123   (Karickhoff 1981)
Boiling Point (°C):
Density (g/cm³ at 20°C):
Molar Volume (cm³/mol):
Molecular Volume (Å³):
Total Surface Area, TSA (Å²):
        266.0   (Yalkowsky & Valvani 1979)
        292   (calculated as per Gavezzotti 1985, Sabljic 1987)
Heat of Fusion, ΔH_{fus}, kcal/mol:
Entropy of Fusion, ΔS_{fus}, cal/mol K (e.u.):
Fugacity Ratio at 25°C (assuming ΔS = 13.5 e.u.), F:
        0.11   (at M.P. = 122°C)

Water Solubility (g/m³ or mg/L at 25°C):
        0.043   (27°C, shake flask-nephelometry, Davis et al. 1942)
        0.061   (shake flask-fluorescence, Mackay & Shiu 1977; quoted, IUPAC 1989)
        0.053   (24°C, shake flask-nephelometry, Hollifield 1979)
        0.025   (24 °C, shake flask-LSC, Means et al. 1979)
        0.0244 (shake flask-LSC, Means et al. 1980)
        0.061, 0.0109 (quoted, calculated-K_{OW}, Mackay et al. 1980)
        0.025   (quoted, Karickhoff 1981)
        0.034   (lit. mean, Yalkowsky et al. 1983)
        0.043, 0.061, 0.053; 0.054 (quoted lit. values; lit. mean, Pearlman et al. 1984)
        0.50   (subcooled liq. value, Banerjee 1985; quoted, Farrington 1991)

Vapor Pressure (Pa at 25°C):

3.84×10^{-7} (extrapolated-Antoine eqn., Stephenson & Malanowski 1987)

6.78×10^{-6} (extrapolated-Antoine eqn., Stephenson & Malanowski 1987)

Henry's Law Constant (Pa m^3/mol):

Octanol/Water Partition Coefficient, log K_{OW}:

6.95 (calculated-TSA, Yalkowsky & Valvani 1979)

5.80 (shake flask-LSC, Means et al. 1980)

6.95 (estimated-f const., Valvani & Yalkowsky 1980)

6.93 (calculated-f const., Yalkowsky et al. 1983)

5.98 (shake flask-LSC, concn. ratio, Means et al. 1979, 1980; quoted, Karickhoff 1981)

5.80, 6.16 (quoted, calculated-UNIFAC, Banerjee & Howard 1988)

5.80 (recommended, Sangster 1989)

Bioconcentration Factor, log BCF:

Sorption Partition Coefficient, log K_{OC}:

5.77 (soil, calculated-K_{OW}, Karickhoff et al. 1979; quoted, Sabljic 1987)

4.61 (soil, calculated-K_{OW}, Kenaga & Goring 1980; quoted, Sabljic 1987)

5.68 (average of 3 soil/sediment samples, shake flask-LSC, sorption isotherms, Means et al. 1979)

5.37 (average of 12 soil/sediment samples, shake flask-LSC, sorption isotherms, Means et al. 1980; quoted, Sabljic 1987)

5.35 (soil, calculated-regression of K_p versus substrate properties, Means et al. 1980)

5.35, 5.59, 5.74, 5.01 (quoted, calculated-K_{OW}, S & M.P., S, Karickhoff 1981)

3.75 (soil, calculated-K_{OW}, Briggs 1981; quoted, Sabljic 1987)

5.66 (soil, calculated-K_{OW}, Means et al. 1982)

4.60 (soil, calculated-K_{OW}, Chiou et al. 1983)

5.66, 4.60; 5.83 (quoted values; calculated-χ, Sabljic 1987)

Half-Lives in the Environment:

Air: 0.32-3.2 hours, based on estimated photooxidation half-life in air (Atkinson 1987; quoted, Howard et al. 1991).

Surface water: 480-672 hours, based on aerobic soil die-away test data (Sims 1990; quoted, Howard et al. 1991).

Groundwater: 960-1344 hours, based on estimated unacclimated aqueous aerobic biodegradation half-life (Howard et al. 1991).

Sediment:

Soil: biodegradation rate constant of 0.0339 day^{-1} with a half-life of 20 days for Kidman sandy loam soil and 0.0252 day^{-1} with half-life of 28 days for McLaurin sandy loam soil (Park et al. 1990); half-lives were estimated to be, 480-672 hours, based on aerobic soil die-away test data (Sims 1990; quoted, Howard et al. 1991).

Environmental Fate Rate Constants or Half-Lives:

Volatilization:

Photolysis:

Hydrolysis: no hydrolyzable groups (Howard et al. 1991).

Oxidation: photooxidation half-life in air estimated to be 0.32-3.2 hours, based on estimated rate constant for the reaction with hydroxyl radicals in air (Atkinson 1987; quoted, Howard et al. 1991); photooxidation half-life in water estimated to be 1.57-157 years, based on measured rate constant for the reaction with singlet oxygen in benzene (Stevens et al. 1974; quoted, Howard et al. 1991).

Biodegradation: biodegradation rate constant of 0.0339 day^{-1} with a half-life of 20 days for Kidman sandy loam soil and 0.0252 day^{-1} with half-life of 28 days for McLaurin sandy loam soil (Park et al. 1990); aerobic half-life estimated to be 480-672 hours, based on aerobic soil die-away test data (Sims 1990; quoted, Howard et al. 1991); anaerobic half-life estimated to be 1920-2688 hours, based on estimated unacclimated aqueous aerobic biodegradation half-life (Howard et al. 1991).

Biotransformation:

Bioconcentration, Uptake (k_1) and Elimination (k_2) Rate Constants:

Common Name: 9,10-Dimethylbenz[a]anthracene
Synonym:
Chemical Name: 9,10-dimethylbenz[a]anthracene
CAS Registry No: 56-56-4
Molecular Formula: $C_{20}H_{16}$
Molecular Weight: 256.35

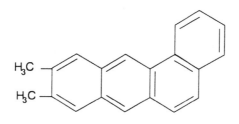

Melting Point (°C):
      122 (Yalkowsky et al. 1983)
Boiling Point (°C):
Density (g/cm³ at 20°C):
Molar Volume (cm³/mol):
      282.7 (LeBas method)
Molecular Volume (Å³):
Total Surface Area, TSA (Å²):
Heat of Fusion, ΔH_{fus}, kcal/mol:
      5.28 (Kelly & Rice 1974)
Entrophy of Fusion, ΔS_{fus}, cal/mol K (e.u.):
      13.1 (Kelly & Rice 1974)
Fugacity Ratio at 25 °C (assuming ΔS = 13.5 e.u.), F:
      0.11

Water Solubility (g/m³ or mg/L at 25°C):
      0.0435 (27 °C, shake flask-nephelometry, Davis et al. 1942)
      0.0435 (quoted, Yalkowsky et al. 1983)
      0.0435 (recommended, IUPAC 1989)

Vapor Pressure (Pa at 25°C):
      3.73×10^{-7} (solid vapor pressure, extrapolated, effusion method, Kelly & Rice 1974)
      3.70×10^{-6} (extrapolated, subcooled liquid value, Kelly & Rice 1974)

Henry's Law Constant (Pa m³/mol):

Octanol/Water Partition Coefficient, log K_{ow}:
  6.93   (calculated-f const., Yalkowsky et al. 1983)

Bioconcentration Factor, log BCF:

Sorption Partition Coefficient, log K_{oc}:

Half-Lives in the Environment:
  Air:
  Surface water:
  Groundwater:
  Sediment:
  Soil:
  Biota:

Environmental Fate Rate Constants or Half-Lives:
  Volatilization:
  Photolysis:
  Oxidation:
  Hydrolysis:
  Biodegradation:
  Biotransformation:
  Bioconcentration, Uptake (k_1) and Elimination (k_2) Rate Constants:

Common Name: 3-Methylcholanthrene
Synonym: 20-methylcholanthrene
Chemical Name: 3-methylcholanthrene
CAS Registry No: 56-49-5
Molecular Formula: $C_{21}H_{16}$
Molecular Weight: 268.38

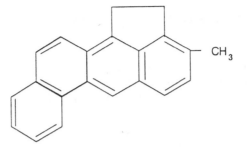

Melting Point (°C):
          178   (Yalkowsky & Valvani 1979, 1980; Mackay et al. 1980; Miller et al. 1985)
          179   (Karickhoff 1981)
          180   (Bjørseth 1983)
Boiling Point (°C):
Density (g/cm³ at 20°C):
Molar Volume (cm³/mol):
          296.0 (LeBas method, Miller et al. 1985)
Molecular Volume (Å³):
Total Surface Area, TSA (Å²):
          282.70   (Yalkowsky & Valvani 1979)
          288.50   (calculated as per Gavezzotti 1985, Sabljic 1987)
Heat of Fusion, ΔH_{fus}, kcal/mol:
Entropy of Fusion, ΔS_{fus}, cal/mol K (e.u.):
Fugacity Ratio at 25°C (assuming ΔS = 13.5 e.u.), F:
          0.0306   (at M.P. = 178°C)
          0.003   (Mackay et al. 1980)

Water Solubility (g/m³ or mg/L at 25°C):
          0.0015   (Weimer & Prausnitz 1965; quoted, Mackay & Shiu 1977)
          0.0029   (shake flask-fluorescence, Mackay & Shiu 1977; quoted, Yalkowsky & Valvani
                    1979; Mackay et al. 1980; Karickhoff 1981; Pearlman et al. 1984; IUPAC
                    1989)
          0.00323   (shake flask-scintillation counting, Means et al. 1980)
          0.0029   (quoted, Valvani & Yalkowsky 1980; Yalkowsky & Valvani 1980)
          0.0022   (calculated-K_{ow}, Valvani & Yalkowsky 1980)
          0.0018   (calculated-K_{ow}, Yalkowsky & Valvani 1980)

227

0.00295 (selected, Miller et al. 1985)

0.0022 (quoted, Banerjee 1985)

0.0022, 0.0774 (quoted, calcd.-UNIFAC, (subcooled liquid value), Banerjee 1985)

Vapor Pressure (Pa at 25°C):

1.03×10^{-7} (extrapolated-Antoine eqn., Stephenson & Malanowski 1987)

Henry's Law Constant (Pa m^3/mol):

Octanol/Water Partition Coefficient, log K_{OW}:

6.42 (shake flask-LSC, Means et al. 1980; quoted, Karickhoff 1981)

7.11 (calculated-f const., Valvani & Yalkowsky 1980; Yalkowsky & Valvani 1979, 1980)

7.11, 6.69 (quoted, calculated-S & M.P., Mackay et al. 1980)

7.11 (Hansch & Leo 1985)

7.11 (calculated, Miller et al. 1985)

7.11, 6.45, 7.07 (quoted, calculated-UNIFAC, f const., Banerjee & Howard 1988)

6.75 (recommended, Sangster 1989)

Bioconcentration Factor, log BCF:

4.12 (*daphnia magna*, McCarthy et al. 1985)

Sorption Partition Coefficient, log K_{OC}:

6.29 (Soil, calculated-K_{OW}, Karickhoff et al. 1979; quoted, Sabljic 1987)

4.89 (soil, calculated-K_{OW}, Kenaga & Goring, quoted, Sabljic 1987)

6.25 (average of 14 soil/sediment samples, shake flask-LSC, sorption isotherms, Means et al. 1980; quoted, Sabljic 1987)

6.09, 6.10 (calculated-regression of K_p versus substrate properties, calcd.-K_{OW}, Means et al. 1980)

4.02 (soil, calculated-K_{OW}, Briggs 1981; quoted, Sabljic 1987)

6.09, 6.03, 6.02, 5.54 (quoted, calculated-K_{OW}, S & M.P., S, Karickhoff 1981)

6.18 (soil, calculated-K_{OW}, Means et al. 1982)

5.07 (soil, calculated-K_{OW}, Chiou et al. 1983)

6.18, 5.07; 6.13 (soil: quoted values, calculated- χ , Sabljic 1987)

Half-Lives in the Environment:

Air: 0.317-3.17 hours, based on estimated photooxidation half-life in air (Atkinson 1987; quoted, Howard et al. 1991).

Surface water: 14616-33600 hours, based on mineralization half-life in fresh water and estuarine ecosystems (Heitkamp 1988; quoted, Howard et al. 1991).

Groundwater: 29232-672000 hours, based on estimated unacclimated aqueous aerobic biodegradation half-life (Howard et al. 1991).

Sediment:

Soil: 14616-33600 hours, based on estimated mineralization half-life in fresh water and estuarine ecosystems (Heitkamp 1988; quoted, Howard et al. 1991).

Biota:

Environmental Fate Rate Constants or Half-Lives:

Volatilization:

Photolysis:

Hydrolysis: no hydrolyzable groups (Howard et al. 1991).

Oxidation: photooxidation half-life in air of 0.317-3.17 hours, based on estimated rate constant for reaction with hydroxyl radical in air (Atkinson 1987; quoted, Howard et al. 1991).

Biodegradation: aerobic half-life of 14616-33600 hours, based on mineralization half-life in fresh water and estuarine ecosystems (HEitkamp 1988; quoted, Howard et al. 1991); anaerobic half-life of 58464-134400 hours, based on estimated unacclimated aqueous aerobic biodegradation half-life (Howard et al. 1991).

Biotransformation:

Bioconcentration, Uptake (k_1) and Elimination (k_2) Rate Constants:

Common Name: Benzo[ghi]perylene
Synonym: 1,12-benzoperylene
Chemical Name: 1,12-benzoperylene
CAS Registry No: 191-24-2
Molecular Formula: $C_{22}H_{12}$
Molecular Weight: 276.34

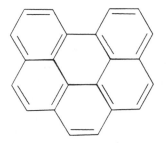

Melting Point (°C):

      275-277, 272-273 (measured, quoted, Murray et al. 1974)

      222 (Cleland & Kingsbury 1977; quoted, Callahan et al. 1979; Mabey et al. 1982)

      277 (Yalkowsky & Valvani 1979,1980; Mackay et al. 1980; Bruggeman et al. 1982; Miller et al. 1985)

      278 (Bjørseth 1983; Pearlman et al. 1984)

Boiling Point (°C):

      525 (Pearlman et al. 1984)

Density (g/cm³ at 20°C):

Molar Volume (cm³/mol):

      277.0 (LeBas method, Miller et al. 1985)

      1.547 (intrinsic volume: $V_I/100$, Kamlet et al. 1988; Hawker 1989)

Molecular Volume (Å³):

      244.3 (Pearlman et al. 1984)

Total Surface Area, TSA (Å²):

      266.9 (Yalkowsky & Valvani 1979)

      266.926 (Pearlman et al. 1984)

Heat of Fusion, ΔH_{fus}, kcal/mol:

Entropy of Fusion, ΔS_{fus}, cal/mol K (e.u.):

Fugacity Ratio at 25°C (assuming ΔS = 13.5 e.u.), F:

      0.00321 (at M.P. = 277°C)

      0.003 (Mackay et al. 1980)

Water Solubility (g/m³ or mg/L at 25°C):

      0.00026 (shake flask-fluorescence, Mackay & Shiu 1977; quoted, API 1978; Callahan et al. 1979; Yalkowsky & Valvani 1979; Mackay et al. 1980; Mabey et al. 1982; Pearlman et al. 1984; Miller et al. 1985; IUPAC 1989)

0.00026 (quoted, Valvani & Yalkowsky 1980; Yalkowsky & Valvani 1980)

0.00024 (calculated-K_{OW}, Valvani & Yalkowsky 1980)

0.00022 (calculated-K_{OW}, Yalkowsky & Valvani 1980)

0.00083 (Wise et al. 1981; quoted, Pearlman et al. 1984)

0.000264 (RP-TLC, Bruggeman et al. 1982)

0.00026 (selected, Miller et al. 1985; quoted, Eastcott et al. 1988; Hawker 1989; Capel et al. 1991)

0.00029 (quoted, Isnard & Lambert 1989)

Vapor Pressure (Pa at 25°C):

1.33×10^{-8} (20°C, estimated, Callahan et al. 1979)

1.39×10^{-8} (effusion method, Murray et al. 1974; quoted, Mabey et al. 1982)

6.69×10^{-7} (Yamasaki et al. 1984; quoted, Capel et al. 1991)

1.38×10^{-8} (extrapolated-Antoine eqn., Stephenson & Malanowski 1987)

Henry's Law Constant (Pa m^3/mol):

0.0146 (calculated-P/C, Mabey et al. 1982)

0.015 (calculated-P/C, Eastcott et al. 19880

0.001 (15°C, calculated, Baker & Eisenreich 1990)

0.709 (calculated-P/C, Capel et al. 1991)

Octanol/Water Partition Coefficient, log K_{OW}:

7.23 (calculated-π const., Callahan et al. 1979)

7.10 (calculated-f const., Yalkowsky & Valvani 1979, 1980; quoted, Eastcott et al. 1988)

7.10, 6.77 (quoted, calculated-S, M.P., Mackay et al. 1980)

7.10 (RP-TLC, Bruggeman et al. 1982)

6.51 (calculated-f const., Mabey et al. 1982)

6.85 (average lit. value, Yalkowsky et al. 1983)

7.10, 6.86 (quoted, calculated-MR, Yoshida et al. 1983)

7.10, 7.05 (quoted, HPLV-RT, Rapaport et al. 1984)

7.19 (calculated, Miller et al. 1985; quoted, Hawker 1989; Landrum et al. 1991)

7.04 (calculated- χ as per Rekker & De Kort 1979, Ruepert et al. 1985)

6.63 (HPLC, Wang et al. 1986)

7.10 (calculated, Newsted & Giesy 1987)

7.19, 6.25 (quoted, calculated-solvatochromic p., Kamlet et al. 1988)

7.23 (quoted, Pavlou 1987; Ryan et al. 1988)

7.10 (selected, Isnard & Lambert 1989)

6.90 (recommended, Sangster 1989)

7.60 (calculated-K_{OC}, Broman et al. 1991)

231

Bioconcentration Factor, log BCF:

    5.54   (microorganisms-water, calculated from K_{ow}, Mabey et al. 1982)

    4.45   (*daphnia magna*, Newsted & Giesy 1987)

Sorption Partition Coefficient, log K_{oc}:

    6.20   (calculated-K_{ow}, Mabey et al. 1982)

    6.26   (calculated, Pavlou 1987)

Half-Lives in the Environment:

    Air: 0.321-3.21 hours, based on estimated photooxidation half-life in air (Atkinson 1987; quoted, Howard et al. 1991); 0.6 hour for adsorption on wood soot particles in an outdoor teflon chamber with an estimated first order rate constant of 0.0179 at 1 cal cm^{-2} $minute^{-1}$, 10 g/m^3 H_2O and 20°C (Kamens et al. 1988).

    Surface water: 14160-15600 hours, based on aerobic soil dieaway test data at 10-30°C (Coover & Sims 1987; quoted, Howard et al. 1991).

    Groundwater: 28320-31200 hours, based on aerobic soil dieaway test data at 10-30°C (Coover & Sims 1987; quoted, Howard et al. 1991).

    Sediment:

    Soil: 14160-15600 hours, based on aerobic soil dieaway test data (Coover & Sims 1987; quoted, Howard et al. 1991); >50 days (Ryan et al. 1988); mean half-life of 9.1 years for Luddington soil (Wild et al. 1991).

    Biota:

Environmental Fate Rate Constants or Half-Lives:

    Volatilization:

    Photolysis: half-lives on different atmospheric particulate substrates (appr. 25 μg/gm on substrate): 7.0 hours on silica gel, 22 hours on alumina and 29 hours on fly ash (Behymer & Hites 1985); first order daytime photodegradation rate constants for adsorption on wood soot particles in an outdoor teflon chamber: 0.0077 $minute^{-1}$ with 1000-2000 ng/mg loading and 0.0116 $minute^{-1}$ with 30-350 ng/mg loading (Kamens et al. 1988).

    Hydrolysis: no hydrolyzable groups (Howard et al. 1991).

    Oxidation: rate constants of <360 M^{-1} hr^{-1} for singlet oxygen and <36 M^{-1} hr^{-1} for peroxy radical (Mabey et al. 1982); photooxidation half-life in air of 0.321-3.21 hours, based on estimated rate constant for reaction with hydroxyl radical in air (Atkinson 1987; quoted, Howard et al. 1991).

    Biodegradation: aerobic half-life of 14160-15600 hours, based on aerobic soil dieaway test data at 10-30°C (Coover & Sims 1987; quoted, Howard et al. 1991);

anaerobic half-life of 56640-62400 hours, based on aerobic soil dieaway test data at 10-30°C (Coover & Sims 1987; quoted, Howard et al. 1991).

Biotransformation: estimated to be 3×10^{-12} ml cell^{-1} hr^{-1} for bacteria (Mabey et al. 1982).

Bioconcentration, Uptake (k_1) and Elimination (k_2) Rate Constants:

Common Name: Dibenz[a,c]anthracene
Synonym: 1,2:3,4-Dibenzanthracene, naphtho-2',3':9,10-phenanthrene
Chemical Name: dibenz[a,c]anthracene
CAS Registry No: 215-58-7
Molecular Formula: $C_{22}H_{14}$
Molecular Weight: 278.35

Melting Point (°C):
        205-207 (Weast 1982-83)
        206 (Miller et al. 1985)
        205 (Bjorseth 1983)
Boiling Point (°C):
        518 (Weast 1982-83)
Density (g/cm³ at 20°C):
Molar Volume (cm³/mol):
        300 (LeBas method, Miller et al. 1985)
Molecular Volume (Å³):
Total Surface Area, TSA (Å²):
Heat of Fusion, ΔH_{fus}, kcal/mol:
Entrophy of Fusion, ΔS_{fus}, cal/mol K (e.u.):
Fugacity Ratio at 25°C (assuming ΔS = 13.5 e.u.), F:

Water Solubility (g/m³ or mg/L at 25°C):
        0.0016 (gen. col.-HPLV/UV, Billington et al. 1988)

Vapor Pressure (Pa at 25°C):
        1.3×10^{-9} (effusion method, De Kruif 1980)

Henry's Law Constant (Pa m³/mol):

Octanol/Water Partition Coefficient, log K_{ow}:
        7.19 (calculated, Miller et al. 1985)
        7.11 (calculated- χ as per Rekker & De Kort 1979, Ruepert et al. 19850
        7.19 (recommended, Sangster 1989)

234

Bioconcentration Factor, log BCF:

Sorption Partition Coefficient, log K_{OC}:

Half-Lives in the Environment:
        Air:
        Surface water:
        Groundwater:
        Sediment:
        Soil:
        Biota:

Environmental Fate Rate Constants or Half-Lives:
        Volatilization:
        Photolysis:
        Oxidation:
        Hydrolysis:
        Biodegradation:
        Biotransformation:
        Bioconcentration, Uptake (k_1) and Elimination (k_2) Rate Constants:

Common Name: Dibenz[a,h]anthracene
Synonym: DB[a,h]A, 1,2,5,6-dibenzanthracene, 1,2:5,6-dibenzanthracene
Chemical Name: 1,2:5,6-dibenzanthracene
CAS Registry No: 53-70-3
Molecular Formula: $C_{22}H_{14}$
Molecular Weight: 278.36

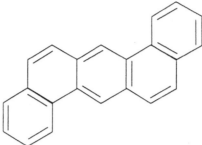

Melting Point (°C):
    270 (Weast 1977; quoted, Callahan et al. 1979; Mabey et al. 1982; Bjørseth 1983)
    269 (Karickhoff 1981)
    267 (Pearlman et al. 1984)
    266 (Lande et al. 1985; Miller et al. 1985; Gobas et al. 1988; Budavari 1989)
Boiling Point (°C):
    524 (Weast 1977; Pearlman 1984)
Density (g/cm³ at 20°C):
Molar Volume (cm³/mol):
    300.0 (LeBas method, Gobas et al. 1988)
    1.539 (intrinsic volume: $V_I/100$, Kamlet et al. 1988; Hawker 1989)
Molecular Volume (Å³):
    244.3 (Pearlman et al. 1984)
Total Surface Area, TSA (Å²):
    286.475 (Pearlman et al. 1984)
    296.4 (calculated as per Gavezzotti 1985, Sabljic 1987)
Heat of Fusion, ΔH_{fus}, kcal/mol:
Entropy of Fusion, ΔS_{fus}, cal/mol K (e.u.):
Fugacity Ratio at 25°C (assuming $\Delta S = 13.5$ e.u.), F:
    0.00403 (at M.P. = 267°C)

Water Solubility (g/m³ or mg/L at 25°C):
    0.0005 (27°C, shake flask-nephelometry, Davis et al. 1942; quoted, Callahan et al.
        1979; Mabey et al. 1982; IUPAC 1989)
    0.0006 (shake flask-UV, Klevens 1950; quoted, IUPAC 1989)
    0.0006 (Callahan et al. 1979)
    0.0025 (shake flask-LSC, Means et al. 1980; quoted, Karickhoff 1981)
    0.0005 (selected, Mills et al. 1982)

236

0.0009 (lit. mean, Yalkowsky et al. 1983)
0.00056 (lit. mean, Pearlman et al. 1984; quoted, Banerjee 1985)
0.0005 (selected, Miller et al. 1985)
0.146 (calculated-UNIFAC for subcooled liq., Banerjee 1985; quoted, Farrington 1991)
0.0009, 0.00029 (quoted, predicted, Lande et al. 1985)
0.00050 (quoted, Gobas et al. 1988; Hawker 1989)
0.0005 (quoted, Isnard & Lambert 1989)
0.0005 (selected, Ma et al. 1990)

Vapor Pressure (Pa at 25°C):
1.33×10^{-8} (20°C, estimated, Callahan et al. 1979; quoted, Mabey et al. 1982; Eastcott et al. 1988)
3.70×10^{-10} (effusion method, De Kruif 1980)
4.25×10^{-10} (extrapolated-Antoine eqn., Stephenson & Malanowski 1987)
1.30×10^{-8} (selected, Ma et al. 1990)

Henry's Law Constant (Pa m^3/mol):
0.0074 (calculated-P/C, Mabey et al. 1982)
0.0076 (calculated-P/C, Eastcott et al. 1988)

Octanol/Water Partition Coefficient, log K_{ow}:
5.97 (calculated-f const., Callahan et al. 1979)
6.50 (shake flask-LSC, Means et al. 1980; quoted, Karickhoff 1981)
6.84 (calculated-f const., Mabey et al. 1982)
7.19 (calculated-f const., Yalkowsky et al. 1983)
6.50, 6.88, 6.84 (quoted, HPLC-RT/MS, calculated-f const., Burkhard et al. 1985)
6.50 (Freitag et al. 1985)
7.11 (calculated- χ as per Rekker & De Kort 1979, Ruepert et al. 1985)
5.80 (Hansch & Leo 1985)
7.19 (calculated, Miller et al. 1985; quoted, Eastcott et al. 1988)
6.50 (quoted, Pavlou 1987)
5.80, 6.16 (quoted, calculated-UNIFAC, Banerjee & Howard 1988)
7.19, 6.52 (quoted, calculated-solvatochromic p., Kamlet et al. 1988)
7.19 (quoted, Isnard & Lambert 1989)
6.75 (recommended, Sangster 1989)
6.20 (selected, Ma et al. 1990)

Bioconcentration Factor, log BCF:
5.84 (microorganisms-water, calculated from K_{ow}, Mabey et al. 1982)
4.63 (activated sludge, Freitag et al. 1984)
3.38, 4.63, 1.0 (algae, activated sludge, fish, Freitag et al. 1985)
4.00 (*daphnia magna*, Newsted & Giesy 1987)

Sorption Partition Coefficient, log K_{OC}:

 6.31 (average of 14 soil/sediment samples, shake flask-LSC, concn. ratio at equilibrium, sorption isotherms, Means et al. 1980)

 6.22, 6.18 (calculated-regression of k_p versus substrate properties, calculated-k_{OW}, Means et al. 1980)

 6.22; 6.11, 5.30, 5.62 (quoted; calculated-K_{OW}, S & M.P., S, Karickhoff 1981)

 6.52 (calculated-K_{OW}, Mabey et al. 1982)

 5.20 (calculated, Pavlou 1987)

Half-Lives in the Environment:

 Air: 0.428-4.28 hours, based on estimated photooxidation half-life in air (Atkinson 1987; quoted, Howard et al. 1991).

 Surface water: 6-782 hours, based on sunlight photolysis half-life in water (Smith et al. 1978; Muel & Saguim 1985; quoted, Howard et al. 1991).

 Groundwater: 17328-45120 hours, based on estimated unacclimated aqueous aerobic biodegradation half-life (Howard et al. 1991).

 Sediment:

 Soil: biodegration rate constant of 0.0019 day^{-1} with a half-life of 361 days for Kidman sandy loam soil and 0.117 day^{-1} with half-life of 420 days for McLaurin sandy loam soil (Park et al. 1990); half-lives were estimated to be 8664-22560 hours, based on aerobic soil dieaway test data (Coover & Sims 1987; Sims 1990; quoted, Howard et al. 1991); mean half-life of 20.607 weeks (quoted, Wild et al. 1991).

 Biota:

Environmental Fate Rate Constants or Half-Lives:

 Volatilization:

 Photolysis: atmospheric and aqueous photolysis half-life of 782 hours, based on measured rate of photolysis in heptane under November sunlight (Muel & Saguim 1985; quoted, Howard et al. 1991) and of 6 hours after adjusting the ratio of sunlight photolysis in water vs. heptane (Smith et al. 1978; Muel & Saguim 1985; quoted, Howard et al. 1991).

 Hydrolysis: not hydrolyzable (Mabey et al. 1982); no hydrolyzable groups (Howard et al. 1991).

 Oxidation: rate constant of $5x10^8$ M^{-1} hr^{-1} for singlet oxygen and $1.5x10^4$ M^{-1} hr^{-1} for peroxy radical (Mabey et al. 1982); photooxidation half-life of 0.428-4.28 hours, based on estimated rate constant for reaction with hydroxyl radical in air (Atkinson 1987; quoted, Howard et al. 1991).

 Biodegradation: aerobic half-life of 8664-22560 hours, based on aerobic soil dieaway test data (Coover & Sims 1987; Sims 1990; quoted, Howard et al. 1991); rate constants of 0.0019 day^{-1} with half-life of 361 days for Kidman sandy loam and 0.0017 day^{-1} with half-life of 420 days for McLarin sandy loam all at -0.33 bar soil moisture (Park et al. 1990).

 Biotransformation: estimated to be $3x10^{-12}$ ml cell^{-1} hr^{-1} for bacteria (Mabey et al. 1982).

 Bioconcentration, Uptake (k_1) and Elimination (k_2) Rate Constants:

238

Common Name: Dibenz[a,j]anthracene
Synonym: 1,2:7,8-dibenzanthracene, 1,2:7,8-dibenzanthracene
Chemical Name: dibenz[a,j]anthracene
CAS Registry No: 58-70-3
Molecular Formula: $C_{22}H_{14}$
Molecular Weight: 278.36

Melting Point (°C):
      198   (Bjørseth 1983)
      196   (Yalkowsky et al. 1983)
Boiling Point (°C):
Density (g/cm³ at 20°C):
Molar Volume (cm³/mol):
      300.0   (LeBas method)
Molecular Volume (Å³):
      255.44 (Pearlman et al. 1984)
Total Surface Area, TSA (Å²):
      286.5   (Yalkowsky et al. 1983)
      286.575 (Pearlman et al. 1984)
Heat of Fusion, ΔH_{fus}, kcal/mol:
Entropy of Fusion, ΔS_{fus}, cal/mol K (e.u.):
Fugacity Ratio at 25°C (assuming ΔS = 13.5 e.u.), F:
      0.0195 (at M.P. = 198°C)

Water Solubility (g/m³ or mg/L at 25°C):
      0.012 (27°C, nephelometry, Davis et al. 1942; quoted, IUPAC 1989)
      0.012 (quoted, Yalkowsky et al. 1983; Pearlman et al. 1984)

Vapor Pressure (Pa at 25°C):

Henry's Law Constant (Pa m³/mol):

Octanol/Water Partition Coefficient, log K_{OW}:

      7.19   (calculated-f const., Yalkowsky et al. 1983)

      7.11   (calculated- χ as per Rekker & De Kort 1979, Ruepert et al. 1985)

Bioconcentration Factor, log BCF:

Sorption Partition Coefficient, log K_{OC}:

Half-Lives in the Environment:

      Air:

      Surface water:

      Groundwater:

      Sediment:

      Soil:

      Biota:

Environmental Fate Rate Constants or Half-Lives:

      Volatilization:

      Photolysis:

      Hydrolysis:

      Oxidation:

      Biodegradation:

      Biotransformation:

      Bioconcentration, Uptake (k_1) and Elimination (k_2) Rate Constants:

Common Name: Pentacene
Synonym: 2,3,6,7-dibenzanthracene, 2,3:6,7-dibenzanthracene
Chemical Name: pentancene
CAS Registry No: 135-48-83
Molecular Formula: $C_{22}H_{14}$
Molecular Weight: 278.35

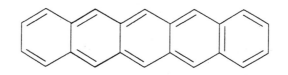

Melting Point (°C):
        270-271 (Weast 1982-83)
Boiling Point (°C):
        290-300 (sublime, Weast 1982-83)
Density (g/cm³ at 20°C):
Molar Volume (cm³/mol):
        300   (LeBas method)
        1.539 (intrinsic molar volume, $V_I/100$)
Molecular Volume (Å³):
Total Surface Area, TSA (Å²):
Heat of Fusion, ΔH_{fus}, kcal/mol:
Entrophy of Fusion, ΔS_{fus}, cal/mol K (e.u.):
Fugacity Ratio at 25 °C (assuming ΔS = 13.5 e.u.), F:
        0.00377 (at M.P. = 270°C)

Water Solubility (g/m³ or mg/L at 25°C):

Vapor Pressure (Pa at 25°C):
        1.0×10^{-13} (effusion method, De Kruif 1980)

Henry's Law Constant (Pa m³/mol):

Octanol/Water Partition Coefficient, log K_{ow}:
        7.19   (calculated-f const., Miller et al. 1985)
        7.19   (recommended, Sangster 1989)

Bioconcentration Factor, log BCF:

Sorption Partition Coefficient, log K_{oc}:

Half-Lives in the Environment:
       Air:
       Surface water:
       Groundwater:
       Sediment:
       Soil:
       Biota:

Environmental Fate Rate Constants or Half-Lives:
       Volatilization:
       Photolysis:
       Oxidation:
       Hydrolysis:
       Biodegradation:
       Biotransformation:
       Bioconcentration, Uptake (k_1) and Elimination (k_2) Rate Constants:

Common Name: Coronene
Synonym: hexabenzobenzene
Chemical Name: coronene
CAS Registry No: 191-07-1
Molecular Formula: $C_{24}H_{12}$
Molecular Weight: 300.36

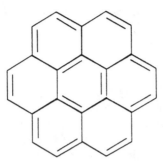

Melting Point (°C):
      440, 438-440 (measured, quoted, Murray et al. 1974)
      439   (Bjørseth 1983; Yalkowsky et al. 1983)
      360   (Miller et al. 1985)
      >36   (Pearlman et al. 1984)
Boiling Point (°C):
      525   (Weast 1982-83)
      > 500 (Bjørseth 1983; Pearlman et al. 1984)
Density (g/cm^3 at 20°C):
Molar Volume (cm^3/mol):
      292.0   (LeBas method, Miller et al. 1985)
Molecular Volume (Å^3):
      260.83 (Pearlman et al. 1984)
Total Surface Area, TSA (Å^2):
      282.4   (Yalkowsky et al. 1983)
      282.41 (Pearlman et al. 1984)
Heat of Fusion, ΔH_{fus}, kcal/mol:
Entropy of Fusion, ΔS_{fus}, cal/mol K (e.u.):
Fugacity Ratio at 25 °C (assuming ΔS = 13.5 e.u.), F:
      4.84×10^{-4} (at M.P. = 442°C)

Water Solubility (g/m^3 or mg/L at 25°C):
      0.00014 (shake flask-fluorescence, Mackay & Shiu 1977; quoted, Futoma et al. 1981;
                 Pearlman et al. 1984; Miller et al. 1985; Billington et al. 1988; IUPAC 1989)
      0.00069 (average lit. value, Yalkowsky et al. 1983)

0.00014   (average lit. value, Pearlman et al. 1984; quoted, Billington et al. 1988)
0.00014   (selected, Miller et al. 1985)
0.00010   (gen. col.-HPLC/UV, Billington et al. 1988)
0.00014   (quoted, Isnard & Lambert 1989)

Vapor Pressure (Pa at 25°C):
1.95x10^{-10} (effusion method, Murray et al. 1974)
2.00x10^{-10} (extrapolated-Antoine eqn., Stephenson & Malanowski 1987)

Henry's Law Constant (Pa m^3/mol):

Octanol/Water Partition Coefficient, log K_{ow}:
7.64   (average lit. value, Yalkowsky et al. 1983)
7.64   (calculated, Miller et al. 1985; quoted, Eastcott et al. 1988)
7.64   (calculated- χ as per Rekker & De Kort 1979, Ruepert et al. 1985)
8.20, 6.70 (HPLC-RP predicted, Brooke et al. 1986)
5.40   (HPLC-RV measured, Brooke et al. 1986)
7.64   (quoted, Isnard & Lambert 1989)
6.50   (recommended, Sangster 1989)
8.0   (calculated-K_{oc}, Broman et al. 1991)

Bioconcentration Factor, log BCF:

Sorption Partition Coefficient, log K_{oc}:
7.80   (specified particulate log K_{oc}, Broman et al. 1991)
5.0   (predicted dissolved log K_{oc}, Broman et al. 1991)

Half-Lives in the Environment:
Air:
Surface water:
Groundwater:
Sediment:
Soil: mean half-life of 16.5 years for Luddington soil (Wild et al. 1991).
Biota:

Environmental Fate Rate Constants or Half-Lives:
Volatilization:
Photolysis:
Hydrolysis:
Oxidation:
Biodegradation:
Biotransformation:
Bioconcentration, Uptake (k_1) and Elimination (k_2) Rate Constants:

2.2 Summary Tables and QSPR Plots

Table 2.1   Summary of physical-chemical properties of PAHs at 25 °C

COMPOUND	CAS no.	formula	MW g/mol	M.P., °C	B.P., °C	fugacity ratio, F at 25 °C	V_M[a] LeBas cm³/mol	V[b] from ρ cm³/mol	$V_I/100$[c] intrinsic cm³/mol	TSA[d] Å²	TSA[e] Å²	TMV[e] Å³
Indan	496-11-7	C_9H_{10}	118.18	-51	178	1	143.7	123.0		151.1	156.70	121.60
Naphthalene	91-20-3	C_8H_{10}	128.19	80.5	218	0.283	147.6	125.0	0.753	155.8	155.837	126.905
1-Methyl-	90-12-0	$C_{11}H_{10}$	142.2	-22	244.6	1	169.8	139	0.851	172.5	174.948	142.558
2-Methyl-	91-57-6	$C_{11}H_{10}$	142.2	34.6	241.9	0.804	169.8	141	0.851	176.3	177.682	142.801
1,2-Dimethyl-	573-98-8	$C_{12}H_{12}$	156.23	-1	266-267	1	192.0		0.949			
1,3-Dimethyl-	575-41-7	$C_{12}H_{12}$	156.23		265	1	192.0	154	0.949	192.9	196.793	158.453
1,4-Dimethyl-	571-58-4	$C_{12}H_{12}$	156.23	7.66	262	1	192.0		0.949	189.2	194.06	158.21
1,5-Dimethyl-	571-61-9	$C_{12}H_{12}$	156.23	81	249	0.279	192.0		0.949		194.06	158.21
2,3-Dimethyl-	581-40-8	$C_{12}H_{12}$	156.23	105	269	0.162	192.0	156	0.949	193.1	196.209	158.331
2,6-Dimethyl-	581-40-2	$C_{12}H_{12}$	156.23	108	262	0.151	192.0		0.949	196.7	199.527	158.697
1-Ethyl-	1127-76-0	$C_{12}H_{12}$	156.23	-13.8	258.7	1	192.0	155	0.949	187.4	192.275	158.579
2-Ethyl-	939-27-5	$C_{12}H_{12}$	156.23	-70	251-2	1	192.0		0.949		199.909	158.892
1,4,5-Trimethyl-	2131-41-1	$C_{13}H_{14}$	176.2	64	185	0.411	214.2	169	1.047		204.393	170.72

COMPOUND	CAS no.	formula	MW g/mol	M.P., °C	B.P., °C	fugacity ratio, F at 25 °C	$V_{M(a)}$ LeBas cm³/mol	$V^{(b)}$ from ρ cm³/mol	$V_I/100^{(c)}$ intrinsic cm³/mol	TSA$^{(d)}$ Å²	TSA$^{(e)}$ Å²	TMV$^{(e)}$ Å³
Biphenyl	92-52-4	$C_{12}H_{10}$	154.21	71	256	0.351	185	178	0.92	182.0	189.60	155.10
4-Methyl-	644-08-6	$C_{13}H_{12}$	168.24	45.5	267-268	0.627	206.8				212.88*	
4,4'-Dimethyl-	613-33-2	$C_{13}H_{14}$	182.27	119	295	0.118	229				233.4*	
Diphenylmethane	101-81-5	$C_{13}H_{12}$	168.24	25	264.3	1	206.8				214.20	171.70
Bibenzyl	103-29-7	$C_{14}H_{14}$	182.27	52.2	285	0.538					237.11	188.00
trans-Stilbene	103-30-0	$C_{14}H_{12}$	180.25	124		0.105	221.6			175.0		
Acenaphthene	83-32-9	$C_{12}H_{10}$	154.21	96.2	277.5	0.198	173	151	0.916	180.8	180.773	148.816
Acenaphthylene	208-96-8	$C_{12}H_8$	150.2	92	265-275	0.217	165.7			193.6		
Fluorene	86-73-7	$C_{13}H_{10}$	166.2	116	295	0.126	188	138	0.96	194	194.012	160.392
1-Methylfluorene	1730-37-6	$C_{14}H_{12}$	180.25	85		0.255	210		1.058			
Phenanthrene	85-01-8	$C_{14}H_{10}$	178.2	101	339	0.177	199	182	1.015	198	197.996	169.482
1-Methyl-	832-69-9	$C_{15}H_{12}$	192.26	123	359	0.107	218.7		1.113		217.104	185.135
Anthracene	120-12-7	$C_{14}H_{10}$	178.2	216.2	340	0.0129	197	139	1.015	202.2	202.164	170.286
9-Methyl-	779-02-2	$C_{15}H_{12}$	192.26	81.5	355	0.276	219	181		215.1	216.142	184.863
2-Methyl-	613-12-7	$C_{15}H_{12}$	192.26	209	359	0.0151	219	106(?)		222.6	224.009	186.182
9,10-Dimethyl-	781-43-1	$C_{16}H_{14}$	206.3	182		0.028	241		1.211	228.0	230.119	199.51

247

COMPOUND	CAS no.	formula	MW g/mol	M.P., °C	B.P., °C	fugacity ratio, F at 25 °C	$V_M^{(a)}$ LeBas cm³/mol	$V^{(b)}$ from ρ cm³/mol	$V_I/100^{(c)}$ intrinsic cm³/mol	$TSA^{(d)}$ Å²	$TSA^{(e)}$ Å²	$TMV^{(e)}$ Å³
Pyrene	129-00-0	$C_{16}H_{10}$	202.3	156	360	0.0506	214	159	1.156	213.0	213.466	186.008
Fluoranthene	206-44-0	$C_{16}H_{10}$	202.3	111	375	0.141	217	162		218.0	218.631	187.741
Benzo[a]fluorene	238-84-6	$C_{17}H_{12}$	216.28	187	407	0.0249	240		1.224	237.4	240.327	203.77
Benzo[b]fluorene	243-17-4	$C_{17}H_{12}$	216.28	209	402	0.0151	240		1.224	239.9	240.335	203.778
Chrysene	218-01-9	$C_{18}H_{12}$	228.3	255	448	0.0053	251	179	1.277	241.0	240.15	212.059
Triphenylene	217-59-4	$C_{18}H_{12}$	228.3	199	438	0.019	251		1.277	236	235.992	211.255
p-Terphenyl	92-94-4	$C_{18}H_{14}$	230.1	213		0.0138	258.2					
Naphthacene	92-24-0	$C_{18}H_{12}$	228.3	357		0.00052	251		1.277	248.0	248.50	213.7
B[a]A	56-55-3	$C_{18}H_{12}$	228.3	160	435	0.0462	248		1.277	244.3	244.32	212.863
Benzo[a]pyrene	50-32-8	$C_{20}H_{12}$	252.3	175	495	0.0328	263	187	1.418	256.0	255.628	228.585
Benzo[e]pyrene	192-97-2	$C_{20}H_{12}$	252.3	178		0.0307					251.457	227.781
Perylene	198-55-0	$C_{20}H_{12}$	252.32	277	495	0.00321	263		1.415	251.5	251.456	227.781
B[b]F	205-99-2	$C_{20}H_{12}$	252.32	168	481	0.0385	268.9				260.783	230.315
B[j]F	205-82-2	$C_{20}H_{12}$	252.32	166	480	0.0403	268.9				259.705	230.405
B[k]F	207-08-9	$C_{20}H_{12}$	252.32	217	481	0.0126	268.9			266.0	264.952	231.121
7,12-DMBA	57-97-6	$C_{20}H_{16}$	256.35	122		0.11	282.7			267.3	267.3	239.4
9,10-DMBA	56-56-35	$C_{20}H_{16}$	256.35	122		0.11	282.7					

COMPOUND	CAS no.	formula	MW g/mol	M.P., °C	B.P., °C	fugacity ratio, F at 25 °C	V_M[a] LeBas cm³/mol	V[b] from ρ cm³/mol	$V_I/100$[c] intrinsic cm³/mol	TSA[d] Å²	TSA[e] Å²	TMV[e] Å³
3-MCA	56-49-5	$C_{21}H_{16}$	268.36	178		0.0307	296			282.7	291.0	250.7
B[ghi]P	191-24-2	$C_{21}H_{16}$	268.36	277		0.00321	277		1.547	266.9	266.926	244.307
D[a,c]A	215-58-7	$C_{22}H_{14}$	278.35	205		0.0166	300					
D[a,h]A	53-70-1	$C_{22}H_{14}$	278.35	267	524	0.0403	300		1.539	286.5	286.475	255.40
D[a,j]A	58-70-3	$C_{22}H_{14}$	278.35	196		0.0195	300		1.539	286.4	286.475	255.40
Pentacene	135-48-8	$C_{22}H_{14}$	278.35	> 300		0.0019	292.5					
Coronene	191-07-1	$C_{24}H_{12}$	300.36	> 350	525		292			282.4	282.409	260.833

Note:

B[a]A	benz[a]anthracene
B[ghi]P	benzo[ghi]perylene
B[b]F	benzo[b]fluoranthene
B[j]F	benzo[j]fluoranthene
B[k]F	benzo[k]fluoranthene
3-MCA	3-methylcholanthrene
7,12-DMBA	7,12-dimethylbenz[a]anthracene
9,10-DMBA	9,10-dimethylbenz[a]anthracene

(a) Miller et al. 1985, Eastcott et al. 1988 and calculated for this work
(b) Lande & Banerjee 1981
(c) Kamlet et al. 1988
(d) Yalkowsky & Valvani 1979
(e) Pearlman et al. 1984
* Doucette and Andren 1988

Table 2.2   Summary of selected physical-chemical properties of PAHs at 25 °C

COMPOUND	Vapor pressure		Solubility			log K_{ow}	Henry's law
	P^S	P_L	S	C^S	C_L		const.,H
	Pa	Pa	g/m^3	mmol/m^3	mmol/m^3		Pa m^3/mol
		(liquid)			(liquid)		calcd. P/C
Indan	197	197	100	846.17	846.17	3.33	232.82
Naphthalene	10.4	36.81	31	241.83	855.92	3.37	43.01
1-Methyl-	8.84	8.84	28	196.91	196.91	3.87	44.90
2-Methyl-	9	11.2	25	175.81	218.78	3.86	51.19
1,2-Dimethyl-	0.87	0.870				4.31	
1,3-Dimethyl-			8	51.21	51.21	4.42	
1,4-Dimethyl-	2.27	2.27	11.4	72.97	72.97	4.37	31.11
1,5-Dimethyl-			3.1	19.84	71.34	4.38	
2,3-Dimethyl-	1	6.17	2.5	16.00	98.95	4.40	62.49
2,6-Dimethyl-	1.4	9.27	1.7	10.88	72.044	4.31	128.66
1-Ethyl-	2.51	2.51	10.1	64.65	64.65	4.40	38.825
2-Ethyl-	4	4.00	8	51.21	51.21		78.12
1,4,5-Trimethyl-	0.681	1.66	2.1	11.92	28.970	5.00	57.14
Biphenyl	1.3	3.706	7	45.393	129.40	3.90	28.64
4-Methyl-			4.05	24.073	38.40	4.63	
4,4'-Dimethyl-			0.175	0.960	8.166	5.09	
Diphenylmethane	0.0885	0.0885	16	95.10	95.10	4.14	0.931
Bibenzyl	0.406	0.754	4.37	23.98	44.544	4.70	16.93
trans-Stilbene	0.065	0.619	0.29	1.609	15.335	4.81	40.40

COMPOUND	P^S Pa	P$_L$ Pa (liquid)	S g/m^3	C^S mmol/m^3	C$_L$ mmol/m^3 (liquid)	log K$_{ow}$	H Pa m^3/mol calcd.
Acenaphthene	0.3	1.52	3.80	24.642	124.70	3.92	12.17
Acenaphthylene	0.9	4.14	16.1	107.19	492.97	4.00	8.40
Fluorene	0.09	0.715	1.90	11.43	90.816	4.18	7.87
1-Methyl-			1.09	6.047	23.713	4.97	
Phenanthrene	0.02	0.113	1.10	6.173	34.847	4.57	3.24
1-Methyl-			0.27	1.404	13.084	5.14	
Anthracene	0.001	0.0778	0.045	0.253	19.65	4.54	3.96
9-Methyl-	0.0024		0.261	1.358	4.915	5.07	
2-Methyl			0.03	0.156	10.31	5.15	
9,10-Dimethyl	1.53x10^{-4}		0.056	0.271	9.70	5.25	
Pyrene	0.0006	0.0119	0.132	0.652	12.89	5.18	0.92
Fluoranthene	0.00123	8.72x10^{-3}	0.26	1.186	8.410	5.22	1.037
Benzo[a]fluorene			0.045	0.208	8.326	5.40	
Benzo[b]fluorene			0.002	0.00925	0.602	5.75	
Chrysene	5.70x10^{-7}	1.07x10^{-4}		0.00876	1.649	5.86	0.065
Triphenylene	2.30x10^{-6}	1.21x10^{-4}	0.043	0.188	9.906	5.49	0.012
p-Terphenyl	4.86x10^{-6}		0.0180	0.0782	5.664	6.03	
Naphthacene	9.30x10^{-9}	1.79x10^{-5}	0.0006	0.00263	5.050	5.76	0.004
B[a]A	2.80x10^{-5}	6.06x10^{-4}	0.011	0.0482	1.045	5.91	0.581
Benzo[a]pyrene	7.00x10^{-7}	2.13x10^{-5}	0.0038	0.0151	0.459	6.04	0.046
Benzo[e]pyrene	7.40x10^{-7}	2.41x10^{-5}	0.004	0.0159	0.517		0.020

COMPOUND	P^S Pa	P_L Pa	S, g/m³	C^S mmol/m³	C_L mmol/m³	log K_{ow}	H Pa m³/mol P/C
Perylene	1.40×10^{-8}		0.0004	0.00159	0.493	6.25	0.003
B[b]F			0.0015	0.00595	0.154	5.80	
B[j]F			0.0025	0.0099	0.246		
B[k]F	5.20×10^{-8}	4.12×10^{-6}	0.0008	0.00317	0.252	6.00	0.016
7,12-DMBA	3.84×10^{-8}		0.0500	0.195	1.776	6.00	
9,10-DMBA	3.73×10^{-7}		0.0435	0.170	1.543	6.00	
3-MCA	1.03×10^{-6}	3.46×10^{-5}	0.0019	0.00708	0.231	6.42	0.145
B[ghi]P		2.25×10^{-5}	0.00026	0.000968	0.301	6.50	0.075
Pentacene	1.0×10^{-13}	5.26×10^{-11}					
DB[a,c]A	1.30×10^{-9}	7.84×10^{-8}	0.0016	0.00575	0.346		
DB[a,h]A	3.70×10^{-10}	9.16×10^{-8}	0.0006	0.00216	0.533	6.75	
DB[a,j]A			0.012	0.0431	2.210		
Coronene	2.0×10^{-10}		0.00014	0.000466		6.75	

Abbreviations:

B[a]A	benz[a]anthracene
B[b]F	benzo[b]fluoranthene
B[j]F	benzo[j]fluoranthene
B[k]F	benzo[k]fluoranthene
7,12-DMBA	7,12-dimethylbenz[a]anthracene
9,10-DMBA	9,10-dimethylbenz[a]anthracene
3-MCA	3-methylcholanthrene
B[ghi]P	benzo[ghi]perylene
DB[a,c]A	dibenz[a,c]anthracene
DB[a,h]A	dibenz[a,h]anthracene
DB[a,j]A	dibenz[a,j]anthracene

Table 2.3 Suggested half-life classes of polynuclear aromatic hydrocarbons (PAHs) in various environmental compartments

Compounds	Air class	Water class	Soil class	Sediment class
Indan	2	4	6	7
Naphthalene	2	4	6	7
1-Methyl-	2	4	6	7
2,3-Dimethyl-	2	4	6	7
1-Ethyl-	2	4	6	7
1,4,5-Trimethyl-	2	4	6	7
Biphenyl	3	4	5	6
Acenaphthalene	3	5	7	8
Fluorene	3	5	7	8
Phenanthrene	3	5	7	8
Anthracene	3	5	7	8
Pyrene	4	6	8	9
Fluoranthene	4	6	8	9
Chrysene	4	6	8	9
Benz[a]anthracene	4	6	8	9
Benzo[k]fluoranthene	4	6	8	9
Benzo[a]pyrene	4	6	8	9
Perylene	4	6	8	9
Dibenz[a,h]anthracene	4	6	8	9

where,

Class	Mean half-life (hours)	Range (hours)
1	5	< 10
2	17 (~ 1 day)	10-30
3	55 (~ 2 days)	30-100
4	170 (~ 1 week)	100-300
5	550 (~ 3 weeks)	300-1,000
6	1700 (~ 2 months)	1,000-3,000
7	5500 (~ 8 months)	3,000-10,000
8	17000 (~ 2 years)	10,000-30,000
9	55000 (~ 6 years)	> 30,000

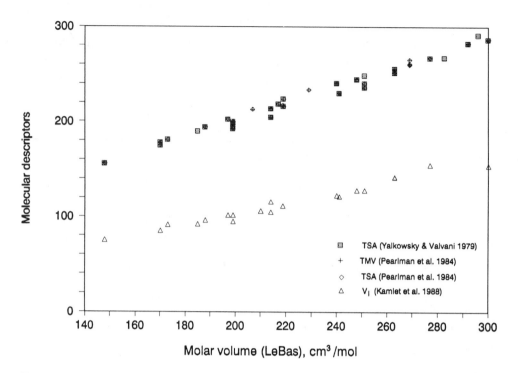

Figure 2.1   Plot of molecular descriptors versus LeBas molar volume.

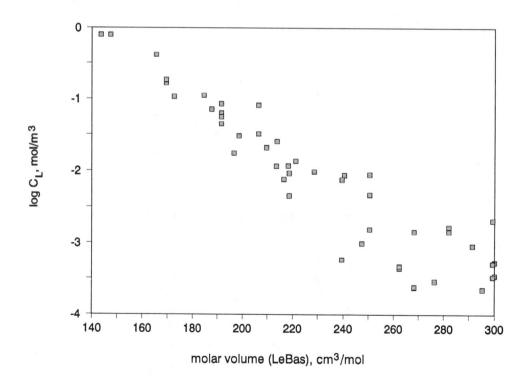

Figure 2.2   Plot of log C_L (liquid solubility) versus molar volume.

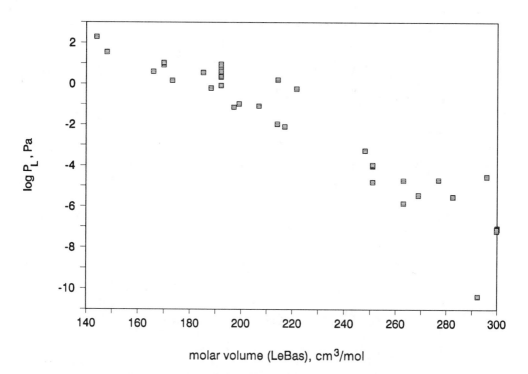

Figure 2.3   Plot of log P_L (liquid vapor pressure) versus molar volume.

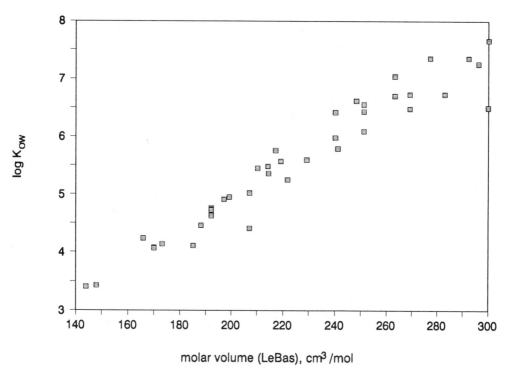

Figure 2.4   Plot of log K_{OW} versus LeBas molar volume.

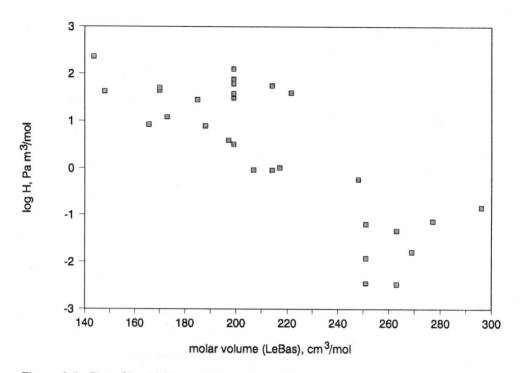

Figure 2.5 Plot of log H (Henry's law constant) versus molar volume.

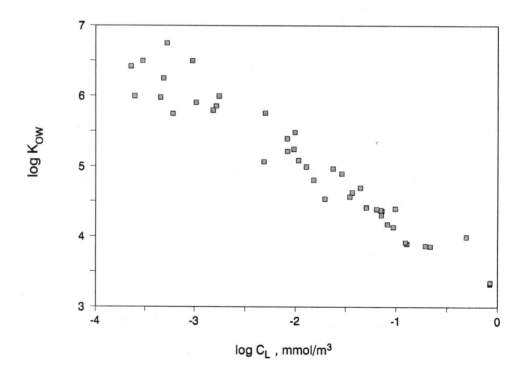

Figure 2.6 Plot of log K_{OW} versus log C_L (liquid solubility).

2.3 Illustrative Fugacity Calculations: Level I, II and III

Chemical name: Indan

Level I calculation: (six compartment model)

100000 kg

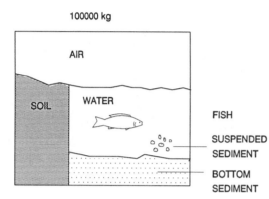

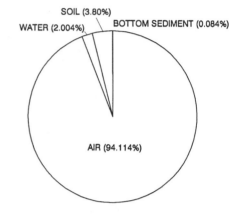

Distribution of mass

physical-chemical properties:

MW: 118.18 g/mol

M.P.: -51°C

Fugacity ratio: 1.0

vapor pressure: 197 Pa

solubility: 100 g/m^3

log K_{OW} : 3.33

Compartment	Z	Concentration			Amount	Amount
	mol/m3 Pa	mol/m3	mg/L (or g/m3)	ug/g	kg	%
Air	4.034E-04	7.964E-09	9.411E-07	7.939E-04	94114	94.114
Water	4.295E-03	8.479E-08	1.002E-05	1.002E-05	2004.1	2.004
Soil	1.807E-01	3.568E-06	4.216E-04	1.757E-04	3794.5	3.795
Biota (fish)	4.592E-01	9.064E-06	1.071E-03	1.071E-03	0.2142	2.14E-04
Suspended sediment	1.130E+00	2.230E-05	2.635E-03	1.757E-03	2.635	2.64E-03
Bottom sediment	3.614E-01	7.135E-06	8.432E-04	3.513E-04	84.323	0.0843
	Total				100000	100

f = 1.974E-05 Pa

258

Chemical name: Indan

Level II calculation: (six compartment model)

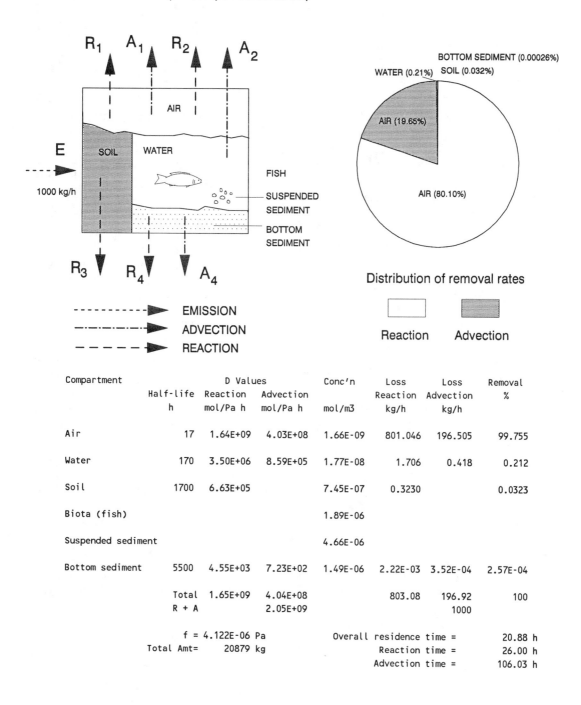

Distribution of removal rates

Reaction   Advection

Compartment	Half-life h	D Values Reaction mol/Pa h	Advection mol/Pa h	Conc'n mol/m3	Loss Reaction kg/h	Loss Advection kg/h	Removal %
Air	17	1.64E+09	4.03E+08	1.66E-09	801.046	196.505	99.755
Water	170	3.50E+06	8.59E+05	1.77E-08	1.706	0.418	0.212
Soil	1700	6.63E+05		7.45E-07	0.3230		0.0323
Biota (fish)				1.89E-06			
Suspended sediment				4.66E-06			
Bottom sediment	5500	4.55E+03	7.23E+02	1.49E-06	2.22E-03	3.52E-04	2.57E-04
	Total	1.65E+09	4.04E+08		803.08	196.92	100
	R + A		2.05E+09			1000	

$f = 4.122E-06$ Pa
Total Amt= 20879 kg

Overall residence time = 20.88 h
Reaction time = 26.00 h
Advection time = 106.03 h

259

Fugacity Level III calculations: (four compartment model)

Chemical name: Indan

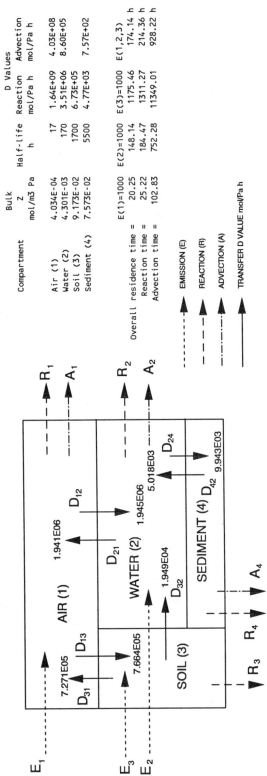

Phase Properties and Rates:

Compartment	Bulk Z mol/m3 Pa	Half-life h	D Values Reaction mol/Pa h	Advection mol/Pa h
Air (1)	4.034E-04	17	1.64E+09	4.03E+08
Water (2)	4.301E-03	170	3.51E+06	8.60E+05
Soil (3)	9.173E-02	1700	6.73E+05	
Sediment (4)	7.573E-02	5500	4.77E+03	7.57E+02

	E(1)=1000	E(2)=1000	E(3)=1000	E(1,2,3)
Overall residence time =	20.25	148.14	1175.46	174.14 h
Reaction time =	25.22	184.47	1311.27	214.36 h
Advection time =	102.83	752.28	11349.01	928.22 h

EMISSION (E)
REACTION (R)
ADVECTION (A)
TRANSFER D VALUE mol/Pa h

Phase Properties, Compositions, Transport and Transformation Rates:

Emission, kg/h

E(1)	E(2)	E(3)		f(1)	f(2)	f(3)	f(4)
1000	0	0		4.128E-06	1.279E-06	2.229E-06	1.206E-06
0	1000	0		1.269E-06	1.341E-03	6.852E-07	1.264E-03
0	0	1000		2.132E-06	1.906E-05	5.961E-03	1.797E-04
600	300	100		3.071E-06	4.049E-04	5.977E-04	3.817E-04

Fugacity, Pa

Concentration, g/m3

	C(1)	C(2)	C(3)	C(4)
	1.968E-07	6.501E-07	2.416E-05	1.079E-05
	6.051E-08	6.815E-04	7.428E-06	1.131E-02
	1.016E-07	9.690E-06	6.463E-02	1.608E-04
	1.464E-07	2.058E-07	6.479E-03	3.416E-03

Amounts, kg

m(1)	m(2)	m(3)	m(4)
1.968E+04	1.300E+02	4.349E+02	5.395E+00
6.051E+03	1.363E+05	1.337E+02	5.656E+03
1.016E+04	1.938E+03	1.163E+06	8.042E+01
1.464E+04	4.116E+04	1.166E+05	1.708E+03

Total Amount, kg

2.025E+04
1.481E+05
1.175E+06
1.741E+05

Emission, kg/h

E(1)	E(2)	E(3)		R(1)	R(2)	R(3)	R(4)
1000	0	0		8.023E+02	5.301E-01	1.77E-01	6.798E-04
0	1000	0		2.467E+02	5.556E+02	5.45E-02	7.126E-01
0	0	1000		4.143E+02	7.900E+00	4.74E+02	1.013E-02
600	300	100		5.968E+02	1.678E+02	4.75E+01	2.152E-01

Loss, Reaction, kg/h

Loss, Advection, kg/h

A(1)	A(2)	A(3)	A(4)
1.968E+02	1.300E-01		1.079E-04
6.051E+01	1.363E+02		1.131E-01
1.016E+02	1.938E+00		1.608E-02
1.464E+02	4.116E+02		3.416E-01

Intermedia Rate of Transport, kg/h

	T12	T13	T21	T31	T32	T24	T42
	air-water	air-soil	water-air	soil-air	soil-water	water-sed	sed-water
	9.491E-01	3.739E-01	2.934E-01	1.915E-01	5.134E-03	1.503E-03	7.151E-04
	2.918E-01	1.150E-01	3.075E+02	5.888E-02	1.578E-01	1.575E+00	7.496E-01
	4.901E-01	1.931E-01	4.372E+00	5.123E+02	1.373E+01	2.240E-02	1.066E-02
	7.060E-01	2.782E-01	9.287E+01	5.136E+01	1.377E+01	4.757E-01	2.264E-01

Level III Distribution

Chemical name: Indan

Distribution of mass

Distribution of removal rates

Emission rates:

E(1) = 1000 kg/h

E(2) = 0

E(3) = 0

Residence time:

t = 20.25 h

E(1) = 0

E(2) = 1000 kg/h

E(3) = 0

t = 148 h

E(1) = 0

E(2) = 0

E(3) = 1000 kg/h

t = 1175.5 h

E(1) = 600 kg/h

E(2) = 300 kg/h

E(3) = 100 kg/h

t = 174 h

261

Chemical name: Naphthalene

Level I calculation: (six compartment model)

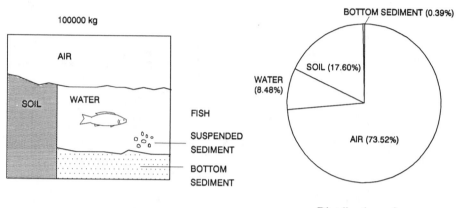

Distribution of mass

physical-chemical properties:

MW: 128.18

M.P.: 80.5°C

Fugacity ratio: 0.284

vapor pressure: 10.4 Pa

solubility: 31 g/m^3

log K_{OW} : 3.37

Compartment	Z	Concentration			Amount	Amount
	mol/m3 Pa	mol/m3	mg/L (or g/m3)	ug/g	kg	%
Air	4.034E-04	5.736E-09	7.352E-07	6.202E-04	73524	73.524
Water	2.325E-02	3.306E-07	4.238E-05	4.238E-05	8475.7	8.476
Soil	1.073E+00	1.525E-05	1.955E-03	8.146E-04	17596.1	17.596
Biota (fish)	2.725E+00	3.875E-05	4.967E-03	4.967E-03	0.9935	9.93E-04
Suspended sediment	6.705E+00	9.532E-05	1.222E-02	8.146E-03	12.219	1.22E-02
Bottom sediment	2.146E+00	3.050E-05	3.910E-03	1.629E-03	391.024	0.3910
	Total				100000	100

f = 1.422E-05 Pa

262

Chemical name: Naphthalene

Level II calculation: (six compartment model)

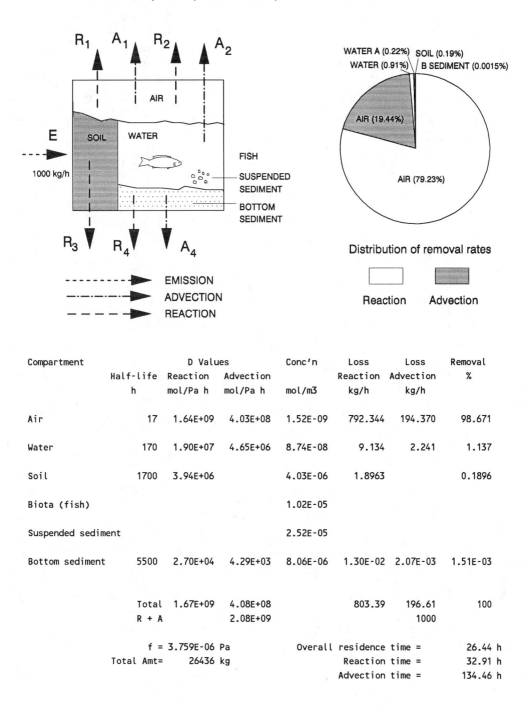

Compartment	Half-life h	D Values Reaction mol/Pa h	D Values Advection mol/Pa h	Conc'n mol/m3	Loss Reaction kg/h	Loss Advection kg/h	Removal %
Air	17	1.64E+09	4.03E+08	1.52E-09	792.344	194.370	98.671
Water	170	1.90E+07	4.65E+06	8.74E-08	9.134	2.241	1.137
Soil	1700	3.94E+06		4.03E-06	1.8963		0.1896
Biota (fish)				1.02E-05			
Suspended sediment				2.52E-05			
Bottom sediment	5500	2.70E+04	4.29E+03	8.06E-06	1.30E-02	2.07E-03	1.51E-03
	Total	1.67E+09	4.08E+08		803.39	196.61	100
	R + A		2.08E+09			1000	

f = 3.759E-06 Pa

Total Amt= 26436 kg

Overall residence time = 26.44 h

Reaction time = 32.91 h

Advection time = 134.46 h

Fugacity Level III calculations: (four compartment model)

Chemical name: Naphthalene

Phase Properties and Rates:

Compartment	Bulk Z mol/m3 Pa	Half-life h	D Values Reaction mol/Pa h	Advection mol/Pa h
Air (1)	4.034E-04	17	1.64E+09	4.03E+08
Water (2)	2.329E-02	170	1.90E+07	4.66E+06
Soil (3)	5.434E-01	1700	3.99E-06	
Sediment (4)	4.477E-01	5500	2.82E+04	4.48E+03

	E(1)=1000	E(2)=1000	E(3)=1000	E(1,2,3)
Overall residence time =	21.15	161.82	2029.01	264.13 h
Reaction time =	26.33	201.50	2101.70	322.38 h
Advection time =	107.38	821.81	58664.77	1461.91 h

EMISSION (E)

REACTION (R)

ADVECTION (A)

TRANSFER D VALUE mol/Pa h

Amounts, kg

m(1)	m(2)	m(3)	m(4)	Total Amount, kg
1.964E-04	5.418E+02	9.419E+02	2.455E+01	2.115E+04
4.664E+03	1.501E+05	2.237E+02	6.802E+03	1.618E+05
3.122E+03	3.361E+02	2.022E+06	1.523E+02	2.029E+06
1.349E+04	4.570E+04	2.029E+05	2.071E+03	2.641E+05

Intermedia Rate of Transport, kg/h

air-water T12	water-air T21	air-soil T13	soil-air T31	soil-water T32	water-sed T24	sed-water T42
3.602E+00	8.578E-01	4.658E-01	7.164E-02	1.017E-02	6.604E-03	3.020E-03
8.554E-01	2.377E-02	1.106E-01	1.702E-02	2.415E-03	1.830E+00	8.369E-01
5.726E-01	5.322E-01	7.405E-02	1.538E+00	2.183E+01	4.097E-02	1.874E-02
2.475E+00	7.236E+01	3.200E-01	1.543E+01	2.190E+00	5.571E-01	2.548E-01

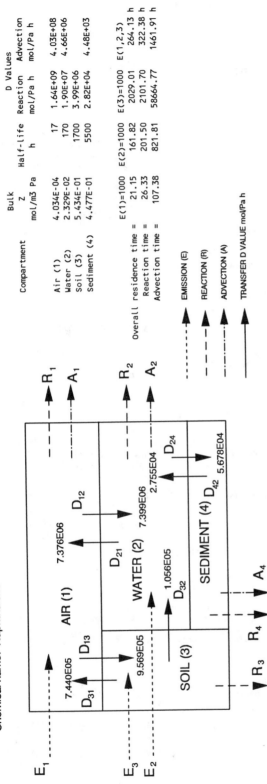

E_1, E_2, E_3, R_1, R_2, R_3, R_4, A_1, A_2, A_4

AIR (1)   WATER (2)   SOIL (3)   SEDIMENT (4)

D_{12} 7.376E06 D_{21} D_{13} 7.440E05 D_{31} 9.569E05 D_{32} 1.056E05 D_{24} 7.399E06   2.755E04 D_{42} 5.678E04

Phase Properties, Compositions, Transport and Transformation Rates:

Emission, kg/h

E(1)	E(2)	E(3)
1000	0	0
0	1000	0
0	0	1000
600	300	100

Fugacity, Pa

f(1)	f(2)	f(3)	f(4)
3.797E-06	9.074E-07	7.511E-07	8.554E-07
9.019E-07	2.514E-04	1.784E-07	2.370E-04
6.038E-07	5.629E-06	1.613E-03	5.307E-06
2.609E-06	7.654E-05	1.618E-04	7.216E-05

Concentration, g/m3

C(1)	C(2)	C(3)	C(4)
1.964E-07	2.709E-06	5.233E-05	4.909E-05
4.664E-08	7.507E-04	1.243E-05	1.360E-02
3.122E-08	1.680E-05	1.124E-01	3.045E-04
1.349E-07	2.285E-04	1.127E-02	4.141E-03

Emission, kg/h

E(1)	E(2)	E(3)
1000	0	0
0	1000	0
0	0	1000
600	300	100

Loss, Reaction, kg/h

R(1)	R(2)	R(3)	R(4)
8.005E+00	2.208E+00	3.84E-01	3.093E-03
1.901E+02	6.120E+02	9.12E-02	8.571E-01
1.273E+02	1.370E+01	8.24E+02	1.919E-02
5.501E+02	1.863E+02	8.27E+02	2.609E-01

Loss, Advection, kg/h

A(1)	A(2)	A(4)
1.964E+02	5.418E-01	4.909E-01
4.664E+01	1.501E+02	1.360E-01
3.122E+01	3.361E+00	3.045E-03
1.349E+02	4.570E+00	4.141E-01

Level III Distribution

Chemical name: Naphthalene

Distribution of mass

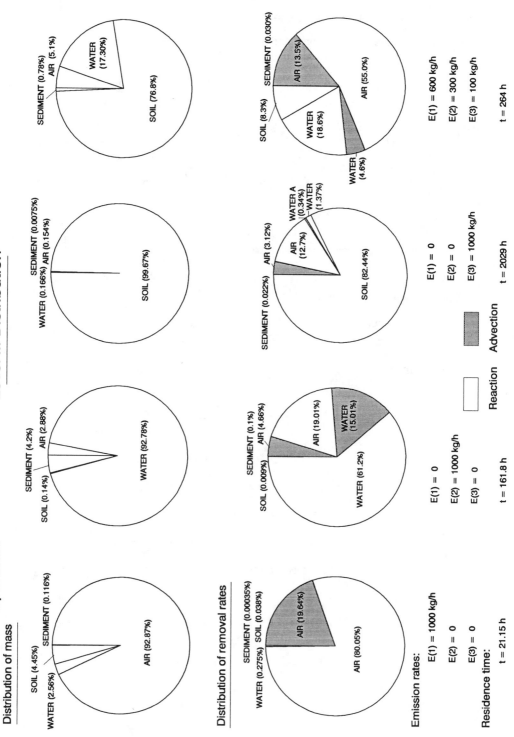

Distribution of removal rates

Emission rates:

E(1) = 1000 kg/h	E(1) = 0	E(1) = 0
E(2) = 0	E(2) = 1000 kg/h	E(2) = 0
E(3) = 0	E(3) = 0	E(3) = 1000 kg/h

Residence time:

t = 21.15 h   t = 161.8 h   t = 2029 h

E(1) = 600 kg/h

E(2) = 300 kg/h

E(3) = 100 kg/h

t = 264 h

Reaction   Advection

265

Chemical name: 1-Methylnaphthalene

Level I calculation: (six compartment model)

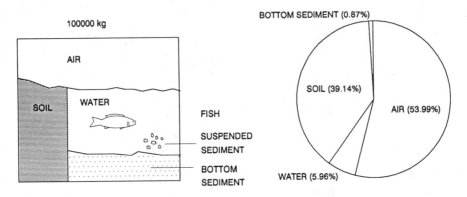

Distribution of mass

physical-chemical properties:

MW: 142.2 g/mol

M.P.: -22 °C

Fugacity ratio: 1.0

vapor pressure: 8.84 Pa

solubility: 28 g/m^3

log K_{OW} : 3.87

Compartment	Z mol/m3 Pa	Concentration mol/m3	mg/L (or g/m3)	ug/g	Amount kg	Amount %
Air	4.034E-04	3.797E-09	5.399E-07	4.555E-04	53994	53.994
Water	2.227E-02	2.097E-07	2.981E-05	2.981E-05	5962.5	5.962
Soil	3.250E+00	3.059E-05	4.349E-03	1.812E-03	39144.0	39.144
Biota (fish)	8.256E+00	7.771E-05	1.105E-02	1.105E-02	2.2100	2.21E-03
Suspended sediment	2.031E+01	1.912E-04	2.718E-02	1.812E-02	27.183	2.72E-02
Bottom sediment	6.499E+00	6.117E-05	8.699E-03	3.624E-03	869.868	0.8699
Total					100000	100

f = 9.412E-06 Pa

Chemical name: 1-Methylnaphthalene

Level II calculation: (six compartment model)

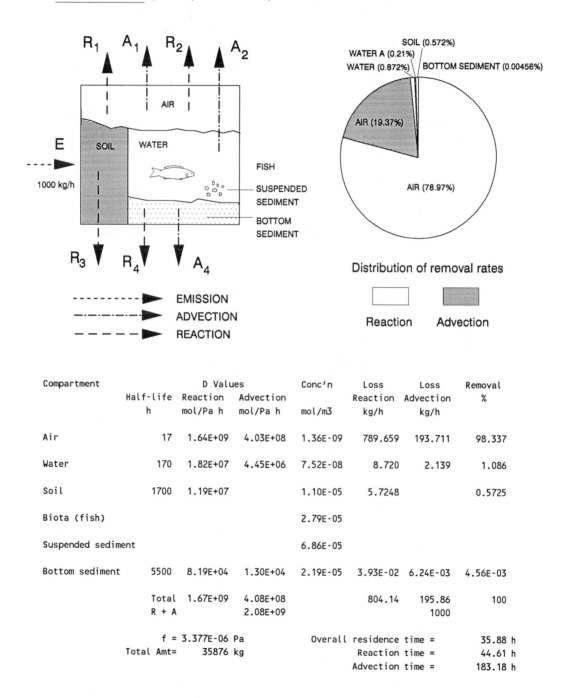

Distribution of removal rates

Reaction   Advection

Compartment	Half-life h	D Values Reaction mol/Pa h	Advection mol/Pa h	Conc'n mol/m3	Loss Reaction kg/h	Loss Advection kg/h	Removal %
Air	17	1.64E+09	4.03E+08	1.36E-09	789.659	193.711	98.337
Water	170	1.82E+07	4.45E+06	7.52E-08	8.720	2.139	1.086
Soil	1700	1.19E+07		1.10E-05	5.7248		0.5725
Biota (fish)				2.79E-05			
Suspended sediment				6.86E-05			
Bottom sediment	5500	8.19E+04	1.30E+04	2.19E-05	3.93E-02	6.24E-03	4.56E-03
	Total	1.67E+09	4.08E+08		804.14	195.86	100
	R + A		2.08E+09			1000	

f = 3.377E-06 Pa
Total Amt= 35876 kg

Overall residence time =   35.88 h
Reaction time =   44.61 h
Advection time =   183.18 h

Fugacity Level III calculations: (four compartment model)

Chemical name: 1-Methylnaphthalene

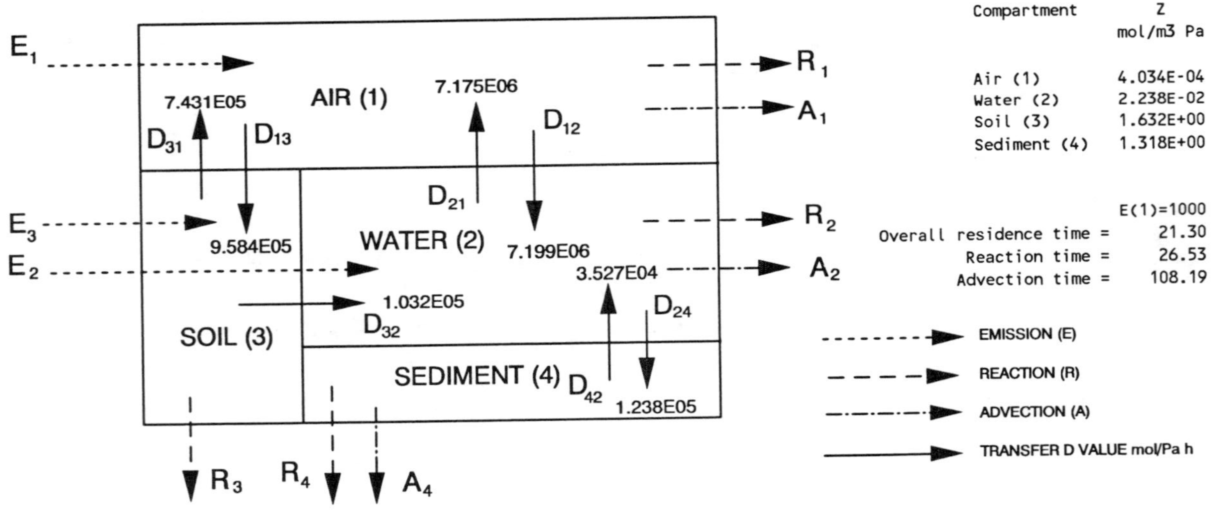

Diagram labels (four-compartment model):

- AIR (1), WATER (2), SOIL (3), SEDIMENT (4)
- Emissions: E_1, E_2, E_3
- Reactions: R_1, R_2, R_3, R_4
- Advections: A_1, A_2, A_4
- D_{12} = 7.199E06
- D_{21} = 7.175E06
- D_{13} = 9.584E05
- D_{31} = 7.431E05
- D_{32} = 1.032E05
- D_{24} = 3.527E04
- D_{42} = 1.238E05

Legend: EMISSION (E); REACTION (R); ADVECTION (A); TRANSFER D VALUE mol/Pa h

Phase Properties and Rates:

Compartment	Bulk Z mol/m3 Pa	Half-life h	D Values Reaction mol/Pa h	Advection mol/Pa h
Air (1)	4.034E-04	17	1.64E+09	4.03E+08
Water (2)	2.238E-02	170	1.82E+07	4.48E+06
Soil (3)	1.632E+00	1700	1.20E+07	
Sediment (4)	1.318E+00	5500	8.30E+04	1.32E+04

	E(1)=1000	E(2)=1000	E(3)=1000	E(1,2,3)	Total Amount, kg
Overall residence time =	21.30	175.05	2293.78	294.68 h	2.130E+04
Reaction time =	26.53	217.94	2323.99	358.69 h	1.751E+05
Advection time =	108.19	889.56	176455.73	1651.07 h	2.294E+06
					2.947E+05

Phase Properties, Compositions, Transport and Transformation Rates:

Emission, kg/h — Fugacity, Pa — Concentration, g/m3

E(1)	E(2)	E(3)	f(1)	f(2)	f(3)	f(4)	C(1)	C(2)	C(3)	C(4)
1000	0	0	3.423E-06	8.226E-07	2.559E-07	7.748E-07	1.964E-07	2.618E-06	5.938E-05	1.452E-04
0	1000	0	8.190E-07	7.470E-04	6.123E-08	2.210E-04	4.698E-08	7.470E-04	1.421E-05	4.142E-02
0	0	1000	2.051E-07	1.936E-06	5.486E-04	1.824E-06	1.176E-08	6.163E-06	1.273E-01	3.417E-04
600	300	100	2.320E-06	7.109E-05	5.504E-05	6.696E-05	1.331E-07	2.263E-04	1.277E-02	1.255E-02

Amounts, kg

m(1)	m(2)	m(3)	m(4)
1.964E+04	5.236E+02	1.069E+03	7.259E+01
4.698E+03	1.494E+05	2.557E+02	2.071E+04
1.176E+03	1.233E+03	2.291E+06	1.709E+02
1.331E+04	4.525E+04	2.298E+05	6.273E+05

Emission, kg/h — Loss, Reaction, kg/h — Loss, Advection, kg/h

E(1)	E(2)	E(3)	R(1)	R(2)	R(3)	R(4)	A(1)	A(2)	A(4)
1000	0	0	8.005E+02	2.135E+00	4.36E-01	9.146E-03	1.964E+02	5.236E-01	1.452E-03
0	1000	0	1.915E+02	6.090E+02	1.04E-01	2.609E+00	4.698E+01	1.494E+02	4.142E-01
0	0	1000	4.795E+01	5.025E+00	9.34E+02	2.153E-02	1.176E+01	1.233E+00	3.417E-03
600	300	100	5.426E+02	1.845E+02	9.37E+01	7.904E-01	1.331E+02	4.525E-01	1.255E-01

Intermedia Rate of Transport, kg/h

T12 air-water	T21 water-air	T13 air-soil	T24 water-sed	T31 soil-air	T32 soil-water	T42 sed-water
3.504E+00	8.393E-01	4.665E-01	1.448E-02	2.705E-02	3.755E-03	3.886E-03
8.384E-01	2.394E+02	1.116E-01	4.132E+00	6.471E-03	8.982E-04	1.109E+00
2.099E-01	1.976E+00	2.795E-02	3.409E-02	5.798E+01	8.048E+00	9.148E-03
2.375E+00	7.253E+01	3.162E-01	1.252E+00	5.816E+00	8.073E-01	3.358E-01

Level III Distribution

Chemical name: 1-Methylnaphthalene

Distribution of mass

Distribution of removal rates

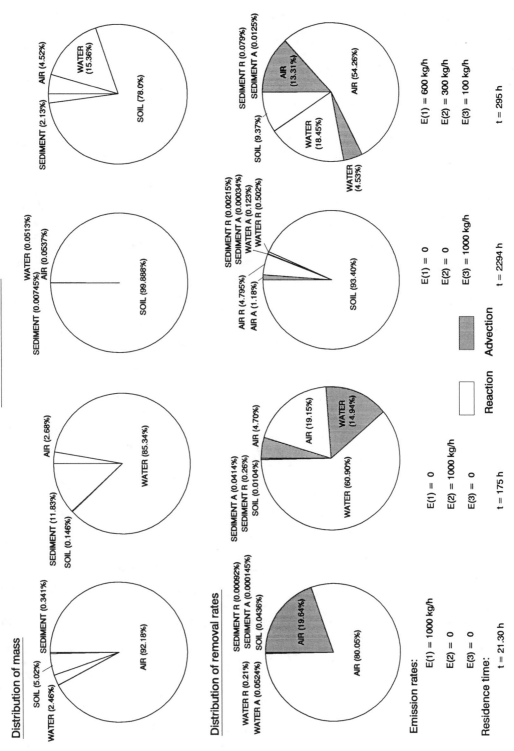

269

Chemical name: 2,3-Dimethylnaphthalene

Level I calculation: (six compartment model)

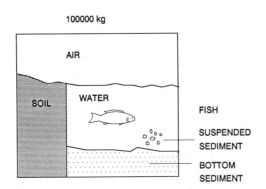

100000 kg

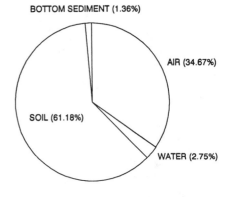

BOTTOM SEDIMENT (1.36%)

AIR (34.67%)

SOIL (61.18%)

WATER (2.75%)

Distribution of mass

physical-chemical properties:

MW: 156.23 g/mol

M.P.: 105 °C

Fugacity ratio: 0.162

vapor pressure: 1.0 Pa

solubility: 2.50 g/m^3

log K_{OW} : 3.87

Compartment	Z	Concentration			Amount	Amount
	mol/m3 Pa	mol/m3	mg/L (or g/m3)	ug/g	kg	%
Air	4.034E-04	2.219E-09	3.467E-07	2.924E-04	34666	34.666
Water	1.600E-02	8.802E-08	1.375E-05	1.375E-05	2750.2	2.750
Soil	7.910E+00	4.351E-05	6.798E-03	2.832E-03	61178	61.178
Biota (fish)	2.010E+01	1.105E-04	1.727E-02	1.727E-02	3.4540	3.45E-03
Suspended sediment	4.944E+01	2.719E-04	4.248E-02	2.832E-02	42.485	4.25E-02
Bottom sediment	1.582E+01	8.702E-05	1.360E-02	5.665E-03	1359.51	1.3595
	Total				100000	100

f = 5.500E-06 Pa

Chemical name: 2,3-Dimethylnaphthalene

Level II calculation: (six compartment model)

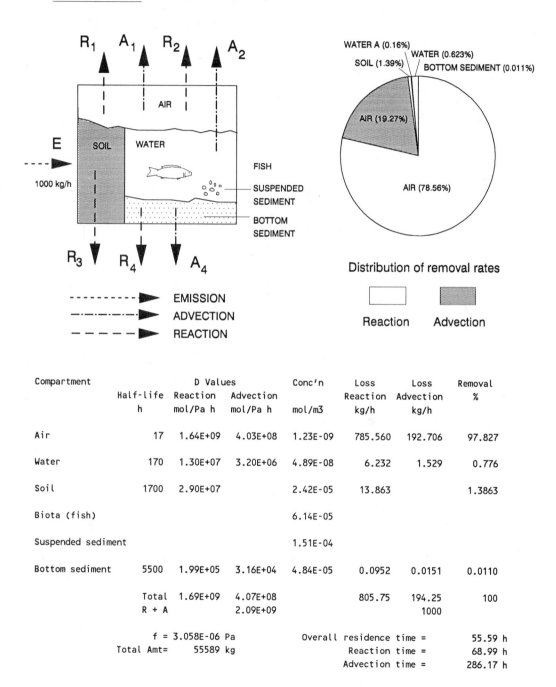

Distribution of removal rates

Reaction   Advection

Compartment	Half-life h	D Values Reaction mol/Pa h	Advection mol/Pa h	Conc'n mol/m3	Loss Reaction kg/h	Loss Advection kg/h	Removal %
Air	17	1.64E+09	4.03E+08	1.23E-09	785.560	192.706	97.827
Water	170	1.30E+07	3.20E+06	4.89E-08	6.232	1.529	0.776
Soil	1700	2.90E+07		2.42E-05	13.863		1.3863
Biota (fish)				6.14E-05			
Suspended sediment				1.51E-04			
Bottom sediment	5500	1.99E+05	3.16E+04	4.84E-05	0.0952	0.0151	0.0110
	Total	1.69E+09	4.07E+08		805.75	194.25	100
	R + A		2.09E+09			1000	

f = 3.058E-06 Pa
Total Amt=   55589 kg

Overall residence time =   55.59 h
Reaction time =   68.99 h
Advection time =   286.17 h

271

Fugacity Level III calculations: (four compartment model)

Chemical name: 2,3-Dimethylnaphthalene

Compartment flow diagram showing AIR (1), WATER (2), SOIL (3), and SEDIMENT (4) with emission (E_1, E_2, E_3), reaction (R_1, R_2, R_3, R_4), advection (A_1, A_2, A_4), and transfer D-value arrows:
D_{12} 5.747E06, D_{21}, D_{13}, D_{31} 7.375E05, D_{31} 9.027E05, D_{32} 7.913E04, D_{24} 4.764E04, D_{42} 2.632E05, 5.729E06.

Legend: EMISSION (E); REACTION (R); ADVECTION (A); TRANSFER D VALUE mol/Pa h

Phase Properties and Rates:

Compartment	Bulk Z mol/m3 Pa	Half-life h	Reaction mol/Pa h	Advection mol/Pa h
Air (1)	4.034E-04	17	1.64E+09	4.03E+08
Water (2)	1.627E-02	170	1.33E+07	3.25E+06
Soil (3)	3.960E+00	1700	2.91E+07	
Sediment (4)	3.177E+00	5500	2.00E+05	3.18E+04

	E(1)=1000	E(2)=1000	E(3)=1000	E(1,2,3)
Overall residence time =	21.29	216.84	2387.14	316.54 h
Reaction time =	26.51	269.83	2400.06	384.90 h
Advection time =	108.13	1104.18	443600.96	1782.39 h

272

Phase Properties, Compositions, Transport and Transformation Rates:

Emission, kg/h

E(1)	E(2)	E(3)
1000	0	0
0	1000	0
0	0	1000
600	300	100

Fugacity, Pa

f(1)	f(2)	f(3)	f(4)
3.118E-06	7.979E-07	9.420E-08	7.512E-07
7.950E-07	2.851E-04	2.402E-08	2.684E-04
7.907E-08	7.749E-07	2.143E-04	7.296E-07
2.117E-06	8.609E-05	2.149E-05	8.106E-05

Concentration, g/m3

C(1)	C(2)	C(3)	C(4)
1.965E-07	2.028E-06	5.828E-05	3.728E-04
5.011E-04	7.247E-04	1.486E-05	1.332E-01
4.984E-09	1.970E-06	1.326E-01	3.621E-04
1.334E-07	2.188E-04	1.330E-02	4.023E-02

Amounts, kg

m(1)	m(2)	m(3)	m(4)
1.965E+04	4.056E+02	1.049E+03	1.864E+04
5.011E+03	1.449E+05	2.675E+02	6.662E+04
4.984E+02	3.939E+02	2.386E+06	1.811E+02
1.334E+04	4.377E+04	2.393E+05	2.012E+04

Total Amount, kg

Total Amount, kg
2.129E+04
2.168E+05
2.387E+06
3.165E+05

Emission, kg/h

E(1)	E(2)	E(3)
1000	0	0
0	1000	0
0	0	1000
600	300	100

Loss, Reaction, kg/h

R(1)	R(2)	R(3)	R(4)
8.010E+02	1.653E+00	4.28E-01	2.349E-02
2.043E+02	5.909E+02	1.09E-01	8.394E+00
2.032E+01	1.606E+00	9.73E+02	2.281E-02
5.439E+02	1.784E+02	9.76E+01	2.535E+00

Loss, Advection, kg/h

A(1)	A(2)	C(3)	A(4)
1.965E+02	4.056E-01	3.728E-02	3.728E-03
5.011E+01	1.449E+02	1.332E+00	1.332E+00
4.984E+00	3.939E-01	2.281E-01	3.621E-01
1.334E+02	4.377E+01	9.76E-01	4.023E+00

Intermedia Rate of Transport, kg/h

T12 air-water	T21 water-air	T13 air-soil	T31 soil-air	T32 soil-water	T24 water-sed	T42 sed-water
2.799E+00	7.141E-01	4.397E-01	1.085E-02	1.165E-02	3.281E-02	5.591E-03
7.138E-01	2.552E+02	1.121E+01	2.768E-04	2.970E-04	1.172E+01	1.998E+00
7.100E-02	6.935E-01	1.115E-02	2.469E+01	2.649E+00	3.186E-02	5.430E-03
1.901E+00	7.705E+01	2.986E-01	2.476E+00	2.657E-01	3.540E-01	6.033E-01

Level III Distribution

Chemical name: 2,3-Dimethylnaphthalene

Distribution of mass

- SOIL (4.93%)
- SEDIMENT (0.876%)
- WATER (1.905%)
- AIR (92.29%)

- AIR (2.31%)
- SEDIMENT (30.7%)
- WATER (66.84%)
- SOIL (0.123%)
- WATER (1.905%)

- WATER (0.017%)
- AIR (0.021%)
- SEDIMENT (0.0076%)
- SOIL (99.955%)

- AIR (4.22%)
- WATER (13.83%)
- SEDIMENT (6.35%)
- SOIL (75.6%)

Distribution of removal rates

- SEDIMENT A (0.000377%)
- SEDIMENT R (0.0024%)
- SOIL (0.00037%)
- WATER R (0.165%)
- WATER A (0.041%)
- AIR R (80.1%)
- AIR (19.65%)

- SEDIMENT A (0.133%)
- SEDIMENT R (0.84%)
- AIR (5.01%)
- SOIL (0.0011%)
- AIR (20.43%)
- WATER (14.49%)
- WATER (59.086%)

- SEDIMENT A (0.023%)
- SEDIMENT R (0.0023%)
- WATER A (0.039%)
- WATER R (0.161%)
- AIR R (2.032%)
- AIR A (0.498%)
- SOIL (97.267%)

- SEDIMENT R (0.253%)
- SEDIMENT A (0.0402%)
- AIR (13.3%)
- AIR (54.4%)
- SOIL (9.76%)
- WATER (17.8%)
- WATER (4.38%)

Reaction | Advection

Emission rates:

E(1) = 1000 kg/h	E(1) = 0	E(1) = 0	E(1) = 600 kg/h
E(2) = 0	E(2) = 1000 kg/h	E(2) = 0	E(2) = 300 kg/h
E(3) = 0	E(3) = 0	E(3) = 1000 kg/h	E(3) = 100 kg/h

Residence time:

t = 21.29 h	t = 217 h	t = 2387 h	t = 316.5 h

Chemical name: 1-Ethylnaphthalene

Level I calculation: (six compartment model)

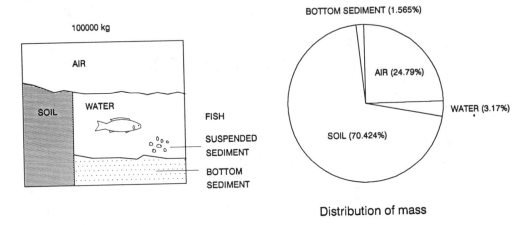

Distribution of mass

physical-chemical properties:

MW: 156.23 g/mol

M.P.: -13.8 °C

Fugacity ratio: 1.0

vapor pressure: 2.51 Pa

solubility: 10.1 g/m^3

log K_{OW} : 4.40

Compartment	Z mol/m3 Pa	Concentration mol/m3	mg/L (or g/m3)	ug/g	Amount kg	Amount %
Air	4.034E-04	1.587E-09	2.479E-07	2.091E-04	24793	24.793
Water	2.576E-02	1.013E-07	1.583E-05	1.583E-05	3165.8	3.166
Soil	1.273E+01	5.009E-05	7.825E-03	3.260E-03	70424	70.424
Biota (fish)	3.235E+01	1.272E-04	1.988E-02	1.988E-02	3.9760	3.98E-03
Suspended sediment	7.958E+01	3.130E-04	4.891E-02	3.260E-02	48.905	4.89E-02
Bottom sediment	2.546E+01	1.002E-04	1.565E-02	6.521E-03	1564.97	1.5650
Total					100000	100

f = 3.934E-06 Pa

Chemical name: 1-Ehtylnaphthalene

Level II calculation: (six compartment model)

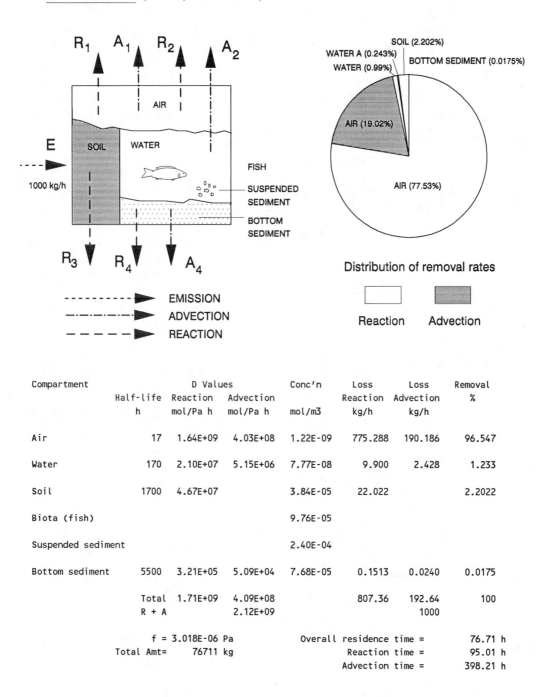

Distribution of removal rates

Reaction   Advection

Compartment	Half-life h	D Values Reaction mol/Pa h	D Values Advection mol/Pa h	Conc'n mol/m3	Loss Reaction kg/h	Loss Advection kg/h	Removal %
Air	17	1.64E+09	4.03E+08	1.22E-09	775.288	190.186	96.547
Water	170	2.10E+07	5.15E+06	7.77E-08	9.900	2.428	1.233
Soil	1700	4.67E+07		3.84E-05	22.022		2.2022
Biota (fish)				9.76E-05			
Suspended sediment				2.40E-04			
Bottom sediment	5500	3.21E+05	5.09E+04	7.68E-05	0.1513	0.0240	0.0175
Total		1.71E+09	4.09E+08		807.36	192.64	100
R + A			2.12E+09			1000	

f = 3.018E-06 Pa

Total Amt= 76711 kg

Overall residence time = 76.71 h
Reaction time = 95.01 h
Advection time = 398.21 h

Fugacity Level III calculations: (four compartment model)

Chemical name: 1-Ethylnaphthalene

Diagram labels: E₁, E₃, E₂ (emissions); R₁, A₁, R₂, A₂, R₃, R₄, A₄; AIR (1) 7.860E06; WATER (2) 7.891E06; SOIL (3); SEDIMENT (4); D₃₁ 1.030E06; D₁₃ 7.463E05; D₁₂; D₂₁; D₃₂ 1.274E05; D₂₄ 7.669E04; D₄₂ 4.236E05.

Legend:
- EMISSION (E)
- REACTION (R)
- ADVECTION (A)
- TRANSFER D VALUE mol/Pa h

Phase Properties and Rates:

Compartment	Bulk Z mol/m3 Pa	Half-life h	D Values Reaction mol/Pa h	Advection mol/Pa h
Air (1)	4.034E-04	17	1.64E+09	4.03E+08
Water (2)	2.619E-02	170	2.13E+07	5.24E+06
Soil (3)	6.374E+00	1700	4.68E+07	
Sediment (4)	5.114E+00	5500	3.22E+05	5.11E+04

	E(1)=1000	E(2)=1000	E(3)=1000	E(1,2,3)
Overall residence time =	21.68	224.58	2409.06	321.29 h
Reaction time =	27.00	279.45	2417.79	390.58 h
Advection time =	110.12	1143.70	667550	1811.08 h

Phase Properties, Compositions, Transport and Transformation Rates:

Emission, kg/h — Fugacity, Pa

E(1)	E(2)	E(3)	f(1)	f(2)	f(3)	f(4)
1000	0	0	3.115E-06	7.065E-07	6.734E-08	6.652E-07
0	1000	0	7.035E-07	1.841E-04	1.521E-08	1.733E-06
0	0	1000	5.066E-08	5.032E-07	1.343E-04	4.738E-07
600	300	100	2.085E-06	5.570E-05	1.348E-05	5.244E-05

Concentration, g/m3

C(1)	C(2)	C(3)	C(4)
1.963E-07	2.891E-06	6.706E-05	5.314E-04
4.434E-08	7.532E-04	1.515E-05	1.385E-01
3.193E-09	2.059E-06	1.338E-01	3.785E-04
1.314E-07	2.279E-04	1.342E-02	4.190E-02

Amounts, kg

m(1)	m(2)	m(3)	m(4)	Total Amount, kg
1.963E+04	5.781E+02	1.207E+03	2.657E+02	2.168E+04
4.434E+03	1.506E+05	2.726E+02	6.923E+04	2.246E+05
3.193E+02	4.117E+02	2.408E+06	1.892E+02	2.409E+06
1.314E+04	4.558E+04	2.416E+05	2.095E+04	3.213E+05

Loss, Reaction, kg/h

E(1)	E(2)	E(3)	R(1)	R(2)	R(3)	R(4)
1000	0	0	8.002E+02	2.357E+00	4.92E-01	3.348E-02
0	1000	0	1.807E+02	6.141E+02	1.11E-01	8.724E+00
0	0	1000	1.302E+01	1.678E+00	9.82E+02	2.384E-02
600	300	100	5.357E+02	1.858E+02	9.85E+01	2.640E+00

Loss, Advection, kg/h

A(1)	A(2)	A(4)
1.963E+02	5.781E+01	5.314E-03
4.434E+01	1.506E+02	1.385E+00
3.193E+00	4.117E-01	3.785E-03
1.314E+02	4.558E+01	4.190E+00

Intermedia Rate of Transport, kg/h

T12 air-water	T21 water-air	T13 air-soil	T31 soil-air	T32 soil-water	T24 water-sed	T42 sed-water
3.840E+00	8.676E-01	5.012E-01	7.851E-03	1.340E-03	4.676E-02	7.969E-03
8.673E-01	2.261E+02	1.132E-01	1.773E-03	3.027E-04	1.218E+01	2.077E+00
6.246E-02	6.179E-01	8.154E-03	1.566E+01	2.673E+00	3.331E-02	5.676E-03
2.570E+00	6.840E+01	3.355E-01	1.572E+00	2.682E-01	3.687E+00	6.283E-01

Level III Distribution

Chemical name: 1-Ethylnaphthalene

Distribution of mass

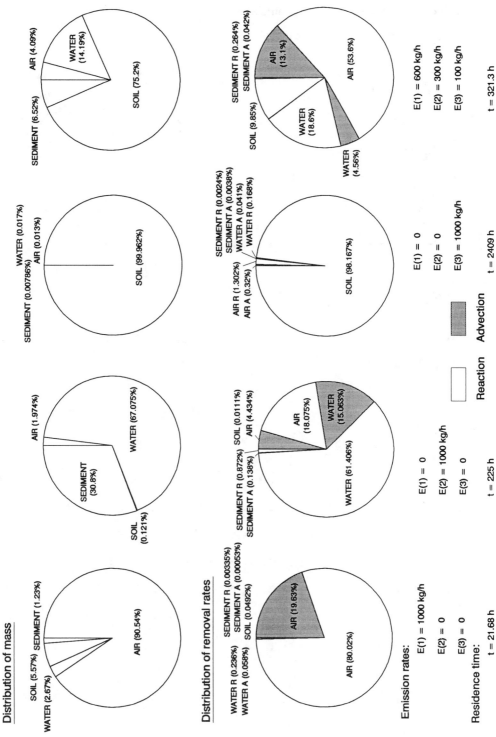

Distribution of removal rates

Emission rates:

E(1) = 1000 kg/h   E(1) = 0   E(1) = 0   E(1) = 600 kg/h

E(2) = 0   E(2) = 1000 kg/h   E(2) = 0   E(2) = 300 kg/h

E(3) = 0   E(3) = 0   E(3) = 1000 kg/h   E(3) = 100 kg/h

Residence time:

t = 21.68 h   t = 225 h   t = 2409 h   t = 321.3 h

☐ Reaction   ▨ Advection

Chemical name: 1,4,5-Trimethylnaphthalene

Level I calculation: (six compartment model)

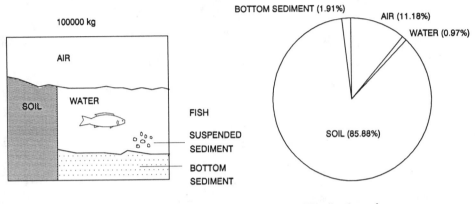

100000 kg

AIR

SOIL WATER

FISH

SUSPENDED SEDIMENT

BOTTOM SEDIMENT

BOTTOM SEDIMENT (1.91%)   AIR (11.18%)

WATER (0.97%)

SOIL (85.88%)

Distribution of mass

physical-chemical properties:

MW: 176.2 g/mol

M.P.: 64.0 °C

Fugacity ratio: 0.4114

vapor pressure: 0.681 Pa

solubility: 2.10 g/m³

log K_{OW} : 5.00

Compartment	Z mol/m3 Pa	Concentration mol/m3	Concentration mg/L (or g/m3)	Concentration ug/g	Amount kg	Amount %
Air	4.034E-04	6.343E-10	1.118E-07	9.429E-05	11177	11.177
Water	1.750E-02	2.752E-08	4.849E-06	4.849E-06	969.7	0.970
Soil	3.444E+01	5.416E-05	9.542E-03	3.976E-03	85881	85.881
Biota (fish)	8.751E+01	1.376E-04	2.424E-02	2.424E-02	4.8487	4.85E-03
Suspended sediment	2.153E+02	3.385E-04	5.964E-02	3.976E-02	59.639	5.96E-02
Bottom sediment	6.888E+01	1.083E-04	1.908E-02	7.952E-03	1908.46	1.9085
Total					100000	100

f = 1.572E-06 Pa

278

Chemical name: 1,4,5-Trimethylnaphthalene

Level II calculation: (six compartment model)

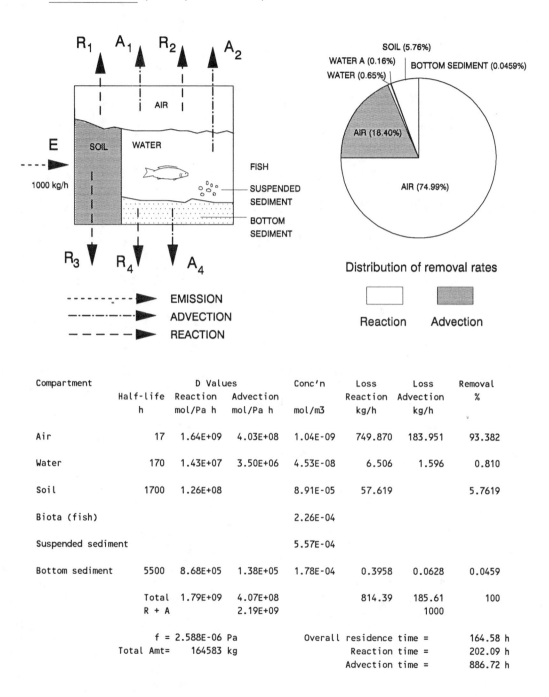

Distribution of removal rates

Reaction   Advection

Compartment	Half-life h	D Values Reaction mol/Pa h	Advection mol/Pa h	Conc'n mol/m3	Loss Reaction kg/h	Loss Advection kg/h	Removal %
Air	17	1.64E+09	4.03E+08	1.04E-09	749.870	183.951	93.382
Water	170	1.43E+07	3.50E+06	4.53E-08	6.506	1.596	0.810
Soil	1700	1.26E+08		8.91E-05	57.619		5.7619
Biota (fish)				2.26E-04			
Suspended sediment				5.57E-04			
Bottom sediment	5500	8.68E+05	1.38E+05	1.78E-04	0.3958	0.0628	0.0459
	Total	1.79E+09	4.07E+08		814.39	185.61	100
	R + A		2.19E+09			1000	

f = 2.588E-06 Pa
Total Amt= 164583 kg

Overall residence time = 164.58 h
Reaction time = 202.09 h
Advection time = 886.72 h

279

Fugacity Level III calculations: (four compartment model)

Chemical name: 1,4,5-Trimethylnaphthalene

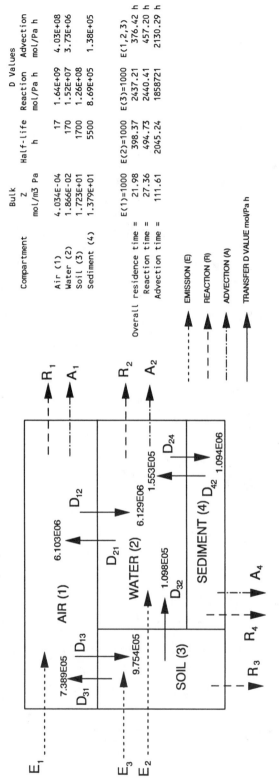

Phase Properties and Rates:

Compartment	Bulk Z mol/m3 Pa	Half-life h	D Values Reaction mol/Pa h	Advection mol/Pa h
Air (1)	4.034E-04	17	1.64E+09	4.03E+08
Water (2)	1.866E-02	170	1.52E+07	3.73E+06
Soil (3)	1.723E+01	1700	1.26E+08	1.38E+05
Sediment (4)	1.379E+01	5500	8.69E+05	

	E(1)=1000	E(2)=1000	E(3)=1000	E(1,2,3)
Overall residence time =	21.98	398.37	2437.21	376.42 h
Reaction time =	27.36	494.73	2440.41	457.20 h
Advection time =	111.61	2045.24	1858721	2130.29 h

EMISSION (E)

REACTION (R)

ADVECTION (A)

TRANSFER D VALUE mol/Pa h

Phase Properties, Compositions, Transport and Transformation Rates:

Emission, kg/h

E(1)	E(2)	E(3)	f(1)	f(2)	f(3)	f(4)
1000	0	0	2.763E-06	6.515E-07	2.118E-06	6.133E-06
0	1000	0	6.486E-07	2.184E-04	4.972E-09	2.056E-04
0	0	1000	1.661E-08	1.922E-07	4.460E-05	1.809E-07
600	300	100	1.854E-06	6.594E-05	6.474E-06	6.207E-05

Concentration, g/m3

C(1)	C(2)	C(3)	C(4)
1.964E-07	2.143E-06	6.429E-05	1.490E-03
4.611E-08	7.184E-04	1.509E-05	4.996E-01
1.180E-09	6.320E-07	1.354E-01	4.396E-04
1.318E-07	2.169E-04	1.358E-02	1.508E-01

Amounts, kg

m(1)	m(2)	m(3)	m(4)
1.964E+04	4.285E+02	1.157E+03	7.451E+02
4.611E+03	1.437E+05	2.716E+02	2.498E+05
1.180E+02	1.264E+02	2.437E+06	2.198E+02
1.318E+04	4.337E+04	2.445E+05	7.541E+04

Total Amount, kg

| 2.198E+04 |
| 3.984E+05 |
| 2.437E+06 |
| 3.764E+05 |

Loss, Reaction, kg/h

R(1)	R(2)	R(3)	R(4)
8.008E+02	1.747E+00	4.72E-01	9.389E-02
1.880E+02	5.857E+02	1.11E-01	3.148E+01
4.812E+00	5.153E-01	9.93E+02	2.769E-02
5.374E+02	1.768E+02	9.96E+01	9.502E+00

Loss, Advection, kg/h

A(1)	A(2)	A(4)
1.964E+02	4.285E-01	1.490E-02
4.611E+01	1.437E+02	4.996E+00
1.180E+00	1.264E-01	4.396E-03
1.318E+02	4.337E+01	1.508E+00

Emission, kg/h

E(1)	E(2)	E(3)
1000	0	0
0	1000	0
0	0	1000
600	300	100

Intermedia Rate of Transport, kg/h

T12	T21	T13	T31	T32	T24
air-water	water-air	air-soil	soil-air	soil-water	water-sed
2.984E+00	7.006E-01	4.749E-01	2.758E-03	4.096E-04	1.256E-01
7.005E-01	2.349E+02	1.115E-01	6.473E-04	9.614E-05	4.210E+01
1.793E-02	2.067E-01	2.854E-03	5.807E+00	8.625E-01	3.704E-02
2.003E+00	7.091E+01	3.187E-01	7.091E+00	8.652E-01	1.271E+01

T42	
sed-water	
1.678E-02	
5.625E+00	
4.949E-03	
1.698E+00	

Chemical name: 1,4,5-Trimethylnaphthalene

Level III Distribution

Distribution of mass

Distribution of removal rates

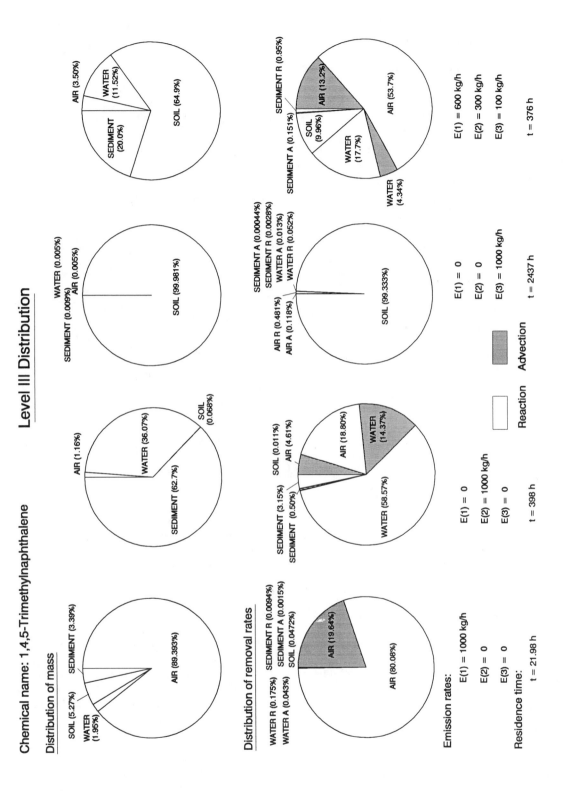

Emission rates:

E(1) = 1000 kg/h	E(1) = 0	E(1) = 0
E(2) = 0	E(2) = 1000 kg/h	E(2) = 0
E(3) = 0	E(3) = 0	E(3) = 1000 kg/h

E(1) = 600 kg/h		
E(2) = 300 kg/h		
E(3) = 100 kg/h		

Residence time:

t = 21.98 h	t = 398 h	t = 2437 h	t = 376 h

Reaction   Advection

281

Chemical name: Biphenyl

Level I calculation: (six compartment model)

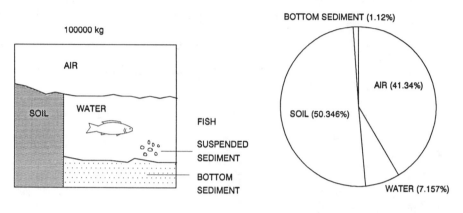

100000 kg

BOTTOM SEDIMENT (1.12%)

AIR (41.34%)

SOIL (50.346%)

WATER (7.157%)

Distribution of mass

physical-chemical properties:

MW: 154.2 g/mol

M.P.: 71.0 °C

Fugacity ratio: 0.3508

vapor pressure: 1.30 Pa

solubility: 7.0 g/m^3

log K_{OW} : 3.90

Compartment	Z mol/m3 Pa	Concentration			Amount kg	Amount %
		mol/m3	mg/L (or g/m3)	ug/g		
Air	4.034E-04	2.681E-09	4.134E-07	3.487E-04	41341	41.341
Water	3.492E-02	2.321E-07	3.578E-05	3.578E-05	7156.9	7.157
Soil	5.459E+00	3.628E-05	5.594E-03	2.331E-03	50346	50.346
Biota (fish)	1.387E+01	9.217E-05	1.421E-02	1.421E-02	2.8425	2.84E-03
Suspended sediment	3.412E+01	2.267E-04	3.496E-02	2.331E-02	34.962	3.50E-02
Bottom sediment	1.092E+01	7.255E-05	1.119E-02	4.662E-03	1118.79	1.1188
Total					100000	100

f = 6.646E-06 Pa

Chemical name: Biphenyl

Level II calculation: (six compartment model)

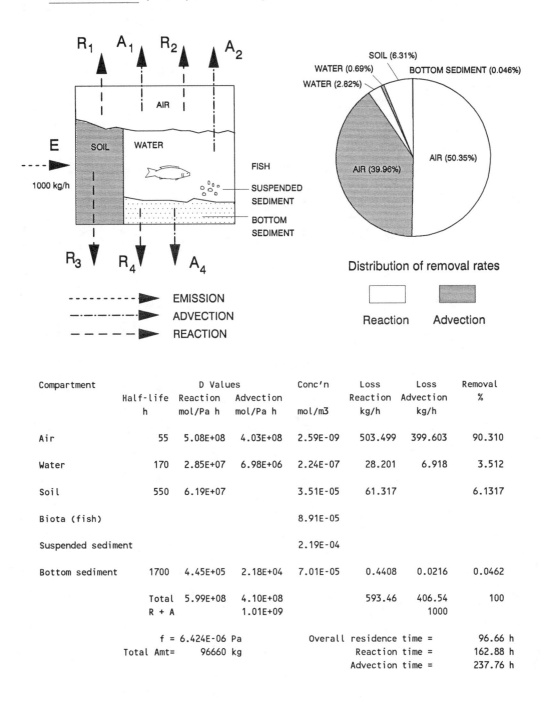

Distribution of removal rates

Reaction   Advection

Compartment	Half-life h	D Values Reaction mol/Pa h	D Values Advection mol/Pa h	Conc'n mol/m3	Loss Reaction kg/h	Loss Advection kg/h	Removal %
Air	55	5.08E+08	4.03E+08	2.59E-09	503.499	399.603	90.310
Water	170	2.85E+07	6.98E+06	2.24E-07	28.201	6.918	3.512
Soil	550	6.19E+07		3.51E-05	61.317		6.1317
Biota (fish)				8.91E-05			
Suspended sediment				2.19E-04			
Bottom sediment	1700	4.45E+05	2.18E+04	7.01E-05	0.4408	0.0216	0.0462
	Total	5.99E+08	4.10E+08		593.46	406.54	100
	R + A		1.01E+09			1000	

f = 6.424E-06 Pa
Total Amt= 96660 kg

Overall residence time = 96.66 h
Reaction time = 162.88 h
Advection time = 237.76 h

283

Fugacity Level III calculations: (four compartment model)

Chemical name: Biphenyl

Diagram (four-compartment model):

- AIR (1): emission E_1; loss R_1, A_1; transfer D values 7.544E05 (D_{31}), 1.104E06 (D_{13}), 9.359E06 (D_{12})
- WATER (2): emission E_2; loss R_2, A_2; D_{21} = 9.398E06, D_{32} = 1.621E05, D_{24} = 5.675E04
- SOIL (3): emission E_3; loss R_3; D_{13}, D_{31}
- SEDIMENT (4): loss R_4, A_4; D_{42} = 2.055E05
- D_{12}, D_{13}, D_{21}, D_{24}, D_{32}, D_{42}

Legend:
- EMISSION (E)
- REACTION (R)
- ADVECTION (A)
- TRANSFER D VALUE mol/Pa h

Phase Properties and Rates:

Compartment	Bulk Z mol/m3 Pa	Half-life h	D Values Reaction mol/Pa h	Advection mol/Pa h
Air (1)	4.034E-04	55	5.08E+08	4.03E+08
Water (2)	3.510E-02	170	2.86E+07	7.02E+06
Soil (3)	2.740E+00	550	6.21E+07	
Sediment (4)	2.211E+00	1700	4.51E+05	2.21E+04

	E(1)=1000	E(2)=1000	E(3)=1000	E(1,2,3)
Overall residence time =	46.46	174.50	783.12	158.54 h
Reaction time =	82.96	231.65	787.77	239.69 h
Advection time =	105.60	707.34	132784	468.25 h

Phase Properties, Compositions, Transport and Transformation Rates:

Emission, kg/h

E(1)	E(2)	E(3)
1000	0	0
0	1000	0
0	0	1000
600	300	100

Fugacity, Pa

f(1)	f(2)	f(3)	f(4)
7.047E-06	1.466E-06	1.234E-07	5.689E-07
1.460E-06	1.438E-04	2.555E-08	5.581E-05
8.806E-08	3.872E-07	1.028E-04	1.502E-07
4.675E-06	4.407E-05	1.037E-05	1.710E-05

Concentration, g/m3

C(1)	C(2)	C(3)	C(4)
4.384E-07	7.936E-06	5.212E-05	1.940E-04
9.080E-08	7.786E-04	1.080E-05	1.903E-02
5.478E-09	2.096E-06	4.345E-02	5.123E-05
2.908E-07	2.385E-04	4.380E-03	5.831E-03

Amounts, kg

m(1)	m(2)	m(3)	m(4)	Total Amount, kg
4.384E+04	1.587E+03	9.382E+02	9.700E+01	4.646E+04
9.080E+03	1.557E+05	1.943E+02	9.516E+03	1.745E+05
5.478E+02	4.192E+02	7.821E+05	2.562E+01	7.831E+05
2.908E+04	4.771E+04	7.883E+04	2.916E+03	1.585E+05

Intermedia Rate of Transport, kg/h

air-water T12	air-soil T13	water-air T21	water-sed T24	soil-air T31	soil-water T32	sed-water T42
1.021E+01	2.116E+01	1.200E+00	4.646E-02	1.435E-02	3.083E-03	4.979E-03
2.115E+00	2.076E+02	2.485E-01	4.558E+00	2.973E-03	6.385E-04	4.884E-01
1.276E-01	1.499E-02	5.587E-01	1.227E-02	1.196E+01	2.570E+00	1.315E-03
6.774E+00	6.360E+01	7.958E-01	1.396E+00	1.206E+00	2.590E-01	1.496E-01

Emission, kg/h

E(1)	E(2)	E(3)
1000	0	0
0	1000	0
0	0	1000
600	300	100

Loss, Reaction, kg/h

R(1)	R(2)	R(3)	R(4)
5.523E+02	6.470E+00	1.18E+00	3.954E+00
1.144E+02	6.348E+02	2.45E-01	3.879E+00
6.902E+00	1.709E+00	9.85E+02	1.044E-02
3.664E+02	1.945E+02	9.93E+01	1.189E+00

Loss, Advection, kg/h

A(1)	A(2)	A(4)
4.384E+02	1.587E+00	1.940E-03
9.080E+01	1.557E+02	1.903E-01
5.478E+00	4.192E-01	5.123E-04
2.908E+02	4.771E+01	5.831E-02

Level III Distribution

Chemical name: Biphenyl

Distribution of mass

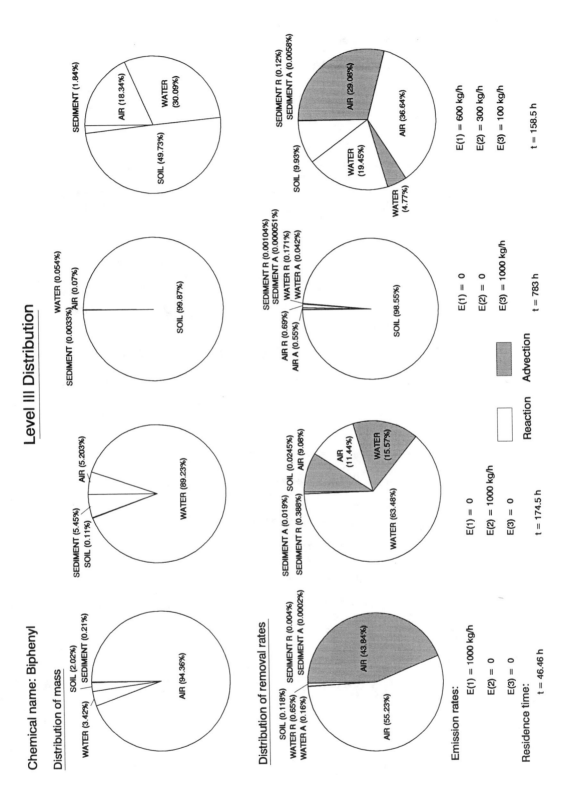

SOIL (2.02%)
SEDIMENT (0.21%)
WATER (3.42%)
AIR (94.36%)

SEDIMENT (5.45%)
SOIL (0.11%)
AIR (5.203%)
WATER (89.23%)

WATER (0.054%)
AIR (0.07%)
SEDIMENT (0.0033%)
SOIL (99.87%)

SEDIMENT (1.84%)
AIR (18.34%)
WATER (30.09%)
SOIL (49.73%)

Distribution of removal rates

SOIL (0.118%)
WATER R (0.65%)
WATER A (0.16%)
SEDIMENT R (0.004%)
SEDIMENT A (0.0002%)
AIR (43.84%)
AIR (55.23%)

SEDIMENT A (0.019%)
SEDIMENT R (0.388%)
SOIL (0.0245%)
AIR (9.08%)
AIR (11.44%)
WATER (15.57%)
WATER (63.48%)

SEDIMENT R (0.00104%)
SEDIMENT A (0.000051%)
WATER R (0.171%)
WATER A (0.042%)
AIR R (0.69%)
AIR A (0.55%)
SOIL (98.55%)

SEDIMENT R (0.12%)
SEDIMENT A (0.0058%)
AIR (29.09%)
AIR (36.64%)
SOIL (9.93%)
WATER (19.45%)
WATER (4.77%)

Reaction   Advection

Emission rates:

E(1) = 1000 kg/h
E(2) = 0
E(3) = 0

E(1) = 0
E(2) = 1000 kg/h
E(3) = 0

E(1) = 0
E(2) = 0
E(3) = 1000 kg/h

E(1) = 600 kg/h
E(2) = 300 kg/h
E(3) = 100 kg/h

Residence time:

t = 46.46 h

t = 174.5 h

t = 783 h

t = 158.5 h

285

Chemical name: Acenaphthene

Level I calculation: (six compartment model)

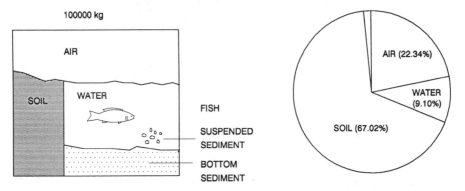

Distribution of mass

physical/chemical properties:

MW: 154.20 g/mol

M.P.: 96.2°C

Fugacity ratio: 0.198

vapor pressure: 0.30 Pa

solubility: 3.80 g/m³

log K_{OW} : 3.92

Compartment	Z	Concentration			Amount	Amount
	mol/m3 Pa	mol/m3	mg/L (or g/m3)	ug/g	kg	%
Air	4.034E-04	1.449E-09	2.234E-07	1.885E-04	22342	22.342
Water	8.214E-02	2.950E-07	4.549E-05	4.549E-05	9098.5	9.098
Soil	1.345E+01	4.829E-05	7.447E-03	3.103E-03	67020	67.020
Biota (fish)	3.416E+01	1.227E-04	1.892E-02	1.892E-02	3.7839	3.78E-03
Suspended sediment	8.404E+01	3.018E-04	4.654E-02	3.103E-02	46.542	4.65E-02
Bottom sediment	2.689E+01	9.658E-05	1.489E-02	6.206E-03	1489.34	1.4893
	Total				100000	100

f = 3.591E-06 Pa

Chemical name: Acenaphthene

Level II calculation: (six compartment model)

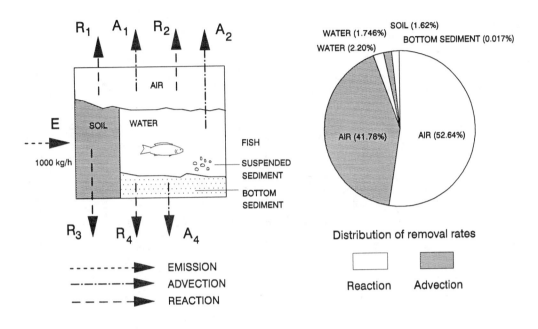

Distribution of removal rates

Reaction   Advection

Compartment	Half-life h	D Values Reaction mol/Pa h	D Values Advection mol/Pa h	Conc'n mol/m3	Loss Reaction kg/h	Loss Advection kg/h	Removal %
Air	55	5.08E+08	4.03E+08	2.71E-09	527.144	418.368	94.551
Water	550	2.07E+07	1.64E+07	5.52E-07	21.468	17.038	3.851
Soil	5500	1.52E+07		9.04E-05	15.813		1.5813
Biota (fish)				2.30E-04			
Suspended sediment				5.65E-04			
Bottom sediment	17000	1.10E+05	5.38E+04	1.81E-04	0.1137	0.0558	0.0169
Total		5.44E+08	4.20E+08		564.54	435.46	100
R + A			9.64E+08			1000	

f = 6.725E-06 Pa

Total Amt= 187259 kg

Overall residence time =   187.26 h
Reaction time =   331.70 h
Advection time =   430.02 h

Fugacity Level III calculations: (four compartment model)
Chemical name: Acenaphthene

Phase Properties and Rates:

Compartment	Bulk Z (mol/m3 Pa)	Half-life (h)	D Values Reaction (mol/Pa h)	Advection (mol/Pa h)
Air (1)	4.034E-04	55	5.08E+08	4.03E+08
Water (2)	8.260E-02	550	2.08E+07	1.65E+07
Soil (3)	6.748E-02	5500	1.53E+07	
Sediment (4)	5.444E+00	17000	1.11E+05	5.44E+04

	E(1)=1000	E(2)=1000	E(3)=1000	E(1,2,3)
Overall residence time =	62.73	428.38	7382.09	904.36 h
Reaction time =	112.37	767.25	7622.80	1509.05 h
Advection time =	142.00	969.89	233769	2256.90 h

↑ EMISSION (E)
↑ REACTION (R)
↑ ADVECTION (A)
↑ TRANSFER D VALUE mol/Pa h

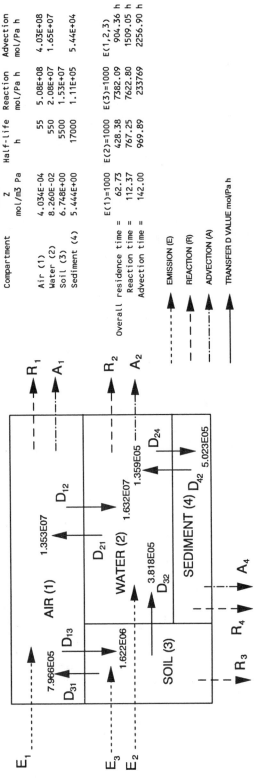

288

Phase Properties, Compositions, Transport and Transformation Rates:

Emission, kg/h — Fugacity, Pa — Concentration, g/m3 — Amounts, kg — Total Amount, kg

E(1)	E(2)	E(3)	f(1)	f(2)	f(3)	f(4)	C(1)	C(2)	C(3)	C(4)	m(1)	m(2)	m(3)	m(4)	Total Amount, kg
1000	0	0	7.023E-06	1.876E-06	6.911E-07	3.127E-06	4.369E-07	2.389E-05	7.192E-04	2.625E-03	4.369E+04	4.778E+03	1.294E+04	1.312E+03	6.273E+04
0	1000	0	1.858E-06	1.273E-04	1.828E-07	2.122E-04	1.156E-07	1.622E-03	1.902E-04	1.782E-01	1.156E+04	3.243E+05	3.424E+03	8.909E+04	4.284E+05
0	0	1000	3.825E-07	3.039E-06	3.935E-04	5.067E-06	2.379E-08	3.871E-05	4.094E-01	4.253E-03	2.379E+03	7.742E+03	7.370E+06	2.127E+03	7.382E+06
600	300	100	4.810E-06	3.962E-06	3.982E-05	6.605E-05	2.992E-07	5.047E-04	4.143E-02	5.545E-02	2.992E+04	1.009E+05	7.458E+05	2.773E+05	9.044E+05

Emission, kg/h — Loss, Reaction, kg/h — Loss, Advection, kg/h

E(1)	E(2)	E(3)	R(1)	R(2)	R(3)	R(4)	A(1)	A(2)	A(4)
1000	0	0	5.505E+02	6.020E+00	1.63E+00	5.350E-02	4.369E+02	4.778E+00	2.625E-02
0	1000	0	1.456E+02	4.086E+02	4.31E-01	3.632E+00	1.156E+02	3.243E+02	1.782E+00
0	0	1000	2.998E+01	9.755E+00	9.29E+02	8.670E-02	2.379E+01	7.742E+00	4.253E-02
600	300	100	3.770E+02	1.272E+02	9.40E+01	1.130E+00	2.992E+02	1.009E+02	5.545E-01

Intermedia Rate of Transport, kg/h

	T12 air-water	T13 air-soil	T21 water-air	T24 water-sed	T31 soil-air	T32 soil-water	T42 sed-water
	1.475E+01	1.757E+00	3.913E+00	1.453E-01	8.489E-02	4.068E-02	6.554E-02
	3.902E+00	4.647E-01	2.656E+02	9.862E+00	2.246E-02	1.076E-02	4.449E+00
	8.032E-02	9.566E-02	6.340E+00	2.354E-01	4.833E+01	2.316E+01	1.062E-01
	1.010E+01	1.203E+00	8.265E+01	3.069E+00	4.891E+00	2.344E+00	1.384E+00

Level III Distribution

Chemical name: Acenaphthene

Distribution of mass

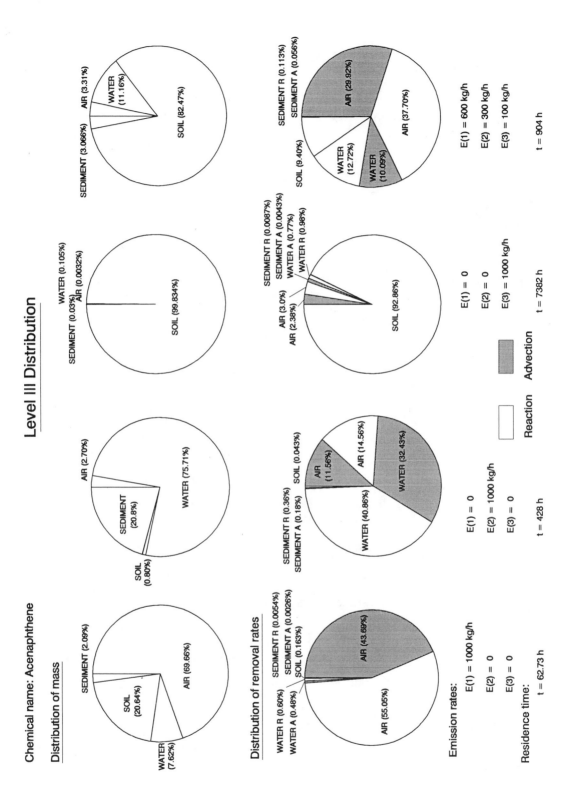

Distribution of removal rates

Emission rates:

E(1) = 1000 kg/h
E(2) = 0
E(3) = 0

E(1) = 0
E(2) = 1000 kg/h
E(3) = 0

E(1) = 0
E(2) = 0
E(3) = 1000 kg/h

E(1) = 600 kg/h
E(2) = 300 kg/h
E(3) = 100 kg/h

Residence time:

t = 62.73 h

t = 428 h

t = 7382 h

t = 904 h

Reaction   Advection

Chemical name: Fluorene

Level I calculation: (six compartment model)

100000 kg

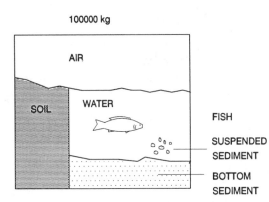

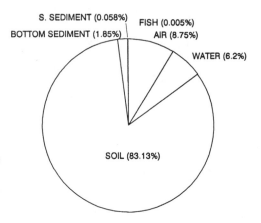

Distribution of mass

physical-chemical properties:

MW: 166.2 g/mol

M.P.: 116°C

Fugacity ratio: 0.126

vapor pressure: 0.080 Pa

solubility: 1.90 g/m^3

log K_{ow} : 4.18

Compartment	Z	Concentration				Amount	Amount
	mol/m3 Pa	mol/m3	mg/L (or g/m3)	ug/g		kg	%
Air	4.034E-04	5.267E-10	8.755E-08	7.385E-05		8755	8.755
Water	1.429E-01	1.866E-07	3.101E-05	3.101E-05		6202.1	6.202
Soil	4.257E+01	5.558E-05	9.237E-03	3.849E-03		83134	83.134
Biota (fish)	1.081E+02	1.412E-04	2.347E-02	2.347E-02		4.6936	4.69E-03
Suspended sediment	2.660E+02	3.474E-04	5.773E-02	3.849E-02		57.732	5.77E-02
Bottom sediment	8.513E+01	1.112E-04	1.847E-02	7.698E-03		1847.41	1.8474
	Total					100000	100

f = 1.306E-06 Pa

Chemical name: Fluorene

Level II calculation: (six compartment model)

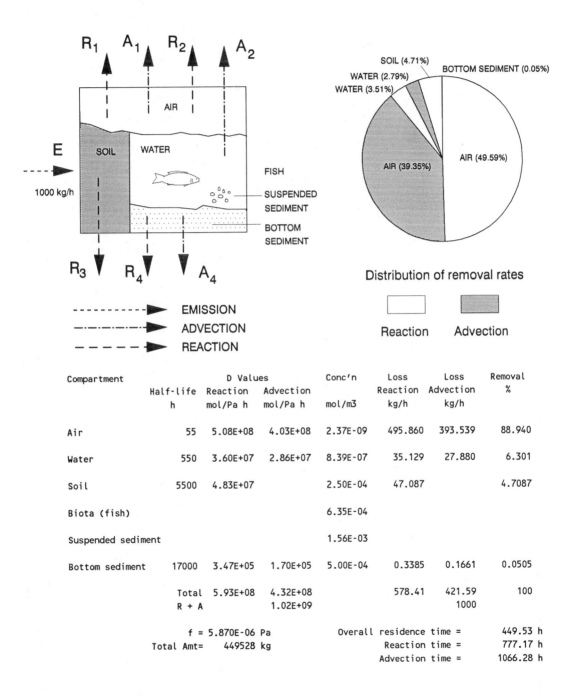

Distribution of removal rates

Reaction   Advection

Compartment	Half-life h	D Values Reaction mol/Pa h	Advection mol/Pa h	Conc'n mol/m3	Loss Reaction kg/h	Loss Advection kg/h	Removal %
Air	55	5.08E+08	4.03E+08	2.37E-09	495.860	393.539	88.940
Water	550	3.60E+07	2.86E+07	8.39E-07	35.129	27.880	6.301
Soil	5500	4.83E+07		2.50E-04	47.087		4.7087
Biota (fish)				6.35E-04			
Suspended sediment				1.56E-03			
Bottom sediment	17000	3.47E+05	1.70E+05	5.00E-04	0.3385	0.1661	0.0505
Total		5.93E+08	4.32E+08		578.41	421.59	100
R + A			1.02E+09			1000	

f = 5.870E-06 Pa

Total Amt= 449528 kg

Overall residence time = 449.53 h
Reaction time = 777.17 h
Advection time = 1066.28 h

Fugacity Level III calculations: (four compartment model)

Chemical name: Fluorene

Phase Properties and Rates:

Compartment	Bulk Z mol/m3 Pa	Half-life h	D Values Reaction mol/Pa h	Advection mol/Pa h
Air (1)	4.035E-04	55	5.08E+08	4.03E+08
Water (2)	1.443E-01	550	3.64E+07	2.89E+07
Soil (3)	2.133E+01	5500	4.84E+07	
Sediment (4)	1.714E+01	17000	3.49E+05	1.71E+05

	E(1)=1000	E(2)=1000	E(3)=1000	E(1,2,3)
Overall residence time =	72.20	551.25	7701.58	978.85 h
Reaction time =	129.24	986.11	7807.35	1627.28 h
Advection time =	163.57	1250.01	568512	2456.49 h

EMISSION (E)
REACTION (R)
ADVECTION (A)
TRANSFER D VALUE mol/Pa h

Diagram

E_1 R_1 A_1

AIR (1)

D_{12} 1.573E07 D_{21} 1.590E07

D_{13} 8.508E05 D_{31} 2.343E06

E_2 R_2 A_2

WATER (2)

D_{32} 6.814E05 D_{24} 3.132E05 D_{42} 1.473E06

SOIL (3)

SEDIMENT (4)

E_3 R_3 R_4 A_4

Phase Properties, Compositions, Transport and Transformation Rates:

Emission, kg/h

E(1)	E(2)	E(3)
1000	0	0
0	1000	0
0	0	1000
600	300	100

Fugacity, Pa

f(1)	f(2)	f(3)	f(4)
6.490E-06	1.262E-06	3.047E-07	2.230E-06
1.247E-06	7.372E-05	5.853E-08	1.302E-04
1.277E-07	1.028E-06	1.206E-04	1.816E-06
4.281E-06	2.298E-05	1.226E-05	4.058E-05

Concentration, g/m3

C(1)	C(2)	C(3)	C(4)
4.353E-07	3.028E-05	1.080E-03	6.352E-03
8.361E-08	1.768E-03	2.074E-04	3.709E-01
8.563E-09	2.466E-05	4.274E-01	5.174E-01
2.871E-07	5.512E-04	4.345E-02	1.156E-01

Amounts, kg

m(1)	m(2)	m(3)	m(4)	Total Amount, kg
4.353E+04	6.057E+03	1.944E+04	3.176E+03	7.220E+04
8.361E+03	3.537E+05	3.734E+03	1.855E+05	5.512E+05
8.563E+02	4.933E+03	7.693E+06	2.587E+03	7.702E+06
2.871E+04	1.102E+05	7.821E+05	5.781E+04	9.788E+05

Intermedia Rate of Transport, kg/h

T12 air-water	T21 water-air	T13 air-soil	T31 soil-air	T32 soil-water	T24 water-sed	T42 sed-water
1.715E+01	3.300E+01	2.527E+00	4.308E-02	3.450E+00	3.091E-01	1.161E-01
3.294E+00	1.927E+02	4.854E-01	8.276E-03	6.628E-03	1.805E+02	6.777E+00
3.373E-01	2.688E+00	4.971E-02	1.705E+01	1.366E+01	2.517E-01	9.452E-02
1.131E+01	6.007E+01	1.667E+00	1.733E+00	1.388E+00	5.625E+00	2.112E+00

Emission, kg/h

E(1)	E(2)	E(3)
1000	0	0
0	1000	0
0	0	1000
600	300	100

Loss, Reaction, kg/h

R(1)	R(2)	R(3)	R(4)
5.484E+02	7.631E+00	2.45E+00	1.295E-01
1.053E+02	4.456E+02	4.70E-01	7.561E+00
1.079E+00	6.215E+00	9.69E+02	1.054E+00
3.617E+02	1.389E+02	9.85E+01	2.356E+00

Loss, Advection, kg/h

A(1)	A(2)	A(4)
4.353E+02	6.057E+00	6.352E-02
8.361E+01	3.537E+02	3.709E+00
8.563E+00	4.933E+00	5.174E-01
2.871E+02	1.102E+02	1.156E+00

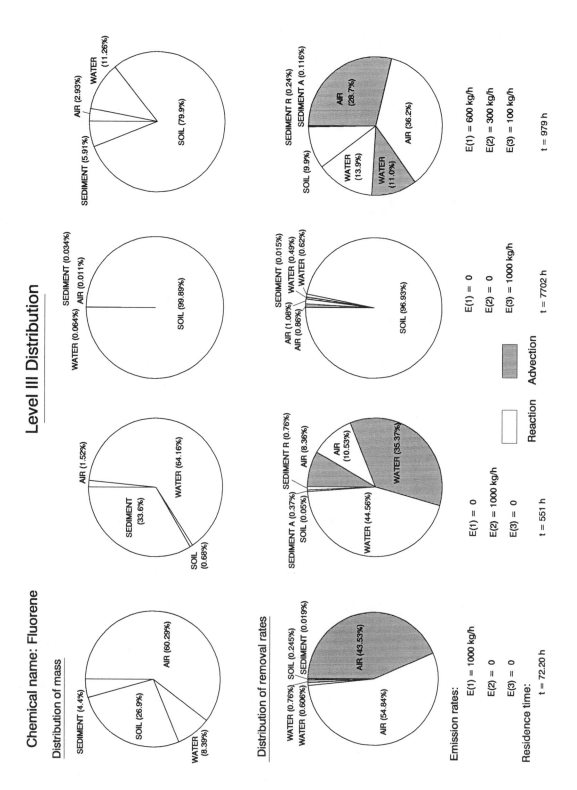

Level III Distribution

Chemical name: Fluorene

Distribution of mass

Distribution of removal rates

Emission rates:

Residence time:

E(1) = 1000 kg/h	E(1) = 0	E(1) = 0	E(1) = 600 kg/h
E(2) = 0	E(2) = 1000 kg/h	E(2) = 0	E(2) = 300 kg/h
E(3) = 0	E(3) = 0	E(3) = 1000 kg/h	E(3) = 100 kg/h
t = 72.20 h	t = 551 h	t = 7702 h	t = 979 h

Reaction   Advection

293

Chemical name: Phenanthrene

Level I calculation: (six compartment model)

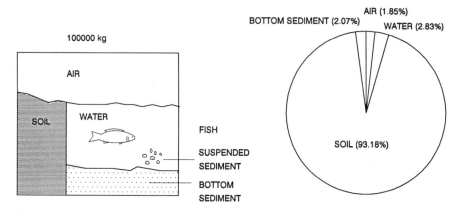

100000 kg

AIR

SOIL   WATER

FISH

SUSPENDED
SEDIMENT

BOTTOM
SEDIMENT

AIR (1.85%)
BOTTOM SEDIMENT (2.07%)   WATER (2.83%)

SOIL (93.18%)

Distribution of mass

physical-chemical properties:

MW: 178.2 g/mol

M.P.: 101 °C

Fugacity ratio: 0.177

vapor pressure: 0.020 Pa

solubility: 1.10 g/m³

log K_{OW} : 4.57

Compartment	Z mol/m3 Pa	Concentration			Amount kg	Amount %
		mol/m3	mg/L (or g/m3)	ug/g		
Air	4.034E-04	1.039E-10	1.851E-08	1.561E-05	1850.72	1.851
Water	3.086E-01	7.946E-08	1.416E-05	1.416E-05	2831.9	2.832
Soil	2.257E+02	5.810E-05	1.035E-02	4.314E-03	93177	93.177
Biota (fish)	5.734E+02	1.476E-04	2.630E-02	2.630E-02	5.2607	5.26E-03
Suspended sediment	1.410E+03	3.631E-04	6.471E-02	4.314E-02	64.706	6.47E-02
Bottom sediment	4.513E+02	1.162E-04	2.071E-02	8.627E-03	2070.60	2.0706
Total					100000	100

f = 2.574E-07 Pa

Chemical name: Phenanthrene

Level II calculation: (six compartment model)

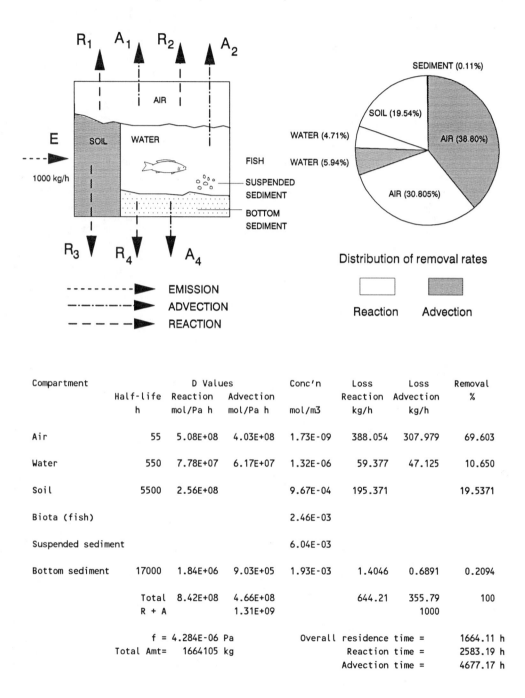

Distribution of removal rates

Reaction   Advection

- - - - - - - - - ▶ EMISSION
- - · - - · - - · ▶ ADVECTION
- - - - - - ▶ REACTION

Compartment	Half-life h	D Values Reaction mol/Pa h	D Values Advection mol/Pa h	Conc'n mol/m3	Loss Reaction kg/h	Loss Advection kg/h	Removal %
Air	55	5.08E+08	4.03E+08	1.73E-09	388.054	307.979	69.603
Water	550	7.78E+07	6.17E+07	1.32E-06	59.377	47.125	10.650
Soil	5500	2.56E+08		9.67E-04	195.371		19.5371
Biota (fish)				2.46E-03			
Suspended sediment				6.04E-03			
Bottom sediment	17000	1.84E+06	9.03E+05	1.93E-03	1.4046	0.6891	0.2094
Total R + A		8.42E+08	4.66E+08 1.31E+09		644.21	355.79 1000	100

f = 4.284E-06 Pa
Total Amt= 1664105 kg

Overall residence time = 1664.11 h
Reaction time = 2583.19 h
Advection time = 4677.17 h

295

Fugacity Level III calculations: (four compartment model)

Chemical name: Phenanthrene

Four-compartment fugacity diagram (AIR (1), WATER (2), SOIL (3), SEDIMENT (4)) with emission, reaction, advection arrows and transfer D values:

- E₁, R₁, A₁ — AIR (1)
- D₁₂ = 1.784E07, D₂₁, 9.984E05, D₃₁, D₁₃ = 4.934E06
- E₂, E₃, R₂, A₂ — WATER (2) = 1.828E07
- D₃₂ = 1.592E06, D₂₄ = 1.211E06, D₄₂ = 7.361E06
- SOIL (3), R₃; SEDIMENT (4), R₄, A₄

Phase Properties and Rates:

Compartment	Bulk Z mol/m3 Pa	Half-life h	D Values Reaction mol/Pa h	Advection mol/Pa h
Air (1)	4.038E-04	55	5.09E+08	4.04E+08
Water (2)	3.163E-01	550	7.97E+07	6.33E+07
Soil (3)	1.129E+02	5500	2.56E+08	
Sediment (4)	9.052E+01	17000	1.84E+06	9.05E+05

	E(1)=1000	E(2)=1000	E(3)=1000	E(1,2,3)
Overall residence time =	102.16	899.16	7862.97	1117.34 h
Reaction time =	182.46	1602.01	7897.70	1850.26 h
Advection time =	232.13	2049.45	1787808	2820.74 h

Legend: EMISSION (E); REACTION (R); ADVECTION (A); TRANSFER D VALUE mol/Pa h

Phase Properties, Compositions, Transport and Transformation Rates:

Emission, kg/h — Fugacity, Pa

E(1)	E(2)	E(3)	f(1)	f(2)	f(3)	f(4)
1000	0	0	6.009E-06	6.630E-07	1.146E-07	1.232E-06
0	1000	0	6.461E-07	3.390E-05	1.232E-08	6.298E-05
0	0	1000	2.716E-08	2.111E-07	2.169E-05	3.923E-07
600	300	100	3.802E-06	1.059E-05	2.242E-06	1.967E-05

Concentration, g/m3

C(1)	C(2)	C(3)	C(4)
4.324E-04	3.737E-05	2.306E-03	1.987E-02
4.650E-08	1.910E-03	2.480E-04	1.016E+00
1.955E-09	1.190E-05	4.365E-01	6.328E-03
2.736E-07	5.967E-04	4.511E-02	3.173E-01

Amounts, kg

m(1)	m(2)	m(3)	m(4)	Total Amount, kg
4.324E+04	7.474E+03	4.151E+04	9.936E+03	1.022E+05
4.650E+03	3.821E+05	4.463E+03	5.080E+05	8.992E+05
1.955E+02	2.380E+03	7.857E+06	3.164E+05	7.863E+06
2.736E+04	1.193E+05	8.120E+05	1.587E+05	1.117E+06

Emission, kg/h — Loss, Reaction, kg/h

E(1)	E(2)	E(3)	R(1)	R(2)	R(3)	R(4)
1000	0	0	5.449E+02	9.417E+00	5.23E+00	4.051E-01
0	1000	0	5.859E+01	4.814E+02	5.62E-01	2.071E+01
0	0	1000	2.463E+00	2.999E+00	9.90E+02	1.290E-01
600	300	100	3.447E+02	1.504E+02	1.02E+02	6.468E+00

Loss, Advection, kg/h

A(1)	A(2)	A(3)	A(4)
4.324E+02	7.474E+00	5.23E+00	1.987E-01
4.650E+01	3.821E+02	5.62E-01	1.016E+01
1.955E+00	2.380E+00	9.90E+02	6.328E-02
2.736E+02	1.193E+02	1.02E+02	3.173E+00

Intermedia Rate of Transport, kg/h

T12 air-water	T13 air-soil	T21 water-air	T24 water-sed	T31 soil-air	T32 soil-water	T42 sed-water
1.957E+01	5.285E+00	2.108E+00	8.697E-01	2.039E-02	3.251E-02	2.659E-01
2.104E+00	5.681E-01	1.078E+02	4.446E+01	2.192E-03	3.496E-03	1.360E+01
8.847E-02	2.388E-02	6.712E-01	2.770E-01	3.859E+00	6.154E+00	8.469E-02
1.238E+02	3.343E+01	3.366E+01	1.389E+01	3.988E-01	6.359E-01	4.247E+00

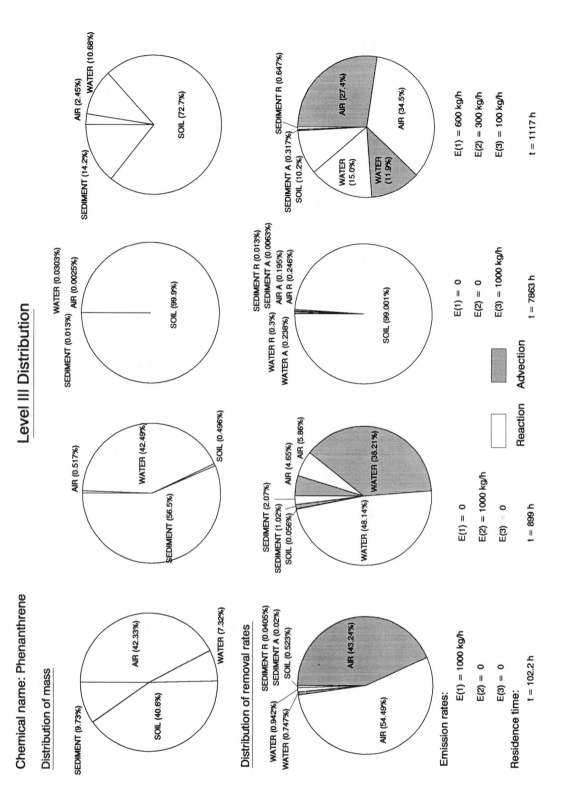

Level III Distribution

Chemical name: Phenanthrene

Distribution of mass

SEDIMENT (9.73%)
AIR (42.33%)
SOIL (40.6%)
WATER (7.32%)

WATER (0.0303%)
SEDIMENT (0.013%)
AIR (0.0025%)
SOIL (99.9%)

AIR (0.517%)
WATER (42.49%)
SOIL (0.496%)
SEDIMENT (56.5%)

WATER (10.68%)
AIR (2.45%)
SEDIMENT (14.2%)
SOIL (72.7%)

Distribution of removal rates

SEDIMENT R (0.0405%)
SEDIMENT A (0.02%)
SOIL (0.523%)
WATER (0.942%)
WATER (0.747%)
AIR (43.24%)
AIR (54.49%)

SEDIMENT (2.07%)
SEDIMENT (1.02%)
SOIL (0.056%)
AIR (4.65%)
AIR (5.86%)
WATER (38.21%)
WATER (48.14%)

SEDIMENT R (0.013%)
SEDIMENT A (0.0063%)
AIR A (0.195%)
AIR R (0.246%)
WATER R (0.3%)
WATER A (0.238%)
SOIL (99.001%)

SEDIMENT R (0.647%)
SEDIMENT A (0.317%)
SOIL (10.2%)
WATER (15.0%)
AIR (27.4%)
WATER (1.9%)
AIR (34.5%)

Reaction   Advection

Emission rates:

E(1) = 1000 kg/h	E(1) = 0	E(1) = 600 kg/h
E(2) = 0	E(2) = 1000 kg/h	E(2) = 300 kg/h
E(3) = 0	E(3) = 0	E(3) = 100 kg/h

E(1) = 0		
E(2) = 0		
E(3) = 1000 kg/h		

Residence time:

t = 102.2 h   t = 899 h   t = 7863 h   t = 1117 h

Chemical name: Anthracene

Level I calculation: (six compartment model)

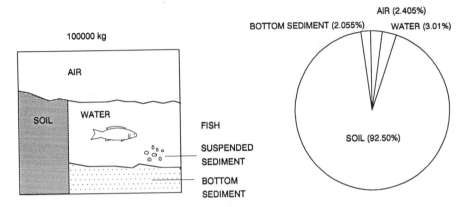

Distribution of mass

physical-chemical properties:

MW: 178.2 g/mol

M.P.: 216.2 °C

Fugacity ratio: 0.0129

vapor pressure: 0.001 Pa

solubility: 0.045 g/m^3

log K_{OW} : 4.54

Compartment	Z	Concentration			Amount	Amount
	mol/m3 Pa	mol/m3	mg/L (or g/m3)	ug/g	kg	%
Air	4.034E-04	1.350E-10	2.405E-08	2.029E-05	2405.11	2.405
Water	2.525E-01	8.448E-08	1.506E-05	1.506E-05	3011.0	3.011
Soil	1.723E+02	5.765E-05	1.027E-02	4.281E-03	92460	92.460
Biota (fish)	4.378E+02	1.465E-04	2.610E-02	2.610E-02	5.2202	5.22E-03
Suspended sediment	1.077E+03	3.603E-04	6.421E-02	4.281E-02	64.208	6.42E-02
Bottom sediment	3.446E+02	1.153E-04	2.055E-02	8.561E-03	2054.66	2.0547
Total					100000	100

f = 3.346E-07 Pa

298

Chemical name: Anthracene

Level II calculation: (six compartment model)

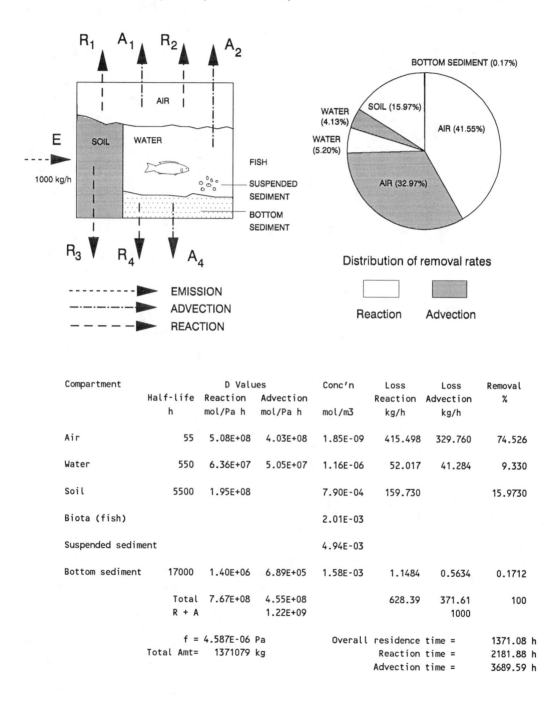

Distribution of removal rates

Reaction   Advection

Compartment	Half-life h	D Values Reaction mol/Pa h	D Values Advection mol/Pa h	Conc'n mol/m3	Loss Reaction kg/h	Loss Advection kg/h	Removal %
Air	55	5.08E+08	4.03E+08	1.85E-09	415.498	329.760	74.526
Water	550	6.36E+07	5.05E+07	1.16E-06	52.017	41.284	9.330
Soil	5500	1.95E+08		7.90E-04	159.730		15.9730
Biota (fish)				2.01E-03			
Suspended sediment				4.94E-03			
Bottom sediment	17000	1.40E+06	6.89E+05	1.58E-03	1.1484	0.5634	0.1712
Total		7.67E+08	4.55E+08		628.39	371.61	100
R + A			1.22E+09			1000	

f = 4.587E-06 Pa

Total Amt= 1371079 kg

Overall residence time = 1371.08 h

Reaction time = 2181.88 h

Advection time = 3689.59 h

299

Fugacity Level III calculations: (four compartment model)
Chemical name: Anthracene

Diagram (four-compartment model): AIR (1), WATER (2), SOIL (3), SEDIMENT (4) with emission (E_1, E_2, E_3), reaction (R_1, R_2, R_3, R_4), advection (A_1, A_2, A_4) and transfer D values.

Transfer D values shown: D_{13} = 9.484E05, D_{31} = (4.901E06), D_{12} = 1.739E07, D_{21}, D_{32} = 1.291E06, D_{24}, D_{42} = 5.637E06, plus 1.783E07 and 9.418E05.

Legend: EMISSION (E), REACTION (R), ADVECTION (A), TRANSFER D VALUE mol/Pa h

Phase Properties and Rates:

Compartment	Bulk Z mol/m3 Pa	Half-life h	D Values Reaction mol/Pa h	D Values Advection mol/Pa h
Air (1)	4.040E-04	55	5.09E+08	4.04E+08
Water (2)	2.583E-01	550	6.51E+07	5.17E+07
Soil (3)	8.623E+01	5500	1.96E+08	
Sediment (4)	6.913E+01	17000	1.41E+06	6.91E+05

	E(1)=1000	E(2)=1000	E(3)=1000	E(1,2,3)
Overall residence time =	100.51	850.88	7852.68	1100.84 h
Reaction time =	179.52	1516.68	7891.95	1823.38 h
Advection time =	228.36	1938.29	1578105	2778.03 h

Phase Properties, Compositions, Transport and Transformation Rates:

Emission, kg/h — Fugacity, Pa — Concentration, g/m3 — Amounts, kg

E(1)	E(2)	E(3)	f(1)	f(2)	f(3)	f(4)	C(1)	C(2)	C(3)	C(4)	m(1)	m(2)	m(3)	m(4)	Total Amount, kg
1000	0	0	6.011E-06	7.778E-07	1.489E-07	1.441E-06	4.328E-07	3.581E-05	2.288E-03	1.776E-02	4.328E+04	7.161E+03	4.119E+04	8.878E+03	1.005E+05
0	1000	0	7.572E-07	4.075E-05	1.876E-08	7.551E-05	5.452E-08	1.876E-03	2.883E-04	9.302E-01	5.452E+03	3.752E+05	5.189E+03	4.651E+05	8.509E+05
0	0	1000	3.376E-08	2.697E-07	2.837E-05	4.998E-07	2.431E-09	1.242E-05	4.359E-01	6.158E-03	2.431E+02	2.484E+03	7.847E+06	3.079E+03	7.853E+06
600	300	100	3.837E-06	1.272E-05	2.932E-06	2.357E-05	2.763E-07	5.855E-04	4.505E-02	2.903E-01	2.763E+04	1.171E+05	8.110E+05	1.452E+05	1.101E+06

Emission, kg/h — Loss, Reaction, kg/h — Loss, Advection, kg/h

E(1)	E(2)	E(3)	R(1)	R(2)	R(3)	R(4)	A(1)	A(2)	A(4)
1000	0	0	5.453E+02	9.023E+00	5.19E+00	3.619E-01	4.328E+02	7.161E+00	1.776E-01
0	1000	0	6.870E+01	4.727E+02	6.54E-01	1.896E+01	5.452E+01	3.752E+02	9.302E-01
0	0	1000	3.063E+00	3.129E+00	9.89E+02	1.255E-01	2.431E+00	2.484E+00	6.158E-03
600	300	100	3.481E+02	1.475E+02	1.02E+02	5.917E+00	2.763E+02	1.171E+02	2.903E+00

Intermedia Rate of Transport, kg/h

T12 air-water	T21 water-air	T13 air-soil	T31 soil-air	T32 soil-water	T24 water-sed	T42 sed-water
1.910E+01	2.411E+00	5.249E+00	2.517E-02	3.427E-02	7.813E-01	2.419E-01
2.406E+00	1.263E+02	6.613E-01	3.171E-03	4.317E-01	4.093E+01	1.267E+01
1.073E-01	8.360E-01	2.949E-02	4.795E+00	6.529E+00	2.710E-01	8.389E-02
1.219E+01	3.941E+01	3.351E+00	4.955E-01	6.747E-01	1.278E+01	3.955E+00

Level III Distribution

Chemical name: Anthracene

Distribution of mass

Distribution of removal rates

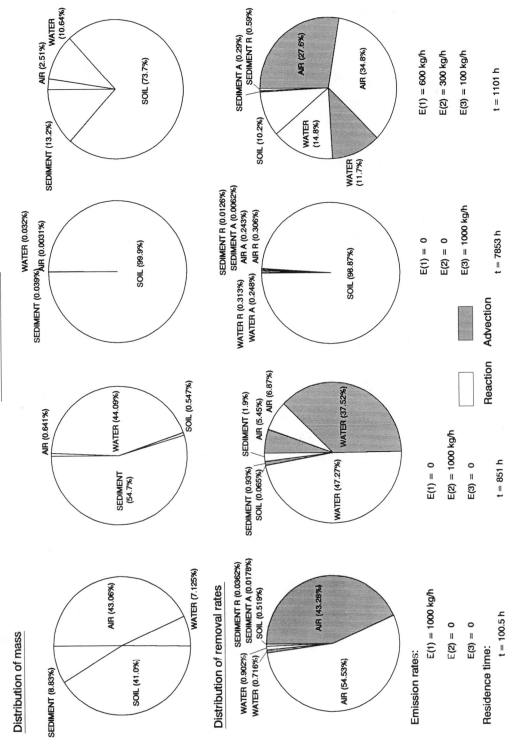

Emission rates:

E(1) = 1000 kg/h	E(1) = 0	E(1) = 0	E(1) = 600 kg/h
E(2) = 0	E(2) = 1000 kg/h	E(2) = 0	E(2) = 300 kg/h
E(3) = 0	E(3) = 0	E(3) = 1000 kg/h	E(3) = 100 kg/h

Residence time:

t = 100.5 h	t = 851 h	t = 7853 h	t = 1101 h

Reaction   Advection

301

Chemical name: Pyrene

Level I calculation: (six compartment model)

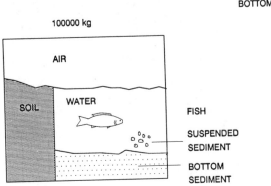

100000 kg

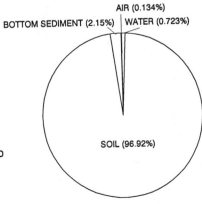

AIR (0.134%)

BOTTOM SEDIMENT (2.15%) | WATER (0.723%)

SOIL (96.92%)

Distribution of mass

physical-chemical properties:

MW: 202.3 g/mol

M.P.: 156 °C

Fugacity ratio: 0.0506

vapor pressure: 0.0006 Pa

solubility: 0.132 g/m^3

log K_{OW} : 5.18

Compartment	Z mol/m3 Pa	Concentration				Amount kg	Amount %
		mol/m3	mg/L (or g/m3)	ug/g			
Air	4.034E-04	6.629E-12	1.341E-09	1.131E-06		134.11	0.1341
Water	1.087E+00	1.787E-08	3.615E-06	3.615E-06		723.0	0.7230
Soil	3.239E+03	5.323E-05	1.077E-02	4.487E-03		96916	96.916
Biota (fish)	8.230E+03	1.352E-04	2.736E-02	2.736E-02		5.4718	5.47E-03
Suspended sediment	2.025E+04	3.327E-04	6.730E-02	4.487E-02		67.303	6.73E-02
Bottom sediment	6.479E+03	1.065E-04	2.154E-02	8.974E-03		2153.70	2.1537
Total						100000	100

f = 1.643E-08 Pa

Chemical name: Pyrene

Level II calculation: (six compartment model)

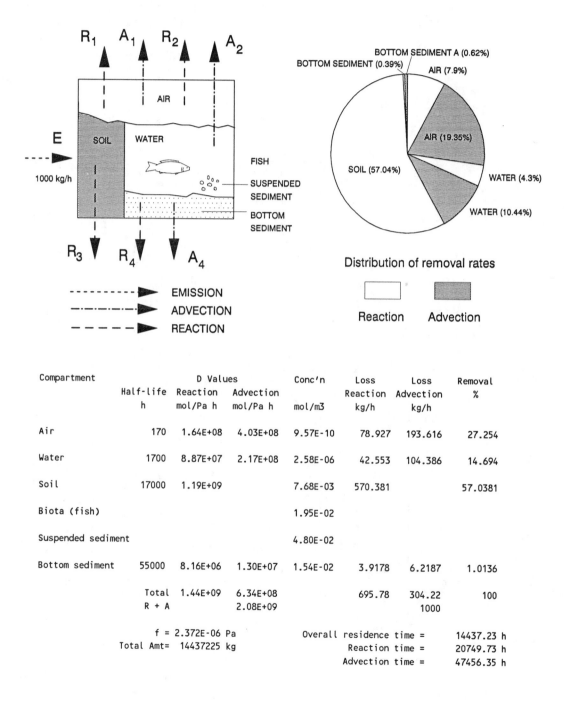

EMISSION

ADVECTION

REACTION

Distribution of removal rates

Reaction   Advection

Compartment	Half-life h	D Values Reaction mol/Pa h	Advection mol/Pa h	Conc'n mol/m3	Loss Reaction kg/h	Loss Advection kg/h	Removal %
Air	170	1.64E+08	4.03E+08	9.57E-10	78.927	193.616	27.254
Water	1700	8.87E+07	2.17E+08	2.58E-06	42.553	104.386	14.694
Soil	17000	1.19E+09		7.68E-03	570.381		57.0381
Biota (fish)				1.95E-02			
Suspended sediment				4.80E-02			
Bottom sediment	55000	8.16E+06	1.30E+07	1.54E-02	3.9178	6.2187	1.0136
Total		1.44E+09	6.34E+08		695.78	304.22	100
R + A			2.08E+09			1000	

f = 2.372E-06 Pa

Total Amt= 14437225 kg

Overall residence time =   14437.23 h

Reaction time =   20749.73 h

Advection time =   47456.35 h

303

Fugacity Level III calculations: (four compartment model)

Chemical name: Pyrene

Four-compartment diagram (AIR (1), WATER (2), SOIL (3), SEDIMENT (4)) with emission (E), reaction (R), advection (A) arrows and transfer D values (mol/Pa h):

- E_1, R_1, A_1 — AIR (1)
- E_2, E_3, R_2, A_2 — WATER (2)
- R_3 — SOIL (3)
- R_4, A_4 — SEDIMENT (4)
- D_{12} = 1.945E07
- D_{13} = 1.689E06
- D_{21} = 2.176E07
- D_{31} = 2.250E07
- D_{32} = 7.809E06
- D_{24} = 1.404E07
- D_{42} = 1.023E08

Phase Properties and Rates:

Compartment	Bulk Z mol/m3 Pa	Half-life h	D Values Reaction mol/Pa h	D Values Advection mol/Pa h
Air (1)	4.075E-04	170	1.66E+08	4.08E+08
Water (2)	1.197E+00	1700	9.76E+07	2.39E+08
Soil (3)	1.620E+03	17000	1.19E+09	
Sediment (4)	1.297E+03	55000	8.17E+06	1.30E+07

	E(1)=1000	E(2)=1000	E(3)=1000	E(1,2,3)
Overall residence time =	1134.55	5137.60	24371.65	4659.17 h
Reaction time =	3592.47	16840.77	24506.28	12246.18 h
Advection time =	1658.25	7392.96	4436326	7520.37 h

Legend:
- EMISSION (E)
- REACTION (R)
- ADVECTION (A)
- TRANSFER D VALUE mol/Pa h

Phase Properties, Compositions, Transport and Transformation Rates:

Emission, kg/h — Fugacity, Pa — Concentration, g/m3

E(1)	E(2)	E(3)	f(1)	f(2)	f(3)	f(4)	C(1)	C(2)	C(3)	C(4)
1000	0	0	8.014E-06	4.201E-07	1.505E-07	1.222E-07	6.606E-07	1.017E-04	4.933E-02	3.205E-02
0	1000	0	3.730E-07	1.185E-05	7.005E-09	3.446E-06	3.075E-08	2.869E-03	2.296E-03	9.039E-01
0	0	1000	1.373E-08	7.781E-08	4.126E-06	2.263E-08	1.132E-09	1.884E-05	1.352E+00	5.936E-03
600	300	100	4.922E-06	3.814E-06	5.050E-07	1.109E-06	4.057E-07	9.236E-04	1.655E-01	2.910E-01

Amounts, kg

m(1)	m(2)	m(3)	m(4)	Total Amount, kg
6.606E+04	2.035E+04	8.879E+05	1.603E+05	1.135E+06
3.075E+03	5.738E+05	4.132E+04	4.519E+06	5.138E+06
1.132E+02	3.768E+03	2.434E+07	2.968E+04	2.437E+07
4.057E+04	1.847E+05	2.979E+06	1.455E+06	4.659E+06

Intermedia Rate of Transport, kg/h

T12 air-water	T21 water-air	T13 air-soil	T31 soil-air	T32 soil-water	T24 water-sed	T42 sed-water
3.528E+01	1.653E+00	3.648E-01	5.143E-02	2.378E-01	8.696E+00	3.472E+00
1.642E+00	4.662E+01	1.698E+00	2.394E-03	1.107E-02	2.452E+02	9.791E+01
6.044E-02	3.062E-01	6.250E-02	1.410E+00	6.518E+00	1.611E+00	6.430E+01
2.167E+01	1.501E+01	2.241E+01	1.726E+01	7.978E-01	7.895E+01	3.152E+01

Emission, kg/h — Loss, Reaction, kg/h — Loss, Advection, kg/h

E(1)	E(2)	E(3)	R(1)	R(2)	R(3)	R(4)	A(1)	A(2)	A(4)
1000	0	0	2.693E+02	8.294E+00	3.62E+01	2.019E+00	6.606E+02	2.035E+01	3.205E+00
0	1000	0	1.253E+01	2.339E+02	1.68E+00	5.694E+01	3.075E+01	5.738E+02	9.039E+01
0	0	1000	4.613E-01	1.536E+00	9.92E+02	3.740E-01	1.132E+00	3.768E+00	5.936E-01
600	300	100	1.654E+02	7.530E+01	1.21E+02	1.833E+01	4.057E+02	1.847E+02	2.910E+01

Level III Distribution

Chemical name: Pyrene

Distribution of mass

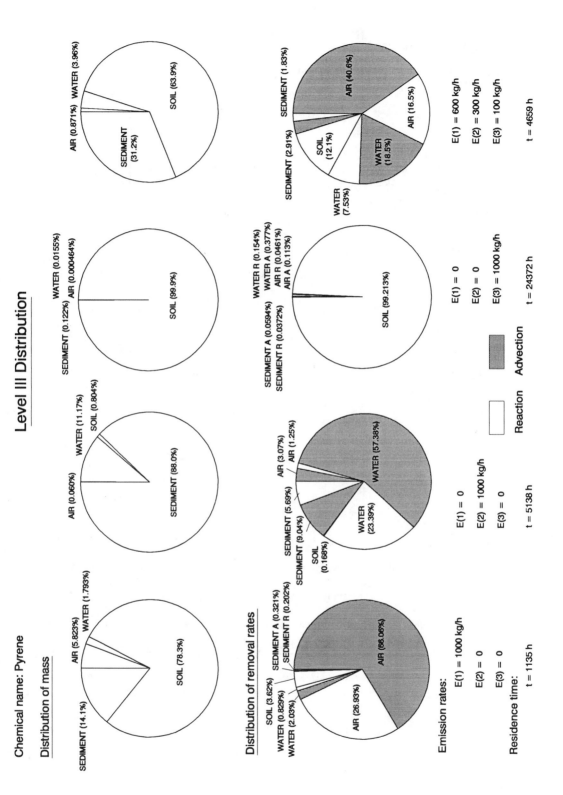

Distribution of removal rates

Emission rates:

Pie 1	Pie 2	Pie 3	Pie 4
E(1) = 1000 kg/h	E(1) = 0	E(1) = 0	E(1) = 600 kg/h
E(2) = 0	E(2) = 1000 kg/h	E(2) = 0	E(2) = 300 kg/h
E(3) = 0	E(3) = 0	E(3) = 1000 kg/h	E(3) = 100 kg/h

Residence time:

t = 1135 h	t = 5138 h	t = 24372 h	t = 4659 h

Reaction ☐   Advection ▨

305

Chemical name: Fluoranthene

Level I calculation: (six compartment model)

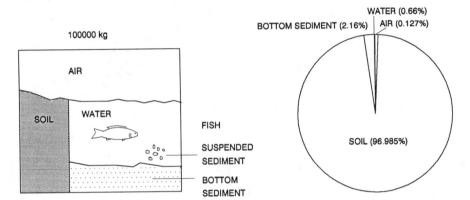

Distribution of mass

physical-chemical properties:

MW: 202.30 g/mol

M.P.: 111 °C

Fugacity ratio: 0.141

vapor pressure: 0.00123 Pa

solubility: 0.260 g/m³

log K_{OW} : 5.22

| Compartment | Z mol/m3 Pa | Concentration | | | Amount kg | Amount % |
		mol/m3	mg/L (or g/m3)	ug/g		
Air	4.034E-04	6.297E-12	1.274E-09	1.075E-06	127.39	0.1274
Water	1.045E+00	1.631E-08	3.299E-06	3.299E-06	659.9	0.6599
Soil	3.413E+03	5.327E-05	1.078E-02	4.490E-03	96985	96.985
Biota (fish)	8.670E+03	1.353E-04	2.738E-02	2.738E-02	5.4756	5.48E-03
Suspended sediment	2.133E+04	3.329E-04	6.735E-02	4.490E-02	67.350	6.74E-02
Bottom sediment	6.825E+03	1.065E-04	2.155E-02	8.980E-03	2155.22	2.1552
Total					100000	100

f = 1.561E-08 Pa

Chemical name: Fluoranthene

Level II calculation: (six compartment model)

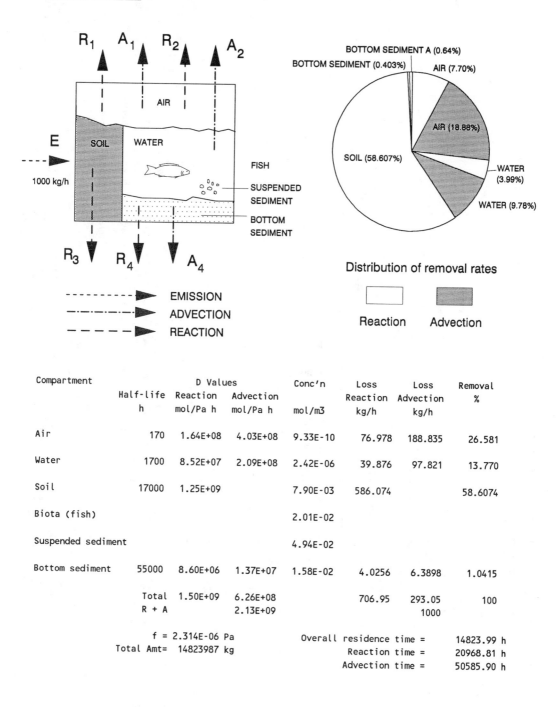

BOTTOM SEDIMENT A (0.64%)
BOTTOM SEDIMENT (0.403%)
AIR (7.70%)
AIR (18.88%)
SOIL (58.607%)
WATER (3.99%)
WATER (9.78%)

Distribution of removal rates

Reaction   Advection

EMISSION
ADVECTION
REACTION

Compartment	Half-life h	D Values Reaction mol/Pa h	Advection mol/Pa h	Conc'n mol/m3	Loss Reaction kg/h	Loss Advection kg/h	Removal %
Air	170	1.64E+08	4.03E+08	9.33E-10	76.978	188.835	26.581
Water	1700	8.52E+07	2.09E+08	2.42E-06	39.876	97.821	13.770
Soil	17000	1.25E+09		7.90E-03	586.074		58.6074
Biota (fish)				2.01E-02			
Suspended sediment				4.94E-02			
Bottom sediment	55000	8.60E+06	1.37E+07	1.58E-02	4.0256	6.3898	1.0415
Total		1.50E+09	6.26E+08		706.95	293.05	100
R + A			2.13E+09			1000	

f = 2.314E-06 Pa
Total Amt= 14823987 kg

Overall residence time = 14823.99 h
Reaction time = 20968.81 h
Advection time = 50585.90 h

307

Fugacity Level III calculations: (four compartment model)

Chemical name: Fluoranthene

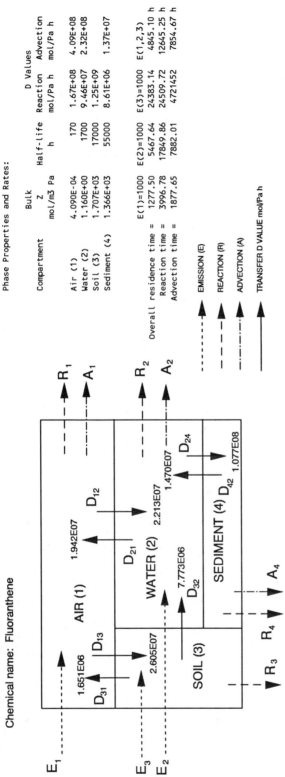

Legend:
- EMISSION (E)
- REACTION (R)
- ADVECTION (A)
- TRANSFER D VALUE mol/Pa h

Phase Properties and Rates:

Compartment	Bulk Z mol/m3 Pa	Half-life h	D Values Reaction mol/Pa h	D Values Advection mol/Pa h
Air (1)	4.090E-04	170	1.67E+08	4.09E+08
Water (2)	1.160E+00	1700	9.46E+07	2.32E+08
Soil (3)	1.707E+03	17000	1.25E+09	
Sediment (4)	1.366E+03	55000	8.61E+06	1.37E+07

	E(1)=1000	E(2)=1000	E(3)=1000	E(1,2,3)
Overall residence time =	1277.50	5467.64	24383.14	4845.10 h
Reaction time =	3996.78	17849.86	24509.72	12645.25 h
Advection time =	1877.65	7882.01	4721452	7854.67 h

Phase Properties, Compositions, Transport and Transformation Rates:

Emission, kg/h

E(1)	E(2)	E(3)
1000	0	0
0	1000	0
0	0	1000
600	300	100

Fugacity, Pa

f(1)	f(2)	f(3)	f(4)
7.937E-06	4.306E-07	1.639E-07	1.255E-06
3.751E-07	1.205E-05	7.744E-09	3.511E-05
1.270E-08	7.480E-08	3.918E-06	2.180E-07
4.876E-06	3.881E-06	4.924E-07	1.131E-05

Concentration, g/m3

C(1)	C(2)	C(3)	C(4)
6.567E-07	1.011E-04	5.657E-02	3.467E-01
3.104E-08	2.828E-03	2.674E-03	9.702E-02
1.051E-09	1.756E-05	1.353E+00	6.023E-02
4.034E-07	9.108E-04	1.700E-01	3.124E-01

Emission, kg/h

E(1)	E(2)	E(3)
1000	0	0
0	1000	0
0	0	1000
600	300	100

Loss, Reaction, kg/h

R(1)	R(2)	R(3)	R(4)
2.677E+02	8.240E+00	4.15E-01	2.184E+00
1.265E+01	2.306E+02	1.96E+00	6.112E+01
4.283E-01	1.431E+00	9.93E+02	3.794E-01
1.645E+02	7.426E+01	1.25E+02	1.968E+01

Loss, Advection, kg/h

A(1)	A(2)	A(4)
6.567E+02	2.021E+01	3.467E+00
3.104E+01	5.656E+02	9.702E+01
1.051E+00	3.511E+01	6.023E+01
4.034E+02	1.822E+02	3.124E+01

Amounts, kg

m(1)	m(2)	m(3)	m(4)	Total Amount, kg
6.567E+04	2.021E+04	1.018E+06	1.733E+05	1.277E+06
3.104E+03	5.656E+05	3.104E+04	4.851E+06	5.468E+06
1.051E+02	3.511E+03	2.435E+07	3.011E+04	2.438E+07
4.034E+04	1.822E+05	3.060E+06	1.562E+06	4.845E+06

Intermedia Rate of Transport, kg/h

air-water T12	water-air T21	air-soil T13	soil-air T31	soil-water T32	water-sed T24	sed-water T42
3.554E+01	1.692E+01	4.182E+01	5.474E-02	2.577E-01	9.381E+00	3.730E+00
1.680E+00	4.734E+01	1.977E+00	2.587E-03	1.218E-02	2.625E+02	1.044E+02
5.686E-02	2.939E-01	6.692E-02	1.309E+00	6.161E+00	1.630E+00	6.480E-01
2.183E+01	1.525E+01	2.569E+01	1.645E+01	7.744E-01	8.455E+01	3.362E+01

Level III Distribution

Chemical name: Fluoranthene

Distribution of mass

Distribution of removal rates

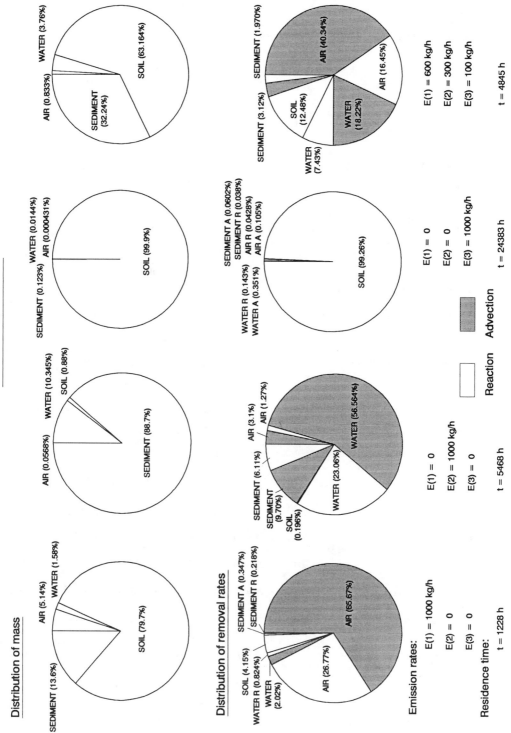

Emission rates:

E(1) = 1000 kg/h   E(1) = 0   E(1) = 0   E(1) = 600 kg/h

E(2) = 0   E(2) = 1000 kg/h   E(2) = 0   E(2) = 300 kg/h

E(3) = 0   E(3) = 0   E(3) = 1000 kg/h   E(3) = 100 kg/h

Residence time:

t = 1228 h   t = 5468 h   t = 24383 h   t = 4845 h

Advection

Reaction

Chemical name: Chrysene

Level I calculation: (six compartment model)

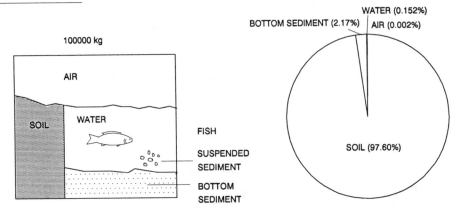

Distribution of mass

physical-chemical properties:

MW: 228.3 g/mol

M.P.: 255 °C

Fugacity ratio: 0.0053

vapor pressure: 5.70×10^{-7} Pa

solubility: 0.002 g/m^3

log K_{OW} : 5.86

Compartment	Z	Concentration			Amount	Amount
	mol/m3 Pa	mol/m3	mg/L (or g/m3)	ug/g	kg	%
Air	4.034E-04	8.746E-14	1.997E-11	1.684E-08	1.9967	0.0020
Water	1.537E+01	3.332E-09	7.607E-07	7.607E-07	152.13	0.1521
Soil	2.191E+05	4.750E-05	1.084E-02	4.519E-03	97604	97.604
Biota (fish)	5.567E+05	1.207E-04	2.755E-02	2.755E-02	5.5106	5.51E-03
Suspended sediment	1.369E+06	2.969E-04	6.778E-02	4.519E-02	67.780	6.78E-02
Bottom sediment	4.382E+05	9.501E-05	2.169E-02	9.037E-03	2168.97	2.1690
Total					100000	100

f = 2.168E-10 Pa

Chemical name: Chrysene

Level II calculation: (six compartment model)

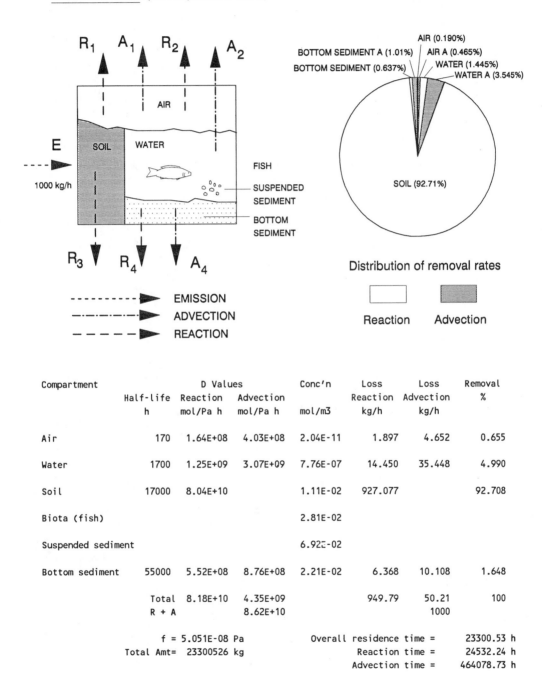

Distribution of removal rates

Compartment	Half-life h	D Values Reaction mol/Pa h	Advection mol/Pa h	Conc'n mol/m3	Loss Reaction kg/h	Loss Advection kg/h	Removal %
Air	170	1.64E+08	4.03E+08	2.04E-11	1.897	4.652	0.655
Water	1700	1.25E+09	3.07E+09	7.76E-07	14.450	35.448	4.990
Soil	17000	8.04E+10		1.11E-02	927.077		92.708
Biota (fish)				2.81E-02			
Suspended sediment				6.92E-02			
Bottom sediment	55000	5.52E+08	8.76E+08	2.21E-02	6.368	10.108	1.648
Total		8.18E+10	4.35E+09		949.79	50.21	100
R + A			8.62E+10			1000	

f = 5.051E-08 Pa
Total Amt= 23300526 kg

Overall residence time =   23300.53 h
Reaction time =   24532.24 h
Advection time =   464078.73 h

Fugacity Level III calculations: (four compartment model)

Chemical name: Chrysene

Diagram (four-compartment model) labels:

- Boxes: AIR (1), WATER (2), SOIL (3), SEDIMENT (4)
- Inputs/outputs: E_1, R_1, A_1, R_2, A_2, E_3, E_2, R_3, R_4, A_4
- Transfer D values: D_{31} 1.348E07, D_{13} 1.370E09, 2.012E07, D_{12}, D_{21}, 1.708E08, 8.918E08, D_{24}, D_{32} 2.664E08, D_{42} 6.863E09

Phase Properties and Rates:

Compartment	Bulk Z mol/m3 Pa	Half-life h	D Values Reaction mol/Pa h	D Values Advection mol/Pa h
Air (1)	8.545E-04	170	3.48E+08	8.54E+08
Water (2)	2.277E+01	1700	1.86E+09	4.55E+09
Soil (3)	1.096E+05	17000	8.04E+10	
Sediment (4)	8.766E+04	55000	5.52E+08	8.77E+08

	E(1)=1000	E(2)=1000	E(3)=1000	E(1,2,3)
Overall residence time =	13044	12614	24490	14060 h
Reaction time =	20205	38378	24546	24004 h
Advection time =	36804	18791	10761260	33938 h

Legend:
- ---- EMISSION (E)
- ---- REACTION (R)
- -·-·- ADVECTION (A)
- → TRANSFER D VALUE mol/Pa h

Phase Properties, Compositions, Transport and Transformation Rates:

Emission, kg/h — Fugacity, Pa — Concentration, g/m3 — Amounts, kg — Total Amount, kg

E(1)	E(2)	E(3)	f(1)	f(2)	f(3)	f(4)	C(1)	C(2)	C(3)	C(4)	m(1)	m(2)	m(3)	m(4)	Total Amount, kg
1000	0	0	1.597E-06	2.627E-08	2.711E-08	7.770E-08	3.115E-07	1.366E-04	6.782E-01	1.555E+00	3.115E+00	2.732E+04	1.221E+07	7.775E+05	1.304E+07
0	1000	0	3.015E-09	4.111E-07	5.119E-11	1.216E-06	5.881E-10	2.137E-03	1.280E-03	2.433E+01	5.881E+01	4.274E+05	2.305E+04	1.216E+07	1.261E+07
0	0	1000	2.768E-10	1.362E-09	5.430E-08	4.027E-09	5.399E-11	7.079E-06	1.358E+00	8.059E-02	5.399E+00	1.416E+03	2.445E+07	4.029E+04	2.449E+07
600	300	100	9.591E-07	1.392E-07	2.171E-08	4.117E-07	1.871E-07	7.238E-04	5.431E-01	8.239E-01	1.871E+04	1.448E+05	9.776E+06	4.120E+06	1.406E+07

Emission, kg/h — Loss, Reaction, kg/h — Loss, Advection, kg/h — Intermedia Rate of Transport, kg/h

E(1)	E(2)	E(3)	R(1)	R(2)	R(3)	R(4)	A(1)	A(2)	A(4)
1000	0	0	1.270E+02	1.114E+01	4.98E+02	9.796E+00	3.115E+02	2.732E+01	1.555E+01
0	1000	0	2.398E-01	1.742E+02	9.39E-01	1.533E+02	5.881E-01	4.274E+02	2.433E+02
0	0	1000	2.201E-02	5.772E-01	9.97E+02	5.077E+00	5.399E-02	1.416E+00	8.059E-01
600	300	100	7.627E+01	5.901E+01	3.99E+02	5.191E+01	1.871E+02	1.448E+02	8.239E+01

Intermedia Rate of Transport, kg/h

T12 air-water	T21 water-air	T13 air-soil	T31 soil-air	T32 soil-water	T24 water-sed	T42 sed-water
6.228E+01	1.207E-01	9.427E-01	8.343E-02	1.649E+00	4.117E+01	1.582E+01
1.176E-01	1.888E+00	9.427E-01	1.575E-04	3.113E-03	6.441E+02	2.475E+02
1.079E-02	6.254E-03	8.654E-02	1.671E-01	3.302E+00	2.133E+00	8.199E-01
3.740E+01	6.394E-01	2.999E+02	6.681E-02	1.320E+02	2.181E+02	8.383E+01

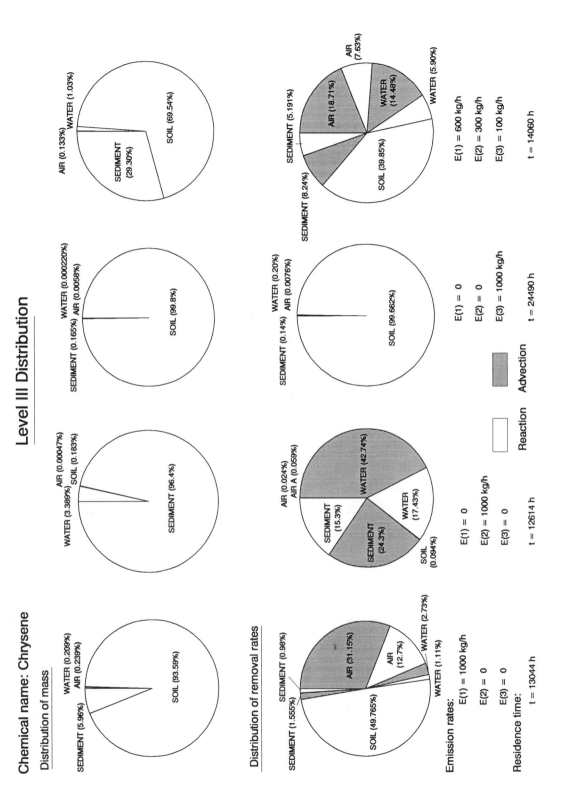

Level III Distribution

Chemical name: Chrysene

Distribution of mass

Pie 1 (top left):
- WATER (0.209%)
- AIR (0.239%)
- SEDIMENT (5.96%)
- SOIL (93.59%)
- t = 13044 h

Pie 2 (middle left):
- WATER (3.389%)
- AIR (0.00047%)
- SOIL (0.183%)
- SEDIMENT (96.4%)
- t = 12614 h

Pie 3 (top right):
- WATER (0.000220%)
- AIR (0.0058%)
- SEDIMENT (0.165%)
- SOIL (99.8%)
- t = 24490 h

Pie 4 (far right top):
- WATER (1.03%)
- AIR (0.133%)
- SEDIMENT (29.30%)
- SOIL (69.54%)
- t = 14060 h

Distribution of removal rates

Pie 5 (bottom left):
- SEDIMENT (0.98%)
- SEDIMENT (1.555%)
- AIR (31.15%)
- AIR (12.7%)
- WATER (2.73%)
- WATER (1.11%)
- SOIL (49.765%)

Pie 6 (bottom middle):
- AIR (0.024%)
- AIR A (0.059%)
- WATER (42.74%)
- WATER (17.43%)
- SEDIMENT (15.3%)
- SEDIMENT (24.3%)
- SOIL (0.094%)

Pie 7 (bottom right):
- WATER (0.20%)
- AIR (0.0076%)
- SEDIMENT (0.14%)
- SOIL (99.662%)

Pie 8 (far right):
- AIR (7.63%)
- WATER (5.90%)
- WATER (14.48%)
- AIR (18.71%)
- SEDIMENT (5.191%)
- SEDIMENT (8.24%)
- SOIL (39.85%)

Emission rates:

E(1) = 1000 kg/h	E(1) = 0	E(1) = 0	E(1) = 600 kg/h
E(2) = 0	E(2) = 1000 kg/h	E(2) = 0	E(2) = 300 kg/h
E(3) = 0	E(3) = 0	E(3) = 1000 kg/h	E(3) = 100 kg/h

Residence time:

Advection (shaded)
Reaction (white)

313

Chemical name: Benz[a]anthracene

Level I calculation: (six compartment model)

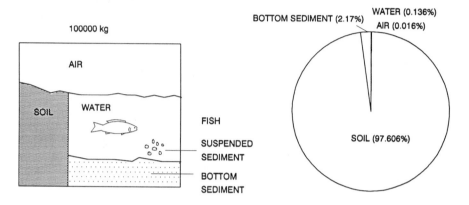

100000 kg

AIR

SOIL   WATER

FISH

SUSPENDED
SEDIMENT

BOTTOM
SEDIMENT

BOTTOM SEDIMENT (2.17%)   WATER (0.136%)
AIR (0.016%)

SOIL (97.606%)

Distribution of mass

physical-chemical properties:

MW: 228.3 g/mol

M.P.: 160 °C

Fugacity ratio: 0.0462

vapor pressure: 0.000028 Pa

solubility: 0.011 g/m^3

log K_{OW} : 5.91

| Compartment | Z | Concentration | | | Amount | Amount |
	mol/m3 Pa	mol/m3	mg/L (or g/m3)	ug/g	kg	%
Air	4.034E-04	6.962E-13	1.589E-10	1.341E-07	15.8941	0.0159
Water	1.721E+00	2.970E-09	6.780E-07	6.780E-07	135.59	0.1356
Soil	2.753E+04	4.750E-05	1.085E-02	4.519E-03	97606	97.606
Biota (fish)	6.994E+04	1.207E-04	2.755E-02	2.755E-02	5.5107	5.51E-03
Suspended sediment	1.720E+05	2.969E-04	6.778E-02	4.519E-02	67.782	6.78E-02
Bottom sediment	5.505E+04	9.501E-05	2.169E-02	9.038E-03	2169.03	2.1690
Total					100000	100

f = 1.726E-09 Pa

Chemical name: Benz[a]anthracene

Level II calculation: (six compartment model)

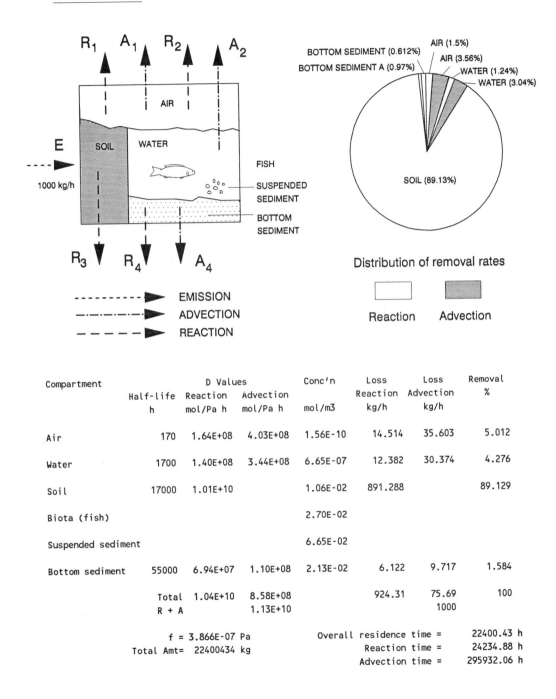

Distribution of removal rates

Reaction   Advection

Compartment	Half-life h	D Values Reaction mol/Pa h	D Values Advection mol/Pa h	Conc'n mol/m3	Loss Reaction kg/h	Loss Advection kg/h	Removal %
Air	170	1.64E+08	4.03E+08	1.56E-10	14.514	35.603	5.012
Water	1700	1.40E+08	3.44E+08	6.65E-07	12.382	30.374	4.276
Soil	17000	1.01E+10		1.06E-02	891.288		89.129
Biota (fish)				2.70E-02			
Suspended sediment				6.65E-02			
Bottom sediment	55000	6.94E+07	1.10E+08	2.13E-02	6.122	9.717	1.584
Total		1.04E+10	8.58E+08		924.31	75.69	100
R + A			1.13E+10			1000	

f = 3.866E-07 Pa
Total Amt= 22400434 kg

Overall residence time =   22400.43 h
Reaction time =   24234.88 h
Advection time =   295932.06 h

315

Fugacity Level III calculations: (four compartment model)

Chemical Name: Benz[a]anthracene

Phase Properties and Rates:

Compartment	Bulk Z mol/m3 Pa	Half-life h	D Values Reaction mol/Pa h	Advection mol/Pa h
Air (1)	4.833E-04	170	1.97E+08	4.83E+08
Water (2)	2.651E+00	1700	2.16E+08	5.30E+08
Soil (3)	1.376E+04	17000	1.01E+10	1.10E+08
Sediment (4)	1.101E+04	55000	6.94E+07	1.10E+08

	E(1)=1000	E(2)=1000	E(3)=1000	E(1,2,3)
Overall residence time =	6630	13069	24490	10348 h
Reaction time =	14290	39373	24546	21660 h
Advection time =	12368	19562	10824422	19812 h

EMISSION (E)
REACTION (R)
ADVECTION (A)
TRANSFER D VALUE mol/Pa h

Phase Properties, Compositions, Transport and Transformation Rates:

Emission, kg/h / Fugacity, Pa / Concentration, g/m3 / Amounts, kg

E(1)	E(2)	E(3)	f(1)	f(2)	f(3)	f(4)	C(1)	C(2)	C(3)	C(4)	m(1)	m(2)	m(3)	m(4)	Total Amount, kg
1000	0	0	4.570E-06	1.626E-07	1.053E-07	4.811E-07	5.043E-07	9.841E-05	3.308E-01	1.209E+00	5.043E+04	1.968E+04	5.955E+06	6.047E+05	6.630E+06
0	1000	0	6.944E-08	3.379E-06	1.600E-09	9.999E-06	7.662E-09	2.045E-03	5.027E-03	2.514E+01	7.662E+02	4.090E+05	9.048E+04	1.257E+07	1.307E+07
0	0	1000	1.236E-09	1.088E-08	4.322E-07	3.219E-08	1.364E-10	6.584E-06	1.358E+00	8.092E-02	1.364E+01	1.317E+03	2.445E+07	4.046E+04	2.449E+07
600	300	100	2.763E-06	1.112E-06	1.069E-07	3.291E-06	3.049E-07	6.733E-04	3.358E-01	8.275E-01	3.049E+04	1.347E+05	6.045E+06	4.137E+06	1.035E+07

Emission, kg/h / Loss, Reaction, kg/h / Loss, Advection, kg/h

E(1)	E(2)	E(3)	R(1)	R(2)	R(3)	R(4)	A(1)	A(2)	A(4)
1000	0	0	2.056E+02	8.023E+00	2.43E+02	7.619E+00	5.043E+02	1.968E+01	1.209E+01
0	1000	0	3.123E+00	1.667E+01	3.69E+00	1.584E+02	7.662E+00	4.090E+02	2.514E+02
0	0	1000	5.559E-02	5.368E-01	9.97E+02	5.098E-01	1.364E-01	1.317E+00	8.092E-01
600	300	100	1.243E+02	5.489E+01	2.46E+02	5.213E+01	3.049E+02	1.347E+02	8.275E+01

Intermedia Rate of Transport, kg/h

	T12 air-water	T21 water-air	T13 air-soil	T31 soil-air	T32 soil-water	T24 water-sed	T42 sed-water
	4.737E+01	7.316E-01	2.436E+02	5.400E-02	7.816E-01	3.200E+01	1.228E+01
	7.197E-01	1.521E+01	3.701E+00	8.205E-04	1.188E-02	6.650E+02	2.553E+02
	1.281E-02	4.895E-02	6.588E-02	2.217E-01	3.209E+00	2.141E+00	8.218E-01
	2.864E+01	5.006E+00	1.473E+02	5.482E-02	7.934E-01	2.189E+02	8.403E+01

Level III Distribution

Chemical name: Benz[a]anthracene

Distribution of mass

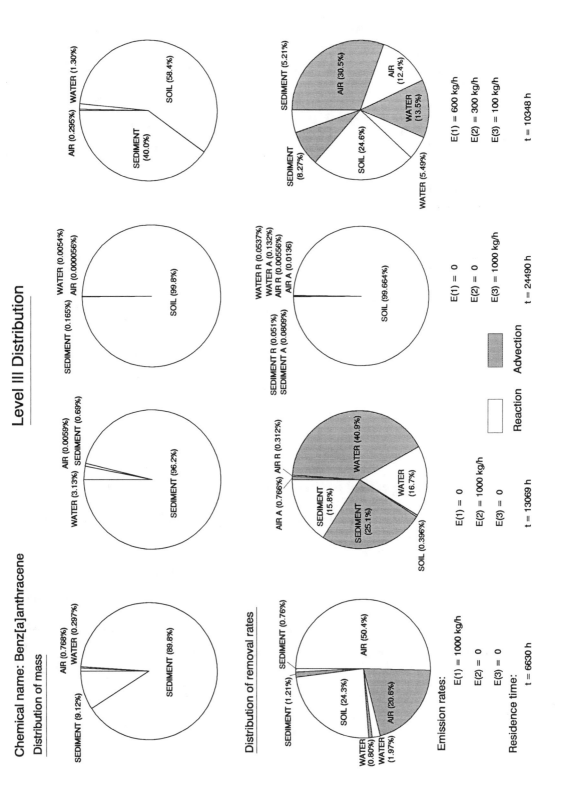

Distribution of removal rates

Chemical name: Benzo[a]pyrene

Level I calculation: (six compartment model)

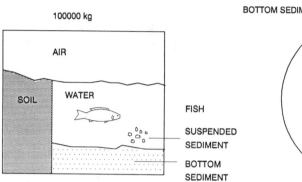

100000 kg

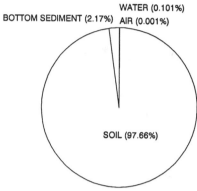

WATER (0.101%)
BOTTOM SEDIMENT (2.17%) AIR (0.001%)

SOIL (97.66%)

Distribution of mass

physical-chemical properties:

MW: 252.32 g/mol

M.P.: 175 °C

Fugacity ratio: 0.0328

vapor pressure: 7.0×10^{-7} Pa

solubility: 0.0038 g/m^3

log K_{OW} : 6.04

Compartment	Z	Concentration			Amount	Amount
	mol/m3 Pa	mol/m3	mg/L (or g/m3)	ug/g	kg	%
Air	4.034E-04	3.737E-14	9.428E-12	7.953E-09	0.9428	9.43E-04
Water	2.152E+01	1.993E-09	5.028E-07	5.028E-07	100.57	0.1006
Soil	4.643E+05	4.301E-05	1.085E-02	4.521E-03	97655	97.655
Biota (fish)	1.180E+06	1.093E-04	2.757E-02	2.757E-02	5.5135	5.51E-03
Suspended sediment	2.902E+06	2.688E-04	6.782E-02	4.521E-02	67.816	6.78E-02
Bottom sediment	9.286E+05	8.601E-05	2.170E-02	9.042E-03	2170.11	2.1701
	Total				100000	100

f = 9.263E-11 Pa

Chemical name: Benzo[a]pyrene

Level II calculation: (six compartment model)

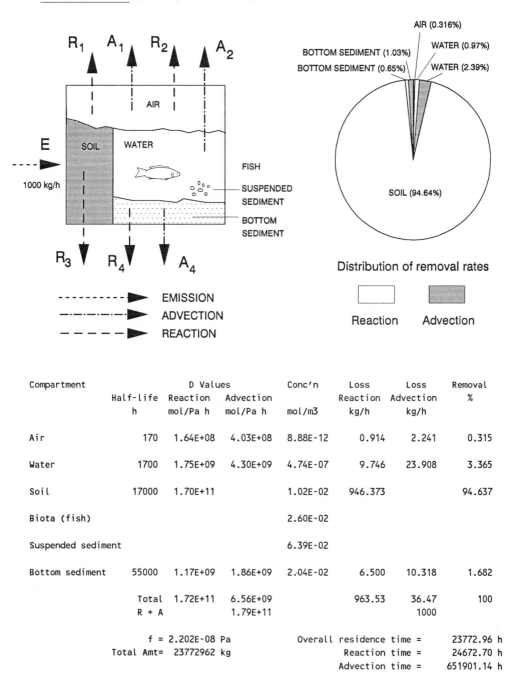

Distribution of removal rates

Reaction   Advection

Compartment	Half-life h	D Values Reaction mol/Pa h	Advection mol/Pa h	Conc'n mol/m3	Loss Reaction kg/h	Loss Advection kg/h	Removal %
Air	170	1.64E+08	4.03E+08	8.88E-12	0.914	2.241	0.315
Water	1700	1.75E+09	4.30E+09	4.74E-07	9.746	23.908	3.365
Soil	17000	1.70E+11		1.02E-02	946.373		94.637
Biota (fish)				2.60E-02			
Suspended sediment				6.39E-02			
Bottom sediment	55000	1.17E+09	1.86E+09	2.04E-02	6.500	10.318	1.682
Total		1.72E+11	6.56E+09		963.53	36.47	100
R + A			1.79E+11			1000	

f = 2.202E-08 Pa
Total Amt= 23772962 kg

Overall residence time =   23772.96 h
Reaction time =   24672.70 h
Advection time =   651901.14 h

Fugacity Level III calculations: (four compartment model)

Chemical Name: Benzo[a]pyrene

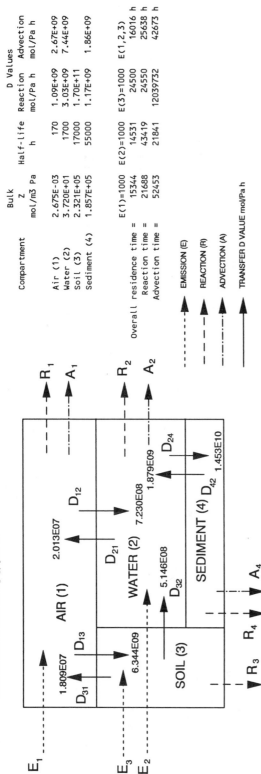

Phase Properties and Rates:

Compartment	Bulk Z mol/m3 Pa	Half-life h	D Values Reaction mol/Pa h	Advection mol/Pa h
Air (1)	2.675E-03	170	1.09E+09	2.67E+09
Water (2)	3.720E+01	1700	3.03E+09	7.44E+09
Soil (3)	2.321E-05	17000	1.70E+11	
Sediment (4)	1.857E+05	55000	1.17E+09	1.86E+09

	E(1)=1000	E(2)=1000	E(3)=1000	E(1,2,3)
Overall residence time =	15344	14531	16016 h	
Reaction time =	21688	43419	25638 h	
Advection time =	52453	21841	42673 h	

EMISSION (E)
REACTION (R)
ADVECTION (A)
TRANSFER D VALUE mol/Pa h

320

Phase Properties, Compositions, Transport and Transformation Rates:

Emission, kg/h

E(1)	E(2)	E(3)
1000	0	0
0	1000	0
0	0	1000
600	300	100

Fugacity, Pa

f(1)	f(2)	f(3)	f(4)
3.659E-07	1.396E-08	1.359E-08	4.133E-08
3.786E-10	2.037E-07	1.406E-11	6.032E-07
3.988E-11	6.150E-10	2.320E-08	1.821E-09
2.197E-07	6.954E-08	1.048E-08	2.060E-07

Concentration, g/m3

C(1)	C(2)	C(3)	C(4)
2.470E-07	1.310E-04	7.958E-01	1.937E+00
2.555E-11	1.912E-03	8.233E-04	2.827E+01
2.691E-11	5.773E-06	1.359E+00	8.535E-02
1.483E-07	6.527E-04	6.136E-01	9.651E-01

Amounts, kg

m(1)	m(2)	m(3)	m(4)	Total Amount, kg
2.470E+04	2.620E+04	1.432E+07	9.684E+05	1.534E+07
2.691E+00	3.824E+05	1.482E+04	1.413E+07	1.453E+07
2.691E-02	1.155E+03	2.446E+07	4.267E+04	2.450E+07
1.483E+04	1.305E+05	1.104E+07	4.826E+06	1.602E+07

Emission, kg/h

E(1)	E(2)	E(3)
1000	0	0
0	1000	0
0	0	1000
600	300	100

Loss, Reaction, kg/h

R(1)	R(2)	R(3)	R(4)
1.007E+02	1.068E+01	5.84E+02	1.220E+01
1.042E-01	1.559E+02	6.04E-01	1.781E+02
1.097E-02	4.706E-01	9.97E+02	5.377E-01
6.043E+01	5.322E+01	4.50E+02	6.080E+01

Loss, Advection, kg/h

A(1)	A(2)	A(4)
2.470E+02	2.620E+01	1.937E+01
2.555E-01	3.824E+02	2.827E+02
2.691E-02	1.155E+00	8.535E-01
1.483E+02	1.305E+02	9.651E+01

Intermedia Rate of Transport, kg/h

T12 air-water	T13 air-soil	T21 water-air	T24 water-sed	T31 soil-air	T32 soil-water	T42 sed-water
6.676E+01	5.858E+02	7.089E-02	5.116E+01	6.201E-02	1.764E+00	1.959E+01
6.907E+02	6.060E-01	1.035E+00	7.467E+02	6.415E-05	1.825E-03	2.859E+02
7.275E-03	6.383E-02	3.124E-03	2.255E+00	1.059E-01	3.012E+00	8.633E-01
4.008E+01	3.516E+02	3.533E-01	2.549E+02	4.781E+01	1.360E+00	9.762E+01

Level III Distribution

Chemical name: Benzo[a]pyrene

Distribution of mass

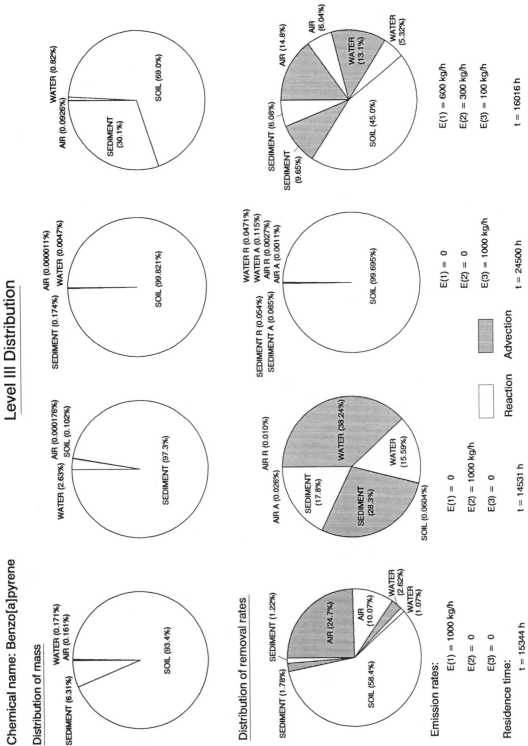

Distribution of removal rates

Emission rates:

E(1) = 1000 kg/h
E(2) = 0
E(3) = 0

Residence time:
t = 15344 h

E(1) = 0
E(2) = 1000 kg/h
E(3) = 0

t = 14531 h

E(1) = 0
E(2) = 0
E(3) = 1000 kg/h

t = 24500 h

E(1) = 600 kg/h
E(2) = 300 kg/h
E(3) = 100 kg/h

t = 16016 h

Reaction   Advection

Chemical name: Perylene

Level I calculation: (six compartment model)

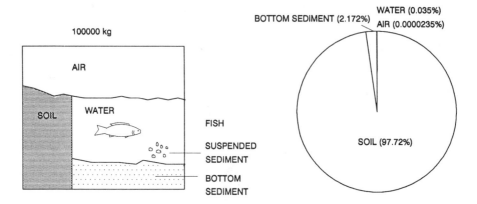

Distribution of mass

physical-chemical properties:

MW: 252.32 g/mol

M.P.: 277 °C

Fugacity ratio: 0.0032

vapor pressure: 1.40×10^{-8} Pa

solubility: 0.0004 g/m^3

log K_{OW} : 6.25

Compartment	Z	Concentration			Amount	Amount
	mol/m3 Pa	mol/m3	mg/L (or g/m3)	ug/g	kg	%
Air	4.034E-04	4.379E-15	1.105E-12	9.322E-10	0.1105	1.11E-04
Water	1.132E+02	1.229E-09	3.102E-07	3.102E-07	62.03	0.0620
Soil	3.963E+06	4.302E-05	1.085E-02	4.523E-03	97694	97.694
Biota (fish)	1.007E+07	1.093E-04	2.758E-02	2.758E-02	5.5157	5.52E-03
Suspended sediment	2.477E+07	2.689E-04	6.784E-02	4.523E-02	67.843	6.78E-02
Bottom sediment	7.926E+06	8.604E-05	2.171E-02	9.046E-03	2170.97	2.1710
	Total				100000	100

f = 1.086E-11 Pa

Chemical name: Perylene

Level II calculation: (six compartment model)

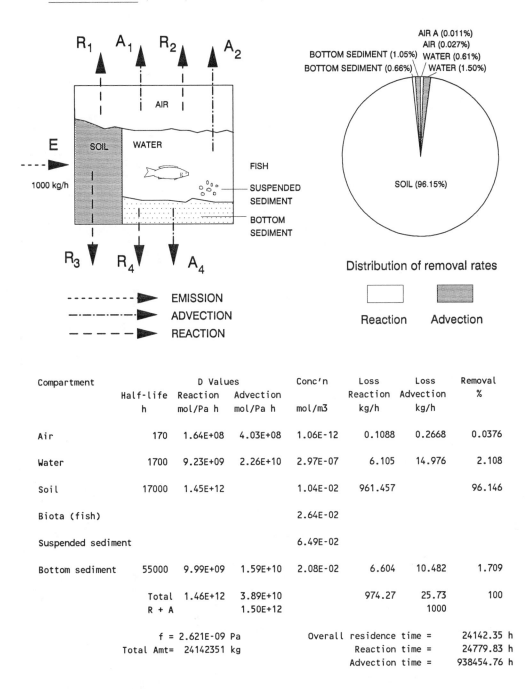

Distribution of removal rates

Reaction   Advection

Compartment	Half-life h	D Values Reaction mol/Pa h	Advection mol/Pa h	Conc'n mol/m3	Loss Reaction kg/h	Loss Advection kg/h	Removal %
Air	170	1.64E+08	4.03E+08	1.06E-12	0.1088	0.2668	0.0376
Water	1700	9.23E+09	2.26E+10	2.97E-07	6.105	14.976	2.108
Soil	17000	1.45E+12		1.04E-02	961.457		96.146
Biota (fish)				2.64E-02			
Suspended sediment				6.49E-02			
Bottom sediment	55000	9.99E+09	1.59E+10	2.08E-02	6.604	10.482	1.709
Total		1.46E+12	3.89E+10		974.27	25.73	100
R + A			1.50E+12			1000	

f = 2.621E-09 Pa
Total Amt= 24142351 kg

Overall residence time =   24142.35 h
Reaction time =   24779.83 h
Advection time =   938454.76 h

Fugacity Level III calculations: (four compartment model)

Chemical name: Perylene

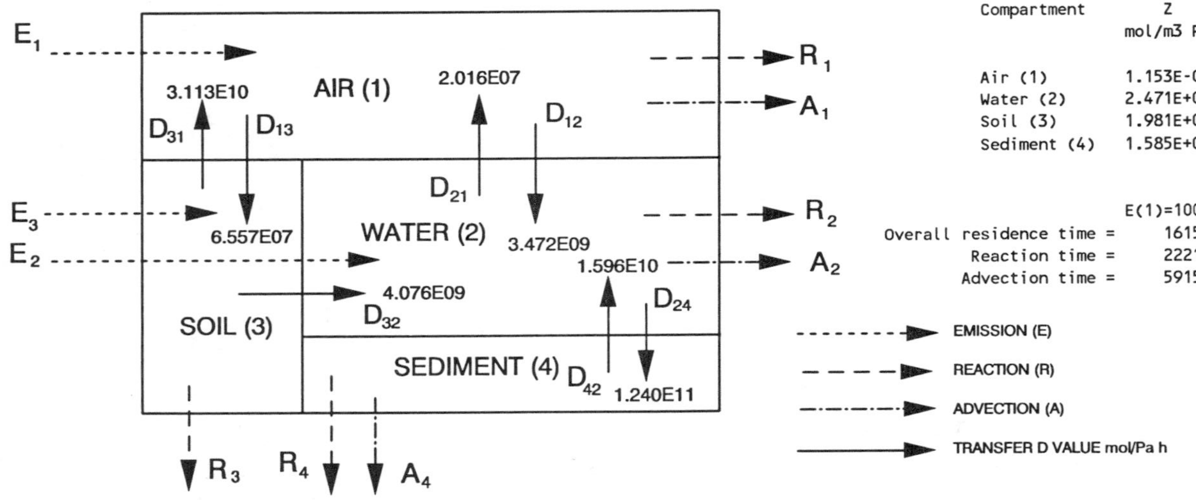
E1
E3
E2
AIR (1)
WATER (2)
SOIL (3)
SEDIMENT (4)
3.113E10
D31
D13
6.557E07
2.016E07
D21
D12
3.472E09
4.076E09
D32
1.596E10
D24
D42
1.240E11
R1
A1
R2
A2
R3
R4
A4
EMISSION (E)
REACTION (R)
ADVECTION (A)
TRANSFER D VALUE mol/Pa h

Phase Properties and Rates:

Compartment	Bulk Z mol/m3 Pa	Half-life h	D Values Reaction mol/Pa h	Advection mol/Pa h
Air (1)	1.153E-02	170	4.70E+09	1.15E+10
Water (2)	2.471E+02	1700	2.01E+10	4.94E+10
Soil (3)	1.981E+06	17000	1.45E+12	
Sediment (4)	1.585E+06	55000	9.99E+09	1.59E+10

	E(1)=1000	E(2)=1000	E(3)=1000	E(1,2,3)
Overall residence time =	16152	16414	24508	17066 h
Reaction time =	22218	48214	24554	26744 h
Advection time =	59159	24886	13203738	47161 h

Phase Properties, Compositions, Transport and Transformation Rates:

Emission, kg/h			Fugacity, Pa				Concentration, g/m3				Amounts, kg				Total Amount, kg
E(1)	E(2)	E(3)	f(1)	f(2)	f(3)	f(4)	C(1)	C(2)	C(3)	C(4)	m(1)	m(2)	m(3)	m(4)	
1000	0	0	7.797E-08	1.898E-09	1.665E-09	5.627E-09	2.269E-07	1.183E-04	8.322E-01	2.251E+00	2.269E+04	2.367E+04	1.498E+07	1.125E+06	1.615E+07
0	1000	0	1.075E-11	2.711E-08	2.296E-13	8.037E-08	3.128E-11	1.690E-03	1.148E-04	3.215E+01	3.128E+00	3.381E+05	2.066E+03	1.607E+07	1.641E+07
0	0	1000	3.536E-12	7.586E-11	2.718E-09	2.249E-10	1.029E-11	4.731E-06	1.359E+00	8.997E-02	1.029E+00	9.461E+02	2.446E+07	4.499E+04	2.451E+07
600	300	100	4.679E-08	9.278E-09	1.271E-09	2.751E-08	1.361E-07	5.786E-04	6.353E-01	1.100E+01	1.361E+04	1.157E+05	1.144E+07	5.502E+06	1.707E+07

Emission, kg/h			Loss, Reaction, kg/h				Loss, Advection, kg/h			Intermedia Rate of Transport, kg/h T12	T21	T13	T31	T32	T24	T42
E(1)	E(2)	E(3)	R(1)	R(2)	R(3)	R(4)	A(1)	A(2)	A(4)	air-water	water-air	air-soil	soil-air	soil-water	water-sed	sed-water
1000	0	0	9.248E+01	9.648E+00	6.11E+02	1.418E+01	2.269E+02	2.367E+01	2.251E+01	6.830E+01	9.655E-03	6.124E+02	2.754E-02	1.712E+00	5.935E+01	2.267E+01
0	1000	0	1.275E-02	1.378E+02	8.42E-02	2.025E+02	3.128E-02	3.381E+02	3.215E+02	9.419E-03	1.379E-01	8.446E-02	3.798E-06	2.361E-04	8.478E+02	3.238E+02
0	0	1000	4.194E-03	3.857E-01	9.97E+02	5.668E-01	1.029E-02	9.461E-01	8.997E-01	3.098E-03	3.860E-04	2.778E-02	4.497E-02	2.796E+00	2.373E+00	9.061E-01
600	300	100	5.549E+01	4.717E+01	4.66E+02	6.932E+01	1.361E+02	1.157E+02	1.100E+02	4.098E+01	4.720E-02	3.675E+02	2.102E-02	1.307E+00	2.902E+02	1.108E+02

Level III Distribution

Chemical name: Perylene

Distribution of mass

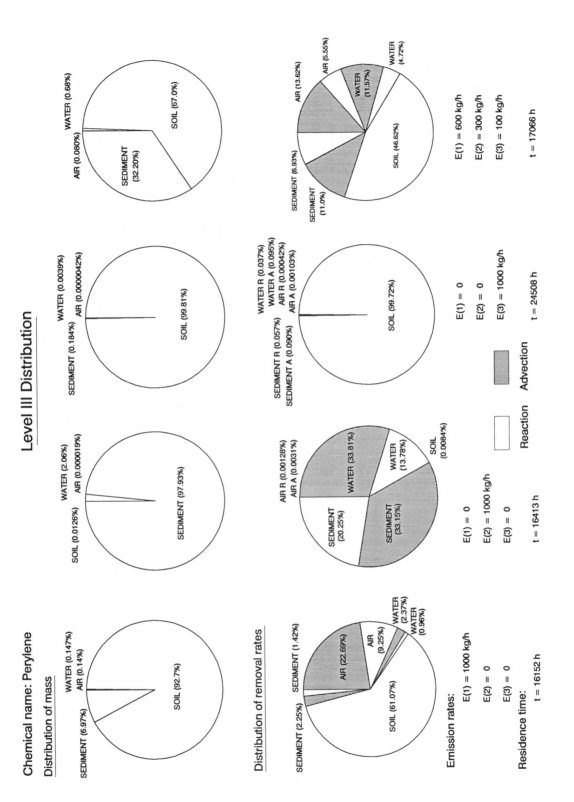

Distribution of removal rates

Emission rates:

E(1) = 1000 kg/h
E(2) = 0
E(3) = 0

Residence time:

t = 16152 h

E(1) = 0
E(2) = 1000 kg/h
E(3) = 0

t = 16413 h

E(1) = 0
E(2) = 0
E(3) = 1000 kg/h

t = 24508 h

E(1) = 600 kg/h
E(2) = 300 kg/h
E(3) = 100 kg/h

t = 17066 h

Chemical name: Benzo[k]fluoranthene

Level I calculation: (six compartment model)

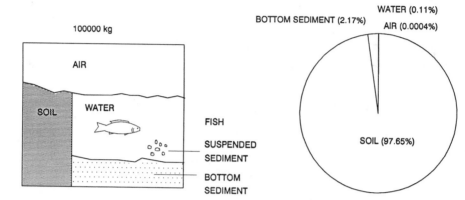

Distribution of mass

physical-chemical properties:

MW: 278.35 g/mol

M.P.: 217°C

Fugacity ratio: 0.0126

vapor pressure: 5.2×10^{-8} Pa

solubility: 0.0008 g/m^3

log K_{OW} : 6.0

Compartment	Z	Concentration			Amount	Amount
	mol/m3 Pa	mol/m3	mg/L (or g/m3)	ug/g	kg	%
Air	4.034E-04	1.446E-14	4.024E-12	3.395E-09	4.02E-01	4.02E-04
Water	5.527E+01	1.981E-09	5.513E-07	5.513E-07	110.2598	0.1103
Soil	1.088E+06	3.898E-05	1.085E-02	4.521E-03	97646.1	97.646
Biota (fish)	2.764E+06	9.903E-05	2.756E-02	2.756E-02	5.5130	5.51E-03
Suspended sediment	6.798E+06	2.436E-04	6.781E-02	4.521E-02	67.810	6.78E-02
Bottom sediment	2.175E+06	7.796E-05	2.170E-02	9.041E-03	2169.913	2.1699
Total					100000	100

f = 3.583E-11 Pa

Chemical name: Benzo[k]fluoranthene

Level II calculation: (six compartment model)

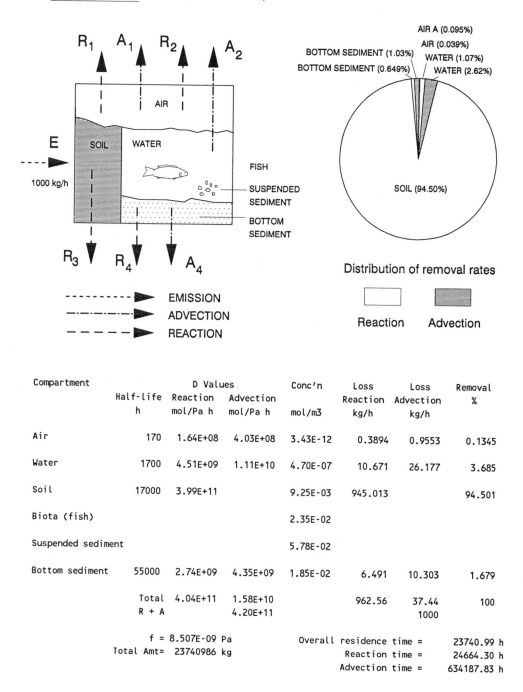

Distribution of removal rates

Reaction   Advection

Compartment	Half-life h	D Values Reaction mol/Pa h	Advection mol/Pa h	Conc'n mol/m3	Loss Reaction kg/h	Loss Advection kg/h	Removal %
Air	170	1.64E+08	4.03E+08	3.43E-12	0.3894	0.9553	0.1345
Water	1700	4.51E+09	1.11E+10	4.70E-07	10.671	26.177	3.685
Soil	17000	3.99E+11		9.25E-03	945.013		94.501
Biota (fish)				2.35E-02			
Suspended sediment				5.78E-02			
Bottom sediment	55000	2.74E+09	4.35E+09	1.85E-02	6.491	10.303	1.679
	Total	4.04E+11	1.58E+10		962.56	37.44	100
	R + A		4.20E+11			1000	

$f = 8.507E-09$ Pa
Total Amt= 23740986 kg

Overall residence time = 23740.99 h
Reaction time = 24664.30 h
Advection time = 634187.83 h

327

Fugacity Level III calculations: (four compartment model)

Chemical name: Benzo[k]fluoranthene

Phase Properties and Rates:

Compartment	Bulk Z mol/m3 Pa	Half-life h	D Values Reaction mol/Pa h	Advection mol/Pa h
Air (1)	1.215E-02	170	4.95E+09	1.22E+10
Water (2)	9.203E+01	1700	7.50E+09	1.84E+10
Soil (3)	5.439E+05	17000	3.99E+11	
Sediment (4)	4.351E+05	55000	2.74E+09	4.35E+09

	E(1)=1000	E(2)=1000	E(3)=1000	E(1,2,3)
Overall residence time =	15904	14126	24549	16230 h
Reaction time =	21968	42402	24549	25595 h
Advection time =	57617	21183	11822671	44359 h

EMISSION (E)
REACTION (R)
ADVECTION (A)
TRANSFER D VALUE mol/Pa h

R_1 A_1
R_2 A_2
E_1 E_3 E_2
R_3 R_4 A_4

AIR (1)   2.016E07
WATER (2)   3.600E09
SOIL (3)
SEDIMENT (4) D_{42} 3.405E10

D_{12} D_{21} D_{13} D_{31} 3.949E07   3.226E10
D_{32} 1.228E09 D_{24} 4.406E09

Phase Properties, Compositions, Transport and Transformation Rates:

Emission, kg/h

E(1)	E(2)	E(3)
1000	0	0
0	1000	0
0	0	1000
600	300	100

Fugacity, Pa

f(1)	f(2)	f(3)	f(4)
6.784E-09	5.347E-09	5.466E-09	1.583E-08
2.914E-11	7.656E-08	2.348E-12	2.267E-07
6.781E-12	2.353E-10	8.974E-09	6.967E-10
4.071E-08	2.620E-08	4.178E-09	7.757E-08

Concentration, g/m3

C(1)	C(2)	C(3)	C(4)
2.295E-07	1.370E-04	8.275E-01	1.918E+00
9.855E-11	1.961E-03	3.554E-04	2.745E+01
2.294E-11	6.027E-06	1.359E+00	8.438E-02
1.377E-07	6.711E-04	6.325E-01	9.395E-01

Amounts, kg

m(1)	m(2)	m(3)	m(4)	Total Amount, kg
2.295E+04	2.739E+04	1.489E+07	9.588E+05	1.590E+07
9.855E+00	3.922E+05	6.397E+03	1.373E+07	1.413E+07
2.294E+00	1.205E+03	2.445E+07	4.219E+04	2.450E+07
1.377E+04	1.342E+05	1.138E+07	4.698E+06	1.623E+07

Emission, kg/h

E(1)	E(2)	E(3)
1000	0	0
0	1000	0
0	0	1000
600	300	100

Loss, Reaction, kg/h

R(1)	R(2)	R(3)	R(4)
9.354E+01	1.117E+01	6.07E+02	1.208E+01
4.018E-02	1.599E+02	2.61E-01	1.730E+02
9.350E-03	4.914E-01	9.97E+02	5.316E-01
5.614E+01	5.471E+01	4.64E+02	5.919E+01

Loss, Advection, kg/h

A(1)	A(2)	A(4)
2.295E+02	2.739E+01	1.918E+01
9.855E-02	3.922E+02	2.745E+02
2.294E-02	1.205E+00	8.438E-01
1.377E+02	1.342E+02	9.395E+01

Intermedia Rate of Transport, kg/h

T12 air-water	T13 air-soil	T21 water-air	T31 soil-air	T32 soil-water	T24 water-sed	T42 sed-water
6.798E-01	6.091E-02	3.000E+02	6.008E-02	1.868E-02	5.067E+01	1.942E+01
2.920E-02	2.616E-01	4.295E-01	2.581E-05	8.023E-04	7.255E+02	2.780E+02
6.795E-03	6.089E-02	1.320E-03	9.865E-02	3.067E+00	2.230E+00	8.544E-01
4.080E+01	3.656E+02	1.470E-01	4.592E+02	1.428E+02	2.483E+02	9.514E+01

Level III Distribution

Chemical name: Benzo[k]fluoranthene

Distribution of mass

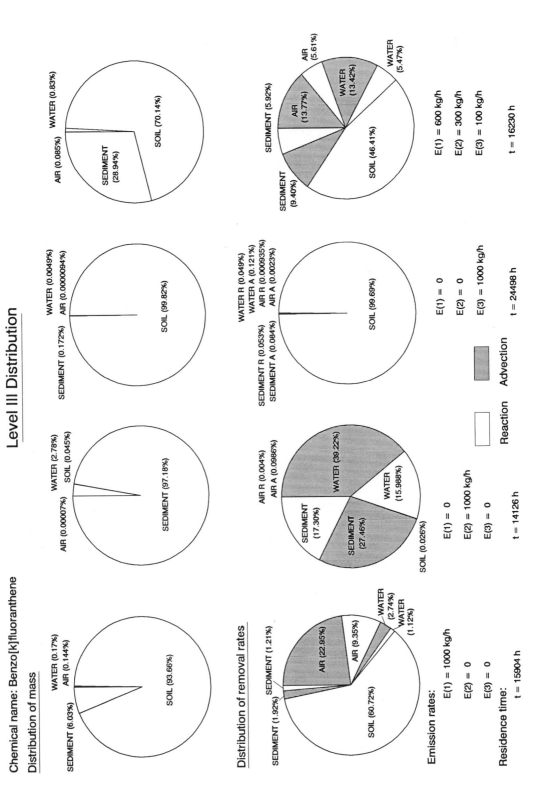

Distribution of removal rates

Emission rates:

E(1) = 1000 kg/h
E(2) = 0
E(3) = 0

E(1) = 0
E(2) = 1000 kg/h
E(3) = 0

E(1) = 0
E(2) = 0
E(3) = 1000 kg/h

E(1) = 600 kg/h
E(2) = 300 kg/h
E(3) = 100 kg/h

Residence time:

t = 15904 h

t = 14126 h

t = 24498 h

t = 16230 h

Reaction   Advection

329

Chemical name: Dibenz[a,h]anthracene

Level I calculation: (six compartment model)

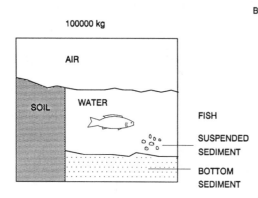

100000 kg

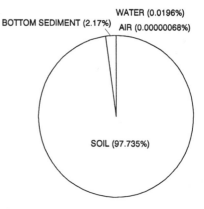

WATER (0.0196%)
BOTTOM SEDIMENT (2.17%) AIR (0.00000068%)
SOIL (97.735%)

Distribution of mass

physical-chemical properties:

MW: 278.35 g/mol

M.P.: 267 °C

Fugacity ratio: 0.0040

vapor pressure: 3.70×10^{-10} Pa

solubility: 0.0006 g/m^3

log K_{OW} : 6.75

Compartment	Z	Concentration			Amount	Amount
	mol/m3 Pa	mol/m3	mg/L (or g/m3)	ug/g	kg	%
Air	4.034E-04	2.441E-17	6.795E-15	5.732E-12	0.0007	6.79E-07
Water	5.826E+03	3.525E-10	9.813E-08	9.813E-08	19.63	0.0196
Soil	6.447E+08	3.901E-05	1.086E-02	4.525E-03	97735	97.735
Biota (fish)	1.638E+09	9.912E-05	2.759E-02	2.759E-02	5.5180	5.52E-03
Suspended sediment	4.030E+09	2.438E-04	6.787E-02	4.525E-02	67.872	6.79E-02
Bottom sediment	1.289E+09	7.803E-05	2.172E-02	9.050E-03	2171.89	2.1719
	Total				100000	100

f = 6.051E-14 Pa

Chemical name: Dibenz[a,h]anthracene

Level II calculation: (six compartment model)

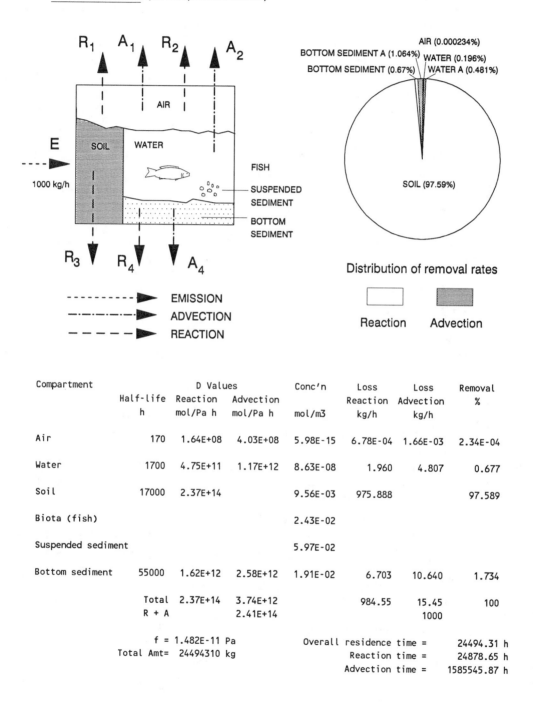

Distribution of removal rates

Reaction   Advection

Compartment	Half-life h	D Values Reaction mol/Pa h	Advection mol/Pa h	Conc'n mol/m3	Loss Reaction kg/h	Loss Advection kg/h	Removal %
Air	170	1.64E+08	4.03E+08	5.98E-15	6.78E-04	1.66E-03	2.34E-04
Water	1700	4.75E+11	1.17E+12	8.63E-08	1.960	4.807	0.677
Soil	17000	2.37E+14		9.56E-03	975.888		97.589
Biota (fish)				2.43E-02			
Suspended sediment				5.97E-02			
Bottom sediment	55000	1.62E+12	2.58E+12	1.91E-02	6.703	10.640	1.734
	Total	2.37E+14	3.74E+12		984.55	15.45	100
	R + A		2.41E+14			1000	

f = 1.482E-11 Pa
Total Amt= 24494310 kg

Overall residence time =   24494.31 h
Reaction time =   24878.65 h
Advection time =   1585545.87 h

Fugacity Level III calculations: (four compartment model)

Chemical name: Dibenz[a,h]anthracene

Phase Properties and Rates:

Compartment	Bulk Z mol/m3 Pa	Half-life h	Reaction mol/Pa h	Advection mol/Pa h	E(1,2,3)
Air (1)	5.291E-01	170	2.16E+11	5.29E+11	5.29E+11
Water (2)	2.761E+04	1700	2.25E+12	5.52E+12	5.52E+12
Soil (3)	3.224E+08	17000	2.37E+14		2.58E+12
Sediment (4)	2.579E+08	55000	1.62E+12	2.58E+12	

D Values (Reaction mol/Pa h, Advection mol/Pa h)

	E(1)=1000	E(2)=1000	E(3)=1000	E(1,2,3)
Overall residence time =	16529	19174	24517	18122 h
Reaction time =	22558	54892	24558	28127 h
Advection time =	61847	29467	14731762	50942 h

Legend:
- EMISSION (E)
- REACTION (R)
- ADVECTION (A)
- TRANSFER D VALUE mol/Pa h

Diagram (four compartment model):

E_1, R_1, A_1 — AIR (1); R_2, A_2 — WATER (2); E_3, R_3 — SOIL (3); E_2, R_4, A_4 — SEDIMENT (4)

Transfer D values:
- D_{12} = 1.645E11
- D_{21} = 2.017E07
- D_{13} = 1.755E08
- D_{31} = 1.480E12
- D_{32} = 6.065E11
- D_{24} = 2.585E12
- D_{42} = 2.015E13

332

Phase Properties, Compositions, Transport and Transformation Rates:

Emission, kg/h

E(1)	E(2)	E(3)
1000	0	0
0	1000	0
0	0	1000
600	300	100

Fugacity, Pa

f(1)	f(2)	f(3)	f(4)
1.504E-09	1.249E-11	9.384E-12	3.708E-11
1.497E-15	1.774E-10	9.346E-18	5.266E-10
1.116E-15	4.536E-13	1.515E-11	1.347E-12
9.021E-10	6.075E-11	7.145E-12	1.804E-10

Concentration, g/m3

C(1)	C(2)	C(3)	C(4)
2.214E-07	9.599E-05	8.421E-01	2.662E-05
2.205E-13	1.363E-03	8.386E-01	3.780E-03
1.644E-13	3.487E-06	1.359E+00	9.668E-06
1.329E-07	4.669E-04	6.412E-01	1.295E-04

Amounts, kg

m(1)	m(2)	m(3)	m(4)	Total Amount, kg
2.214E+04	1.920E+04	1.516E+07	1.331E+06	1.653E+07
2.205E-02	2.727E+05	1.509E+01	1.890E+07	1.917E+07
1.644E-02	6.973E+02	2.447E+07	4.834E+04	2.452E+07
1.329E+04	9.339E+04	1.154E+07	6.474E+07	1.812E+07

Emission, kg/h

E(1)	E(2)	E(3)
1000	0	0
0	1000	0
0	0	1000
600	300	100

Loss, Reaction, kg/h

R(1)	R(2)	R(3)	R(4)
9.027E+01	7.826E+00	6.18E+02	1.67E+01
8.990E-05	1.111E+02	6.15E-04	2.382E+02
6.702E-05	2.843E-01	9.97E+02	6.091E-02
5.416E+01	3.807E+01	4.70E+02	8.157E+01

Loss, Advection, kg/h

A(1)	A(2)	A(4)
2.214E+02	1.920E+01	2.662E+01
2.205E-04	2.727E+02	3.780E+02
1.644E-04	6.973E-02	9.668E-01
1.329E+02	9.339E+02	1.295E+02

Intermedia Rate of Transport, kg/h

T12 air-water	T21 water-air	T13 air-soil	T31 soil-air	T32 soil-water	T24 water-sed	T42 sed-water
6.883E+01	7.012E-05	6.195E+02	4.583E-04	1.584E+00	7.007E+01	2.668E+01
6.855E-05	9.959E-04	6.169E-04	4.564E-01	1.578E-06	9.951E+02	3.789E+02
5.110E-05	2.547E-06	4.599E-04	7.399E-04	2.557E+00	2.545E+00	9.689E-01
4.130E+01	3.411E-04	3.717E+02	3.490E-04	1.206E+00	3.408E+02	1.298E+02

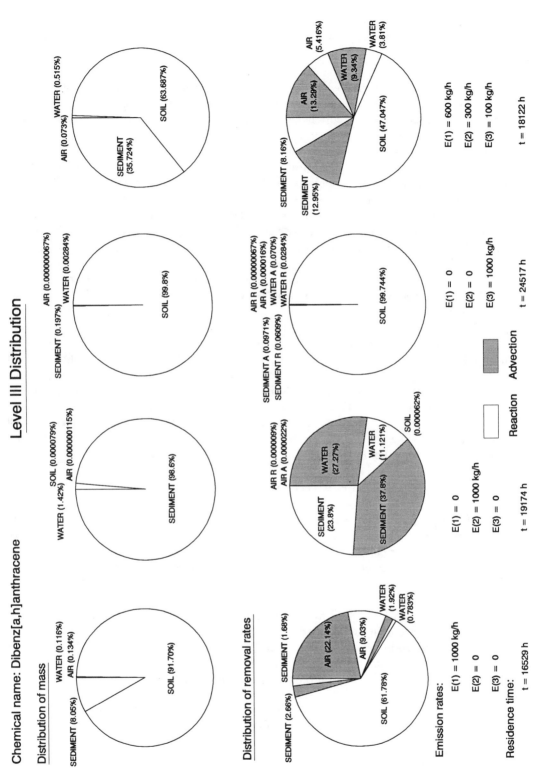

Chemical name: Dibenz[a,h]anthracene

Level III Distribution

Distribution of mass

Distribution of removal rates

333

2.4 COMMENTARY ON THE PHYSICAL-CHEMICAL PROPERTIES AND ENVIRONMENTAL FATE

QSPR Plots

The QSPR plots of the PAHs indicate that total surface area (TSA) and molar volume (V_M) are well correlated, although it is probable that TSA is fundamentally more accurate as a descriptor. There may be loss of correlation by about half a log-unit by using the simpler V_M as descriptor.

Figure 2.2 shows the expected steady drop in subcooled liquid solubility as a function of V_M with a slope of about 0.025 log-units per cm³/mol, or about a factor of 4.2 in solubility (0.625 log-units) for every 25 cm³/mol added. This is about the volume of a methyl group. It must be emphasized that some of the properties of larger PAHs with very low solubilities and high melting points may be subject to considerable error, especially in the fugacity ratio. It seems likely that these solubilities are predictable within a factor of about 4.

The vapor pressure QSPR plot in Figure 2.3 is similar but with greater scatter in the data, and with a steeper slope of about 0.05 log-units per cm³/mol, corresponding to a factor of 18 for a molar volume increment of 25 cm³/mol.

The log K_{OW} plot in Figure 2.4 does not contain the fugacity ratio correction, and more data have been determined, thus the relationship with V_M is better established. The slope is about 0.025 log-units per cm³/mol. The implication is that a 25 cm³/mol increase in molar volume causes K_{OW} to increase by approximately 0.625 log-units or a factor of 4.2 while water solubility also falls by a factor of 4.2. These slopes are different in magnitude than those discussed in Volume I of this series for chlorinated aromatics, thus there is clearly a fundamentally different dependence of these hydrophobic descriptors on LeBas molar volume. This is shown in the plot of log K_{OW} versus solubility in Figure 2.6 in which the slope is about 1.0. This is steeper than for chlorinated aromatics and suggests that it may be unwise to apply solubility - K_{OW} correlations derived for one homologous series of chemicals to others.

The log H plot in Figure 2.5 shows considerable scatter but a marked decrease in H with molar volume, i.e., the larger PAHs have much lower air-water partition coefficients. The slope of about 0.025 units per cm³/mol, or a factor of about 4.2 per 25 cm³/mol is attributable to the fact that increasing molar volumes causes a more rapid decrease in vapor pressure than in solubility.

In summary, a 25 cm³/mol increase in LeBas molar volume causes approximately
(1) a decrease in log solubility of 0.625 units (factor of 4.2).
(2) a decrease in log vapor pressure of 1.25 units (factor of 18).
(3) a decrease in log H of 0.625 units (factor of 4.2)
(4) an increase in log K_{OW} of 0.625 units (factor of 4.2).

The net result is a marked increase in hydrophobicity with molecular weight and also an increased tendency to partition from air into water. This latter behavior contrasts with chlorinated benzenes and biphenyls in which H is relatively constant for the series.

Selected Values

The physical-chemical properties listed for each chemical were inspected, examined in the light of the QSPR plots and values selected as reported in Tables 2.1 and 2.2. The sparse reactivity data were also examined and half-lives selected in Table 2.4. It must be emphasized that these values are subject to considerable variation, thus presenting single values is over-simplistic. PAHs are fairly reactive chemicals, being particularly subject to oxidation and photolysis, but it is suspected that in many situations the chemical is "shielded" from reaction by sorption to organic material. The most reliable half-lives are measured under laboratory conditions of "ideal" exposure to reactants, but these experiments may overestimate environmental rates. For example, a typical laboratory half-life of 6 hours would imply loss of 94% in 1 day, thus little of the chemical is likely to be observed in the environment. From an examination of the literature it was decided that a reasonable approach was to assign half-lives in air of 17 hours (class 2) to the 2-ring compounds, 55 hours (class 3) to 3-ring compounds, and 170 hours (class 4) to higher ring compounds. In surface water where photolysis is slower, half-lives were increased by two classes (i.e., a factor of 10). In soils, in which sorption is likely to retard reaction, classes were assigned of 1700 hours (class 6) to 2 rings, 5500 hours (class 7) to 3 rings and 17000 hours (class 8) to higher PAHs. For sediments, a further one class increase from that of soils was assigned.

It is clear that these reactivity class assignments are fairly arbitrary, but they do reflect observations that PAHs are well preserved in soils and sediments. Obviously, conditions will differ from day to night, summer to winter, with climate and with oxidation status of the environment. In many cases it transpires that even an approximate half-life is adequate for assessment purposes. For detailed site-specific assessments reliable, locally applicable estimates of rates are essential.

Evaluative Calculations

Level I, II and III calculations are shown for 18 selected PAHs. The aim was to cover the range of properties from the two-ring naphthalene to the five-ring dibenz[a,h]anthracene (1,2:5,6-dibenzanthracene). It is possible to infer the likely environmental behavior of other PAHs for those of similar structure and hence properties.

The Level I calculation shows that air is the predominant medium of partitioning for the two-ring PAHs with soil being the other important medium. The higher ring compounds increasingly partition into soil as a result of the lower Henry's law constant (HLC) and the high K_{OW}, with typically 92% in soil and similar amounts in air and water for three-ring PAHs. Four-ring and high PAHs partition typically 98% into soil with most of the remainer in sediment. Very little is present in air or water. Clearly, as the molecular weight increases these chemicals become

335

so sparingly soluble in water, and so involatile, that they are destined to accumulate primarily in organic matter in soils and sediments.

The Level III calculations, which are regarded as the most realistic and indicative of true environmental behavior, are best summarized in the set of 8 pie charts. For the 2-ring PAHs, the chemical is found predominantly in the medium into which it is discharged. Most removal also takes place in that medium, however, since there is appreciable evaporation, removal from air can become important. The 3-ring PAHs show the effect of reduced volatility and increased hydrophobicity. If discharged to air, about equal amounts are found in air and soil, but most loss is from air. If discharged to water, similar amounts are found in water and sediment indicating the increased importance of deposition to sediment. If discharged to soil, almost all chemical remains in soil. The favored intermedia transport processes are thus deposition from the atmosphere to soil and water and deposition from water to sediment. The chemical is thus striving to partition into media with solid organic carbon phases.

The higher PAHs show even more pronounced behavior of this type. There is appreciable deposition from the atmosphere, thus even if discharged to air, only some 6% of the pyrene and less than 1% of the benzo[a]pyrene (BaP) is found in the air. Deposition is very rapid because of the high partitioning to aerosols. If discharged to water, most chemical is found in sediment, although reaction and advection from water are important mechanisms of removal. For BaP, 97% is found in sediments. If the discharge is to soil, the chemical is essentially immobile, remaining there until it reacts.

In summary, the 2-ring PAHs are sufficiently volatile and water soluble that they partition appreciably into air and water. They are environmentally mobile. At the other end of the molecular mass spectrum the behavior of the 4-ring and high PAHs is dominated by a strong tendency to deposit with aerosols and suspended aquatic matter into soils and sediments. They will rapidly migrate into these solid media and tend to remain there. The intermediate PAHs such as pyrene show intermediate behavior, a tendency to partition into air and be transformed or transported there but also a tendency to be deposited to solid media.

The most important fate processes for this series of chemicals are reactivities in air, soil and sediments, and the deposition rates which are largely a function of air-aerosol and water-particle partitioning, as well as the deposition rates in solid dispersed phases.

2.5 REFERENCES

Abernethy, S., Bobra, A.M., Shiu, W.Y., Wells, P.G., Mackay, D. (1986) Acute lethal toxicity of hydrocarbons and chlorinated hydrocarbons to two planktonic crustaceans: The key role of organisms-water partitioning. *Aquatic Toxicology* 8, 163-174.

Abernethy, S., Mackay, D. (1987) A discussion of correlations for narcosis in aquatic species. In: *QSAR in Environmental Toxicology* - II, Kaiser, K.L.E., Ed., pp. 1-16, D. Reidel Publ. Co., Dordrecht, Netherlands.

Abernethy, S., Mackay, D., McCarty, L.S. (1988) "Volume fraction" correlation for narcosis in aquatic organisms: The key role of partitioning. *Environ. Toxicol. & Chem.* 7, 469-481.

Aihara, A. (1959) Estimation of the energy of hydrogen bonds formed in crystals. I.Sublimation pressure of some organic molecular crystals and the additivity of lattice energy. *Bull. Chem. Soc. Japan* 32, 1242-1248.

Akiyoshi, M., Deguchi, T., Sanemasa, I. (1987) The vapor saturation method for preparing aqueous solutions of solid aromatic hydrocarbons. *Bull. Chem. Soc. Jpn.* 60, 3935-3939

Ambrose, M., Lawrenson, I.J., Sprake, C.H.S. (1975) The vapour pressure of naphthalene. *J. Chem. Thermodyn.* 7, 1173-1176.

Ambrose, D., Sprake, H.S. (1976) The vapour pressure of indane. *J. Chem. Thermodyn.* 8, 601-602.

Amidon, G.L., Anik, S.T. (1980) Hydrophobicity of polycyclic aromatic compounds. Thermodynamic partitioning analysis. *J. Phys. Chem.* 84, 970-974.

Amidon, G.L., Anik, S.T. (1981) Application of the surface area approach to the correlation and estimation of aqueous solubility and vapor pressure. Alkyl aromatic hydrocarbons. *J. Chem. Eng. Data* 26, 28-33.

Amidon, G.L., Pearlman, R.S., Anik, S.T. (1979) The solvent contribution to the free energy of protein-ligand interactions. *J. Theor. Biol.* 77, 161-170.

Amidon, G.L., Williams, N.A. (1982) A solubility equation for nonelectrolytes in water. *Int'l. J. Pharma.* 11, 249-256.

Amidon, G.L., Yalkowsky, S.H., Anik, S.T., Valvani, S.C. (1975) Solubility of nonelectrolytes in polar solvents. V. Estimation of the solubility of aliphatic monofunctional compounds in water using a molecular area approach. *J. Phys. Chem.* 79, 2239-2245.

Amidon, G.L., Yalkowsky, S.H., Leung, S. (1974) Solubility of nonelectrolytes in polar solvents. II. Solubility of aliphatic alcohols in water. *J. Pharm. Sci.* 63, 1858-1866.

Andrews, L.J., Keefer, R.M. (1949) Cation complexed of compounds containing carbon-carbon double bonds. IV. The argentation of aromatic hydrocarbons. *J. Am. Chem. Soc.* 71, 3644-3647.

Andrews, L.J., Keefer, R.M., (1950) Cation complexed of compounds containing carbon-carbon double bonds. VI. The argentation of susbstituted benzene. *J. Am. Chem. Soc.* 72, 3110-3116.

API (1978) *Fate and Effects of Polynuclear Aromatic Hydrocarbons in the Aquatic Environment.* Publication No. 4297. American Petroleum Institute, Washington, D.C..

API (1979) *Monograph Series 707, Naphthalene; 708, Anthracene and Phenanthrene; and 709, A Ring Condensed Series.* American Petroleum Institute, Washington, D.C..

API (1981) *Selected Values of Properties of Hydrocarbons and Related Compounds.* (Table 23-2) API Research Project 44, Supplementary Vol. No. A-81., Thermodynamic Research Center, Texas A & M University, College Station, Texas.

Aquan-Yuen, M., Mackay, D., Shiu, W.Y. (1979) Solubility of hexane, phenanthrene, chlorobenzene, and p-chlorobenzene in electrolyte solutions. *J. Chem. Eng. Data* 24, 30-34.

Arbuckle, W.B. (1983) Estimating activity coefficients for use in calculating environmental parameters. *Environ. Sci. Technol.* 17(9), 537-542.

Arbuckle, W.B. (1986) Using UNIFAC to calculate aqueous solubilities. *Environ. Sci. Technol.* 20, 1060-1064.

Arey, J., Atkinson, R., Aschmann, S.M., Schuetzle, D. (1990) Experimental investigation of the atmospheric chemistry of 2-methyl-1-nitronaphthalene and a comparison of predicted nitroarene concentrations with ambient air data. In: *Polycyclic Aromatic Compounds.* Vol. 1(1-2), pp.33-50. Gordon and Breach Science Publishers, Inc., United Kingdom.

Armstrong, D.E., Hurley, J.P., Swackhamer, D.L., Shafer, M.M. (1987) Cycles of nutrient elements, hydrophobic organic compounds, and metals in Crystal Lake. Role of particle-mediated processed in regulation. In: *Sources and Fates of Aquatic Pollutants.* Hites, R.A., Eisenreich, S.J., Eds., *Advances in Chemistry Series* 216, pp.491-518, Amer. Chem. Soc., Washington, D.C.

Ashworth, R.A., Howe, G.B., Mullins, M.E., Rogers, T.N. (1988) Air-water partitioning coefficients of organics in dilute aqueous solutions. *J. Hazardous Materials* 18, 25-36.

Atkinson, R. (1985) Kinetics and mechanisms of the gas phase reactions of hydroxy radical with organic compounds under atmospheric conditions. *Chem. Rev.* 85, 69-201.

Atkinson, R. (1987) Structure-activity relationship for the estimation of rate constants for the gas phase reactions of OH radicals with organic compounds. *Int'l J. Chem. Kinetics* 19, 799-828.

Atkinson, R. (1990) Kinetics and mechnaisms of the gas-phase reactions of the hydroxy radical with organic compounds. *J. Phys. Chem. Ref. Data* (in press)

Atkinson, R., Aschmann, S.M. (1985) Rate constants for the gas-phase reaction of hydroxyl radicals with biphenyl and the monochlorobiphenyls at 295°K. *Environ. Sci. Technol.* 19, 462-464.

Atkinson, R., Aschmann, S.M. (1986) Kinetics of the reactions of naphthalene, 2-methylnaphthalene and 2,3-dimethylnaphathalene with OH radicals and with O_3 at 295 $\pm$ 1 K. *Intl. J. Chem. Kinetics* 18, 569.

Atkinson, R., Aschmann, S.M. (1987) Kinetics of the gas-phase reactions of alkylnaphthalenes with O_3, N_2O_5 and OH radicals at 298 $\pm$ 2 K. *Atmos. Environ.* 21, 2323-2326.

Atkinson, R., Aschman, S.M., Pitts Jr., J.N. (1984) Kinetics of the reactions of naphthalene and biphenyl with OH radicals and O_3 at 294 $\pm$ 1 K. *Environ. Sci. Technol.* 18, 110-113.

Atkinson, R., Arey, J., Zielinska, B., Aschmann, S.M. (1987) Kinetics and products of the gas-phase reactions of OH radicals and N_2O_5 with naphthalene and biphenyl. *Environ. Sci. Technol.* 21, 1014-1022.

Atkinson, R., Carter, W.P.L. (1984) Kinetics and mechanisms of gas-phase ozone reaction with organic compounds under atmospheric conditions. *Chem. Rev.* 84, 437-470.

Augood, D.R., Hey, D.H., Williams, G.H. (1953) Homolytic aromatic substitution. Part III. Ratio of isomerides formed in the phenylation of chlorobenzene. Competitive experiments on the phenylation of p-dichlorobenzene and 1,3,5-trichlorobenzene. Partial rate factors of chlorobenzene. *J. Chem. Soc.* (London), pp.44-50.

Bahnick, D.A., Doucette, W.J. (1988) Use of molecular indices to estimate soil sorption coefficients for organic chemicals. *Chemosphere* 17(9), 1703-1715.

Bailey, R.E., Gonslor, S.J., Rhinehart, W.L. (1983) Biodegratation of the monochlorbiphenyls and biphenyl in river water. *Environ. Sci. Technol.* 17, 617-621.

Baker, J.E., Eisenreich, S.J. (1990) Concentrations and fluxes of polycyclic aromatic hydrocarbons and polychlorinated biphenyls across the air-water interface of Lake Superior. *Environ. Sci. Technol.* 24(3), 342-352.

Banerjee, S. (1985) Calculation of water solubility of organic compounds with UNIFAC-derived parameters. *Environ. Sci. Technol.* 19(4), 369-370.

Banerjee, S., Baughman, G.L. (1991) Bioconcentration factors and lipid solubility. *Environ. Sci. Technol.* 25(3), 536-539.

Banerjee, S., Howard, P. (1988) Improved estimation of solubility and partitioning through correction of UNIFAC-derived activity coefficients. *Environ. Sci. Technol.* 22(7), 839-841.

Banerjee, S., Howard, P., Lande, S.S. (1990) General structure-vapor pressure relationships for organics. *Chemosphere* 21(10-11), 1173-1180.

Banerjee, S., Yalkowsky, S.H., Valvani, S.C. (1980) Water solubility and octanol/water partition coefficients of organics. Limitations of the solubility-partition coefficient correlation. *Environ. Sci. Technol.* 14, 1227-1229.

Barnsley, E.A. (1975) The bacterial degradation of fluoranthene and benzo(a)pyrene. *Can. J. Microbiol.* 21, 1004-1008.

Barone, G., Crescenzi, V., Liquori, A.M., Quadrifoglio, F. (1967) Solubilization of polycyclic aromatic hydrocarbons in poly(methacrylic acid) aqueous solutions. *J. Phys. Chem.* 71, 2341-2345.

Barrows M.E., Petrocelli, S.R., Macek, K.J., Carroll, J.J. (1980) Bioconcentration and elimination of selected water pollutants by bluegill sunfish (*lepomis macrochirus*). In: *Dynamic, Exposure, Hazard Assessment of Toxic Chemicals.* Ann Arbor Press, pp.379-392, Ann Arbor, Michigan.

Baughman, G.L., Paris, D.F. (1981) Microbial bioconcentration of organic pollutants from aquatic systems-a critical review. *CRC Critical Reviews in Microbiology.* pp. 205-228.

Bayona, J.M., Fernandez, P., Porte, C., Tolosa, I., Valls, M., Albaiges, J. (1991) Partitioning of urban wastewater organic microcontaminants among coastal compartments. *Chemosphere* 23(3), 313-326.

Behymer, T.D., Hites, R.A. (1985) Photolysis of polycyclic aromatic hydrocarbons adsorbed on simulated atmospheric particulates. *Environ. Sci. Technol.* 19(10), 1004-1006.

Bender, R., Bieling, V., Maurer, G. (1983) The vapour pressures of solids: anthracene, hydroquinone, and resorcinol. *J. Chem. Thermodyn.* 15, 585-594.

Bennet, D., Canady, J. (1984) Thermodynamics of solution of naphthalene in various water-ethanol mixtures. *J. Am. Chem. Soc.* 106, 910-915.

Bidleman T.F. (1984) Estimation of vapor pressures for nonpolar organic compounds by capillary gas chromatography. *Anal. Chem.* 56, 2490-2496.

Bidleman T.F., Foreman, W.T. (1987) Vapor-particle partitioning of semivolatile organic compounds. In: *Sources, and Fates of Aquatic Pollutants.* Hite, R.A., Eisenreich, S.J., Editors, *Advances in Chemistry Series* 216, American Chemical Society, Washington, D.C.

Biermann, H.W., MacLeod, H., Atkinson, R., Winer, A.M., Pitts Jr., J.N. (1985) Kinetics of the gas-phase reactions of the hydroxyl radical with naphthalene, phenanthrene and anthracene. *Environ. Sci. Technol.* 19(3), 244-248.

Billington, J.W., Huang, G.L., Szeto, F., Shiu, W.Y., Mackay, D. (1988) Preparation of sparingly soluble organic substances: I. Single component systems. *Environ. Toxicol. & Chem.* 7, 117-124.

Bjørseth, A. Editor (1983) *Handbook of Polycyclic Aromatic Hydrocarbons.* Marcel Dekker, Inc., New York, and Basel.

Bohon, R.L., Claussen, W.F. (1951) The solubility of aromatic hydorcarbons in water. *J. Am. Chem. Soc.* 73, 1571-1578.

Bondi, A. (1964) van der Waals volumes and radii. *J. Phys. Chem.* 68, 441-451.

Bopp, R.F. (1983) Revised parameters for modelling the transport of PCB compounds across the air-water interface. *J. Geophys. Res.* 88, 2521-2529.

Bossert, I., Kachel, W.M., Bartha, R. (1984) Fate of hydrocarbons during oily sludge disposal in soil. *Appl. Environ. Microbiol.* 47, 763-767.

Boublik, T., Fried, V., Hala, E. (1973) *The Vapour Pressures of Pure Substances.* Elsevier, Amsterdam.

Boublik, T., Fried, V., Hala, E. (1984) *The Vapour Pressures of Pure Substances.* 2nd revised ed., Elsevier, Amsterdam.

Bradley, R.S., Cleasby, T.G. (1953) The vapor pressure & lattice energy of some aromatic ring compounds. *J. Chem. Soc.* 1690-1692.

Briggs, G.G. (1981) Theorectical and experimental relationships between soil adsorption, octanol-water partition coefficients, water solubilities, bioconcentration factors, and the Parachor. *J. Agric Food Chem.* 29, 1050-1059.

Bright, N.F.H. (1951) The vapor pressure of diphenyl, dibenzyl and diphenylethane. *J. Chem. Soc.* Part I, 624-625.

Broman, D., Naf, C., Rolff, C., Zebuhr, Y. (1990) Occurance and dynamics of polychlorinated dibenzo-p-dioxins and dibenzofurans and polycyclic aromatic hydrocarbons in the mixed surface layer of remote coastal and offshore waters of the Baltic. *Environ. Sci. Technol.* 25, 1850-1864.

Brooke, D.N., Dobbs, A.J., Williams, N. (1986) Octanol/water partition coefficients (P): Measurements, estimation, and interpartition particularly for chemicals with P > 10^5. *Ecotoxicol. & Environ. Safety* 11, 251-260.

Bruggeman, W.A., Van Der Steen, J., Hutzinger, O. (1982) Reversed-phase thin-layer chromatography of polynuclear aromatic hydrocarbons and chlorinated biphenyls. Relationship with hydrophobicity as measured by aqueous solubility and octanol-water partition coefficient. *J. Chromatogr.* 238, 335-346.

Budavari, S., Editor (1989) The Merck Index. *An Encyclopedia of Chemicals, Drugs and Biologicals.* 11th edition, Merck & Co., Rahway, New Jersey, U.S.A.

Bulman, T.L., Lesage, S., Fowlie, P., Webber, M.D. (1987) The fate polynuclear aromatic hydrocarbons in soil. In: *Oil in Fresh Water: Chemistry, Biology, Countermeasure Technology.* Vandermeulan, J.H., Hurley, S.E., Editors, Pergamon Press, New York.

Burkhard, L.P. (1984) *Physical-chemical Properties of the Polychlorinated Biphenyls: Measurement, Estimation, and Application to Environmental Systems.* Ph.D. Thesis, University of Wisconsin-Madison.

Burkhard, L.P., Andren, A.W., Armstrong, D.E. (1985) Estimation of vapor pressures for polychlorinated biphenyls: A comparison of eleven predictive methods. *Environ. Sci. Technol.* 19, 500-507.

Burkhard, L.P., Armstrong, D.E., Andren, A.W. (1985) Henry's law constants for the polychlorinated biphenyls. *Environ. Sci. Technol.* 19, 590-596.

Burkhard, L.P., Kuehl, D.W., Veith, G.D. (1985) Evaluation of reverse phase liquid chromatography/mass spectrometry for estimation of n-octanol/water partition coefficients for organic chemicals. *Chemosphere* 14(10), 1551-1560.

Burkhard, L.P., Kuehl, D.W. (1986) n-Octanol/water partition coefficients by reverse phase liquid chromatography/mass spectrometry for eight tetrachlorinated planar molecules. *Chemosphere* 15, 163-167.

Bysshe, S.E. (1982) Bioconcentration factor in aquatic organisms. In: *Handbook of Chemical Property Estimation Methods.* Lyman, W.J., Reehl, W.F., Rosenblatt, D.H., Editors. Chapter 5. Ann Arbor Sci., Ann Arbor, Michigan.

Callahan, M.A.,Slimak, M.W., Gabel, N.W., May, I.P., Fowler, C.F., Freed, J.R., Jennings, P., Durfee, R.L., Whitmore, F.C., Maestri, B., Mabey, W.R., Holt, B.R., Gould, C. (1979) *Water Related Environmental Fate of 129 Priority Pollutants.* EPA-440-4-79-029a,b.

Camin, D.L., Rossini, F.D. (1955) *J. Phys. Chem.* 59, 1173.

Capel, P.D., Leuenberger, C., Giger, W. (1991) Hydrophobic organic chemicals in urban fog. *Atm. Environment* 25A(7), 1335-1346.

Campbell, J.R., Luthy, R.G. (1985) Prediction of aromatic solute partition coefficients using the UNIFAC group contribution model. *Environ. Sci. Technol.* 19(10), 980-985.

Carlson, A.R., Kosian, P.A. (1987) Toxicity of chlorobenzenes to fathead minnows (*pimephales promelas*). *Arch. Environ. Contam. Toxicol. 16, 129-135.*

Caron, G., Suffet, I.H., Belton, T. (1984) Effect of dissolved organic carbon on the environmental distribution of nonpolar organic compounds. *Chemosphere* 14, 993-1000.

Casellato, F., Vecchi, C., Grielli, A., Casu, B. (1973) *Thermochim. Acta* 6, 361.

Casserly, D.M., Davis, E.M., Downs, T.D., Guthrie, R.K. (1983) Sorption of organics by *selenastrum capricornutum*. *Water Res.* 17(11), 1591-1594.

Chao, J., Lin, C.T., Chung, T.H. (1983) Vapor pressure of coal chemicals. *J. Phys. Chem. Ref. Data* 12, 1033-1063.

Chin, Y.P., Weber Jr., W.J., Voice, T.C. (1986) Determination of partition coefficient and water solubilities by reversed phase chromatography. II. Evaluation of partitioning and solubility models. *Water Res.* 20, 1443-1450.

Chiou, C.T. (1981) Partition coefficient and water solubility in environmental chemistry. In: *Hazard Assessment of Chemicals.* Saxena, J., Fisher, F., Editors, Academic Press, New York, pp.117-153.

Chiou, C.T. (1985) Partition coefficients of organic compounds in liqid-water systems and correlations with fish bioconcentration factors. *Environ. Sci. Technol.* 19, 57-62.

Chiou, C.T., Block, J.B. (1986) Parameters affecting the partition coefficients of organic compounds in solvent-water and lipid-water systems. In: *Partition Coefficient, Determination and Estimation.* Dunn III, W.J., Block, J.H., Pearlman, R.S., Editors, Pergamon Press, New York. pp.36-60.

Chiou, C.T., Freed, V.H., Schmedding, D.W., Kohnert, R.L. (1977) Partition coefficient and bioaccumulation of selected organic chemicals. *Environ. Sci. Technol.* 11(5), 475-478.

Chiou, C.T., Porter, P.E., Schmedding, D.W. (1983) Partition equilibria of nonionic organic compounds between soil organic matter and water. *Environ. Sci. Technol.* 17, 227-231.

Chiou, C.T., Schmedding, D.W. (1981) Measurement and interrelation on octanol-water partition coefficient and water solubility of organic chemicals. In: *Test Protocols for Environmental Fate and Movement of Toxicants. J. Assoc. Anal. Chem., pp.28-42., Arlington, Virginia.*

Chiou C.T., Schmedding, D.W., Manes, M. (1982) Partitioning of organic compounds in octanol-water systems. *Environ. Sci. Technol.* 16(1), 4-10.

Chou, S.F.J., Griffin, R.A. (1987) Solubility and soil mobility of polychlorinated biphenyls. In: *PCBs and the Environment.* Waid, J.S., Editor, CRC Press, Inc., Boca Raton, Florida. pp. 101-120.

Chou, J.T., Jurs, P.C. (1979) Computation of partition coefficients from molecular structures by a fragment addition method. In: *Physical Chemical Properties of Drugs.* Medical Research Series, Vol. 10, Yalkowsky, S.H., Sindula, A.A., Valvani, S.C., Editors, Marcel Dekker, Inc., New York. pp. 163-199.

Cleland, J.G., Kingsbury, G.L. (1977) *Multimedia Environmental Goals for Environmental Assessment, Vol. II MEG Charts and Background Information.* U.S. Environmental Protection Agency, (Office Research and Development), Washington D.C., 451p. (EPA-600/7-77-1366).

Colomina, M., Jimenez, P., Turrion, C. (1982) Vapor pressures and enthalpies of sublimation of naphthalene and benzoic acid. *J. Chem. Thermodynamics* 14, 779-784.

Connell, D.W., Hawker, D.W. (1988) Use of polynomial expressions to describe the bioconcentration of hydrophobic chemicals by fish. *Ecotoxicol. Environ. Safety* 16, 242-257.

Coover, M.P., Sims, R.C.C. (1987) The effects of temperature on polycyclic aromatic hydrocarbon persistence in an unacclimated agricultural soil. *Haz. Waste Haz. Mat.* 4, 69-82.

Cope, V.W., Kalkwarf, D.R. (1987) Photooxidation of selected polycyclic aromatic hydrocarbons and pyrenequinones coated on glass surfaces. *Environ. Sci. Technol.* 21(7), 643-648.

D'Amboise, M., Hanai, T. (1982) Hydrophobicity and retention in reverse phase liquid chromatography. *J. Liq. Chromatogr.* 229-244.

Davies, R.P., Dobbs, A.J. (1984) The prediction of bioconcentration in fish. *Water Res.* 18 (10), 1253-1262.

Davis, W.W., Krahl, M.E., Clowes, H.A. (1942) Solubility of carcinogenic and related hydrocarbons in water. *J. Am. Chem. Soc.* 64, 108-110.

Davis, W.W., Parker, Jr., T.V. (1942) A nephelometric method for determination of solubilities of extremely low order. *J. Am. Chem. Soc.* 64, 101.

Dean, J.D., Editor (1985) *Lange's Handbook of Chemistry.* 13th ed., McGraw-Hill, Inc., New York.

De Bruijn, J., Busser, F., Seinen, W., Hermens, J. (1989) Determination of octanol/water partition coefficients for hydrophobic organic chemicals with the "slowing-stirring" method. *Environ. Toxicol. Chem.* 8, 499-512.

De Bruijn, J., Hermens, J. (1990) Relationships between octanol/water partition coefficients and total molecular surface area and total molecular volume of hydrophobic organic chemicals. *Quant. Strut.-Act. Relat.* 9, 11-21.

De Kock, Lord, D.A. (1987) A simple procedure for determining octanol-water partition coefficients using reverse phase high performance liquid chromatography (RPHPLC). *Chemosphere* 16(1), 133-142.

De Kruif, C.G. (1980) Enthalpies of sublimation and vapor pressures of 11 polycyclic hydrocarbons. *J. Chem. Thermodynamics* 12, 243-248.

De Kruif, C.G., Kuipers, T., Van Miltenburg, J.C., Schaake, R.C.F., Stevens, G. (1981) The vapour pressure of solid and liquid naphthalene. *J. Chem. Thermodynam.* 13, 1081-1086.

Deno, N.C., Berkheimer, H.E. (1960) Phase equilibria molecular transport thermodynamics: Activity coefficients as a function of structure and media. *J. Chem. Eng. Data* 5, 1-5.

Dobbs, A.J., Cull, M.R. (1982) Volatilization of chemicals-relative loss rates and the estimation of vapor pressures. *Environ. Pollut.* (series B) 3, 289-298.

Doucette, W.J. (1985) *Measurement and estimation of Octanol/water Partition Coefficients and Aqueous Solubility for Halogenated Aromatic Hydrocarbons.* Ph.D. Thesis, University of Wisconsin, Madison, Wisconsin.

343

Doucette, W.J., Andren, A.W. (1987) Correlation of octanol/water partition coefficients and total molecular surface area for highly hydrophobic aromatic compounds. *Environ. Sci. Technol.* 21, 621-624.

Doucette, W.J., Andren, A.W. (1988) Aqueous solubility of biphenyl, furan, and dioxin congeners. *Chemosphere* 17, 243-252.

Doucette, W.J., Andren, A.W. (1988) Estimation of octanol/water partition coefficients: Evaluation of six methods for highly hydrophobic aromatic hydrocarbons. *Chemosphere* 17(2), 345-359.

Dragoescu, C., Friedlander, S. (1989) Dynamics of the aerosol products of incomplete combustion in urban atmospheres. *Aerosol Sci. Technol.* 10, 249-257.

Eadie, B.J., Landrum, P.F., Faust, W. (1982) Polycyclic aromatic hydrocarbons in sediments, pore water and the *amphipod pontoporeia hoy*i from Lake Michigan. *Chemosphere* 11(9), 847-858.

Eadie, B.J., Morehead, N.R., Landrum, P.F. (1990) Three-phase partitioning of hydrophobic organic compounds in great lakes waters. *Chemosphere* 20(1-2), 161-178.

Eadsforth, C.V., Moser, P.(1983) Assessment of reverse phase chromatographic methods for determining partition coefficients. *Chemosphere* 12, 1459-1475.

Eastcott, L., Shiu, W.Y., Mackay, D. (1988) Environmentally relevant physical-chemical properties of hydrocarbons: a review of data and development of simple correlations. *Oil & Chem. Pollut.* 4, 191-216.

Edwards, D.A., Luthy, R.G., Liu, Z. (1991) Solubilization of polycyclic aromatic hydrocarbons in micellar nonionic surfactant solutions. *Environ. Sci. Technol.* 25(1), 127-133.

Eganhouse, R.P., Calder, J.A. (1976) The solubility of medium molecular weight aromatic hydrocarbons and the effects of hydrocarbon co-solutes and salinity. *Geochim. Cosmochim. Acta* 40, 555-561.

Eichler, W. (Ed.) (1965) Handbuch der insektizidkunde. Veb. Verlag Volk. Gesundheit, Berlin, pp. 1-84.

Eisenbrand, J., Baumann, K. (1970) Über die bestimmung der wasserlöslichkeit von coronen, fluoranthen, perylen, picen, tetracen und triphenylen und über die bildung wasserlöslicher komplexe dieser kohlenwasserstoffe mit coffein. *Zeitschrift für Lebensmittel Untersuchung und Forschung* 144, 312-317.

Eisenreich, S.J., Looney, B.B., Thornton, J.D. (1981) Airborne organic contaminants in the Great Lakes ecosystem. *Environ. Sci. Technol.* 15, 30-38.

Erickson, M.D. (1986) *Analytical Chemistry of PCB's.* Ann Arbor Science Book, Butterworth Publishers, Stoneham, MA.

Ernst, W. (1977) Determination of the bioconcentration potential of marine organisms - a steady-state approach. *Chemosphere* 6(11), 731-740.

Evans, M.S., Landrum, P.F. (1989) Toxicokinetics of DDE, benzo(a)pyrene, and 2,4,5,2',4',5'-hexachlorobiphenyl in *pontoporeia hoy*i and *mysis relicta*. *J. Great Lakes Res.* 15(5), 589-600.

Farrington, J.W. (1991) Biogeochemical processes governing exposure and uptake of organic pollutant compounds in aquatic organicsms. *Environ. Health Perspec.* 90, 75-84.

344

Fendinger, N.J., Goltfelty, D.E. (1988) A laboratory method for the experimental determination of air-water Henry's law constants for several pesticides. *Environ. Sci. Technol.* 22, 1289-1293.

Fendinger, N.J., Goltfelty, D.E. (1990) Henry's law constants for selected pesticides, PAHs and PCBs. *Environ. Toxicol. Chem.* 9, 731-735.

Foreman, W.T., Bidleman, T.F. (1985) Vapor pressure estimates of individual polychlorinated biphenyls and commercial fluids using gas chromatographic retention data. *J. Chromatogra.* 330, 203-216.

Formica, S.J., Baron, J.A., Thibodeaux, L.J., Valsaraj, K.T. (1988) PCB transport into lake sediments. Conceptual model and laboratory simulation. *Environ. Sci. Technol.* 22, 1435-1440.

Fowler, L., Trump, W.N., Vogler, C.E. (1968) Vapor pressure of naphthalene. *J. Chem. Eng. Data* 13(2), 209-210.

Frank, A.P., Landrum, P.F., Eadie, B.J. (1986) Polycyclic aromatic hydrocarbon rates of uptake, depuration, and biotransformation by Lake Michigan *stylodrilus heringianus*. *Chemosphere* 15(3), 317-330.

Freed, V.H., Chiou, C.T., Haque, R. (1977) Chemodynamics: Transport and behavior of chemicals in the environment-A problem in environmental health. *Environ. Health Pespectives* 20, 55-70.

Freitag, D., et al. (1984) Environmental hazard profile-test results as related structures and translation into the environment. In: *QSAR in Environmental Toxicology*. Kaiser, K.L.E., Editor, D. Reidel Publishing Company, Dordrecht, Netherlands.

Freitag, D., Ballhorn, L., Geyer, H., Korte, F. (1985) Environmental hazard profile of organic chemicals. An experimental method for the assessment of the behaviour of chemicals in the ecosphere by simple laboratory tests with C-14 labelled chemicals. *Chemosphere* 14, 1589-1616.

Frez, W.A., Landrum, P.F. (1986) Species dependent uptake of PAH in Great Lakes invertebrates. In: *Polynuclear Aromatic Hydrocarbons: Ninth International Symposium on Chemistry, Characterization and Carcinogenesis.* pp.291-304. Cooke, M.W., Dennis, A.J., Editors, Battelle Press, Columbus, Ohio.

Fu, J.-K., Luthy, R.G. (1985) Aromatic compound solubility in solvent/water mixtures. *J. Environ. Eng.* 112, 328-346.

Fu, J.-K., Luthy, R.G. (1985) *Pollutant Sorption to Soils and Sediments in Organic/Aqueous Solvent Systems. NTIS* PB85-242535. EPA/600/3-85/050.

Fujita, T., Iwasa, J., Hansch, C. (1964) A new substituent constant, "pi" derived from partition coefficients. *J. Am. Chem. Soc.* 86, 5175-5180.

Fukuda, K., Inagaki, Y., Maruyama, T., Kojima, H.I., Yoshida, T. (1988) On the photolysis of akylated naphthalenes in aquatic systems. *Chemosphere* 17(4), 651-659.

Futoma, D.J., Smith, S.R., Smith, T.E., Tanaka, J. (1981) *Polycyclic Aromatic Hydrocarbons in Water Systems.* CRC Press, Inc., Boca Raton, Florida, U.S.A.

Garst, J.E. (1984) Accurate, wide-range, automated high-performance liquid chromatographic method for the estimation of octanol/water partition coefficients. II: Equilibruim in partition coefficient measurements, additivity of substituent constants, and correlation of biological data. *J. Pharm. Sci.* 73, 1623-1629.

Garst, J.E., Wilson, W.C. (1984) Accurate, wide-range, automated high- performance liquid chromatographic method for the estimation of octanol/water partition coefficients. I: Effect of chromatographic conditions and procedure variables on accuracy and reproducibility of the method. *J. Pharm. Sci.* 73, 1616-1622.

Garten, Jr., C.T., Trabalka, J.R. (1983) Evaluation of models for predicting terrestrial food chain behavior of xenobiotics. *Environ. Sci. Technol.* 17, 590-595.

Gauthier, T.D., Shane, E.C., Guerin, W.F., Seltz, W.R., Grant, C.L. (1986) Fluorescence quenching method for determining equilibrium constants for polycyclic aromatic hydrocarbons binding to dissolved humic materials. *Environ. Sci. Technol.* 20(11), 1162-1166.

Gauthier, T.D., Selfz, W.R., Grant, C.L. (1987) Effects of structural and compositional variations of dissolved humic materials on pyrene K_{oc} values. *Environ. Sci. Technol.* 21, 243-248.

Gavezzotti, A. (1985) Molecular free surface: A model method of calculation and its uses in conformational studies and in organic crystal chemistry. *J. Am. Chem. Soc.* 107, 962-967.

Geyer, H., Kraus, A.G., Klein, W., Richter, E., Korte, F. (1980) Relationship between water solubility and bioaccumulation potential of organic chemicals in rats. *Chemosphere* 9, 277-291.

Geyer, H., Sheehan, P., Kotzias, D., Freitag, D., Korte, F. (1982) Prediction of ecotoxicological behaviour of chemicals: Relationship between physico-chemical properties and bioaccumulation of organic chemicals in the mussel mytilus edulis. *Chemosphere* 11(11), 1121-1134.

Geyer, H., Politzki, G., Freitag, D. (1984) Prediction of ecotoxicological behaviour of chemicals: Relationship between n-octanol/water partition coefficient and bioaccumulation of organic chemicals by *alga chlorella*. *Chemosphere* 13(2), 269-284.

Geyer, H. J., Scheunert, I., Korte, F. (1987) Correlation between the bioconcentration potential of organic environmental chemicals in humans & their n-octanol/water partition coefficients. *Chemosphere* 16(1), 239-252.

Geyer, H., Viswanathan, R., Freitag, D., Korte, F. (1981) Relationship between water solubility of organic chemicals and their bioaccumulation by the *Alga Chlorella*. *Chemosphere* 10, 1307-1313.

Gobas, F.A.P.C., Shiu, W.Y., Mackay, D. (1987) Factorsdetermining partitioning of hydrophobic organic chemicals in aquatic organisms. In: *QSAR in Environmental Toxicology* II. Kaiser, K.L.E., Editor, D. Reidel Publ. Co., Dordrecht, Holland.

Gobas, F.A.P.C., Lahittete, J.M., Garofalo, G., Shiu, W.Y., Mackay, D. (1988) A novel method for measuring membrane-water partition coefficients of hydrophobic organic chemicals: Comparison with 1-octanol-water partitioning. *J. Pharma. Sci.* 77(3), 265-272.

Gordon, J.E., Thorne, R.L. (1967) Salt effects on the activity coefficient of naphthalene in mixed aqueous electrolyte solutions. I. Mixtures of two salts. *J. Phys. Chem.* 71, 4390-4399.

Goursot, P., Girdhar, H.L., Westrum, Jr., E.F. (1970) Thermodynamics of polynuclear aromatic molecules. III. Heat capacities and enthalpies of fusion of anthracene. *J. Phys. Chem.* 74, 2538-2541.

Govers, H., Ruepert, C., Aiking, H. (1984) Quantitative structure-activity relationships for polycyclic aromatic hydrocarbons: Correlation between molecular connectivity, physico-chemical properties, bioconcentration and toxicity in *daphnia pulex*. *Chemosphere* 13(2), 227-236.

Grayson, B.T., Fosbraey, L.A. (1982) Determination of the vapor pressure of pesticides. *Pestic. Sci.* 13, 269-278.

Groenewegen, D., Stolp, H. (1975) Microbial degradation of polycyclic aromatic hydrocarbons. *Erdoel Kohle, Erdgas, Pentrachem. Bremst. Chem.* 28(4), 206.

Groenewegen, D., Stolp, H. (1976) Microbial breakdown of polycyclic aromatic hydrocarbons. *Zentralbl. Bakteriol. Parasitenkd. Infekitionskr. Hyg. Abt:* 1, Orig., Reihe, B. 162, 225-232.

Gückel, W., Synnatschke, G., Rittig, R. (1973) A method for determining the volatility of active ingredients used in plant protection. *Pest. Sci.* 4, 137-147.

Gückel, W., Kstel, R., Lewerenz, J., Synnatschke, G. (1982) A method for the determining the volatility of active ingredients used in plant protection. Part III: The temperature relationship between vapor pressure and evaporation rate. *Pest. Sci.* 13, 161-168.

Günther, F.A., Westlake, W.E., Jaglan, P.S. (1968) Reported solubilities of 738 pesticide chemicals in water. *Residue Review* 20,1-148.

Hafkenscheid, T.L., Tomlinson, E. (1983) Correlations between alkane/water and octan-1-ol/water distribution coefficients and isocratic reversed-phase liquid chromatographic capacity factors of acids, bases and neutrals. *Int'l J. Pharmaceutics* 16, 225-239.

Haky, J.E., Young, A.M. (1984) Evaluation of a simple HPLC correlation method for the determination of the octanol-water partition coefficients of organic compounds. *J. Liq. Chrom.* 7(4), 675-689.

Halfon & Reggiani (1986) On ranking chemicals for environmental hazard. *Environ. Sci. Technol.* 20, 1173-1179.

Hallett, D.J., Brecher, R.W. (1984) Cycling of polynuclear aromatic hydrocarbons in the Great Lakes ecosystem. In: *Toxic Contaminants in the Great Lakes. Advances in Environment Sciences and Technology.* Nriagu, J.O., Simmons, M.S., Editors, John Wiley & Sons, New York, New York. pp.213-237.

Hambrick, G.A., Delaune, R.D., Patrick, W.H., Jr. (1980) Effects of estuarine sediment, pH and oxidation-reduction potential on microbial hydrocarbon degradation. *Appl. Environ. Microbiol.* 40, 365-9.

Hammers, W.E., Meurs, G.J., De Ligny, C.L. (1982) Correlations betweeen liquid chromatographic capacity ratio data on Lichrosorb RP-18 and partition coefficients in the octanol-water system. *J. Chromatogr.* 247, 1-13.

Hanai, T., Tran, C., Hrbert, J. (1981) An approach to the prediction of retention times in liquid chromatography. *J. High Resolution Chromatography & Chromatography Communication* (J. HRC & CC) 4, 454-460.

Hansch, C., Fujita, T. (1964) ρ-σ-π Analysis; method for the correlation of biological activity and chemical structure. *J. Am. Chem. Soc.* 86, 1616-1626.

Hansch, C., Leo, A. (1979) *Substituent Constants for Correlation Analysis in Chemistry and Biology.* Wiley, New York.

Hansch, C., Leo, A. (1985) *Medchem Project Issue No. 26.* Pomona College, Claremont, Calofornia.

Hansch, C., Leo, A., Unger, S.H., Kim, K.H., Nikaitani, D., Lien, E.J. (1973) Aromatic substituent constants for structure-activity correlations. *J. Med. Chem.* 16, 1207.

Haque, R., Falco, J., Cohen, S., Riordan, C. (1980) Role of transport and fate studies in the exposure, assessment and screening of toxic chemicals. In: *Dynamics, Exposure and Hazard Assessment of Toxic Chemicals.* R. Haque, Editor. Ann Arbor Science Publishers Inc., Ann Arbor, Michigan.

Haque, R., Schmedding, D. (1975) A method of measuring the water solubility of hydrophobic chemicals. *Bull. Environ. Contam. Toxicol.* 14, 13-18.

Harris, J.C. (1982) Rate of aqueous photolysis. Chapter 8, In: *Handbook of Chemical Property Estimation Methods.* Lyman, W.J., Reehl, W.F., Rosenblatt, D.H., Editors, McGraw-Hill Book Company, New York.

Harris, C.R., Mazurek, J.H. (1964) Comparison of the toxicity to insects of certain insecticides applied by contact and in soil. *J. Econ. Entomol.* 57, 698.

Hashimoto, Y., Tokura, K., Ozaki, K., Strachan, W.M.J. (1982) A comparison of water solubilities by the flask and micro-column methods. *Chemosphere* 11(10), 991-1001.

Hawker, D.W. (1989) The relationship between octan-1-ol/water partition coefficient and aqueous solubility in terms of solvatochromic parameters. *Chemosphere* 19(10/11), 1585-1593.

Hawker, D.W. (1990) Description of fish bioconcentration factors in terms of solvatochromic parameters. *Chemosphere* 20, 467-477.

Hawker, D.W., Connell, D.W. (1985) Relationships between partition coefficient, uptake rate constant, clearance rate constant and time to equilibration for bioaccumulation. *Chemoshere* 14, 1205-1219.

Hawker, D.W., Connell, D.W. (1986) Bioconcentration of lipophilic compounds by some aquatic organisms. *Ecotox. Environ. Safety* 11, 184-197.

Hawker, D.W., Connell, D.W. (1988) Octanol-water partition coefficients of polychlorinated biphenyl congeners. *Environ. Sci. Technol.* 22, 382-387.

Hawker, D.W., Connell, D.W. (1988) Influence of partition coefficient of lipophilic compounds on bioconcentration kinetics with fish. *Water Res.* 22, 701-707.

Heitkamp, M.A. (1988) Environmental and biological factors affecting the biodegradation and detoxification of polycyclic aromatic hydrocarbons. *Diss. Abstr. Int'l. B.* 48, 1926.

Herbes, S.E. (1977) Partitioning of polycyclic aromatic hydrocarbons between dissolved and particulate phases in natural waters. *Water Res.* 11, 493-496.

Herbes, S.E. (1981) Rates of microbial transformation of polycyclic aromatic hydrocarbons in water and sediments in the vicinity of coal-coking waste water discharge. *Appl. Environ. Microbiol.* 41, 20-28.

Herbes, S.E., Risi, G.F. (1978) Metabolic alteration and excretion of anthracene by daphnia pulex. *Bull. Environ. Contam. Toxicol.* 19, 147-155.

Herbes, S.E., Schwall, L.R. (1978) Microbial transformation of polycyclic aromatic hydrocarbons in pristine and petroleum contaminated sediments. *Appl. Environ. Microbiol.* 35, 306-316.

Herbes, S.E., Southworth, G.R., Shaeffer, D.L., Griest, W.H., Maskarinec, M.P. (1980) Critical parthways of polycyclic aromatic hydrocarbons in aquatic environments. In: *The Scientific Basis of Toxidity Assessment.* Witschi, H. Editor, pp. 113-128, Elsevier/North-Holland Biomedical Press, Amsterdam.

Hilpert, S. (1916) The solubility of naphthalene in ammonia. A possible cause for naphthalene stoppages. *Angew. Chem.* 29, 57-59.

Hinckley, D.A. (1989) *Vapor Pressures, Henry's Law Constants and Air-Sea Gas Exchange of Selected Organochlorine Pollutants.* Ph.D. Thesis, University of South Carolina.

Hinckley, D.A., Bidleman, T.F., Foreman, W.T. (1990) Determination of vapor pressures for nonpolar and semipolar organic compounds from gas chromatographic retention data. *J. Chem. Eng. Data* 35, 232-237.

Hine, J., Mookerjee, P.K. (1975) The intrinsic hydrophilic character of organic compounds. Correlations in terms of structural contributions. *J. Org. Chem.* 40, 292-298.

Hodson, J., Williams, N.A. (1988) The estimation of the adsorption (K_{OC}) for soils by high performance liquid chromatography. *Chemosphere* 17(1), 67-77.

Hollifield, H.C. (1979) Rapid nephelometric estimate of water solubility of highly insoluble organic chemicals of environmental interest. *Bull. Environ. Contam. Toxicol.* 23, 579-586.

Howard, P.H., Editor (1989) *Handbook of Environmental Fate and Exposure Data for Organic Chemicals.* Volume I, Lewis Publishers, Inc., Chelsea, Michigan, U.S.A.

Howard, P.H., Boethling, R.S., Jarvis, W.F., Meylan, W.M., Michalenco, E.M., Editors (1991) *Handbook of Environmental Degradation Rates.* Lewis Publishers, Inc., Chelsea, Michigan, U.S.A.

Howard, P.H., Hueber, A.E., Mulesky, B.C., Crisman, J.S., Meylan, W., Crosbie, E., Gray, D.A., Sage, G.W., Howard, K.P., LaMacchia, A., Boethling, R., Troast, R. (1986) Biology, biodegradation and fate/expos: New files on microbial degradation and toxicity as well as environmental fate/exposure of chemicals. *Environ. Toxicol. & Chem.* 5, 977-988.

Hoyer, H., Peperle, W. (1958) Dampfdruckmessungen an organischen substanzen und ihre sublimationswarmen. *Z. Elektrochem.* 62, 61-66.

Hussan, A., Carr, P.W. (1985) A study of a rapid and precise methodology for the measurement of vapor liquid equilibria by headspace gas chromatography. *Anal. Chem.* 57, 793-801.

Hutchinson, T.C., Hellebust, J.A., Tam, D., Mackay, D., Mascarenhas, R.A., Shiu, W.Y. (1980) The correlation of the toxicity to algae of hydrocarbons and halogenated hydrocarbons with their physical-chemical properties. In: *Hydrocarbons and Halogenated Hydrocarbons in Aquatic Environment.* Afghan, B.K., Mackay, D., Editors, pp. 577-586. Plenum Press, New York.

Hutzinger, O., Safe, S., Zitko, V. (1974) *The Chemistry of PCBs.* CRC Press Inc., Cleveland, Ohio.

IARC (1983) *Monographs on the Evaluation of the Carcinogenic Risk of Chemicals to Humans: Polynuclear Aromatic Compounds.* Part I., IARC Press, Lyon.

Inokuchi, H., Shiba, S., Handa, T., Akamatsu, H. (1952) Heats of sublimation of condensed polynuclear aromatic hydrocarbons. *Bull. Chem. Soc. Japan* 25, 299-302.

Irwin, K.C. (1982) SRI International unpublished analysis.

Isnard, P., Lambert, S. (1988) Estimating bioconcentraion factors from octanol-water partition coefficient and aqueous solubility. *Chemosphere* 17, 21-34.

Isnard, P., Lambert, S. (1989) Aqueous solubility and n-octanol/water partition coefficient correlations. *Chemosphere* 18, 1837-1853.

IUPAC Solubility Data Series, Volume 38 (1989) *Hydrocarbons (C_8-C_{36}) with Water and Seawater.* Shaw, D.G., Editor, Pergamon Press, Oxford, England.

Jaber, H.M., Smith, J.H., Cwirla, A.N. (1982) Evaluation of gas saturation methods to measure vapor pressure. (EPA Contract No. 68-01-5117) SRI International, Menlo Park, CA.

Johnson, H. (1982) In: *Aquatic Fate Process Data for Organic Priority Pollutants.* Mabey, W.R., Ed., EPA Final Report on Contract 68-01-3867; U.S. Government Printing Office, Washington, D.C.

Jury, W.A., Spencer, W.F., Farmer, W.J. (1984) Behavior assessment model for trace organics in soil: III.Application of screening model. *J. Environ. Qual.* 13(4), 573-579.

Jury, W.A., Russo, D., Streile, G., EI Abd, H. (1990) Evaluation of volatilization by organic chemicals residing below the soil surface. *Water Resources Res.* 26, 13-26.

Kaiser, K.L.E. (1983) A non-linear function for the calculation of partition coefficients of aromatic compounds with multiple chlorine substitution. *Chemosphere* 12, 1159-1165.

Kamens, R.M., Guo, Z., Fulcher, J.N., Bell, D.A. (1988) Influence of humidity, sunlight, and temperature on daytime decay of polyaromatic hydrocarbons on atmospheric soot particles. *Environ. Sci. Technol.* 22(1), 103-108.

Kamlet, M.J., Doherty, R.M., Abraham, M.H., Carr, P.W., Doherty, R.F., Taft, R.W. (1987) Linear solvation energy relationships. Important differences between aqueous solubility relationships for aliphatic and aromatic solutes. *J. Phys. Chem.* 91, 1996-.

Kamlet, M.J., Doherty, R.M., Carr, P.W., Mackay, D., Abraham, M.H., Taft, R.W. (1988) Linear solvation energy relationships. 44. Parameter estimation rules that allow accurate prediction of octanol/water partition coefficients and other solubility and toxicity properties of polychlorinated biphenyls and polycyclic aromatic hydrocarbons. *Environ. Sci. Technol.* 22(5), 503-509.

Kan, A.T., Tomson, M.B. (1990) Ground water transport of hydrophobic organic compounds in the presence of dissolved organic matter. *Environ. Sci. Technol.* 9, 253-263.

Kanazawa, J. (1981) Measurement of the bioconcentration factors of pesticides by freshwater fish and their correlation with physicochemical properties of acute toxicities. *Pestic. Sci.* 12, 417-424.

Kappeler, T., Wuhrmann, K. (1978) Microbial degradation of water soluble fraction of gas oil. *Water Res.* 12, 327-333.

Karickhoff, S.W. (1981) Semi-empirical estimation of sorption of hydrophobic pollutants on natural sediments and soils. *Chemosphere* 10(8), 833-846.

Karickhoff, S.W., Brown, D.S., Scott, T.A. (1979) Sorption of hydrophobic pollutants on natural water sediments. *Water Res.* 13, 241-248.

Karickhoff, S.W., Morris, K.R. (1985) Sorption dynamics of hydrophobic pollutants in sediment suspensions. *Environ. Toxicol. & Chem.* 4, 469-479.

Kayal, S.I., Connell, D.W. (1990) Partitioning of unsubstituted polycyclic aromatic hydrocarbons between surface sediments and the water column in the Brisbane River estuary. *Aust. J. Mar. Freshwater Res.* 41, 443-456.

Kearney, P.C., Nash, R.G., Isensee, A.R. (1969) Persistence of pesticides in soil. Chemical Fallout: Current research on persistant pesticides. Miller, M.W. & Berg, C.C., Eds., Springfield, Ill. Charles C Thomas. chpt. 3., pp 54-67.

Kelley, J.D., Rice, F.O. (1964) The vapor pressures of some polynuclear aromatic hydrocarbons. *J. Phys. Chem.* 68, 3794-3796.

Kenaga, E.E. (1980a) Predicted bioconcentration factors and soil sorption coefficients of pesticides and other chemicals. *Ecotoxicol. Environ. Safety* 4, 26-38.

Kenaga, E.E. (1980b) Correlation of bioconcentration factors of chemicals in aquatic and terrestrial organisms with their physical and chemical properties. *Environ. Sci. Technol.* 14, 553-556.

Kenaga, E.E., Goring, C.A.I. (1980) In: *Aquatic Toxicology.* Eaton, J.G., Parrish, P.R., Hendrick, A.C. (eds.). Am. Soc. for Testing and Materials, STP 707, pp 78-115.

Kier, L.B., Hall, L.H., Murray, W.J., Randic, M. (1971) Molecular connectivity I:Relationship to nonspecific local anesthesia. *J. Pharm. Sci.* 64, 1971-1981.

Kilzer, L., Scheunert, I., Geyer, H., Klein, W., Korte, F. (1979) Laboratory screening of the volatilization rates of organic chemicals from water and soil. *Chemosphere* 8, 751-761.

Kincannon, D.F., Lin, Y.S. (1985) Microbial degradation of hazardous wastes by land treatment. In: *Proc. Indust. Waste Conf.* 40, 607-619.

Kishi, H., Hashimoto, Y. (1989) Evaluation of the procedures for the measurements of water solubility and n-octanol/water partition coefficient of chemicals results of a ring test in Japan. *Chemosphere* 18(9/10), 1749-1759.

Kishi, H., Kogure, N., Hashimoto, Y. (1990) Contribution of soil constituents in adsorption coefficient of aromatic compounds, halogenated alicyclic and aromatic compounds to soil. *Chemosphere* 21(7), 867-876.

Klein, W., Geyer, H., Freitag, D., Rohleder, H. (1984) Sensitivity of schemes for ecotoxicological hazard ranking of chemicals. *Chemosphere* 13, 203-211.

Klevens, H.B. (1950) Solubilization of polycyclic hydrocarbons. *J. Phys. Colloid Chem.* 54, 283-298.

Klöpffer, W., Rippen, G., Frische, R. (1982) Physicochemical properties as useful tools for predicting the environmental fate of organic chemicals. *Ecotoxicol. & Environ. Safety* 6, 294-301.

Koch, R. (1983) Molecular connectivity index for assessing ecotoxicological behavior of organic chemicals. *Toxicol. Environ. Chem.* 6, 87-96.

Konneman, W.H. (1981) Quantitative structure-activity relationships in fish toxicity studies. Part I: Relationship for 50 industrial pollutants. *Toxicology* 19, 209-221.

Kringstad, K.P., De Sousa, F., Strömberg, L.M. (1984) Evaluation of lipophilic properties of mutagens present in the spent chlorination liquor from pulp bleaching. *Environ. Sci. Technol.* 18(3), 200-203.

Krishnamurthy, T., Wasik, S.P. (1978) Fluorometric determination of partition coefficient of naphthalene homologues in octanol-water mixtures. *J. Environ. Sci. & Health* A13(8), 595-602.

Lande, S.S., Banerjee, S. (1981) Predicting aqueous solubility of organic nonelectrolytes from molar volume. *Chemosphere* 10, 751-759.

Lande, S.S., Hagen, D.F., Seaver, A.E. (1985) Computation of total molecular surface area from gas phase ion mobility data and its correlation with aqueous solubilities of hydrocarbons. *Environ. Toxicol. & Chem.* 4, 325-334.

Landrum, P.F. (1988) Toxicokinetics of organic xenobiotics in the *amphipod, pontoporeia hoy*i: role of physiological and environmental variables. *Aqua. Toxicol.* 12, 245-271.

Landrum, P.F., Stubblefield, C.R. (1991) Role of respiration in accumulation of organic xenophobics by the *amphipod diporeia sp.*. *Environ. Toxicol. & Chem.* 10, 1019-1028.

Landrum, P.F., Eadie, B.J., Faust, W.R. (1991) Toxicokinetics and toxicity of a mixture of sediment-associated polycyclic aromatic hydrocarbons to the amphipod diporeia sp. *Environ. Toxicol. and Chem.* 10, 35-46.

Landrum,, R.F., Nihart, S.R., Edie, B.J., Gardner, W.S. (1984) Reverse-phase separation method for determining pollutant binding to Aldrich humic acid and dissolved organic carbon of natural waters. *Environ. Sci. Technol.* 18, 187-192.

Landrum, P.F., Nihart, S.R., Eadie, B.J., Herche, L.R. (1987) Reduction in bioavailability of organic contaminants to the *amphipod pontoporeia hoy*i by dissolved organic matter of sediment interstitial waters. *Environ. Toxicol. and Chem.* 6, 11-20.

Landrum, P.F., Poore, R. (1988) Toxicokinetics of selected xenobiotics in hexagenia limbata. *J. Great Lakes Res.* 14(4), 427-437.

Landrum, P.F., Reinhold, M.D., Nihart, S.R., Eadie, B.J. (1985) Predicting the bioavailability of organic xenobiotics to *pontoporeia hoy*i in the presence of humic and fulvic materials and natural dissolved organic matter. *Environ. Toxicol. & Chem.* 4, 459-467.

Lane, D.A., Katz, M. (1977) The photomodification of benzo(a)pyrene, benzo(b)fluranthene, and benzo(k)fluranthene under simulated atmospheric conditions. *Adv. Environ. Sci. Technol.* 8, 137-154.

Leahy, D.E. (1986) Instrinsic molecular volume as a measure of the cavity term in linear solvation energy relationships: octanol-water partition coefficients and aqueous solubilties. *J. Pharm. Sci.* 75, 629-636.

Lee, R.F. (1977) Oil Spill Conf., Am. Petrol. Inst. pp.611-616.

Lee, R.F., Anderson, J.W. (1977) Fate and effect of naphthalenes: controlled ecosystem pollution experiment. *Bull. Mar. Sci.* 27, 127.

Lee, R.F., Ryan, C. (1976) Biodegradation of petroleum hydrocarbons by marine microbes. In: *Proc. of the Third International Biodegradation Symposium of 1975*, pp.119-125.

Lee, R.F., Gardner, W.S., Anderson, J.W., Blaylock, J.W. (1978) Fate of polycyclic aromatic hydrocarbons in controlled ecosystem enclosures. *Environ. Sci. Technol.* 12(7), 832-838.

Lee, R.F., Sauerheber, R., Benson, A.A. (1972) Petroleum hydrocarbons: Uptake and discharge by the marine mussel *mytilus edulis*. *Science* 177, 344-345.

Lee, M.D., Wilson, J.T., Ward, C.H. (1984) Microbial degradation of selected aromatics in a hazardous waste site. *Devel. Indust. Microbiol.* 25, 557-565.

Lemaire, P., Mathieu, A., Carriere, S., Drai, P., Giudicelli, J., Lafaurie, M. (1990) The uptake mechanism and biological half-life of benzo(a)pyrene in different tissues of sea bass, *dicentrarchus labrax*. *Ecotoxicol. & Environ. Safety* 20, 223-233.

Leo, A.J. (1975) Calculation of partition coefficients useful in the evaluation of relative hazards of various chemicals in the environment. In: *Symposium on Structure-Activity Correlations in Studies of Toxicity and Bioconcentration with Aquatic Organisms*. G.D. Veith and D.E. Konasewich, Editors, International Joint Commission, Ontario, Canada.

Leo, A.J. (1986) CLOGP-3.42 *Medchem. Software*, Medicinal Chemistry Project, Pomona College, Claremont, California.

Leo, A., Hansch, C., Elkins, D. (1971) Partition coefficients and their uses. *Chemical Reviews* 71, 525-616.

Leversee, G.J., Giesy, J.P., Landrum, P.F., Bartell, S., Gerould, S. Bruno, M., Spacie, A., Bowling, J., Haddock, J., Fannin, T. (1981) Disposition of benzo(a)pyrene in aquatic systems components: *Periphyton, Chironomids, Daphania*, Fish. In: *Polynuclear Aromatic Hydrocarbons: Chemical Analysis and Biological Fate*. Cooke, M., Dennis, A.J. (Eds), pp. 357-367. Battelle Press, Columbus, Ohio.

Leyder, F., Boulanger, P. (1983) Ultraviolet absorption , aqueous solubility, and octanol-water partition for several phthalates. *Bull. Environ. Contam. Toxiocol.* 30 (2), 152.

Linder, G., Bergman, H.L., Meyer, J.S. (1985) Anthracene bioconcentration in rainbow trout during single-compound and complex mixture exposures. *Environ. Toxicol. & Chem.* 4, 549-558.

Lu, P.Y., Metcalf, R.L., Carlson, E.M. (1978) Environmental fate of five radiolabelled coal conversion by-products evaluated in a laboratory model ecosystems. *Environ. Health Perspectives* 24, 201.

Lu, P.Y., Metcalf, R.L., Plummmer, N., Mandel, D. (1977) The environmental fate of three carcinogens: Benzo[a]pyrene, benzidine, and vinyl chloride evaluated in laboratory model ecosystems. *Arch. Environ. Contam. Toxicol.* 6, 129-142.

Lyman, W.J., Reehl, W.F., Rosenblatt, D.H., Editors (1982) *Handbook on Chemical Property Estimation Methods, Environmental Behavior of Organic Compounds*. McGraw-Hill, New York, pp.960.

Ma, K.C., Shiu, W.Y., Mackay, D. (1990) *A Critically Reviewed Compilation of Physical and Chemical and Persistence Data for 110 Selected EMPPL Substances.* A report prepared for the Ontario Ministry of Environment, Water Resources Branch, Toronto, Ontario.

Mabey, W., Mill, T. (1978) Critical review of hydrolysis of organic compounds in water under environmental conditions. *J. Phys. Chem. Ref. Data* 7, 838-415.

Mabey, W., Smith, J.H., Podoll, R.T., Johnson, H.L., Mill, T., Chou, T.W., Gate, J., Waight-Partridge, I., Jaber, H., Vandenberg, D. (1982) *Aquatic Fate Process for Organic Priority Pollutants.* EPA Report, No. 440/4-81-14.

Mackay, D. (1980) Solubility, partition coefficients, volatility and evaporation rates. In: *The Handbook of Environmental Chemistry.* Vol.2/Part A, Hutzinger, O., Ed., pp. 31-45, Springer-Verlag, Berlin, Germany.

Mackay, D. (1981) Basic properties of material. In: *Environmental Risk Analysis for Chemicals.* Conway, R.A., Ed., pp. 33-60, Van Nostrand Reinhold Co. New York, N.Y.

Mackay, D. (1982) Correlation of bioconcentration factors. *Environ. Sci. Technol.* 16, 274-278.

Mackay, D. (1986) Personal Communication.

Mackay, D., Hughes, A.I. (1984) Three parameter equation describing the uptake of organic compounds by fish. *Environ. Sci. Technol.* 18, 439-444.

Mackay, D., Leinonen, P.J. (1975) Rate of evaporation of low-solubility contaminants from water bodies to atmosphere. *Environ. Sci. Technol.* 9, 1178-1180.

Mackay, D., Shiu, W.Y. (1977) Aqueous solubility of polynuclear aromatic hydrocarbons. *J. Chem. Eng. Data* 22, 399-402.

Mackay, D. Shiu, W.Y., Chau, E. (1983) Calculation of diffusion resistances controlling volatilization rates of organic contaminants from water. *Can. J. Fish. & Aqua. Sci.* 40, 295-303.

Mackay, D., Shiu, W.Y., Sutherland, R.P. (1979) Determination of air-water Henry's law constants for hydrophobic pollutants. *Environ. Sci. Technol.* 13, 333-337.

Mackay, D., Bobra, A.M., Shiu, W.Y., Yalkowsky, S.H. (1980) Relationships between aqueous solubility and octanol-water partition coefficient. *Chemosphere* 9, 701-711.

Mackay, D., Bobra, A.M., Chan, D.W., Shiu, W.Y. (1982) Vapor pressure correlation for low-volatility environmental chemicals. *Environ. Sci. Technol.* 16, 645-649.

Mackay, D., Shiu, W.Y., Bobra, A., Billington, J., Chau, E., Yeun, A., Ng, C., Szeto, F. (1982) *Volatilization of Organic Pollutants from Water.* EPA600/3-82-019. National Technical Information Service, Springfield, Virginia.

Mackay, D., Shiu, W.Y. (1981) A critical review of Henry's law constants for chemicals of environmental interest. *J. Phys. Chem. Ref. Data* 10, 1175-1199.

Mackay, D., Shiu, W.Y. (1984) Relationships between physical-chemical and environmental partitioning coefficients. In: *QSAR in Environmental Toxicology.* Kaiser, K.L.E., Ed., pp. 261-278, D. Reidel Publishing Co. Amsterdam.

Mackay, D., Wolkoff, A.W. (1973) Rate of evaporation of low-solubility contaminants from water bodies to atmosphere. *Environ. Sci. Technol.* 7, 611-614.

Macknick, A.B., Prausnitz, J.M. (1979) Vapor pressures of high-molecular-weight hydrocarbons. *J. Chem. Eng. Data* 24, 175-178.

Mailhot, H. (1987) Prediction of algae bioaccumulation and uptake rate of nine organic compounds by ten physicochemical properties. *Environ. Sci. Technol.* 21, 1009-1013.

Mailhot, H., Peters, R.H. (1988) Empirical relationships between the 1-octanol/water partition coefficient and nine physicochemical properties. *Environ. Sci. Technol.* 22(12), 1479-1488.

Malaspina, L., Gigli, R., Bardi, G. (1973) Microcalorimetric determination of the enthalpy of sublimation of benzoic acid and anthracene. *J. Chem. Phys.* 59, 387.

Mallon,B.J., Harris, F. (1984) Octanol-water partition coefficient of benzo(a)pyrene, measurement, calculation and environmental implication. *Bull. Environ. Contam. Toxicol.* 32, 316-323.

Manilal, V.B., Alexander, M. (1991) Factors affecting the microbial degradation of phenanthrene in soil. *Appl. Microbiol. Biotechnol.* 35, 401-405.

Mantoura, R.F.C. (1984) *Speciation and Transformation Model of Hydrocarbons in an Estuary.* Research Report , Institute for Marine Environmental Research, Plymouth, U.K.

Martin, H. (1961) *Guides to the Chemicals used in Crop Protection.* 4th ed., Canadian Dept. Agriculture Publication. pp. 1093.

May, W.E. (1980) The solubility behavior of polycyclic aromatic hydrocarbnons in aqueous systems. In: *Petroleum in the Marine Environment,* Petrakis, L., Weiss, F.T., Eds., *Advances in Chemistry Serie*s No. 85, pp. 143-192, Am. Chem. Soc., Washington, D.C.

May, W.E., Brown, J.M., Chesler, S.N., Guenther, F., Hilpert, L.R., Hertz, H.S., Wise, S.A. (1979) Development of an aqueous polynuclear aromatic hydrocarbon standard reference material. In: *Polynuclear Aromatic Hydrocarbons.* Jones, P.W., Leber, P., Editors, Ann Arbor Science Publishers, Inc., Ann Arbor, Michigan, U.S.A.

May, W.E., Wasik, S.P., Freeman, D.H. (1978a) Determination of aqueous solubility of polynuclear aromatic hydrocarbons by coupled column liquid chromatographic technique. *Anal. Chem.* 50, 175-179.

May, W.E, Wasik, S.P., Freeman, D.H. (1978b) Determination of solubility behaviour of some polycyclic aromatic hydrocarbons in water. *Anal. Chem.* 50, 997-1000.

May, W.E., Wasik, S.P., Miller, M.M., Tewari, Y.B., Brown-Thomas, J.M., Goldberg, R.N. (1983) Solution thermodynamics of some slightly soluble hydrocarbons in water. *J. Chem. Eng. Data* 28, 197-200.

McAuliffe, C. (1966) Solubility in water of paraffin, cycloparaffin, olefin, acetylene, cycloolefin, and aromatic hydrocarbons. *J. Phys. Chem.* 76, 1267-1275.

McCarthy,J.F. (1983) Role of particulate organic matter in decreasing accumulation of polynuclear aromatic hydrocarbons by *Daphania Magna. Arch. Environ. Contam. Toxicol.* 12, 559-568.

McCarthy, J.F., Jimenez, B.D. (1985) Reduction in bioavailability to bluegills of polycyclic aromatic hydrocarbons bound to dissolved humic material. *Environ. Toxicol. & Chem.* 4, 511-521.

McCarthy, J.F., Jimenez, B.D., Barbee, T. (1985) Effect of dissolved humic material on accumulation of polycyclic aromatic hydrocarbons: structure-activity relationships. *Aqua. Toxicol.* 7, 15-24.

McDuffie, D. (1981) Estimation of octanol/water partition coefficients for organic pollutants using reverse-phase HPLC. *Chemosphere* 10, 73-83.

Mclachlan, M., Mackay, D., Jones, P.H. (1990) A conceptual model of organic chemical volatilization at waterfalls. *Environ. Sci. Technol.* 24, 252-257.

Means, J.C., Hassett, J.J., Wood, S.G., Banwart, W.L. (1979) Sorption properties of energy-related pollutants and sediments. In: *Polynuclear Aromatic Hydrocarbons.* P.W. Jones and P. Leber Editors, Ann Arbor Science Publishers, Inc., Ann Arbor, Michigan, U.S.A.

Means, J.C., Wood, S.G., Hassett, J.J., Banwart, W.L. (1980) Sorption of polynuclear aromatic hydrocarbons by sediments and soils. *Environ. Sci. Technol.* 14(12), 1524-1528.

Means, J.C., Wood, S.G., Hassett, J.J., Banwart, W.L. (1982) Sorption of amino-and carboxy-substituted polynuclear aromatic hydrocarbons by sediments and soils. *Environ. Sci. Technol.* 16, 93-98.

Melancon, M.J. Jr., Lech, J.J. (1978) Distribution and elimination of naphthalene and 2-methylnaphthalene in rainbow trout during short and long-term exposures. *Arch. Environ. Contam. Toxicol.* 7, 207.

Metcalfe, D.E., Zukovs, G., Mackay, D., Paterson, S. (1988) Polychlorinated biphenyls (PCBs): physical and chemical property data. In: *Hazards, Decontamination and Replacement of PCB: A Comprehensive Guide.* Crine, J.P., Ed., pp. 3-33, Plenum Press, New York.

Meylan, W.M., Howard, P.H. (1991) Bond contribution method for estimating Henry's law constants. *Environ. Toxicol. and Chem.* 10, 1283-1293.

Mill, T., Mabey, W.R., Lan, B.Y., Baraze, A. (1981) Photolysis of polycyclic aromatic hydrocarbons in water. *Chemosphere* 10, 1281-1290.

Mill, T., et al. (1982) *Laboratory Protocols for Evaluating the Fate of Organic Chemicals in Air and Water.* p. 255 US-EPA-600/3-82-022.

Miller, M.M., Ghodbane, S., Wasik, S.P., Tewari, Y.B., Martire, D.E. (1984) Aqueous solubilities, octanol/water partition coefficients and entropies of melting of chlorinated benzenes and biphenyls. *J. Chem. Eng. Data* 29, 184-190.

Miller, M.M., Wasik, S.P., Huang, G.L., Shiu, W.Y., Mackay, D. (1985) Relationships between octanol-water partition coefficient and aqueous solubility. *Environ. Sci. Technol.* 19, 522-529.

Mills, W.B., Dean, J.D., Porcella, D.B., Gherini, S.A., Huson, R.J.M., Frick, W.E., Rupp, G.L. (1982) *Water Quality Assessment: A Screening Procedure for Toxic and Conventional Pollutants.* Part 1. U.S. EPA Report EPA-600/6-82-004a.

Monsanto Co. (1972) Presentation to the interdepartmental task force on PCB. May 15, 1972, Washington, D.C.

Muel, B., Saguem, S. (1985) Determination of 23 polycyclic hydrocarbons in atmospheric particulate matter of the Paris area and photolysis by sunlight. *Int'l. J. Environ. Anal. Chem.* 19, 111-131.

Murphy, E.M., Zachara, J.M., Smith, S.C. (1990) Influence of mineral-bound humic substances on the sorption of hydrophobic organic compounds. *Environ. Sci. Technol.* 24(10), 1507-1516.

Murray, J.M., Pottie, R.F., Pupp, C. (1974) The vapor pressures and enthalpies of sublimation of five polycyclic aromatic hydrocarbons. *Can. J. Chem.* 52, 557-563.

Neely, W.B. (1980) A method for selecting the most appropriate environmental experiments on a new chemical. In: *Dynamics, Exposure and Hazard Assessment of Toxic Chemicals.* Haque, R., Editor, Ann Arbor Science Publishers, Ann Arbor, Michigan.

Neely, W.B. (1981) Complex problems-simple solutions. *Chemtech.* 11, 249.

Neely, W.B. (1982) Organizing data for environmental studies. *Environ. Toxicol. & Chem.* 1, 259-266.

Neely, W.B. (1983) Reactivity and environmental persistence of PCB isomers. In: *Physical Behavior of PCBs in the Great Lakes.* Mackay, D., Paterson, S., Eisenreich, S.J., Editors, pp.71-88. Ann Arbor Science Publishers, Ann Arbor, Michigan.

Neely, W.B., Branson, D.R., Blau, G.E. (1974) Partition coefficient to measure bioconcentration potential of organic chemicals in fish. *Environ. Sci. Technol.* 8, 1113-1115.

Neff, J.M. (1979) *Polycyclic Aromatic Hydrocarbons in the Aquatic Environment.* Applied Sci. Publisher, London, England, 262pp.

Neuhauser, E.F., Loehr, R.C., Malecki, M.R., Milligan, D.L., Durkin, R.P. (1985) The toxicity of selected organic chemicals to earthworm *eisenia fetida.* *J. Environ. Qual.* 14(3), 383-388.

Newsted, J.L., Giesy, J.P. (1987) Predictive models for photoinduced acute toxicity of polycyclic aromatic hydrocarbons to *daphnia magna,* strauss *(cladocera, crustacea).* *Environ. Toxicol. & Chem.* 6, 445-461.

Nirmalakhandan, N.N., Speece, R.E. (1988) QSAR model for predicting Henry's constant. *Environ. Sci. Technol.* 22(11), 1349-1357.

Nirmalakhandan, N.N., Speece, R.E. (1989) Prediction of aqueous solubility of organic chemicals based on molecular structure. 2. Application to PNAs, PCBs, PCDDs, etc. *Environ. Sci. Technol.* 23, 708-713.

Nkedi-Kizza, P., Rao, P.S.C., Hornsby, A.G. (1985) Influence of organic cosolvent on sorption of hydrophobic organic chemicals by soils. *Environ. Sci. Technol.* 19, 975-979.

OECD (1981) *OECD Gudielines for Testing of Chemicals.* Organization for Economic Co-operation and Development. OECD, Paris.

Ogata, M., Fujisawa, K., Ogino, Y., Mano, E. (19884) Partition coefficients as a measure of bioconcentration potential of crude oil compounds in fish and shellfish. *Bull. Environ. Contam. Toxicol.* 33, 561-567.

Oliver, B.G. (1985) Desorption of chlorinated hydrocarbons for spiked and anthropogenically contaminated sediments. *Chemosphere* 14, 1087-1106.

Oliver, B.G. (1987a) Biouptake of chlorinated hydrocarbons from laboratory-spiked and field sediments by oligochaete worms. *Environ. Sci. Technol.* 21, 785-790.

Oliver, B.G. (1987b) Fate of some chlorobenzenes from the Niagara River in Lake Ontario. In: *Sources and Fates of Aquatic Pollutants.* Hite, R.A., Eisenreich, S.J., Eds., pp. 471-489, Advances in Chemistry Series 216, Am. Chem. Soc., Washington, D.C.

Oliver, B.G., Charlton, C.N. (1984) Chlorinated organic contaminants on settling particulates in the Niagara River vicinity of Lake Onatrio. *Environ. Sci. Technol.* 18, 903-908.

Oliver, B.G., Niimi A.J. (1983) Bioconcentration of chlorobenzenes from water by rainbow trout: Correlations with partition coefficients and environmental residues. *Environ. Sci. Technol.* 17, 287-291.

Oliver, B.G., Niimi, A.J. (1985) Bioconcentration factors of some halogenated organics for rainbow trout: Limitations in their use for prediction of environmental residues. *Environ. Sci. Technol.* 19, 842-849.

Opperhuizen, A., van der Velde, E.W., Govas, F.A.P.C., Liem, D.A.K., van der Steen, J.M. (1985) Relationship between bioconcentration in fish and steric factors of hydrophobic chemicals. *Chemosphere* 14, 1871-1896.

Opperhuizen, A., Gobas, F.A.P.C., Van der Steen, J.M.D., Hutzinger, O. (1988) Aqueous solubility of polychlorinated biphenyls related to molecular structure. *Environ. Sci. Technol.* 22, 638-646.

Opperhuizen, A., Sijm, D.T.H.M. (1989) Bioaccumulation and biotransformation of polychlorinated dibenzo-p-dioxins and dibenzofurans in fish. *Environ. Toxicol. & Chem.* 9, 175-186.

Osborn, A.G., Douslin, D.R. (1975) Vapor pressures and derived enthalpies of vaporization for some condensed-ring hydrocarbons. *J. Chem. & Eng. Data* 20, 229-231.

Pankow, J.F. (1990) Minimization of volatilization losses during sampling and analysis of volatile organic compounds in water. In: *Significance and Treatment of Volatile Organic Compounds in Water Systems.* Ram, N.M., Christman, R.F., Cantor, K.P., Editors, pp.73-86, Lewis Publishers, Inc., Chelsea, Michigan.

Pankow, J.F., Isabelle, L.M., Asher, W.E. (1984) Trace organic compounds in rain. I. Sample design and analysis by adsorption/thermodesorption (ATD). *Environ. Sci. Technol.* 18, 310-318.

Pankow, J.F., Rosen, M.E. (1988) Determination of volatile compounds in water by purging directly to a capillary column with whole column cryotrapping. *Environ. Sci. Technol.* 22, 398-405.

Paris, D.F., Steen, W.C., Barnett, J.T., Bates, E.H. (1980) Kinetics of degradation of xenobiotics by microorganisms. *Paper ENVR-21, 180th National Meeting of American Chemical Society*, San Francisco.

Park, K.S., Sims, R.C., Dupont, R.R., Doucette, W.J., Matthews, J.E. (1990) Fate of PAH compounds in two soil types: Influence of volatilization, abiotic loss and biological activity. *Environ. Toxicol. & Chem.* 9, 187-195.

Parks, G.S., Huffman, H.M. (1931) Some fusion and transition data for hydrocarbons. *Ind. Eng. Chem.* 23(10), 1138-1139.

Passino, D.R.M., Smith, S.B. (1987) Quantitative structure-activity relationships (QSAR) and toxicity data in hazard assessment. In: *QSAR in Environmental Toxicology-II.* Kaiser, K.L.E., Editor, D. Reidel Publishing Company. pp.261-270.

Patton, J.S., Stone, B., Papa, C., Abramowitz, R., Yalkowsky, S.H. (1984) Solubility of fatty acids and other hydrophobic molecules in liquid trioleoylglycerol. *J. Lipid Res.* 25, 189-197.

358

Paul, M.A. (1951) The solubility of naphthalene and biphenyl in aqueous solutions of electrolytes. *J. Am. Chem. Soc.* 74, 5274-5277.

Pavlou, S.P. (1987) The use of equilibrium partition approach in determining safe levels of contaminants in marine sediments. p.388-412. In: *Fate and Effects of Sediments-Bound Chemicals in Aquatic Systems.* Dickson, K.L., Maki, A.W., Brungs, W.A., Editors. Proceedings of the Sixth Pellston Workshop, Florissant, Colorado, August 12-17, 1984. SETAC Special Publication Series, Ward, C.H., Walton, B.T., Editors, Pergamon Press, New York.

Pavlou, S.P., Weston, D.P. (1983, 1984) Initial evaluation of alternatives for development of sediment related criteria for toxic contaminants in marine waters (Puget Sound) phase I & II. EPA Contract No. 68-01-6388.

Pearlman, R.S. (1986) Molecular surface area and volume: their calculation and use in predicting solubilities and free energies of desolvation. In: *Partition Coefficient, Determination and Estimation.* Dunn III, W.J., Block, J.H., Pearlman, R.S., Editors, Pergamon Press, New York. pp.3-20.

Pearlman, R.S., Yalkowsky, S.H., Banerjee, S. (1984) Water solubilities of polynuclear aromatic and heteroaromatic compounds. *J. Phys. Chem. Ref. Data* 13(2), 555-562.

Perkow, J., Eschenroeder, A., Goyer, M., Stevens, J., Wechsler, A. (1980) *An Exposure and Risk Assessment for 2,3,7,8-Tetrachlorodibenzo-p-Dioxin.* Office of Water Regulations and Standards, U.S. EPA, Washington, D.C.

Perez-Tejeda, P., Yanes, C., Maestre, A. (1990) Solubility of naphthalene in water and alcohol solutions at various temperatures. *J. Chem. Eng. Data* 35, 244-246.

Pierotti, C., Deal, C., Derr, E. (1959) Activity coefficient and molecular structure. *Ind. Eng. Chem. Fundam.* 51, 95-101.

Pinal, R. (1988) *Estimation of Aqueous Solubility of Organic Compounds.* Ph.D. Dissertation, University of Arizona, Tucson, Arizona.

Pinal, R., Lee, L.S., Rao, P.S.C. (1991) Prediction of the solubility of hydrophobic compounds in nonideal solvent mixtures. *Chemosphere* 22(9-10), 939-951.

Pinal, R., Rao, S.C., Lee, L.S., Cline, P.V. (1990) Cosolvency of partially miscible organic solvents on the solubility of hydrophobic organic chemicals. *Environ. Sci. Technol.* 24(5), 639-647.

Podoll, R.T., Irwin, K.C., Parish, H.J. (1989) Dynamic studies of naphthalene sorption on soil from aqueous solution. *Chemosphere* 18, 2399-2412.

Pomona (1982) Pomona College Medicinal Data Base, June 1982.

Power, W.H., Woodworth, C.L., Loughary, W.G. (1977) Vapor pressure determination by gas chromatography in the microtorr range-anthracene and triethylene glycol di-2-ethyl butyrate. *J. Chromatogr. Sci.* 15, 203-207.

Price, L.C. (1976) Aqueous solubility of petroleum as applied to its origin and primary migration. *Am. Assoc. Petrol. Geol. Bull.* 60, 213-244.

Pupp, C., Lao, R.C., Murray, J.J., Pottie, R.F. (1974) Equilibrium vapor concentrations of some polycyclic aromatic hydrocarbons, arsenic trioxide (As_4O_6) and selenium dioxide, and the collection efficiencies of these air pollutants. *Atoms. Environ.* 8, 915-925.

Pussemier, L., Szabo, G., Bulman, R.A. (1990) Prediction of the soil adsorption coefficient K_{OC} for aromatic pollutants. *Chemosphere* 21(10-11), 1199-1212.

Radchenko, L.G., Kitiagorodskii, A.I. (1974) Vapor pressure and heat of sublimation of naphthalene, biphenyl, octafluoronaphthalene, decafluorobiphenyl, acenaphthene and α-nitronaphthalene. *Zhur. Fiz. Khim.* 48, 2702-2704.

Radding, S.B., Mill, T., Gould, C.W., Lin, D.H., Johnson, H.L., Bomberger, D.C., Fojo, C.V. (1976) *The Environmental Fates of Selected Polycyclic Aromatic Hydrocarbons.* U.S. Environmental Protection Agency Report No. EPDA-560/5-75-009.

Radding, S.B., Liu, D.H., Johnson, H.L., Mill, T. (1977) *Review of the Environmental Fate of Selected Chemicals.* USEPA report No. EPA-560/5-77-003.

Rapaport, R.A., Eisenreich, S.J. (1984) Chromatographic determination of octanol-water partition coefficients (K_{ow}'s) for 58 polychlorinated biphenyl congeners. *Environ. Sci. Technol.* 18, 163-170.

Riddick, J.A. et al., (1986) Organic Solvents: Physical Properties and Methods of Purification, 4th Edn. J. Wiley & Sons, New York, N.Y.

Reichardt, P.B., Chadwick, B.L., Cole, M.A., Robertson, B.R., Button, D.K. (1981) Kinetic study of biodegradation of biphenyl and its monochlorinated analogues by a mixed marine microbial community. *Environ. Sci. Technol.* 15(1), 75-79.

Reid, R., Prausnitz, J., Sherwood, T. (1979) *The Properties of Gases and Liquids.* 3rd. edition, McGraw Hill, New York., pp.57-58.

Rekker, R.F. (1977) The hydrophobic fragmental constant. In: *Pharmacochemistry Library.* Nauta, W.T., Rekker, R.F., Editors, Elsevier Sci. Publishers Co., New York.

Rekker, R.F., De Kort, N.N. (1979) The hydrophobic fragment constant; an extension to a 1000 data point set. *Eur. J. Med. Chem.-Chim. Ther.* 14(6), 479-488.

Riederer, M.(1990) Estimating partitioning and transport of organic chemicals in the foliage/atmosphere system: Discussion of a fugacity-based model. *Environ. Sci. Technol.* 24, 829-837.

Rogers, K.S., Cammarata, A. (1969) Superdelocalizability and charge density. A correlation with partition coefficients. *J. Med. Chem.* 12, 692.

Rordorf, B.F. (1986) Thermal properties of dioxins, furans and related compounds. *Chemosphere* 15(9-12), 1325-1332.

Rossi, S.S. (1977) Bioavailability of petroleum hydrocarbons from water, sediments and detritus to the marine annelid, *neanthes arenaceodentata. Proc. Oil Spill Conf.*, pp.621-625. Am. Petrol. Inst., Washington D.C.

Rossi, S.S., Thomas, W.H. (1981) Solubility behavior of three aromatic hydrocarbons in distilled water and natural seawater. *Environ. Sci. Technol.* 15, 715-716.

Ruepert, C., Grinwis, A., Govers, H. (1985) Prediction of partition coefficients of unsubstituted polycyclic aromatic hydrocarbons from C_{18} chromatographic and structural properties. *Chemosphere* 14(3/4), 279-291.

Ryan, J.A., Bell, R.M., Davidson, J.M., O'Connor, G.A. (1988) Plant uptake of non-ionic organic chemicals from soils. *Chemosphere* 17, 2299-2323.

Ryan, P.A., Cohen, Y. (1986) Multimedia transport of particle-bound organics: bezo(a)pyrene test case. *Chemosphere* 15(1), 21-47.

Sabljic, A. (1984) Predictions of the nature and strength of soil sorption of organic pollutants by molecular topology. *J. Agric. Food Chem.* 32, 243-246.

Sabljic, A. (1987a) On the prediction of soil sorption coefficients of organic pollutants from molecular structure: application of molecular topology model. *Environ. Sci. Technol.* 21, 358-366.

Sabljic, A. (1987b) Nonempirical modeling of environmental distribution and toxicity of major organic pollutants. In: *QSAR in Environmental Toxicology* - II. Kaiser, K.L.E., Ed., pp 309-322, D. Reidel Publ. Co., Dordrecht, Netherlands.

Sahyun, M.R.V. (1966) Binding of aromatic compounds to bovine serum albumin. *Nature* 209, 613-614.

Sangster, J. (1989) Octanol-water partition coefficients of simple organic compounds. *J. Phys. Chem. Ref. Data* 18, 1111-1230.

Sarna, L.P., Hodge, P.E., Webster, G.R.B. (1984) Octanol-water partition coefficients of chlorinated dioxins and dibenzofurans by reversed-phase HPLC using several C_{18} columns. *Chemosphere* 13, 975-983.

Schmidt-Bleek, F., Haberland, W., Klein, A.W., Caroli, S. (1982) Steps towards environmental hazard assessment of new chemicals (including a hazard ranking scheme, based upon Directive 79/831/EEC). *Chemosphere* 11, 383-415.

Schüürmann, G., Klein, W. (1988) Advances in bioconcentration prediction. *Chemosphere* 17(8), 1551-1574.

Schwarz, F.P. (1977) Determination of temperature dependence of solubilities of polycyclic aromatic hydrocarbons in aqueous solutions by a fluorescence method. *J. Chem. Eng. Data* 22, 273-277.

Schwarz, F.P., Wasik, S.P. (1976) Fluorescence measurements of benzene, naphthalene, anthracene, pyrene, fluoranthene, and benzo[a]pyrene in water. *Anal. Chem.* 48, 524-528.

Schwarz, F.P., Wasik, S.P. (1977) A fluorescence method for the measurement of the partition coefficients of naphthalene, 1-methylnaphthalene, and 1-ethylnaphthalene in water. *J. Chem. Eng. Data* 22, 270-273.

Schwarzenbach, R.P., Westall, J. (1981) Transport of nonpolar organic compounds from surface water to groundwater. Laboratory sorption studies. *Environ. Sci. Technol.* 15(11), 1360-1367.

Sears, G.W., Hopke, E.R. (1949) Vapor pressures of naphthalene, anthracene, and hexachlorobenzene in a low pressure region. *J. Am. Chem. Soc.* 71, 1632-1634.

Seki & Suzuki (1953) Physical chemical studies on molecular compounds. III. Vapor pressures of diphenyl, 4,4'-dinitrophenyl and molecular compound between them. *Bull. Chem. Soc. Japan* 26, 209-213.

Shiu, W.Y., Doucette, W., Gobas, F.A.P.C., Mackay, D., Andren, A.W. (1988) Physical-chemical properties of chlorinated dibenzo-p-dioxins. *Environ. Sci. Technol.* 22, 651-658.

Shiu, W.Y., Gobas, F.A.P.C., Mackay, D. (1987) Physical-chemical properties of three congeneric series of chlorinated hydrocarbons. In: *QSAR in Environmental Toxicology-II*. Kaiser, K.L.E., Editor, D. Reidel Publishing Company. pp.347-362.

Shiu, W.Y., Mackay, D. (1986) A Critical review of aqueous solubilities, vapor pressures, Henry's law constants and octanol-water partition coefficients of the polychlorinated biphenyls. *J. Phys. Chem. Data* 15, 911-929.

Shiu, W.Y., Ma, K.C., Mackay, D., Seiber, J.N., Wauchope, R.D. (1990) Solubilities of Pesticide Chemicals in Water. Part I. Environmental Physical Chemistry, Part II. Data Compilation. *Review Environ. Contam. Toxicol.* 116, 1-187.

Shorten, C.V., Elzerman, A.W., Mills, G.L. (1990) Methods for the determination of PAH desorption kinetics in coal fines and coal contaminated sediments. *Chemosphere* 20(1-2) 137-159.

Sims, R.C.C. (1990) Fate of PAH compounds in soil loss mechanisms. accepted for *Environ. Toxicol. Chem.* 9.

Sims, R.C.C., Overcash, M.R. (1983) Fate of polynuclear aromatic compounds (PNA's) in soil plant systems. *Residue Review* 88.

Sinke, G.C. (1974) A method for measurement of vapor pressures of organic compounds below 0.1 torr, naphthalene as a reference substance. *J. Chem. Thermodynamics* 6, 311-316.

Sklarew, D.S., Girvin, D.C. (1987) Attenuation of polychlorinated biphenyls in soil. *Rev. Environ. Contam. Toxicol.* 98, 1-41.

Smith, J.H., Mabey, W.R., Bahonos, N., Holt, B.R., Lee, S.S., Chou, T.W., Venberger, D.C., Mill, T. (1978) *Environmental Pathways of Selected Chemicals in Fresh Water Systems: Part II. Laboratory Studies*. Interagency Energy-Environment Research Program Report. EPA-600/7-78-074. Environmental Research Laboratory Office of Research and Development. U.S. Environment Protection Agency, Athens, Georgia 30605, p. 304.

Smith, J.H., Mackay, D., Ng, C.W.K. (1983) Volatilization of pesticides from water. *Residue Review* 85, 73-88.

Sonnefeld, W.J., Zoller, W.H., May, W.E. (1983) Dynamic coupled-column liquid chromatographic determination of ambient temperature vapor pressures of polynuclear aromatic hydrocarbons. *Anal. Chem.* 5, 275-280.

Southworth, R.G. (1977) Transport and transformations of anthracene in natural waters. In: *Aquatic Toxicology*. ASTM ATP 667, Marking, L.L., Kimerle, R.A., Editors, American Society for Testing and Materials, pp.359-380, Philadelphia.

Southworth, G.R., Beauchamp, J.J., Schmieders, P.K. (1978) Bioaccumulation potential of polycyclic aromatic hydrocarbons in *Daphnia Pulex*. *Water Res.* 12, 973-977.

Southworth, G.R. (1979) The role of volatilization in removing polycyclic aromatic hydrocarbons from aquatic environments. *Bull. Environ. Contam. Toxicol.* 21, 507-514.

Spacie, A., Landrum, R.F., Leversee, G.J. (1983) Uptake, depuration and biotransformation of anthracene and benzo-a-pyrene in bluegill sunfish. *Ecotox. Environ. Safety* 7, 330-341.

Stauffer, T.B., MacIntyre, W.G., Wickman, D.C. (1989) Sorption of nonpolar organic chemicals on low-carbon-content aquifer materials. *Environ. Toxicol. & Chem.* 8, 845-852.

Steen, W.C., Karickhoff, S.W. (1981) Biosorption of hydrophobic organic pollutants by mixed microbial populations. *Chemosphere* 10, 27-32.

Stephen, H., Stephen, D., Editors (1963) *Solubility of Inorganic and Organic Compounds.* Macmillan Co., New York.

Stephenson, R.M., Malanowski, A. (1987) *Handbook of the Thermodynamics of Organic Compounds.* Elsevier, New York.

Stevens, B., Perez, S.R., Ors, J.A. (1974) Photoperoxidation of unsaturated organic molecules O_2 delta G acceptor properties and reactivity. *J. Am. Chem. Soc.* 96, 6846-6850.

Stucki, G., Alexander, M. (1987) Role of dissolution rate and solubility in biodegradation of aromatic compounds. *Appl. Environ. Microbiol.* 53, 292-297.

Swann, R.L., Laskowski, D.A., McCall, P.J., Vander Kuy, K., Dishburger, H.J. (1983) A rapid method for the estimation of the environmental parameters octanol/water partition coefficient, soil sorption constant, water to air ratio, and water solubility. *Residue Rev.* 85, 17-28.

Szabo, G., Prosser, S.L., Bulman, R.A. (1990a) Prediction of the adsorption coefficient (K_{OC}) for soil by a chemically immobilized humic acid column using RP-HPLC. *Chemosphere* 21(6), 729-739.

Szabo, G., Prosser, S.L., Bulman, R.A. (1990b) Determination of the adsorption coefficient (K_{OC}) of some aromatics for soil by RP-HPLC on two immobilized humic acid phases. *Chemosphere* 21(6), 777-788.

Tabak, H.H., Quave, S.A., Mashni, C.I., Barth, E.F. (1981) Biodegradability studies with organic priority pollutant compounds. *J. Water Pollut. Control. Fed.* 53, 1503-1518.

Taylor, J.W., Crooks, R.J. (1976) *J. Chem. Soc. Faraday Trans.* 72, 723-729.

Thomann, R.V. (1989) Bioaccumulation model of organic chemical distribution in aquatic food chains. *Environ. Sci. Technol.* 23, 699-707.

THOR (1986) *Database of Medicinal Chemistry Project*, Pomona College, Claremont, California.

Tomlinson, E., Hafkenscheid, T.L. (1986) Aqueous solubility and partition coefficient estimation from HPLC data. In: *Partition Coefficient Determination and Estimation.* Dunn III, W.J., Block, J.H., Pearlman, R.S., Editors, Pergamon Press, New York. pp.101-141.

Trabalka, J.R., Garten, C.T., Jr. (1982) *Development of Predictive Models for Xenobiotic Bioaccumulation in Terrestrial Ecosystems.* National Technical Information Service, ORNL-5869. Springfield, Virginia.

Travis, C.C., Arms, A.D. (1988) Bioconcentration of organics in beef, milk, and vegetation. *Environ. Sci. Technol.* 22, 271-174.

Tsai, W., Cohen, Y., Sakugawa, H., Kaplan, I.R. (1991) Dynamic partitioning of semivolatile organics in gas/particle/rain phases during rain scavenging. *Environ. Sci. Technol.* 25(12), 2012-2023.

Tsonopoulos, C., Prausnitz, J.M. (1971) Activity coefficients of aromatic solutes in dilute aqueous solutions. *Ind. Eng. Chem. Fundam.* 10, 593-600.

Ubbelohde, A.R. (1978) *The Molten State of Matter.* p.148, Wiley, New York, New York.

Vadas, G.G., MacIntyre, W.G., Burris, D.R. (1991) Aqueous solubility of liquid hydrocarbon mixtures containing dissolved solid components. *Environ. Toxicol. & Chem.* 10, 633-639.

Vaishnav, D.D., Babeu, L. (1987) Comparison of occurence and rates of chemical biodegradation in natural waters. *Bull. Environ. Contam. Toxicol.* 39, 237-244.

Valerio, F., Pala, M., Brescianini, C., Lazaazrotto, A., Balducci, D. (1991) Effect of sunlight and temperature on concentration of pyrene and benzo(a)pyrene adsorbed on airborne particulate. *Toxicol. & Environ. Chem.* 31-32, 113-118.

Valsaraj, K.T. (1988) On the physico-chemical aspects of partitioning of non-polar hydrophobic organics at the air-water interface. *Chemosphere* 17(5), 875-887.

Valsaraj, K.T., Thibodeaux, L.J. (1989) Relationships between micelle-water and octanol-water partition constants for hydrophobic organics of environmental interest. *Water Res.* 23(2), 183-189.

Valvani, S.C., Yalkowsky, S.H. (1980) Solubility and partitioning in drug design. In:*Physical Chemical Properties of Drug. Med. Res.* series volume 10., pp.201-229, Yalkowsky, S.H., Sinkula, A.A., Valvani, S.C., Editors, Marcel Dekker Inc., New York.

Valvani, S.C., Yalkowsky, S.H., Amidon, G.L. (1976) Solubility of nonelectrolytes in polar solvents. VI. Refinements in molecular surface area computations. *J. Phys. Chem.* 80, 829-836.

Van Ekeren, P.J., Jacobs, M.H.G., Offringa, J.C.A., De Kruif, C.G. (1983) Vapour-pressure measurements on *trans*-diphenylethene and naphthalene using a spinning-rotor friction gauge. *J. Chem. Thermodyn.* 15, 409-417.

Veith, G.D., Austin, N.M., Morris, R.T. (1979a) A rapid method for estimation log P for organic chemicals. *Water Res.* 13, 43-47.

Veith, G.D., Defor, D.L. Bergstedt, B.V. (1979b) Measuring and estimating the bioconcentration factor of chemicals in fish. *J. Fish Res. Board Can.* 26, 1040-1048

Veith, G.D., Kosian, P. (1983) Estimating bioconcentration potential from octanol/water partition coefficients. In: *Physical Behavior of PCBs in the Great Lakes.* Mackay, D., Paterson, S., Eisenreich, S.J., Simmons, M.S., Editors, Ann Arbor Science Publishers, Ann Arbor, Michigan.

Veith, G.D., Macek, K.J., Petrocelli, S.R., Caroll, J. (1980) An evaluation of using partition coefficients and water solubilities to estimate bioconcentration factors for organic chemicals in fish. In: *Aquatic Toxicology.* Eaton, J.G, Parrish, P.R., Hendricks, A.C., Editors, ASTM STP 707, Am. Soc. for Testing and Materials, pp.116-129.

Verschueren, K. (1977) *Handbook of Environmental Data on Organic Chemicals.* Van Nostrand Reinhold Co., New York.

Verschueren, K. (1983) *Handbook of Environmental Data on Organic Chemicals*, 2nd Edition, Van Nostrand Reinhold Co., New York.

Vesala, A. (1974) Thermodynamics of transfer of nonelectrolytes from light to heavy water. I.Linear free energy correlations of free energy of transfer with solubility and heat of melting of a nonelectrolyte. *Acta Chemica Scand.* A28, 839-845.

Voice, T.C., Rice, C.P., Weber, Jr., W.J. (1983) Effects of solids concentration on the sorptive partitioning of hydrophobic pollutants in aqueous systems. *Environ. Sci. Technol.* 17(9), 513-518.

Vowles, P.D., Mantoura, R.F.C. (1987) Sediment-water partition coefficients and HPLC retention factors of aromatic hydrocarbons. *Chemosphere* 16(1), 109-116.

Vozňáková, Z., Popl, M., Berka, M. (1978) Recovery of aromatic hydrocarbons from water. *J. Chromatograph Sci.* 16, 123-127.

Wakayama, N., Inokuchi, H. (1967) Heats of sublimation of polycyclic aromatic hydrocarbons and their molecular packings. *Bull. Chem. Soc. Japan* 40, 2267-2271.

Wakeham S.G., Davis, A.C., Karas, J. (1983) Mesocosm experiments to determine the fate and persistence of volatile organic compounds in coastal seawater. *Environ. Sci. Technol.* 17, 611-617.

Walker, J.D., Colwell, R.R. (1976) Measuring the potential activity of hydrocarbon-degrading bacteria. *Appl. Environ. Microbiol.* 31, 187-197.

Walters, R.W., Luthy, R.G. (1984) Equilibrium adsorption of polycyclic aromatic hydrocarbons from water onto activated carbon. *Environ. Sci. Technol.* 18(6), 395-403.

Wang, L., Wang, X., Xu, O., Tian, L. (1986) *Huanjing Kexue Xuebao* 6, 491.

Wang, L., Kong, L., Chang, C. (1991) Photodegradation of 17 PAHs in methanol (or acetonitrile)-water solution. *Environ. Chem.* 10(2), 15-20.

Warne, M. St. J., Connell, D.W., Hawker, D.W. (1991) Comparison of the critical concentration and critical volume hypotheses to model non-specific toxicity of individual compounds. *Toxicology* 66, 187-195.

Warne, M., St. J., Connell, D.W., Hawker, D.W., Schüürmann, G. (1990) Prediction of aqueous solubility and the octanol-water partition coefficient for lipophilic organic compounds using molecular descriptors and physicochemical properties. *Chemosphere* 21(7), 877-888.

Warner, P.H., et al. (1980) *Determination of Henry's Law Constants of Selected Priority Pollutants.* MERL, Cincinnati.

Warner, P.H., Cohen, J.M., Ireland, J.C. (1987) *Determination of Henry's Law Constants of Selected Priority Pollutants.* NTIS PB87-212684, EPA/600/D-87/229, U.S. Environmental Protection Agency, Cincinnati, Ohio.

Wasik, S.P., Miller, M.M., Teware, Y.B., May, W.E., Sonnefeld, W.J., DeVoe, H., Zoller, W.H. (1983) Determination of the vapor pressure, aqueous solubility, and octanol/water partition coefficient of hydrophobic substances by coupled generator column/liquid chromatographic methods. *Residue Rev.* 85, 29-42.

Wasik, S.P., Tewari, Y.B., Miller, M.M., Martire, D.E. (1981) *Octanol/Water Partition Coefficients and Aqueous Solubilities of Organic Compounds.* NBSIR 81-2406., U.S. Departmnet of Commerce, National Bureau of Standards, Washington, D.C.

Wauchope, R.D., Getzen, F.W. (1972) Temperature dependence of solubilities in water and heats of fusion of solid aromatic hydorcarbons. *J. Chem. Eng. Data* 17, 38-41.

Weast, R. (1972-73) *Handbook of Chemistry and Physics.* 53rd ed., CRC Press, Cleveland.

Weast, R. (1976-77) *Handbook of Chemistry and Physics.* 57th ed., CRC Press, Boca Raton, Florida.

Weast, R. (1983-84) *Handbook of Chemistry and Physics.* 64th ed., CRC Press, Boca Raton, Florida.

Weast, R.C., Astle, M.J., Beyer, W.H., Editors (1984) *CRC Handbook of Chemistry and Physics.* 65th edition, CRC Press, Boca Raton, Florida.

Weimer, R.F., Prausnitz, J.M. (1965) Complex formation between carbon tetrachloride and aromatic hydrocarbons. *J. Chem. Phys.* 42, 3643-3644.

WERL (1989) U.S. *EPA Treatability Database.*

Whitehouse, B.G., Cooke, R.C. (1982) Estimating the aqueous solubility of aromatic hydrocarbons by high performance liquid chromatography. *Chemosphere* 11(8), 689-699.

Wiederman, H.G., Vaguhan, H.P. (1969) Application of thermogravimetry for vapor pressure determinations. In: *Proc. Toronto Symp. Therm. Anal.* (3rd), pp.233-249.

Wild, S.R., Berrow, M.L., Jones, K.C. (1991) The persistence of polynuclear aromatic hydrocarbons (PAHs) in sewage sludge amended agricultural soils. *Envirn. Pollt.* 72(2), 141-157.

Windholtz, M., Editor (1976) *The Merck Index.* 9th edition, Merck & Co., Inc., Rahway, New Jersey, U.S.A.

Windholtz, M., Budavari, S., Blumetti, R.F., Otterbein, E.S., Editors (1983) *The Merck Index,* 10th edition, Merck and Co., Inc., Rahway, New Jersey, U.S.A.

Wise, S.A., Bonnett, W.J., Guenther, F.R., May, W.E. (1981) A relationship between reversed phase C_{18} liquid chromatographic retention and the shape of polycyclic aromatic hydrocarbons. *J. Chromatographic Sci.* 19, 457-465.

Wong, P.T.S., Kaiser, K.L.E. (1975) Bacterial degradation of polychlorinated biphenyls. II. Rate studies. *Bull. Contam. Toxiocol.* 3, 249.

Wood, A.L., Bouchard, D.C., Brusseau, M.L., Rao, P.S.C. (1990) Cosolvent effects on sorption and mobility of organic contaminants in soil. *Chemosphere* 21(4-5), 575-587.

Woodburn, K.B. (1982) *M.Sc. Thesis,* University of Wisconsin, Madison, Wisconsin.

Woodburn, K.B., Doucette, W.J., Andren, A.W. (1984) Generator column determination of octanol/water partition coefficients for selected polychlorinated biphenyl congeners. *Environ. Sci. Technol.* 18, 457-459.

Woodburn, K.B., Lee, L.S., Rao, P.S.C., Delfino, J.J. (1989) Comparison of sorption energetics for hydrophobic organic chemicals by synthetic and natural sorbents from Methanol/water solvent mixtures. *Environ. Sci. Technol.* 23(4), 407-412.

Yalkowsky, S.H. (1981) Solubility and partitioning V:Dependence of solubility on melting point. *J. Pharmaceutical Sci.* 70(8) 971-973.

Yalkowsky, S.H. (1986) Project completion report, CR 811852-01. U.S. Environmental Protection Agency, Ada, Oklahoma, U.S.A.

Yalkowsky, S.H., Mishra, D.S. (1990) Comment on "Prediction of aqueous solubility of organic chemicals based on molecular structure. 2. Application to PNAs, PCBs, PCDDs, etc.". *Environ. Sci. Technol.* 24(6), 927-929.

Yalkowsky, S.H., Orr, R.J., Valvani, S.C. (1979) Solubility and partitioning. 3. The solubility of halobenzenes in water. *Ind. Eng. Chem. Fundam.* 18, 351-353.

Yalkowsky, S.H., Valvani, S.C. (1979) Solubilities and partitioning 2.Relationships between aqueous solubilities, partition coefficients, and molecular surface areas of rigid aromatic hydrocarbons. *J. Chem. Eng. Data* 24, 127-129.

Yalkowsky, S.H., Valvani, S.C. (1980) Solubility and Partitioning. I:Solubility of nonelectrolytes in water. *J. Pharmaceutical Sci.* 69(8), 912-922.

Yalkowsky, S.H., Valvani, S.C., Mackay, D. (1983a) Estimation of the aqueous solubility of some aromatic compounds. *Residue Rev.* 85, 43-55.

Yalkowsky, S.H., Valvani, S.C., Roseman, T.J. (1983b) Solubility and partitioning. VI: Octanol solubility and octanol-water partition coefficients. *J. Pharmaceutical Sci.* 72(8), 866-870.

Yamasaki, H., Kuwata, K., Yoshio, K. (1984) *Nippon Kagaka Kaish.* 1324-1329.

Yoshida, K., Shigeoka, T., Yamauchi, F. (1983) Relationship between molar refraction and n-octanol/water partition coefficient. *Ecotox. Environ. Safety* 7, 558-565.

Yuteri, C., Ryan, D.F., Callow, J.J., Gurol, M.D. (1987) The effect of chemical composition of water on Henry's law constant. *J. Water Pollut. Control Fed.* 59, 950-956.

Zepp, R.G. (1980) In: *Dynamics, Exposure and Hazard Assessment of Toxic Chemicals.* pp.69-110. Haque, R., Editor, Ann Arbor Science, Ann Arbor, Michigan.

Zepp, R.G. (1991) Photochemical fate of agrochemicals in natural waters. In: *Pesticide Chemistry.* Advances in International Research, Development, and Legislation. pp.329-345, Frechse, H., Editor, VCH, Weinheim.

Zepp, R.G., Scholtzhauer, P.F. (1979) Photoreactivity of selected aromatic hydrocarbons in water. In: *Polynuclear Aromatic Hydrocarbons.* Jones, P.W. , Leber, P., Editors, Ann Arbor Sci. Publ. Inc., Ann Arbor, Michigan, pp. 141-58.

Zoeteman, B.C.J., Harmsen, K., Linders, J.B.H. (1980) Persistent organic pollutants in river water and ground water of the Netherlands. *Chemosphere* 9, 231-249.

Zoeteman, B.C.J., De Greef, E., Brinkmann, F.J.J. (1981) Persistency of organic contaminants in groundwater, lessons from soil pollution incidents in the Netherlands. *Sci. Total Environ.* 21, 187-202.

Zwolinski, B.J., Wilhoit, R.C. (1971) *Handbook of Vapor Pressures and Heats of Vaporization of Hydrocarbons and Related Compounds.* American Petroleum Institute Project 44, API 44-TRC Publications in Science and Engineering, Texas A & M University, College Station, Texas.

3. Chlorinated Dibenzo-*p*-Dioxins

3.1   List of Chemicals and Data Compilations

Common Name: Dibenzo-*p*-dioxin
Synonym:
Chemical Name: dibenzo-*p*-dioxin
CAS Registry No: 262-12-4
Molecular Formula: $C_6H_4O_2C_6H_4$
Molecular Weight: 184.0

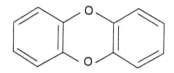

Melting Point (°C):
  120-122 (Gilman & Dietrich 1957)
  122-123 (Pohland & Yang 1972; quoted, Rordorf 1986,1987,1989)
  119-122 (quoted lit., Pohland & Yang 1972)
  124.2   (Shiu et al. 1987)
  123.0   (quoted, Shiu et al. 1988)
Boiling Point (°C):
  266.0   (Rordorf 1986)
  283.5   (calculated, Rordorf 1986,1987,1989)
Density (g/cm³ at 20°C):
Molar Volume (cm³/mol):
  192.0   (LeBas method, Shiu et al. 1987,1988)
  143.82, 142.06 (calculated-liquid density, crystalline volume, Govers et al. 1990)
Molecular Volume (Å³):
Total Surface Area, TSA (Å²):
  185.6   (Doucette & Andren 1987, 1988b)
Entropy of Fusion, ΔS_{fus}, cal/mol K (e.u.):
  13.62   (Rordorf 1986,1987,1989)
Fugacity Ratio at 25 °C (assuming ΔS = 13.5 e.u.):
  0.105   (Shiu et al. 1987)
  0.107   (Shiu et al. 1988)

Water Solubility (g/m³ or mg/L at 25°C):
  0.842   (gen. col.-HPLC/UV, Shiu et al. 1987)
  0.842, 0.90 (gen. col.-HPLC/UV, gen. col.-GC/ECD, Shiu et al. 1988)
  0.861   (selected, Shiu et al. 1988)
  0.90   (gen. col.-GC/ECD, Doucette & Andren 1988a)
  1.26, 0.924 (quoted, calculated- χ , Nirmalakhandan & Speece 1989)

0.861, 1.84, 1.57   (quoted, calculated- χ , log K_{ow}, Yalkowsky & Mishra 1990)

0.861, 0.068, 1.57   (quoted, calculated- χ , log K_{ow}, Speece 1990)

Vapor Pressure (Pa at 25°C):

0.055   (gas saturation, Rordorf 1985ab, 1986, 1987, 1989; quoted, Shiu et al. 1987, 1988)

0.050   (gas saturation, estimated from extrapolated vapor pressure vs. halogen substitution no. plot, Rordorf et al. 1990)

Henry's Law Constant (Pa m³/mol):

12.39   (calculated-P/C, Shiu et al. 1987)

12.29   (calculated-P/C, Shiu et al. 1988)

Octanol/Water Partition Coefficient, log K_{ow}:

3.40   (calculated, Kaiser 1983; quoted, Shiu et al. 1988)

4.31, 4.48, 4.38; 4.38, 4.46, 4.52 (HPLC-RT, linear; quadratic regressions, Sarna et al. 1984)

4.26, 4.65 (Doucette 1985; quoted, Shiu et al. 1988)

4.17, 4.46 (HPLC-RT, Sarna et al. 1984; quoted, Shiu et al. 1988)

4.26, 4.01; 4.34, 4.17 (HPLC-RT, linear; quadratic regressions, Webster et al. 1985)

4.20   (HPLC-RT, Burkhard & Kuehl 1986; quoted, Shiu et al. 1988)

4.20   (selected, Shiu et al. 1987)

4.37, 4.65, 4.26 (gen. col.-GC, calculated-π, TSA, Doucette & Andren 1987)

4.37   (quoted, Doucette & Andren 1988b)

4.65, 4.14, 4.18, 4.37, 4.70, 4.35 (calculated: π-const., f-const., HPLC-RT, MW, χ, TSA, Doucette & Andren 1988b)

4.37   (gen. col.-GC, Shiu et al. 1988)

4.30   (selected, Shiu et al. 1988)

4.20   (quoted, De Voogt et al. 1990)

4.64   (calculated-f const., Yalkowsky & Mishra 1990)

4.37, 4.28 (quoted: gen. col. method, HPLC-RT, Chessells et al. 1991)

4.30, 4.65 (averaged lit. value, calculated-f const., Chessells et al. 1991)

Bioconcentration Factor, log BCF:

Sorption Partition Coefficient, log K_{oc}:

Half-Lives in the Environment:

    Air: room temperature gas-phase reaction rate constant with OH radicals was calculated to be 4.2×10^{-11} cm^3 molecule^{-1} sec^{-1} corresponding to an atmospheric lifetime of about 7 hours (Atkinson 1987a) and 4.0×10^{-11} cm^3 molecule^{-1} sec^{-1} (Atkinson 1987b).

    Surface water:

    Groundwater:

    Sediment:

    Soil:

    Biota:

Environmental Fate Rate Constants or Half-Lives:

    Volatilization:

    Photolysis:

    Oxidation: room temperature gas-phase reaction rate constant with OH radicals was calculated to be 4.2×10^{-11} cm^3 molecule^{-1} sec^{-1} corresponding to an atmospheric lifetime of about 7 hours (Atkinson 1987a) and 4.0×10^{-11} cm^3 molecule^{-1} sec^{-1} (Atkinson 1987b).

    Hydrolysis:

    Biodegradation:

    Biotransformation:

    Bioconcentration, Uptake (k_1) and Elimination (k_2) Rate Constants:

Common Name: 1-Chlorodibenzo-*p*-dioxin
Synonym: 1-CDD, 1-MCDD
Chemical Name: 1-chlorodibenzo-*p*-dioxin
CAS Registry No: 39227-53-7
Molecular Formula: $C_6H_4O_2C_6H_3Cl$
Molecular Weight: 218.5

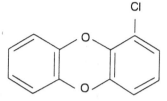

Melting Point (°C):
    104.5-105.5 (Pohland & Yang 1972; quoted, Rordorf 1986,1987,1989)
    105.5 (quoted, Shiu et al. 1988)
Boiling Point (°C):
    296.0 (Rordorf 1986)
    315.5 (calculated, Rordorf 1986,1987,1989)
Density (g/cm³ at 20°C):
Molar Volume (cm³/mol):
    212.9 (LeBas method, Shiu et al. 1988)
    157.41, 158.07 (calculated-liq. density, crystalline volume, Govers et al. 1990)
Molecular Volume (Å³):
Total Surface Area, TSA (Å²):
Entropy of Fusion, ΔS_{fus}, cal/mol K (e.u.):
    13.38 (Rordorf 1986,1987,1989)
Fugacity Ratio at 25 °C (assuming ΔS = 13.5 e.u.):
    0.160 (Shiu et al. 1988)

Water Solubility (g/m³ or mg/L at 25°C):
    417 (gen. col.-HPLC/UV, Shiu et al. 1988)

Vapor Pressure (Pa at 25°C):
    0.012 (gas saturation, Rordorf 1985a,b, 1986,1987,1989; quoted, Shiu et al. 1988)

Henry's Law Constant (Pa m³/mol):
    8.38 (calculated-P/C, Shiu et al. 1988)

Octanol/Water Partition Coefficient, log K_{OW}:

    4.20   (calculated, Kaiser 1983; quoted, Shiu et al. 1988)

    4.97, 5.20, 5.25; 5.05, 5.18, 5.23 (HPLC-RT, linear; quadratic regressions, Sarna et al. 1984)

    4.91, 5.18 (HPLC-RT, Sarna et al. 1984; quoted, Shiu et al. 1988)

    4.81, 4.52; 4.91, 5.74 (HPLC-RT, linear; quadratic regressions, Webster et al. 1985)

    4.75 (HPLC-RT, Burkhard & Kruhl 1986; quoted, Shiu et al. 1988)

    5.30, 4.75 (HPLC-RT, selected, Shiu et al. 1988)

    5.04 (quoted, HPLC-RT, Chessells et al. 1991)

    5.04, 5.47 (mean lit. value, calculated-f const., Chessells et al. 1991)

Bioconcentration Factor, log BCF:

Sorption Partition Coefficient, log K_{OC}:

Half-Lives in the Environment:
    Air:
    Surface water:
    Groundwater:
    Sediment:
    Soil:

Environmental Fate Rate Constants or Half-Lives:
    Volatilization:
    Photolysis:
    Hydrolysis:
    Oxidation:
    Biodegradation:
    Biotransformation:
    Bioconcentration, Uptake (k_1) and Elimination (k_2) Rate Constants:

Common Name: 2-Chlorodibenzo-*p*-dioxin
Synonym: 2-CDD, 2-MCDD
Chemical Name: 2-chlorodibenzo-*p*-dioxin
CAS Registry No: 39227-54-8
Molecular Formula: $C_6H_4O_2C_6H_3Cl$
Molecular Weight: 218.5

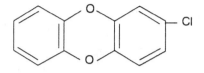

Melting Point (°C):
    87-90   (Gilman & Dietrich 1957; quoted, Pohland & Yang 1972)
    88-89   (Pohland & Yang 1972; quoted, Rordorf 1986,1987,1989)
    89.0   (quoted, Shiu et al. 1988)

Boiling Point (°C):
    298.0   (Rordorf 1986)
    316.0   (calculated, Rordorf 1986,1987,1989)
Density (g/cm³ at 20°C):
Molar Volume (cm³/mol):
    212.9   (LeBas method, Shiu et al. 1988)
    154.85, 155.15 (calculated-liq. density, crystalline volume, Govers et al. 1990)
Molecular Volume (Å³):
Total Surface Area, TSA (Å²):
    203.6   (Doucette & Andren 1987, 1988b)
Entropy of Fusion, ΔS_{fus}, cal/mol K (e.u.):
    12.19   (Rordorf 1987, 1989)
Fugacity Ratio at 25 °C (assuming ΔS = 13.5 e.u.), F:
    0.233   (Shiu et al. 1988)

Water Solubility (g/m³ or mg/L at 25°C):
    0.278, 0.319 (gen. col.-HPLC/UV, gen. col.-GC/ECD, Shiu et al. 1988)
    0.319   (gen. col.-GC/ECD, Doucette & Andren 1988a; quoted, Choudhry & Webster
             1989)
    1.26, 0.932 (quoted, calculated-χ, Nirmalakhandan & Speece 1989)
    0.257, 0.275 (calculated-QSAR, quoted, Fiedler & Schramm 1990)

Vapor Pressure (Pa at 25°C):
    0.017 (gas saturation, Rordorf 1985a,b,1986,1987,1989; quoted, Shiu et al. 1988)
    0.016 (gas saturation, estimated from extrapolated vapor pressure vs. halogen substitution no. plot, Rordorf et al. 1990)

Henry's Law Constant (Pa m^3/mol):
    14.82 (calculated-P/C, Shiu et al. 1988)

Octanol/Water Partition Coefficient, log K_{ow}:
    4.20 (calculated, Kaiser 1983; quoted, Shiu et al. 1988)
    5.36, 5.66, 5.71; 5.45, 5.64, 5.69 (HPLC-RT, linear; quadratic regressions, Sarna et al. 1984)
    4.76, 5.47 (calculated, Doucette 1985; quoted, Shiu et al. 1988)
    5.45, 5.64 (HPLC-RT, Sarna et al. 1984; quoted, Shiu et al. 1988)
    5.33, 5.00; 5.45, 5.29 (HPLC-RT, linear; quadratic regressions, Webster et al. 1985)
    5.08 (HPLC-RT, Burkhard & Kruhl 1986; quoted, Shiu et al. 1988)
    4.94, 5.47, 4.69 (gen. col.-GC/ECD, calculated: π-const., TSA, Doucette & Andren 1987)
    4.94; 5.47, 4.89, 4.98, 4.89, 5.14, 4.79 (selected; calculated: π-const., f-const., HPLC-RT, MW, χ, TSA, Doucette & Andren 1988b)
    5.00 (selected, Shiu et al. 1988; quoted, Gobas & Schrap 1990)
    4.94 (gen. col.-GC, Shiu et al. 1988)
    5.47, 5.00 (calculated-QSAR, quoted, Fiedler & Schramm 1990)
    4.94, 5.39 (quoted: gen. col. method, HPLC-RT, Chessells et al. 1991)
    5.28, 5.47 (mean lit. value, calculated-f const., Chessells et al. 1991)

Bioconcentration Factor, log BCF:

Sorption Partition Coefficient, log K_{oc}:
    3.92 (organic carbon, calculated-QSAR, Fiedler & Schramm 1990)

Half-Lives in the Environment:
    Air:
    Surface water: half-life of 10 minutes irradiated under simulated sunlight in aerated aqueous suspension of semiconductor TiO_2 at 4.0 g/liter and pH 3 (Pelizzetti et al. 1988).
    Groundwater:
    Sediment:

Soil:
Biota:

Environmental Fate Rate Constants or Half-Lives:
    Volatilization:
    Photolysis: photodegradation half-life of 10 minutes with simulated sunlight in aerated
        aqueous suspension of semiconductor TiO_2 at 4.0 g/liter and pH 3 (Pelizzetti et
        al. 1988).
    Hydrolysis:
    Oxidation:
    Biodegradation:
    Biotransformation:
    Bioconcentration, Uptake (k_1) and Elimination (k_2) Rate Constants:

377

Common Name: 2,3-Dichlorodibenzo-*p*-dioxin
Synonym: 2,3-DCDD
Chemical Name: 2,3-dichlorodibenzo-*p*-dioxin
CAS Registry No: 29446-15-9
Molecular Formula: $C_6H_4O_2C_6H_2Cl_2$
Molecular Weight: 253.0

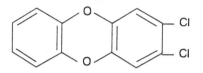

Melting Point (°C):
     163-164 (Pohland & Yang 1972; quoted, Rordorf 1986,1987,1989)
     164.0   (quoted, Shiu et al. 1988)
Boiling Point (°C):
     358   (calculated, Rordorf 1986,1987,1989)
Density (g/cm³ at 20°C):
Molar Volume (cm³/mol):
     233.8   (LeBas method, Shiu et al. 1988)
     164.07 (calculated-liquid density, Govers et al. 1990)
Molecular Volume (Å³):
Total Surface Area, TSA (Å²):
Entropy of Fusion, ΔS_{fus}, cal/mol K (e.u.):
     14.58   (Rordorf 1986,1987,1989)
Fugacity Ratio at 25 °C (assuming ΔS = 13.5 e.u.):
     0.0421 (Shiu et al. 1988)

Water Solubility (g/m³ or mg/L at 25°C):
     0.0149 (gen. col.-HPLC/UV, Shiu et al. 1988)
     0.0341, 0.0149 (calculated-QSAR, quoted, Fiedler & Schramm 1990)

Vapor Pressure (Pa at 25°C):
     0.00039 (gas saturation, Rordorf 1985a,b, 1986,1989; quoted, Shiu et al. 1988)
     0.0004   (gas saturation, estimated from extrapolated vapor pressure vs. halogen substitution no. plot, Rordorf et al. 1990)

Henry's Law Constant (Pa m³/mol):
$$6.61 \quad \text{(calculated-P/C, Shiu et al. 1988)}$$

Octanol/Water Partition Coefficient, log K_{OW}:
4.70   (calculated, Kaiser 1983)
5.60   (selected, Shiu et al. 1988)
6.23, 5.60 (calculated-QSAR, quoted, Fiedler & Schramm 1990)

Bioconcentration Factor, log BCF:

Sorption Partition Coefficient, log K_{OC}:
4.73 (organic carbon, calculated-QSAR, Fiedler & Schramm 1990)

Half-Lives in the Environment:
    Air:
    Surface water:
    Groundwater:
    Sediment:
    Soil:
    Biota: mean biological half-life in rainbow trout, about 2 days (Niimi 1986); biological half-life of 7 days in rainbow trout (Niimi & Oliver 1986).

Environmental Fate Rate Constants or Half-Lives:
    Volatilization:
    Photolysis:
    Hydrolysis:
    Oxidation:
    Biodegradation:
    Biotransformation:
    Bioconcentration, Uptake (k_1) and Elimination (k_2) Rate Constants:
        k_2: 0.092 day^{-1} (rainbow trout, Niimi & Oliver 1986; quoted, Opperhuizen & Sijm 1990)

Common Name: 2,7-Dichlorodibenzo-*p*-dioxin
Synonym: 2,7-DCDD
Chemical Name: 2,7-dichlorodibenzo-*p*-dioxin
CAS Registry No: 33857-26-0
Molecular Formula: $ClC_6H_3O_2C_6H_3Cl$
Molecular Weight: 253.0

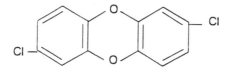

Melting Point (°C):
      201-203 (Gilman & Dietrich 1957; quoted, Pohland & Yang 1972)
      209-210 (Pohland & Yang 1972; quoted, Rordorf 1986,1987,1989)
      210.0   (quoted, Shiu et al. 1988)
Boiling Point (°C):
      374.5   (calculated, Rordorf 1987, 1989)
Density (g/cm³ at 20°C):
Molar Volume (cm³/mol):
      233.8   (LeBas method, Shiu et al. 1988)
      165.88, 162.89 (calculated-liq. density, crystalline volume, Govers et al. 1990)
Molecular Volume (Å³):
Total Surface Area, TSA (Å²):
Entropy of Fusion, ΔS_{fus}, cal/mol K (e.u.):
      13.15   (Rordorf 1986,1987,1989)
Fugacity Ratio at 25 °C (assuming ΔS = 13.5 e.u.):
      0.0148   (Shiu et al. 1988)

Water Solubility (g/m³ or mg/L at 25°C):
      0.00375 (gen. col.-HPLC/UV, Shiu et al. 1988)
      0.253, 0.311 (quoted, calculated-χ , Nirmalakhanden & Speece 1989)
      0.0038   (quoted, Gobas & Schrap 1990)

Vapor Pressure (Pa at 25°C):
      0.00012 (gas saturation, Rordorf 1985a,b, 1986,1987,1989; quoted, Shiu et al. 1988)
      0.00013   (gas saturation, estimated form extrapolated vapor pressure vs. halogen substitution no. plot, Rordorf et al. 1990)

Henry's Law Constant (Pa m^3/mol):
  8.11   (calculated-P/C, Shiu et al. 1988)

Octanol/Water Partition Coefficient, log K$_{OW}$:
  4.70   (calculated, Kaiser 1983; quoted, Shiu et al. 1988)
  6.28, 6.64, 6.72; 6.38, 6.62, 6.71   (HPLC-RT, linear; quadratic regressions, Sarna et
      al. 1984)
  6.62, 6.39   (HPLC-RT, Sarna et al. 1984; quoted, Shiu et al. 1988)
  6.27, 5.86; 6.39, 6.17   (HPLC-RT, linear; quadratic regressions, Webster et al. 1985)
  5.75   (HPLC-RT, Burkhard & Kruhl 1986; quoted, Shiu et al. 1988)
  5.75   (selected, Shiu et al. 1988, quoted, Gobas et al. 1987, 1988; Gobas & Schrap
      1990)
  6.16   (quoted, HPLC-RT, Chessells et al. 1991)
  6.16, 6.23   (mean lit. value, calculated-f const., Chessells et al. 1991)

Bioconcentration Factor, log BCF:
  1.7, 2.9   (guppy, in whole fish, in lipid, Gobas et al. 1987)
  2.56, 3.68   (guppy: wet weight base, lipid weight base, Gobas & Schrap 1990)

Sorption Partition Coefficient, log K$_{OC}$:

Half-Lives in the Environment:
  Air:
  Surface water: photodegradation half-life of 1 hour with irradiation under simulated
      sunlight in aerated aqueous suspension of semiconductor TiO$_2$ at 4.0 g/liter and
      pH 3 (Pelizzetti et al. 1988).
  Groundwater:
  Sediment:
  Soil:
  Biota: mean biological half-life in rainbow trout, about 2 days (Niimi 1986); biological
      half-life of 2 days in rainbow trout (Niimi & Oliver 1986).

Environmental Fate Rate Constants or Half-Lives:
  Volatilization:
  Photolysis: photodegradation half-life of 46 minutes in an aqueous solution assisted by
      TiO$_2$ at 2.0 g/liter under simulated sunlight (Barbeni et al. 1986);
      photodegradation half-life of 1 hour with irradiation under simulated sunlight in

aerated aqueous suspension of semiconductor TiO_2 at 4.0 g/liter and pH 3 (Pelizzetti et al. 1988).

Hydrolysis:

Oxidation:

Biodegradation:

Biotransformation:

Bioconcentration, Uptake (k_1) and Elimination (k_2) Rate Constants:

k_2: 0.462 day^{-1} (rainbow trout, Niimi & Oliver 1986; quoted, Opperhuizen & Sijm 1990)

k_1: 543 day^{-1} (guppy, Gobas & Schrap 1990)

k_2: 1.5 day^{-1} (guppy, Gobas & Schrap 1990)

Common Name: 2,8-Dichlorodibenzo-*p*-dioxin
Synonym: 2,8-TCDD
Chemical Name: 2,8-dichlorodibenzo-*p*-dioxin
CAS Registry No: 38964-22-6
Molecular Formula: $ClC_6H_3O_2C_6H_3Cl$
Molecular Weight: 253.0

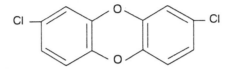

Melting Point (°C):
     150.5-151 (Pohland & Yang 1972; Rordorf 1986,1987,1989)
     151.0   (quoted, Shiu et al. 1988)
Boiling Point (°C):
Density (g/cm³ at 20°C):
Molar Volume (cm³/mol):
     233.8   (LeBas method, Shiu et al. 1988)
     165.88, 167.2 (calculated-liq. density, crystalline volume, Govers et al. 1990)
Molecular Volume (Å³):
Total Surface Area, TSA (Å²):
Entropy of Fusion, ΔS_{fus}, cal/mol K (e.u.):
     13.15   (Rordorf 1986,1987,1989)
Fugacity Ratio at 25 °C (assuming ΔS = 13.5 e.u.), F:
     0.0567   (Shiu et al. 1988)

Water Solubility (g/m³ or mg/L at 25°C):
     0.0167   (gen. col.-HPLC/UV, Shiu et al. 1988)

Vapor Pressure (Pa at 25°C):
     0.00014   (gas saturation, Rordorf 1985a,b, 1986,1987,1989; quoted, Shiu et al. 1988)
     0.00013   (gas saturation, estimated from extrapolated vapor pressure vs. halogen
          substitution no. plot, Rordorf et al. 1990)

Henry's Law Constant (Pa m³/mol):
     2.13   (calculated-P/C, Shiu et al. 1988)

Octanol/Water Partition Coefficient, log K_{OW}:
    4.70   (calculated, Kaiser 1983; quoted, Shiu et al. 1988)
    5.60   (selected, Shiu et al. 1988)

Bioconcentration Factor, log BCF:
    2.77, 2.82   (goldfish: treated with metabolic inhibitor PBO, untreated, Sijm & Opperhuizen 1988)

Sorption Partition Coefficient, log K_{OC}:

Half-Lives in the Environment:
    Air:
    Surface water:
    Groundwater:
    Sediment:
    Soil:
    Biota: mean biological half-life in rainbow trout, about 2 days (Niimi 1986).

Environmental Fate Rate Constants or Half-Lives:
    Volatilization:
    Photolysis:
    Hydrolysis:
    Oxidation:
    Biodegradation:
    Biotransformation: metabolic elimination rate constant from goldfish was estimated to be 0.35 day^{-1} (Sijm & Opperhuizen 1988).
    Bioconcentration, Uptake (k_1) and Elimination (k_2) Rate Constants:
        k_1: 390 L/kg/day   (goldfish, Opperhuizen & Sijm 1990)
        k_2: 0.23 day^{-1} (goldfish, Opperhuizen & Sijm 1988)

Common Name: 1,2,4-Trichlorodibenzo-*p*-dioxin
Synonym: 1,2,4-TCDD
Chemical Name: 1,2,4-trichlorodibenzo-*p*-dioxin
CAS Registry No: 39227-58-2
Molecular Formula: $ClC_6H_3O_2C_6H_2Cl_2$
Molecular Weight: 287.5

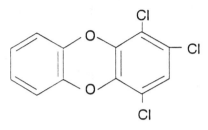

Melting Point (°C):
       128-129 (Pohland & Yang 1972; quoted, Rordorf 1986,1987,1989)
       129.0 (quoted, Shiu et al. 1988)
Boiling Point (°C):
       375.0 (calculated, Rordorf 1986,1987,1989)
Density (g/cm³ at 20°C):
Molar Volume (cm³/mol):
       254.7 (LeBas method, Shiu et al. 1988)
       179.66 (calculated-liquid density, Govers et al. 1990)
Molecular Volume (Å³):
Total Surface Area, TSA (Å²):
Entropy of Fusion, ΔS_{fus}, cal/mol K (e.u.):
       20.07 (Rordorf 1986,1987,1989)
Fugacity Ratio at 25 °C (assuming ΔS = 13.5 e.u.):
       0.0936 (Shiu et al. 1988)

Water Solubility (g/m³ or mg/L at 25°C):
       0.00841 (gen. col.-HPLC/UV, Shiu et al. 1988; quoted, Gobas & Schrap 1990)
       0.0085 (quoted, Fiedler & Schramm 1990)

Vapor Pressure (Pa at 25°C):
       0.0001 (gas saturation, Rordorf 1985a,b, 1986,1987,1989; quoted, Shiu et al. 1988)
       0.0001 (gas saturation, estimated from extrapolated vapor pressure vs. halogen substitution no. plot, Rordorf et al. 1990)

Henry's Law Constant (Pa m^3/mol):
   3.84   (calculated-P/C, Shiu et al. 1988)

Octanol/Water Partition Coefficient, log K_{OW}:
   5.10   (calculated, Kaiser 1983; quoted, Shiu et al. 1988)
   7.40, 7.77, 7.69; 7.47, 7.76, 7.68   (HPLC-RT, linear; quadratic regressions, Sarna et al. 1984)
   7.45, 7.76 (HPLC-RT, Sarna et al. 1984; quoted, Shiu et al. 1988)
   7.36, 6.86; 7.45, 7.11   (HPLC-RT, linear; quadratic regressions, Webster et al. 1985)
   6.45   (HPLC-RT, Burkhard & Kuehl 1986; quoted, Shiu et al. 1988)
   5.57   (shake flask-GC, Shiu et al. 1988)
   6.35   (selected, Shiu et al. 1988; quoted, Fiedler & Schramm 1990; Gobas & Schrap 1990)
   6.45   (quoted, Gobas et al., 1987, 1988)
   7.13   (quoted, HPLC-RT, Chessells et al. 1991)
   6.82, 6.97 (mean lit. value, calculated-f const., Chessells et al. 1991)

Bioconcentration Factor, log BCF:
   1.90, 3.10 (guppy, in whole fish, in lipid, Gobas et al. 1987)
   2.82, 3.95 (guppy: wet weight base, lipid weight base, Gobas & Schrap 1990)

Sorption Partition Coefficient, log K_{OC}:

Half-Lives in the Environment:
   Air:
   Surface water:
   Groundwater: oxidative degradation rate of water dissolved PCDD by ozone is 1.27×10^6 liter gram^{-1} minute^{-1} under alkaline conditions at pH 10 and 20°C (Palauschek & Scholz 1987).
   Sediment:
   Soil:
   Biota: mean biological half-life in rainbow trout, about 10 days (Niimi 1986); 12 days in rainbow trout (Niimi & Oliver 1986).

Environmental Fate Rate Constants or Half-Lives:
   Volatilization:
   Photolysis:

Hydrolysis:

Oxidation: oxidative degradation rate of water dissolved PCDD by ozone is 1.27×10^6 liter gram^{-1} minute^{-1} under alkaline conditions at pH 10 and 20°C (Palauschek & Scholz 1987).

Biodegradation:

Biotransformation:

Bioconcentration, Uptake (k_1) and Elimination (k_2) Rate Constants:

    k_2: 0.058 day^{-1} (rainbow trout, Niimi & Oliver 1986; quoted, Opperhuizen & Sijm 1990)

    k_1: 601   day^{-1} (guppy, Gobas & Schrap 1990)

    k_2: 0.91 day^{-1} (guppy, Gobas & Schrap 1990)

Common Name: 1,2,3,4-Tetrachlorodibenzo-*p*-dioxin
Synonym: 1,2,3,4-TCDD
Chemical Name: 1,2,3,4-tetrachlorodibenzo-*p*-dioxin
CAS Registry No: 30746-58-8
Molecular Formula: $C_6H_4O_2C_6Cl_4$
Molecular Weight: 322.0

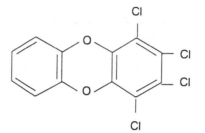

Melting Point (°C):
> 189   (Pohland & Yang 1972; Rordorf 1986,1987,1989)
> 190.0 (quoted, Shiu et al. 1988)
Boiling Point (°C): 419 (calculated, Rordorf 1987, 1989)
Density (g/cm³ at 20°C):
Molar Volume (cm³/mol):
> 275.6 (LeBas method, Shiu et al. 1988)
> 186.71 (calculated-liquid density, Govers et al. 1990)
Molecular Volume (Å³):
Total Surface Area, TSA (Å²):
> 250.3 (Doucette & Andren 1987, 1988b)
Entropy of Fusion, ΔS_{fus}, cal/mol K (e.u.):
> 16.25 (Rordorf 1986,1987,1989)
Fugacity Ratio at 25 °C (assuming ΔS = 13.5 e.u.):
> 0.0223 (Shiu et al. 1988)

Water Solubility (g/m³ or mg/L at 25°C):
> 0.00063, 0.00047 (gen. col.-HPLC/UV, gen. col.-GC/ECD, Shiu et al. 1988)
> 0.00047 (gen. col.-GC/ECD, Doucette & Andren 1988a; quoted, Lodge 1989; Choudhry & Webster 1989)
> 0.0233, 0.0256 (quoted, calculated-χ , Nirmalakhanden & Speece 1989)
> 0.0005 (quoted, Gobas & Schrap 1990)

Vapor Pressure (Pa at 25°C):
> 0.0000064 (gas saturation, Rordorf 1985a,b, 1986,1987,1989; quoted, Shiu et al. 1988)

1.00x10^{-5} (gas saturation, estimated from vapor pressure vs. temperature plot, Rordorf
et al. 1986)

1.044x10^{-4} (subcooled liquid value, GC-RT, Eitzer & Hites 1988)

6.30x10^{-6} (gas saturation, estimated from extrapolated vapor pressure vs. halogen
substitution no. plot, Rordorf et al. 1990)

Henry's Law Constant (Pa m^3/mol):
3.77   (calculated-P/C, Shiu et al. 1988)

Octanol/Water Partition Coefficient, log K_{OW}:
5.50   (calculated, Kaiser 1983; quoted, Shiu et al. 1988)

8.65, 8.90, 8.96; 8.66, 8.91, 8.97 (HPLC-RT, linear; quadratic regressions, Sarna et
al. 1984)

5.8, 7.7 (calculated, Doucette 1985; quoted, Shiu et al. 1988)

8.64, 8.91 (HPLC-RT, Sarna et al. 1984; quoted, Shiu et al. 1988)

8.63, 8.02; 8.64, 8.07 (HPLC-RT, linear; quadratic regressions, Webster et al. 1985)

7.18, 7.08 (HPLC-RT, Burkhard & Kuehl 1986; quoted, Shiu et al. 1988; Sijm et al.
1989a)

6.20, 7.70, 5.80 (gen. col.-GC/ECD, calculated: π-const., TSA, Doucette & Andren
1987)

6.20; 7.70, 7.12, 6.85, 6.26, 6.08, 5.91 (selected; calculated: π-const., f-const., HPLC-
RT, MW, χ , TSA, Doucette & Andren 1988b)

5.77   (shake flask-GC/ECD, Shiu et al. 1988)

6.20   (gen. col.-GC, Shiu et al. 1988)

6.60   (selected, Shiu et al. 1988, quoted, Sijm et al. 1989a; Gobas & Schrap 1990)

6.70   (quoted, Gobas et al. 1987,1988)

5.77, 6.20, 7.97 (quoted: shake flask, gen. col. method, HPLC-RT, Chessells et al.
1991)

7.40, 7.70 (mean lit. value, calculated-f const., Chessells et al. 1991)

Bioconcentration Factor, log BCF:
2.20, 3.40 (guppy, in whole fish, in lipid, Gobas et al. 1987)

2.90, 4.02 (guppy: wet weight base, lipid weight base, Gobas & Schrap 1990)

Sorption Partition Coefficient, log K_{OC}:

Half-Lives in the Environment:

> Air: half-life of photodegradation in a rotary photo-reactor adsorbed to clean silica gels by filtered < 290 nm of light, 88 hours (Koester & Hites 1992).

> Surface water: solution photolysis half-life in n-hexadecane at 1.0 m from a GE Model RS sunlamp with a half-life of 1294 minutes (Nestrick et al. 1980); sunlight induced photolysis half-life in isooctane of 380 minutes and dispersed as solid films with half-life of 65 hours (Buser 1988).

> Groundwater:

> Sediment:

> Soil:

> Biota: mean half-life in rainbow trout, about 43 days (Niimi 1986); biological half-life of 43 days in rainbow trout (Niimi & Oliver 1986).

Environmental Fate Rate Constants or Half-Lives:

> Volatilization:

> Photolysis: solution photolysis half-life in n-hexadecane at 1.0 m from a GE Model RS sunlamp with a half-life of 1294 minutes and surface photolysis half-life on a clean glass surface under the same conditions of 560 minutes (Nestrick et al. 1980); sunlight induced photolysis half-life in isooctane of 380 minutes and dispersed as solid films with a half-life of 65 hours (Buser 1988); half-life of photodegradation adsorbed to clean silica gels in a rotary photo-reactor by filtered < 290 nm of light, 88 hours (Koester & Hites 1992) .

> Hydrolysis:

> Oxidation: oxidative degradation rate of water dissolved PCDD by ozone is 4.73×10^5 liter gram^{-1} minute^{-1} under alkaline condition at pH 10 and 20°C (Palauschek & Scholz 1987).

> Biodegradation:

> Biotransformation:

> Bioconcentration, Uptake (k_1) and Elimination (k_2) Rate Constants:

>> k_2: 0.016 day^{-1} (rainbow trout, Niimi & Oliver 1986; quoted, Opperhuizen & Sijm 1990)

>> k_1: 953   day^{-1} (guppy, Gobas & Schrap 1990)

>> k_2: 1.2   day^{-1} (guppy, Gobas & Schrap 1990)

Common Name: 1,2,3,7-Tetrachlorodibenzo-*p*-dioxin
Synonym: 1,2,3,7-TCDD
Chemical Name: 1,2,3,7-tetrachlorodibenzo-*p*-dioxin
CAS Registry No: 67028-18-6
Molecular Formula: $ClC_6H_4O_2C_6HCl_3$
Molecular Weight: 322.0

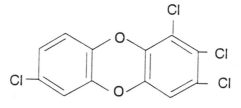

Melting Point (°C):
    172-175 (Friesen et al. 1985; quoted, Rordorf 1987,1989)
    175.0   (quoted, Shiu et al. 1988; Friesen & Webster 1990)
Boiling Point (°C):
    438.3   (calculated, Rordorf 1987, 1989)
Density (g/cm³ at 20°C):
Molar Volume (cm³/mol):
    275.6   (LeBas method, Shiu et al. 1988)
    137.08 (calculated as per Pearlman 1986, De Voogt et al. 1990)
    134.40 (calculated as per Govers & De Voogt et al. 1989, De Voogt et al. 1990)
    186.52 (calculated-liquid density, Govers et al. 1990)
Molecular Volume (Å³): 210 (Friesen & Webster 1990)
Total Surface Area, TSA (Å²): 297 (Friesen & Webster 1990)
Heat of Fusion, ΔH_{fus}, kcal/mol: 6.05 (Friesen & Webster 1990)
Entropy of Fusion, ΔS_{fus}, cal/mol K (e.u.):
    19.60   (Rordorf 1987, 1989)
Fugacity Ratio at 25 °C (assuming ΔS = 13.5 e.u.), F:
    0.0328 (Shiu et al. 1988)

Water Solubility (g/m³ or mg/L at 25°C):
    0.00042, 0.00067 (20°C, ¹⁴C labeled, gen. col.-HPLC/LSC, calculated, Friesen et al. 1985)
    0.000280 (20°C, ¹⁴C labeled, gen. col.-HPLC/LSC, Webster et al. 1986)
    0.000042 (selected, Shiu et al. 1988)
    0.00048, 0.000728 (21°C, 26°C, ¹⁴C-labeled, gen. col.-HPLC/LSC, Friesen & Webster 1990)
    0.000430 (20°C, quoted, De Voogt et al. 1990)

Vapor Pressure (Pa at 25°C):

 0.000001 (calculated, Rordorf 1985a,b, 1986,1989; quoted, Shiu et al. 1988)

 7.0×10^{-6} (gas saturation, estimated from vapor pressure vs. temperature plot, Rordorf et al. 1986)

 5.0×10^{-6} (gas saturation, estimated from extrapolated vapor pressure vs. halogen substitution no. plot, Rordorf et al. 1990)

Henry's Law Constant (Pa m³/mol):

 0.77 (calculated-P/C, Shiu et al. 1988)

Octanol/Water Partition Coefficient, log K_{OW}:

 5.50 (calculated, Kaiser 1983; quoted, Shiu et al. 1988)

 8.19, 8.59, 8.81; 8.22, 8.60, 8,81 (HPLC-RT, linear; quadratic regressions, Sarna et al. 1984)

 8.18, 8.60 (HPLC-RT, Sarna et al. 1984; quoted, Shiu et al. 1988)

 8.15, 7.58; 8.19, 7.72 (HPLC-RT, linear; quadratic regressions, Webster et al. 1985; quoted, Sijm et al. 1989a)

 6.91 (HPLC-RT, Burkhard & Kuehl 1986; quoted, Shiu et al. 1988)

 6.90 (selected, Shiu et al. 1988; quoted, Gobas et al. 1988; Sijm et al. 1989a; Gobas & Schrap 1990)

 6.91 (quoted, De Voogt et al. 1990)

 7.82 (quoted, HPLC-RT, Chessells et al. 1991)

 7.82, 7.70 (mean lit. value, calculated-f const., Chessells et al. 1991)

Bioconcentration Factor, log BCF:

 3.09 (rainbow trout, steady-state, wet weight, Muir et al. 1985)

 3.35 (fathead minnows, steady-state, wet weight, Muir et al. 1985; quoted, Adams et al. 1986)

 3.44, 4.44 (fathead minnows, wet wt. based, lipid based, quoted, Gobas & Schrap 1990)

 3.17, 4.17 (rainbow trout, wet wt. based, lipid based, quoted, Gobas & Schrap 1990)

 3.18, 2.90 (rainbow trout, quoted, Opperhuizen & Sijm 1990)

 3.30, 3.38 (fathead minnows, quoted, Opperhuizen & Sijm 1990)

 3.09, 3.02-3.14; 3.35, 3.30-3.37 (rainbow trout, range; fathead minnows, range, De Voogt et al. 1990)

Sorption Partition Coefficient, log K_{OC}:

 4.26 (DOC, Muir et al. 1985; quoted, De Voogt et al. 1990)

 5.39, 5.98, 6.55 (20°C, fulvic acid, humic acid, Aldrich humic acid, Webster et al. 1986)

Half-Lives in the Environment:

    Air:

    Surface water: solution photolysis half-life in n-hexadecane at 1.0 m from a GE Model RS sunlamp was 563 minutes (Nestrick et al. 1980); direct sunlight photolysis half-lives in aquatic bodies at latitude 40°N for various seasons: 2.08 days in spring, 1.77 days in summer, 3.20 days in fall, 5.42 days in winter, and 8.73 days averaged over a full year (Choudhary & Webster 1986); half-life of 10.69 hours in water-acetonitrile (2:3, v/v) at 313 nm and the calculated midday, midseason direct phototransformation half-lives near water bodies at 40°N latitude: 2.1 days in spring, 1.8 days in summer, 3.2 days in fall and 5.4 days in winter (Choudhary & Webster 1986).

    Groundwater:

    Sediment:

    Soil:

    Biota: half-life of elimination, 2.7 days (Adams et al. 1986); 4 days in whole body of rainbow trout (Muir & Yarechewski 1988; quoted, Muir et al. 1990)

Environmental Fate Rate Constants or Half-Lives:

    Volatilization:

    Photolysis: solution photolysis in n-hexadecane at 1.0 m from a GE Model RS sunlamp with a half-life of 563 minutes and surface photolysis on a clean soft glass surface under the same conditions, 156 minutes (Nestrick et al. 1980); first order photolysis rate constant of 18.13×10^{-6} sec^{-1} with a half-life of 10.69 hours in water-acetonitrile (2:3. v/v) at 313 nm and the calculated midday, midseason direct sunlight photolysis first-order rate constant in aquatic bodies for various seasons: 33.52×10^2 day^{-1} (spring), 39.41×10^2 day^{-1} (summer), 21.77×10^2 day^{-1} (fall), 12.86×10^2 day^{-1} (winter) and calculated direct sunlight photolysis half-lives in aquatic bodies at latitude 40°N for various seasons: 2.08 days in sping, 1.77 days in summer, 3.20 days in fall, 5.42 days in winter and 8.73 days averaged over full year (Choudhry & Webster 1986,1989).

    Hydrolysis:

    Oxidation:

    Biodegradation:

    Biotransformation: rate constant of 0.096 day^{-1} for rainbow trout (Sijm et al. 1990).

    Bioconcentration, Uptake (k_1) and Elimination (k_2) Rate Constants:

        k_1: 317 day^{-1} (rainbow trout, Muir et al. 1985; quoted, Sijm & Opperhuizen 1988)

        k_2: 0.26 day^{-1} (rainbow trout, Muir et al. 1985; quoted, Sijm & Opperhuizen 1988)

        k_1: 529 day^{-1} (fathead minnows, Muir et al. 1985; quoted, Adams et al. 1986; Sijm & Opperhuizen 1988; Sijm et al. 1990)

k_2 0.25 day^{-1} (fathead minnows, Muir et al. 1985, quoted, Adams et al. 1986; Sijm & Opperhuizen 1988; Sijm et al. 1990)

k_2: 0.178, 0.163 day^{-1} (rainbow trout, fathead minnows exposed to dioxins in their diets, Muir & Yarechewski 1988; quoted, Opperhuizen & Sijm 1990; Sijm et al. 1990)

k_1:   420, 213 mL gram^{-1} day^{-1} (rainbow trout exposed to different concentrations, quoted, Opperhuizen & Sijm 1990)

k_2: 0.278, 0.250 day^{-1} (rainbow trout exposed to different concentrations, quoted, Opperhuizen & Sijm 1990)

k_1:   650, 408 mL gram^{-1} day^{-1} (fathead minnows exposed to different concentrations, quoted, Opperhuizen & Sijm 1990)

k_2 0.322, 0.170 day^{-1} (fathead minnows exposed to different concentrations, quoted, Opperhuizen & Sijm 1990)

k_2: 9.5x10^{-2}, 19.1x10^{-2} day^{-1} (rainbow trout for 2-21 days exposure: metabolic inhibitor PBO-treated, control, Sijm et al. 1990)

Common Name: 1,3,6,8-Tetrachlorodibenzo-*p*-dioxin
Synonym: 1,3,6,8-TCDD
Chemical Name: 1,3,6,8-tetrachlorodibenzo-*p*-dioxin
CAS Registry No: 30746-58-8
Molecular Formula: $Cl_2C_6H_2O_2C_6H_2Cl_2$
Molecular Weight: 322.0

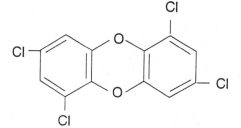

Melting Point (°C):
 219-219.5 (Pohland & Yang 1972; Rordorf 1987,1989)
 219.0 (quoted, Shiu et al. 1988)
Boiling Point (°C):
 438.3 (calculated, Rordorf 1987, 1989)
Density (g/cm³ at 20°C):
Molar Volume (cm³/mol):
 275.6 (LeBas method, Shiu et al. 1988)
 137.32 (calculated as per Pearlman 1986, De Voogt et al. 1990)
 134.40 (calculated as per Govers & De Voogt 1989, De Voogt et al. 1990)
 1.508 (intrinsic V_I/100, Hawker 1990)
 192.34 (calculated-liquid density, Govers et al. 1990)
Molecular Volume (Å³):
Total Surface Area, TSA (Å²):
Entropy of Fusion, ΔS_{fus}, cal/mol K (e.u.):
 17.68 (Rordorf 1987, 1989)
Fugacity Ratio at 25 °C (assuming ΔS = 13.5 e.u.):
 0.0121 (Shiu et al. 1988)

Water Solubility (g/m³ or mg/L at 25°C):
 0.00032 (20°C, ^{14}C-labeled, gen. col.-HPLC/LS, Webster et al. 1985, quoted, Corbet
  et al. 1988; Choudhry & Webster 1989)
 0.00032 (selected, Shiu et al. 1988)
 0.000317 (20°C, ^{14}C-labeled, gen. col.-HPLC/LSC, Friesen et al. 1985; quoted, Servos
  & Muir 1989a; Servos & Muir 1989b)
 0.000168 (calculated, Friesen et al. 1985)

0.000283, 0.000326, 0.000366, 0.000328 (^{14}C-labeled/LSC, Milli-Q treated water, lake water, simulated lake water, Milli-Q water, Servos & Muir 1989a)

0.0262, 0.0256 (quoted, calculated-χ , Nirmalakhanden & Speece 1989)

0.000317 (20°C, quoted, De Voogt et al. 1990)

0.0264 (quoted, Servos et al. 1992a)

Vapor Pressure (Pa at 25°C):

0.000537 (20°C, gas saturation, Webster et al. 1985; quoted Shiu et al. 1988)

0.0000007 (calculated, Rordorf 1985a,b, 1986,1987,1989; quoted, Shiu et al. 1988)

0.0000580 (quoted, Servos et al. 1992a)

Henry's Law Constant (Pa m^3/mol):

6.90 (23°C, batch stripping, Webster et al. 1985; quoted, Corbet et al. 1988; Servos & Muir 1989b)

0.71 (calculated-P/C, Shiu et al. 1988)

0.60 (10°C, calculated-P/C, Servos & Muir 1989b)

Octanol/Water Partition Coefficient, log K_{OW}:

5.50 (calculated, Kaiser 1983; quoted, Shiu et al. 1988)

8.72, 9.00, 9.42; 8.72, 9.02, 9.43 (HPLC-RT, linear; quadratic regressions, Sarna et al. 1984)

8.70, 9.02 (HPLC-RT, Sarna et al. 1984; quoted, Shiu et al. 1988)

8.70, 8.08; 8.70. 8.12 (HPLC-RT, linear; quadratic regressions, Webster et al. 1985; quoted, Sijm et al. 1989a)

7.20, 7.13 (HPLC-RT, Burkhard & Kuehl 1986; quoted, Shiu et al. 1988; Sijm et al. 1989a)

7.13 (corrected HPLC-RT value, Burkhard & Kuehl 1986, quoted, Corbet et al. 1988; Servos & Muir 1989b)

7.10 (selected, Shiu et al. 1988; quoted, Gobas et al. 1988, Sijm et al. 1989a; Gobas & Schrap 1990; Hawker 1990; Servos et al. 1992a,b)

7.20 (quoted, Servos & Muir 1989a; De Voogt 1990)

6.29 (slow stirring-GC/MSD from fly ash extract, Sijm et al. 1989)

8.03 (quoted, HPLC-RT, Chessells et al. 1991)

8.03, 7.70 (mean lit. value, calculated-f const., Chessells 1991)

Bioconcentration Factor, log BCF:

3.03 (fathead minnows, Corbet et al. 1983, 1988)

2.67 (rainbow trout, Corbet et al. 1983, 1988)

3.32 (rainbow trout, steady-state, wet weight, Muir et al. 1985)

3.76 (fathead minnows, steady-state, wet weight, Muir et al. 1985; quoted, Adams et al. 1986)

3.32 (rainbow trout average, Muir et al. 1986)

3.76 (fathead minnows average, Muir et al. 1986)

3.32, 3.76 (quoted, fish, Hawker 1990)

1.08-3.75, 2.14, 1.76-1.93 (Amphipod in lake water, in simulated lake water, in water with Aldrich humic acid, Servos & Muir 1989a)

4.20, 4,20 (goldfish after 8 hours exposure: metabolic inhibitor PBO-treated, control, Sijm et al. 1989b)

5.00 (goldfish after 6 days exposure: metabolic inhibitor PBO-treated, Sijm et al. 1989b)

3.76, 3.54-3.90; 3.32, 3,23-3.40 (fathead minnows, range; rainbow trout, range, quoted, De Voogt et al. 1990)

3.83, 4.83 (fathead minnows, wet wt. based, lipid based, quoted, Gobas & Schrap 1990)

3.39, 4.39 (rainbow trout, wet wt. based, lipid based, quoted, Gobas & Schrap 1990)

3.46, 3.30, 3.15 (quoted, rainbow trout exposed to different concentrations, Opperhuizen & Sijm 1990)

3.76, 3.75 (fathead minnows exposed to different concentrations, quoted, Opperhuizen & Sijm 1990)

3.70 (10.1 ng/L for 0-10 days, predicted for biota held in lake enclosures, Servos et al. 1992b)

3.12 (17.7 ng/L for 0-10 days, invertebrates, Servos et al. 1992b)

3.20 (21.4 ng/L for 0-10 days, unionid clams & white suckers gill, Servos et al. 1992b)

2.36 (3.1 ng/L for 0-10 days, white suckers carcass, Servos et al. 1992b)

3.21 (1.6 ng/L for 14-24 days, invertebrates, Servos et al. 1992b)

2.95 (0.9 ng/L for 14-24 days, unionid clams, Servos et al. 1992b)

3.32 (2.1 ng/L for 14-24 days, white suckers gill, Servos et al. 1992b)

2.70 (1.0 ng/L for 14-24 days, white suckers carcass, Servos et al. 1992b)

4.54 (3.5 ng/L for 0-104 days, white suckers gill, Servos et al. 1992b)

4.26 (1.8 ng/L for 0-104 days, white suckers carcass, Servos et al. 1992b)

Sorption Partition Coefficient, log K_{OC}:

4.36 (DOC, De Voogt et al. 1990)

6.74 (calculated-K_{OW}, Corbet et al. 1988)

2.11-3.75, 2,39, 2.05-2.38 (DOC partition coeff., lake water, simulated lake water, Aldrich humic acid, Servos & Muir 1989a)

5.98-6.23 (sediment, Servos & Muir 1989b)

Half-Lives in the Environment:
     Air:

Surface water: solution photolysis in n-hexadecane at 1.0 m from a GE Model RS sunlamp with half-life of 507 minutes (Nestrick et al. 1980); direct sunlight photolysis in aquatic bodies at latitude 40°N for various seasons with half-lives: 0.35 day in spring, 0.31 day in summer, 0.53 day in fall, 0.84 day in winter and 1.47 days averaged over full year (Choudhary & Webster 1986); 14.0 to 28.5 hours for outdoor pool and 6.3-8.0 days for natural water in a Pyrex flask under sunlight (Corbet et al. 1988); 3.24 hours for photolysis in water-acetonitrile (2:3, v/v) at 313 nm and the calculated midday, midseason direct phototransformation half-lives near water bodies at 40° N latitude calculated to be: 0.35 day in spring, 0.31 day in summer, 0.53 day in fall, and 0.84 day in winter (Choudhry & Webster 1989); calculated transformation rate constant in simulated lake enclosure of 9.4×10^{-2} hour^{-1} (Servos et al. 1992a).

Groundwater:

Sediment: 51.2 to 69.2 days (Corbet et al. 1988); calculated half-life of 10 years with a transformation rate constant of 7.9×10^{-6} hour^{-1} (Servos et al. 1992a).

Soil:

Biota: 6.90 days in fathead minnows (Adams et al. 1986); 4 days in rainbow trout (Neely 1979; quoted, Niimi & Oliver, 1983,1986); 4 days in rainbow trout (Corbet et al. 1983; quoted, Niimi & Oliver 1986); 2 days in rainbow trout (Muir et al. 1984); 15 days in whole body of rainbow trout (Muir et al. 1986; quoted, Muir et al. 1990); 41 to 44 days in rooted vegetable (Corbet et al. 1988).

Environmental Fate Rate Constants or Half-Lives:

Volatilization: 3.9 d in water of 0.5 m depth in a small pond (Corbet et al. 1988)

Photolysis: solution photolysis in n-hexadecane at 1.0 m from a GE Model RS sunlamp with half-life of 507 minutes and surface photolysis on a clean glass surface under the same conditions with half-life of 264 minutes (Nestrick et al. 1980; quoted, Muto et al. 1991); first-order rate constant of 59.57×10^{-6} sec^{-1} in water-acetonitrile (2:3, v/v) at 313 nm and calculated direct sunlight photolysis in aquatic bodies at latitude 40°N for various seasons: 0.35 day in spring, 0.31 day in summer, 0.53 day in fall, 0.84 day in winter and 1.47 day averaged over full year (Choudhry & Webster 1986; quoted, Muto et al. 1991); half-life of 14.0 to 28.5 hours for outdoor pool and 6.3 to 8.0 days for natural water in a Pyrex flask under sunlight (Corbet et al. 1988); half-life of 25 hours in water column (Corbet et al. 1983); half-life of 6.3 to 8.0 days for natural water under sunlight and photodegradation half-life of 0.3 day in summer sunlight at 40°N and 7 days in a 1-liter flask (Corbet et al. 1988); photolysis rate constant of 59.57×10^{-6} sec^{-1} with a half-life of 3.24 hours in water-acetonitrile solution (2:3, v/v) at 313 nm and the calculated midday, midseason direct sunlight photolysis first-order rate constant in aquatic bodies for various seasons: 198.13×10^2 day^{-1} for spring, 226.99×10^2 day^{-1} for summer, 130.91×10^2 day^{-1} for fall, 82.85×10^2 day^{-1} for winter with half-lives: 0.35 day in spring, 0.31 day in summer, 0.53 day in fall, and 0.84 days in winter; while experimentally determined sunlight photolysis half-life of ^{14}C labelled 1,3,6,8-TCDD in pond water was 3.5 days (Choudhry & Webster 1989).

Oxidation: photooxidation may be an important path of transformation (Corbet et al. 1988); oxidative degradation rate of water dissolved PCDD by ozone is 3.21 x 10^5 liter gram^{-1} minute^{-1} at pH 10 and 20°C (Palauschek & Scholz 1987).

Hydrolysis:

Biodegradation:

Biotransformation:

Bioconcentration, Uptake (k_1) and Elimination (k_2) Rate Constants:

k_1: 184 day^{-1} (rainbow trout, Muir et al. 1985,1986)

k_2: 0.10 day^{-1} (rainbow trout, Muir et al. 1985,1986)

k_1: 574 day^{-1} (fathead minnows, Muir et al. 1985,1986; quoted, Adams et al. 1986)

k_2: 0.10 day^{-1} (fathead minnows, Muir et al. 1985,1986; quoted, Adams et al. 1986)

k_1: 225 , 97 L kg^{-1} day^{-1} (goldfish after 8 hours exposure: metabolic inhibitor PBO-treated, control, Sijm et al. 1989)

k_2: 0.211 day^{-1} (fathead minnows, calculated, Corbet et al. 1983; quoted, Opperhuizen & Sijm 1990)

k_2: 0.158 day^{-1} (rainbow trout, calculated, Corbet et al. 1983; quoted, Opperhuizen & Sijm 1990)

k_1: 225, 221, 106 mL/gm/day (rainbow trout exposed to different concentrations in a flow system, quoted, Opperhuizen & Sijm 1990)

k_2: 0.074, 0.110, 0.110 day^{-1} (rainbow trout exposed to different concentratons in a flow system, quoted, Opperhuizen & Sijm 1990)

k_1: 526, 621 mL/gm/day (fathead minnows exposed to different concentrations in a flow system, quoted, Opperhuizen & Sijm 1990)

k_2: 0.080, 0.122 day^{-1} (fathead minnows exposed to different concentrations in a flow system, quoted, Opperhuizen & Sijm 1990).

k_1: 1200 day^{-1} (filter-feeder, estimated from Muir et al. 1985 and Servos et al. 1989, Servos et al. 1992b)

k_2: 0.06 day^{-1} (filter-feeder, estimated from Muir et al. 1985 and Servos et al. 1989, Servos et al. 1992b)

k_1: 285 day^{-1} (small fish, estimated from Muir et al. 1985 and Servos et al. 1989, Servos et al. 1992b)

k_2: 0.12 day^{-1} (small fish, estimated from Muir et al. 1985 and Servos et al. 1989, Servos et al. 1992b)

Common Name: 2,3,7,8-Tetrachlorodibenzo-*p*-dioxin

Synonym: 2,3,6,7-tetrachlorodibenzo-p-dioxin, TCDD, TCDBD, 2,3,6,7-TCDD, 2,3,7,8-TCDD, dioxin

Chemical Name: 2,3,7,8-tetrachlorodibenzo-*p*-dioxin

CAS Registry No: 1746-01-6

Molecular Formula: $Cl_2C_6H_2O_2C_6H_2Cl_2$

Molecular Weight: 322.0

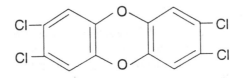

Melting Point (°C):

    305-306 (Pohland & Yang 1972; quoted, Ward & Matsumura 1978; Verschueren 1983; Rordorf 1986,1987,1989; Shiu et al. 1988)

    320-325 (quoted, Pohland & Yang 1972; Merck Index 1989)

    303-305 (Crummett & Stehl 1973; quoted, Callahan et al. 1979; Mabey et al. 1982; Mill 1985)

    306 (Branson et al. 1985; Crosby 1985)

    305.0 (quoted, Schroy et al. 1985a; Shiu et al. 1988)

Boiling Point (°C):

    421.2 (Schroy et al. 1985a)

    446.5 (calculated, Rordorf 1986,1987,1989)

Density (g/cm³ at 20°C):

    1.827 (solid at 25°C, Schroy et al. 1985a)

    1.021 (liquid at normal boiling point, Schroy et al. 1985a)

Molar Volume (cm³/mol):

    275.6 (LeBas method, Shiu et al. 1988)

    137.10 (calculated as per Pearlman 1986, De Voogt et al. 1990)

    134.40 (calculated as per Govers & De Voogt 1989, De Voogt et al. 1990)

    1.508 (intrinsic $V_I/100$, Hawker 1990)

    184.32, 184.97 (calculated-liquid density, crystalline volume, Govers et al. 1990)

Molecular Volume (Å³):

Total Surface Area, TSA (Å²):

Heat of Fusion, ΔH_{fus} kcal/mol:

    9.30 (quoted, Schroy et al. 1985a,b)

Entropy of Fusion, ΔS_{fus}, cal/mol K (e.u.):

    16.49 (Rordorf 1986,1987,1989)

Fugacity Ratio at 25°C (assuming ΔS = 13.5 e.u.):

0.0017   (Shiu et al. 1988)

Water Solubility (g/m^3 or mg/L at 25°C):

0.0002   (GC/ECD, Crummett & Stehl 1973; quoted, Callahan et al. 1979; Mabey et al. 1982; Mackay et al. 1985; Mill 1985; Jackson et al. 1986; Shiu et al. 1988)

0.0002   (quoted, Matsumura & Benezet 1973; Isensee & Jones 1975)

0.0002   (quoted, Neely 1979)

0.0002   (quoted, Kenaga 1980)

0.000317 (^{14}C-labeled, gen. col.-HPLC/LSC, Webster et al. 1983, quoted, Schroy et al. 1985a; Shiu et al. 1988)

0.0002   (quoted, Branson et al. 1985; Crosby 1985; Mill 1985)

1.93×10^{-5} (shake flask-GC/MS, Marple et al. 1986a; quoted, Travis & Hattemer-Frey 1987, Shiu et al. 1988; Walters & Guiseppi-Elie 1988; Kieatiwong et al. 1990)

7.91×10^{-6} (^{14}C-labeled, Adams & Blaine 1986, quoted, Schroy et al. 1985a,b; Mehrle et al. 1988; Shiu et al. 1988; Mackay 1991)

0.00032   (quoted, Srinivasan & Fogler 1987)

1.93×10^{-5} (selected, Shiu et al. 1988, quoted, Paterson et al. 1990)

$(1.25-1.93) \times 10^{-4}$ (quoted, Walters & Giuseppi-Elie 1988)

4.83×10^{-4} (17.3°C, gen. col.-GC/MS, Lodge 1989)

7.90×10^{-6} (20-22°C, quoted, Lodge 1989)

7.20×10^{-6} (quoted, Puri et al. 1989)

0.000192   (20 °C, quoted, De Voogt et al. 1990)

0.000688; 0.0081-0.0194 (calculated-QSAR; quoted, Fiedler & Schramm 1990)

Vapor Pressure (Pa at 25°C):

1.33×10^{-4} (calculated from structure, Mabey et al. 1982)

1.00×10^{-7} (^{14}C-gas saturation method, Jaber & Podoll 1983; quoted, Mill 1985)

9.33×10^{-8} (quoted, Crosby 1985)

6.00×10^{-5} (quoted, Mackay et al. 1985)

9.60×10^{-8} (quoted, Mill 1985)

$(3.5-6.3) \times 10^{-6}$ (gas saturation method, Rordorf 1985a,b, 1986,1987,1989; quoted, Shiu et al. 1988)

4.61×10^{-7} (30.1°C, gas saturation, Schroy et al. 1985a,b)

2.02×10^{-7} (gas saturation method, Schroy et al. 1985b; quoted, Rordorf 1985a,b; Rordorf et al. 1986a,b, 1987,1989,1990; Shiu et al. 1988)

1.51×10^{-7} (Schroy et al. 1985a,b; quoted, Shiu et al. 1988)

4.50×10^{-6} (Webster et al. 1985; quoted, Shiu et al. 1988)

9.86×10^{-8} (^{14}C-gas saturation, Podoll et al. 1986; quoted, Shiu et al. 1988; Kieatiwong et al. 1990)

8.14×10^{-8}, 6.0×10^{-5} (20°C, quoted: solid, subcooled value, Bidleman & Foreman 1987)

4.50x10^{-6} (quoted, Travis & Hattemen-Frey 1987)
6.20x10^{-7} (calculated, Rordorf 1987, 1989)
2.00x10^{-7} (selected, Shiu et al. 1988; quoted, Paterson et al. 1990)

Henry's Law Constant (Pa m^3/mol):
    0.0021   (calculated-P/C, Mabey et al. 1982)
    0.152   (calculated-P/C, Crosby 1985)
    0.212   (calculated-P/C, Schroy et al. 1985)
    1.64   (calculated-P/C, Podoll et al. 1986)
    1.63, 3.34, 10.34 (calculated-P/C, Shiu et al. 1988)
    0.196   (calculated-P/C form data of Schroy et al. 1985, 1988)
    7.93   (calculated-P/C, Jury et al. 1990)

Octanol/Water Partition Coefficient, log K_{OW}:
    5.38   (Crummett & Stehl 1973; quoted, Birnbaum 1985)
    6.19   (quoted, Neely 1979; Veith et al. 1979; Corbet et al. 1983)
    6.15   (quoted, Kenaga 1980; Schroy et 1985b; Marple et al. 1986)
    7.16   (calculated-f const, Perkow et al. 1980; quoted, Marple et al. 1986; Shiu et al. 1988)
    6.84   (calculated-f const, Johnson 1982; quoted, Mill 1985; Marple et al. 1986; Shiu et al. 1988)
    6.84   (calculated-f const, Mabey et al. 1982)
    6.15   (quoted, Garten & Trabalka 1983)
    5.50   (calculated, Kaiser 1983; quoted, Shiu et al. 1988)
    8.93   (HPLC-RT, Sarna et al. 1984)
    6.15   (quoted, Branson et al. 1985)
    6.845   (quoted, Crosby 1985)
    6.839   (quoted, Mill 1985)
    7.14   (quoted, Mackay et al. 1985)
    8.48   (quoted, Podoll et al. 1986)
    7.16, 6.15-7.28 (calculated-f const., quoted range, Jackson et al. 1986)
    7.02 (HPLC-RT, Burkhard & Kuehl 1986; quoted, Shiu et al. 1988; Sijm et al. 1989a; De Voogt et al. 1990)
    6.64   (quoted average, Marple et al. 1986a; quoted, Geyer et al. 1987)
    6.64   (stirring-GC/MS, Marple et al. 1986b; quoted, Geyer et al. 1987; Shiu et al. 1988; Walters & Guiseppi-Elie 1988)
    7.16   (quoted, Jackson et al. 1986)
    6.85   (quoted, Travis & Hattemer & Frey 1987)
    6.80   (selected, Shiu et al. 1988; quoted, Gobas et al. 1988; Sijm et al. 1989a; Goba & Schrap 1990; Paterson et al. 1990)
    6.80   (quoted, Hawker 1990)

6.42   (slow stirring-GC/MSD, from fly ash extract, Sijm et al. 1989a)

7.70, 6.60-7.70   (calculated-QSAR; quoted, Fiedler & Schramm 1990)

6.15   (quoted, Kieatiwong et al. 1990)

7.10   (calculated, Broman et al. 1991)

Bioconcentration Factor, log BCF:

1.69, 2.34, 2.08   (daphnia, ostracod, brine shrimp, ^{14}C-labeled-LSC, Matsumura & Benezet 1973)

4.53   (calculated-K_{ow}, Neely et al. 1974)

4.30-4.41; 3.6-3.95   (snail, *Gambusia*, daphnids; duckweed, algae, catfish; Isensee & Jones 1975)

5.38   (calculated-K_{ow}, Chiou et al. 1977; quoted, Branson et al. 1985)

4.52   (fish, calculated-K_{ow}, Veith et al. 1979; quoted, Corbet et al. 1983)

3.30   (calculated-K_{ow}, Banerjee et al. 1980; quoted, Branson et al. 1985)

3.90   (fathead minnows, steady-state, wet weight, Adams et al. 1986, quoted, De Voogt et al. 1990)

3.73, 4.55   (fish: flowing water test, static ecosystem test, Kenaga & Goring 1980, Kenaga 1980)

5.84   (microorganism, calculated-K_{ow}, Mabey et al. 1982)

3.97   (rainbow trout, Branson et al. 1983; quoted, Adams et al. 1986)

-0.15, 0.54, 3.02   (rodents, cow, fish, Garten & Trabalka 1983)

3.30, 5.38, 4.53   (fish: calculated-K_{ow}, Branson et al. 1985)

3.97, 3.67   (rainbow trout: whole body, muscle, Branson et al. 1985)

1.39-1.037, 1.39-0.568   (rats fat, liver, Geyer et al. 1986)

0.544, -0.155   (beef fat, liver, Geyer et al. 1986)

1.394   (cattle, calculated-steady state, Geyer et al. 1986)

1.38-1.60   (rhesus monkey, Geyer et al. 1986)

4.11   (guppy, Opperhuizen et al. 1986; quoted, Opperhuizen & Sijm 1990)

2.18   (human fat, calculated, Geyer et al. 1986)

1.079   (dry fodder to milk fat, Connett & Webster 1987; quoted, Webster & Connett 1990)

1.398   (dry fodder to milk fat, Travis & Hattemer-Frey 1987; quoted, Webster & Connett 1990)

2.23, 2.18   (calculated-K_{ow} for human fat: lipid basis, wet weight basis, Geyer et al.1987)

1.65-4.46, 4.69   (fish: quoted values, average, Travis & Hattemer-Frey 1987)

4.43, 4.59   (rainbow trout: measured average, estimated BCF at steady-state, for 28-days exposure, Mehrle et al. 1988)

0.833   (dry fodder to milk fat, McKone & Ryan 1989; quoted, Webster & Connett 1990)

0.699   (dry fodder to milk fat, Michaels 1989; quoted, Webster & Connett 1990)

403

4.30, 5.0 (goldfish after 8 hours exposure: metabolic inhibitor PBO-treated, control, Sijm et al. 1989b)

3.89 (fathead minnows, quoted, De Voogt et al. 1990)

4.59 (rainbow trout, quoted, De Voogt et al. 1990)

5.80, 5.90 (goldfish after 6 days exposure: PBO-treated, control, Sijm et al. 1989b)

4.11, 5.64 (guppy, wet wt. based, lipid based, quoted, Gobas & Schrap 1990)

3.97, 4.70; 4.97, 5.70 (rainbow trout, wet wt. based, lipid based, quoted, Gobas & Schrap 1990)

4.11 (fish, quoted, Hawker 1990)

4.63, 4.40 (pine needle/air BCF values, Reissinger et al. 1990)

4.59, 4.58, 4.93, 4.57, 3.97 (rainbow trout, quoted, Opperhuizen & Sijm 1990)

4.11, 3.90, 3.78 (guppy, fathead minnow, mosquito fish, quoted, Opperhuizen & Sijm 1990)

2.62 (human, Webster & Connett 1991)

Sorption Partition Coefficient, log K_{OC}:

5.67 (organic carbon soil, calculated-K_{OW}, Kenaga 1980; quoted, Puri et al. 1989)

6.52 (sediment, calculated-K_{OW}, Mabey et al. 1982)

6.95, 7.39-7.58 (calculated-k_{OW}, 10 soils from Missouri & New Jersey, Jackson et al. 1985)

6.04 (calculated-S, Mill 1985)

6.90 (calculated-K_{OW}, Mill 1985)

6.22-6.54 (red clay soil from Missouri, Marple et al. 1986)

5.96-6.09 (Alluvial soil from Missouri, Marple et al. 1986)

6.95 (calculated-K_{OW}, Jackson et al. 1986)

4.83 (hydroxy aluminum-clay, Srinivasan & Fogler 1987)

6.60 (^{14}C-labeled, soil, batch equilibrium, GC/ECD, Walters & Guiseppi-Elie 1988)

6.7-7.0 (calculated from aqueous solubility, Walters & Guiseppi-Elie 1988)

3.06 (soil, Eduljee 1987)

6.30 (sediment from Lake Ontario, Lodge & Cook 1989)

7.59, 7.25 (solids, organic carbon, Lodge & Cook 1989)

6.24, 6.10, 5.10 (Eglin Air Force Base soil/water with 0.01% surfactant from Florida at pH 4, 7, 8.5, Puri et al. 1989)

6.50, 5.86, 4.81 (Time Beach soil/water with 0.01% surfactant from Missouri at pH 4, 7, 8.5, Puri et al. 1989)

5.7, 5.09, 4.76 (Visalia soil/water with 0.01% surfactant from California at pH 4, 7, 8.5, Puri et al. 1989)

6.44, 6.66 (uncontaminated Time Beach soil/regression analysis of 2-day & 10-day isotherm, Walters et al. 1989)

6.14 (soil, Jury et al. 1990)

6.8 (calculated, Broman et al. 1991)

Half-Lives in the Environment:

Air: dominant transformation process in the atmosphere (Atkinson et al. 1982); 200 hours for OH radical oxidation (Podoll et al. 1986); room temperature gas-phase OH radical reaction rate constant was calculated to be 9×10^{-12} cm^3 molecule^{-1} sec^{-1} corresponding to an atmospheric lifetime of about 3 days (Atkinson 1987a); 22.3-223 hours, based on estimated photooxidation half-life in air (Atkinson 1987b; quoted, Howard et al. 1991); half-life in vapor phase undergo rapid photolysis with an upper limit of 1 hour (Travis & Hattemer-Frey 1987); atmospheric rate constant in summer sunlight at 40°N latitude of 0.012 minute^{-1} with half-life of 58 minutes (Buser 1988); rate constant for photolysis in air at 150-350°C of 5.9×10^{-3} sec^{-1} (Orth et al. 1989); reaction rate constant estimated to be 0.02 hour^{-1} (Paterson et al. 1990).

Surface water: photolysis half-life in methanol solution, 3 hours in sunlight (Plimmer et al. 1973); half-life in a model aquatic environment, 600 days (Ward & Matsumura 1978); photolysis in n-hexadecane at 1.0 m from a GE model RS sunlamp, with half-life of 56.8 minutes (Nestrick et al. 1980); reaction rate constant estimated to be 2.6×10^{-5} hour^{-1} (Mackay et al. 1985); calculated sunlight photolysis half-lives over four seasons at 40°N latitude averaged over for 24 hours exposure per day: 130 hours in winter, 28 hours in spring, 20 hours in summer and 52 hours in fall (Mill et al. 1982; quoted, Mill 1985); volatilization half-lives of about 32 days for water in ponds, lakes and about 16 days for rivers; calculated photolysis half-lives in sunlight at 40°N latitude: 118 hours in winter, 27 hours in spring, 21 hours in summer and 51 hours in fall (Podoll et al. 1986); photolysis in near-surface waters is an important degradative pathway with half-life of 40 hours (Travis & Hattemer-Frey 1987); sunlight-induced photolysis half-life in isooctane, 14 minutes (Buser 1988); reaction rate constant estimated to be 0.008 hour^{-1} (Paterson et al. 1990); 10032-14160 hours, based on estimated unacclimated aqueous aerobic biodegradation half-life (Howard et al. 1991).

Groundwater: 20064-28320 hours, based on estimated unacclimated aerobic biodegradation half-life (Howard et al. 1991).

Sediment: half-life in aquatic sediment 12000 to 14400 hours (Ward & Matsumura 1978; quoted, Quensen & Matsumura 1983); reaction rate constant estimated to be 8.0×10^{-6} hour^{-1} (Macaky et al. 1985); and 1.5×10^{-5} hour^{-1} (Paterson et al. 1990); > 1 year (O'Keefe et al. 1986).

Soil: 10032-14160 hours, based on soil dieaway test data for two soils (Kearney et al. 1971; quoted, Howard et al. 1991); about one year (Kearney et al. 1973; quoted, Quensen & Matsumura 1983); degradation half-life of 10-12 years in soil (De Domenico et al. 1980, Kimbrough et al. 1984); about 1 year if applied to surface on soil with 2,4-D (Nash & Beall 1980); reaction rate constant estimated to be 8.0×10^{-6} hour^{-1} (Mackay et al. 1985); and 1.1×10^{-5} hour^{-1} (Paterson et al. 1990); photodegradation is a rapid process at the soil surface with a half-life of 10 minutes during the day (Facchetti et al. 1986); degradation

405

reaction rate coefficient in soil was considered to be zero (Travis & Hattemer-Frey 1987); half-life in soil about 10 years if TCDD is on or near the surface and 100 years if TCDD is buried at greater depth (Nauman & Schaum 1987); calculated half-life, 10 years (Eduljee 1987); 10 years or longer (Boddington et al. 1990); photolysis in soil is slow (Kieatiwong et al. 1990); half-life for volitilization to atmosphere below surface soil, 365 days (Jury et al. 1990).

Biota: estimated half-life in rat, 31 days (Rose et al. 1976; quoted, Birnbaum 1985); estimated half-life in hamster, 11 days (Olsen et al. 1980; quoted, Birnbaum 1985); estimated half-life in guinea pig, 30 days (Decad et al. 1981a; quoted, Birnbaum 1985); estimated half-life in mouse, 11-24 days (Gasiewicz et al. 1983; quoted, Birnbaum 1985); total body burden depuration half-life of ^{14}C-TCDD in whole rainbow trout is 58 days (Branson et al. 1983,1985; quoted, Adams et al. 1986); half-lives: 17-37 days in mouse, 31 days in rat and 30 days in guinea pig (quoted, Van den Berg et al. 1985); elimination half-life of 14.5 days from fathead minnows (Adams et al. 1986; quoted, Niimi & Oliver 1986); 105 days in whole body of rainbow trout (Kleeman et al. 1986; quoted, Muir et al. 1990); 300-325 days in carp (Kuehl et al. 1986); biological half-life for rainbow trout, 58 days (Niimi & Oliver 1986); estimated half-life in human, 5.8 years (Poiger & Schlatter 1986; quoted, Travis & Hattemer-Frey 1987; Webster & Connett 1991); >336 days for carps in Lake Superior (Kuehl et al. 1987); estimated half-life in human, 6.7 years (Kissel & Robarge 1988; quoted, Webster & Connett 1991); elimination half-lives of 15-48 days from rainbow trout for exposures of different concentrations (Mehrle et al. 1988); estimated half-life in human, 7.1 years (Pirkle et al. 1989; quoted, Webster & Connett 1991); half-lives of 5 to 8 years for human, 17.4-31 days for rats, 9.6-24.4 days for mice, 22-93.7 days for guinea pigs, 12.0-150 days for hamsters, 1 year for monkeys (Boddington et al. 1990); 40.3 days for lactating cows (Olling et al. 1991); assumed half-life for human, 5 to 10 years (Schecter & Ryan 1991); elimination half-life from lake trout sac fry, 35-37 days (Walker et al. 1991); 4.4 years for a 70 kg non-lactating "reference" human (Webster & Connett 1991).

Environmental Fate Rate Constants or Half-Lives:

Volatilization: probably not an important process (Callahan et al. 1979); half-life of 20-200 days from water column which will be slowed down further by the fact that it is sorbed to the sediment and biota (Mill 1985); half-lives of about 32 days for ponds and about 16 days for rivers (Podoll et al. 1986); half-life for volatilization from soil was 104 days by calculation assuming diffusion of TCDD in soil is vapor-dominated up to volumetric water content of 0.3 m^3/m^3, and then liquid-dominated to saturation (Eduljee 1987); half-life of 190 days (Thibodeaux & Lipsky 1985; quoted, Eduljee 1987); half-life for volatilization from below surface soil is 365 days (Jury et al. 1990).

Photolysis: stable to sunlight with a half-life of 14 days in distilled water (Crosby et al.
1971; quoted, Dougherty et al. 1991); TCDD in methanol solution has a half-life
of about 3 hours in sunlight (Plimmer et al. 1973); half-life for vapor in
sunlight, 56 minutes (Peterson 1976; quoted, Mill 1985); thin film of TCDD on
glass plates showed transformation at about 6 hours (Crosby & Wong 1977);
estimated half-life was 320 hours for the reaction with 3×10^{-15} mole OH radicals
in vapor phase in atmosphere with rate constant of 6×10^{-7} sec^{-1} (Singh 1977;
quoted, Mill 1985); surface photolysis half-life on a clean soft glass surface at
1.0 m from a GE Model RS sunlamp with half-life of 8400 minutes (Nestrick
et al. 1980); solution photolysis in n-hexadecane with half-life of 56.8 minutes
(Nestrick et al. 1980; quoted, Mamantov 1984; Dougherty et al. 1991); TCDD
extracted from the aqueous sludge with hexane can be continuously degraded by
a mercury arc of UV radiation (Exner et al. 1982; quoted, Crosby 1985);
reaction rate constant with OH radicals, 2×10^8 $mole^{-1}$ sec^{-1} (Davenport et al.
1984; quoted, Mill 1985); half-lives: about 1 day in water, 0.1 day in vapor and
1-100 days in soil with 50 days for a small fraction in water column in
equilibrium with sediment sorbed with TCDD (Mill 1985); atmospheric and
aqueous photolysis half-life of 27 hours, based on measured rate constant for
photolysis in a 90:10 mixture of distilled water and acetonitrile under summer
sunlight (Dulin et al. 1986; quoted, Howard et al. 1991; Muto et al. 1991) and
81 hours after adjusting for relative winter sunlight intensity (Lyman et al. 1982;
quoted, Howard et al. 1991); half-life of aqueous dissolved TCDD in sunlight
over four seasons at 40°N latitude is calculated to be: 118 hours in winter, 27
hours in spring, 21 hours in summer, 51 hours in fall and with rate constants:
0.14 day^{-1} for winter, 0.61 day^{-1} for spring, 0.78 day^{-1} for summer, and 0.32
day^{-1} for autumn (Podoll et al. 1986); photodegradation is a very rapid process
at soil surface with a half-life of 10 minutes during the day (Facchetti et al.
1986); sunlight induced photolysis in isooctane with half-life of 14 minutes and
dispersed as solid films with half-life of 300 hours (Buser 1988); first-order
photolysis rate constant in isooctane was determined to be 0.15 $minute^{-1}$ and
over 90% was lost in 21 minutes of irradiation in isooctane whereas only greater
than 55% TCDD remained in soil after 15 days of irradiation (Kieatiwong et al.
1990); photolytic degradation in extract from fly ash exposed to UV light from
a distance of 20 cm with half-life of 4.5 hours and 5.2 hours for native and ^{13}C-
labelled congeners in tetradecane solution, whereas half-lives for solid phase
photolysis were relatively longer in fly ash (Tysklind & Rappe 1991); half-life
of 31 minutes in hexadecane and 27 minutes in ethyl oleate (Dougherty et al.
1991); half-life from direct sunlight photolysis in water-acetonitrile in midday
of mid-summer at 40°N was estimated to be 52 hours (quoted, Zeep 1991).

Oxidation: laboratory tests shown that 99.5% TCDD was oxidized in 21 seconds at
800°C while only 50% reacted at 700°C (Esposito et al. 1980; quoted, Crosby
1985); half-life for oxidation in vapor, about 13 days (Mill 1985); rate constant
for reaction of OH radicals with aromatics are large, e.g. > 1.0×10^8 $mole^{-1}$

sec^{-1}, 1.7x10^{-13} molecule^{-1} cm^3 sec^{-1}; half-life in OH oxidation in the atmosphere is 200 hours, applied only to TCDD vapor (Podoll et al. 1986); room temperature gas-phase OH radical reaction rate constant was calculated to be 9x10^{-12} cm^3 molecule^{-1} sec^{-1} corresponding to an atmospheric lifetime of about 3 days (Atkinson 1987a) and 8.0x10^{-12} cm^3 molecule^{-1} sec^{-1} (Atkinson 1987b); photooxidation half-life of 22.3-223 hours, based on estimated rate constant for reaction with hydroxyl radical in air (Atkinson 1987; quoted, Howard et al. 1991); the oxidative degradation of water dissolved TCDDs by ozone takes place only under alkaline conditions at pH 10 and 20°C, the reaction rate constant is 1.33x10^5 liter gram^{-1} minute^{-1} (Palauschek & Scholz 1987).

Hydrolysis: hydrolysis is not likely under environmental conditions (Callahan et al. 1979; Mabey et al. 1982); no hydrolyzable groups (Howard et al. 1991).

Biodegradation: aerobic half-life of 10032 hours, based on soil dieaway test data (Kearney et al. 1971; quoted, Howard et al. 1991) and 14160 hours, based on lake water and sediment dieaway test data (Ward & Matsumara 1978; quoted, Howard et al. 1991); half-life > 1.0 year (Callahan et al. 1979); anaerobic half-life of 40128-56640 hours, based on estimated unacclimated aqueous aerobic biodegradation half-life (Howard et al. 1991).

Biotransformation: rate constant for bacterial transformation in water estimated to be 1x10^{-10} ml cell^{-1} hour^{-1} (Mabey et al. 1982).

Bioconcentration, Uptake (k_1) and Elimination (k_2) Rate Constants:

k_1: 6.30 mL gram^{-1} hour^{-1} (rainbow trout, calculated, Neely 1979)

k_1: 4.64 mL gram^{-1} hour^{-1} (rainbow trout, experimental, Neely 1979)

k_2: 2.12x10^{-4} hour^{-1} (rainbow trout, calculated, Neely 1979)

k_2: 5.00x10^{-4} hour^{-1} (rainbow trout, experimental, Neely 1979)

k_1: 108 mL gram^{-1} day^{-1} (rainbow trout, Branson et al. 1983,1985; quoted, Adams et al. 1986; Sijm & Opperhuizen 1988; Opperhuizen & Sijm 1990)

k_2: 0.012 day^{-1} (rainbow trout, Branson et al. 1983,1985; quoted, Adams et al. 1986; Sijm & Opperhuizen 1988; Opperhuizen & Sijm 1990)

k_1: 476 mL gram^{-1} day^{-1} (fathead minnows, Adams et al. 1986)

k_2: 0.120 day^{-1} (fathead minnows, Adams et al. 1986)

k_1: 1832, 1543, 1337, 1591 day^{-1} (rainbow trout, exposed to 38 pg/L, 176 pg/L, 382 pg/L, 702 pg/L for 28 days, Mehrle et al. 1988; quoted, Opperhuizen & Sijm 1990)

k_2: 0.047, 0.041, 0.015, 0.043 day^{-1} (rainbow trout, exposed to 38, 176, 382, 702 pg/L for 28 days, Mehrle et al. 1988; quoted, Opperhuizen & Sijm 1990)

k_1: 216, 604 L kg^{-1} day^{-1} (goldfish after 8 hours exposure: metabolic inhibitor PBO-treated, control, Sijm et al. 1989)

k_1: 600 mL gram^{-1} day^{-1} (guppy, Opperhuizen et al. 1986; quoted, Opperhuizen & Sijm 1990)

k_2:　　0.046 day^{-1}　　　　(guppy, Opperhuizen et al. 1986; quoted, Opperhuizen & Sijm 1990)

k_2:　　0.008 day^{-1}　　　　(rainbow trout, quoted, Opperhuizen & Sijm 1990)

k_1:　　381 mL gram^{-1} day^{-1} (fathead minnows, quoted, Opperhuizen & Sijm 1990)

k_2:　　0.048 day^{-1}　　　　(fathead minnows, quoted, Opperhuizen & Sijm 1990)

k_1:　　100 mL gram^{-1} day^{-1} (mosquito fish, quoted, Opperhuizen & Sijm 1990)

Common Name: 1,2,3,4,7-Pentachloro-dibenzo-*p*-dioxin
Synonym: 1,2,3,4,7-P$_5$CDD
Chemical Name: 1,2,3,4,7-pentachloro-dibenzo-*p*-dioxin
CAS Registry No: 39227-61-7
Molecular Formula: ClC$_6$H$_3$OC$_6$Cl$_4$
Molecular Weight: 356.4

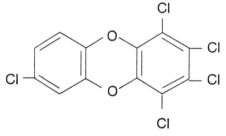

Melting Point (°C):
     195-196 (Pohland & Yang 1972; Rordorf 1987,1989)
     196   (quoted, Shiu et al. 1988)
     188   (Friesen & Webster 1990)
Boiling Point (°C):
     464.7   (calculated, Rordorf 1987, 1989)
Density (g/cm^3 at 20°C):
Molar Volume (cm^3/mol):
     296.5   (LeBas method, Shiu et al. 1988)
     146.22 (calculated as per Pearlman 1986, De Voogt et al. 1990)
     142.92 (calculated as per Govers & De Voogt 1989, De Voogt et al. 1990)
     197.74 (calculated-liquid density, Govers et al. 1990)
Molecular Volume (Å^3):
     224   (Friesen & Webster 1990)
Total Surface Area, TSA (Å^2):
     309   (Friesen & Webster 1990)
Heat of Fusion, ΔH_{fus}, kcal/mol:
     6.214   (Friesen & Webster 1990)
Entropy of Fusion, ΔS_{fus}, cal/mol K (e.u.):
     21.53   (Rordorf 1987, 1989)
Fugacity Ratio at 25 °C (assuming ΔS = 13.5 e.u.), F:
     0.0204 (Shiu et al. 1988)

Water Solubility (g/m^3 or mg/L at 25°C):
     0.000118, 0.0000855 (20°C, ^{14}C-labeled, gen. col.-HPLC/LSC, calculated, Friesen et
          al. 1985; quoted, Shiu et al. 1988)
     0.0000955 (20°C, ^{14}C labeled, gen. col.-HPLC/LSC, Webster et al. 1986)

0.000118 (selected, Shiu et al. 1988; quoted, Fiedler & Schramm 1990)
0.00578, 0.00816 (quoted, calculated-χ , Nirmalakhanden & Speece 1989)
0.000123; 0.000165 (21°C; 26 °C, gen. col.-HPLC/LSC, Friesen & Webster 1990)
0.000120 (20 °C, quoted, De Voogt et al. 1990)

Vapor Pressure (Pa at 25°C):
8.80x10⁻⁸ (calculated, Rordorf 1985a,b, 1987,1989)
1.00x10⁻⁶ (gas saturation, estimated from vapor pressure vs. temperature plot, Rordorf et al. 1986)
1.0x10⁻⁶ (gas saturation, estimated form extrapolated vapor pressure vs. halogen substitution no. plot, Rordorf et al. 1990)

Henry's Law Constant (Pa m³/mol):
0.264 (calculated-P/C, Shiu et al. 1988)

Octanol/Water Partition Coefficient, log K_{OW}:
9.44, 9.62, 10.02; 9.39, 9.65, 10.05 (HPLC-RT, linear; quadratic regressions, Sarna et al. 1984)
9.40, 9.65 (HPLC-RT, Sarna et al. 1984; quoted, Shiu et al. 1988)
9.48, 8.80; 9.40, 8.64 (HPLC-RT, linear; quadratic regressions, Webster et al. 1985; quoted, Sijm et al. 1989a)
9.65, 7.44 (HPLC-RT, Burkhard & Kuehl 1986)
7.44 (quoted, De Voogt et al. 1990)
7.40 (selected, Shiu et al. 1988; quoted, Gobas et al. 1988, Sijm et al. 1989a; Fiedler & Schramm 1990; Gobas & Schrap 1990)
6.60 (slow stirring-GC/MS, from mixture of fly-ash extract, Sijm et al. 1989)

Bioconcentration Factor, log BCF:
3.16 (fathead minnows, steady-state, wet weight, Muir et al. 1985, quoted, Adams et al. 1986)
2.91 (rainbow trout, steady-state, wet weight, Muir et al. 1985)
3.16, 2.91 (fathead minnows, rainbow trout, De Voogt et al. 1990)
3.50, 4.50 (fathead minnows, wet wt. based, lipid based, quoted, Gobas & Schrap 1990)

Sorption Partition Coefficient, log K_{OC}:
4.85, 5.80, 6.38 (20°C, fulvic acid, humic acid, Aldrich humic acid, Webster et al. 1986b)
5.02 (DOC, De Voogt et al. 1990)

411

Half-Lives in the Environment:

Air: photodegradation half-life in a rotary photo-reactor adsorbed to clean silica gel by filtered < 290 nm of light, 92 hours (Koester & Hites 1992).

Surface water: under conditions of variable sunlight intensity at 40°N latitude in aqueous acetonitrile solution (4:6, v/v): half-life of 18.29 days in spring with rate constant of 0.466×10^{-6} sec^{-1}, 15.16 days in summer with rate constant of 0.562×10^{-6} sec^{-1}, 28.59 days in fall with rate constant of 0.298×10^{-6} sec^{-1}, 52.37 days in winter with rate constant of 0.163×10^{-6} sec^{-1}, and 76.82 days averaged over full year with rate constant of 0.111×10^{-6} sec^{-1} (Choudhry & Webster 1985a,1986); photolysis in water-acetonitrile (2:3, v/v) at 313 nm with half-life of 45.86 hours and the calculated midday, midseason direct phototransformation half-lives near water bodies at 40°N latitude: 18.4 days in spring, 15 days in summer, 29 days in fall and 52 days in winter (Choudhry & Webster 1989); 27 days in sunlit surface water and 0.94 days in surface water of actual pond (Friesen et al. 1990).

Groundwater:

Sediment:

Soil:

Biota: 3.1 days in fathead minnows (Adams et al. 1986);2 days in whole body of rainbow trout (Muir & Yarechewski 1988; quoted, Muir et al. 1990)

Environmental Fate Rate Constants or Half-Lives:

Volatilization:

Photolysis: rate constant of 4.31×10^{-6} sec^{-1} in water-acetonitrile (2:3, v/v) at 313 nm and calculated half-lives under conditions of variable sunlight intensity at 40°N latitude: 18.29 days in spring with rate constant of 0.466×10^{-6} sec^{-1}, 15.16 days in summer with rate constant of 0.562×10^{-6} sec^{-1}, 28.59 days in fall with rate constant of 0.298×10^{-6} sec^{-1}, 52.37 days in winter with rate constant of 0.163×10^{-6} sec^{-1}, and 76.82 days averaged over full year with rate constant of 0.111×10^{-6} sec^{-1} (Choudhry & Webster 1985a, 1986); rate constant of 4.3115×10^{-6} sec^{-1} in water-acetonitrile (3:3 v/v) at 313 nm (Choudhry & Webster 1985b); photolysis rate constant of 4.31×10^{-6} sec^{-1} with a half-life of 45.86 hours in water-acetonitrile (2:3, v/v) at 313 nm and the calculated midday, midseason direct sunlight photolysis first-order rate constant in aquatic bodies for various seasons: 4.03×10^2 day^{-1} in spring, 4.86×10^2 day^{-1} in summer, 2.58×10^2 day^{-1} in fall, 1.41×10^2 day^{-1} in winter with half-livers: 18 days in spring, 15 days in summer, 29 days in fall and 52 days in winter (Choudhry & Webster 1989); rate constant of 0.74 day^{-1} under mid-summer sunlight at 50°N latitude in filtered-sterilized natural water and 0.058 day^{-1} in (2:3, v/v) distilled water-acetonitrile (Friesen et al. 1990); photolytic half-life of 38 hours in fly-ash extract (Tysklind & Rappe 1991); half-life of 92 hours for photodegradation in a rotary photoreactor adsorbed to clean silica gel by filtered < 290 nm of light (Koester & Hites 1992).

Oxidation:

Hydrolysis:

Biodegradation:

Biotransformation: biotransformation rate constant of 0.014 day^{-1} for rainbow trout (Sijm et al. 1990).

Bioconcentration, Uptake (k_1) and Elimination (k_2) Rate Constants:

k_1:   285 day^{-1} (fathead minnows, Muir et al. 1985; quoted, Adams et al. 1986)

k_2:   0.22 day^{-1} (fathead minnows, Muir et al. 1985, quoted, Adams et al. 1986)

k_1:   204 day^{-1} (rainbow trout, Muir et al. 1985; quoted, Sijm et al. 1990)

k_2:   0.28 day^{-1} (rainbow trout, Muir et al. 1985; quoted, Sijm et al. 1990)

k_2: 2.5×10^{-2}, 3.9×10^{-2} day^{-1} (rainbow trout for 2 to 21 days exposure: metabolic inhibitor PBO-treated, control, Sijm et al. 1990)

Common Name: 1,2,3,4,7,8-Hexachloro-dibenzo-*p*-dioxin
Synonym: 1,2,3,4,7,8-H$_6$CDD
Chemical Name: 1,2,3,4,7,8-hexachloro-dibenzo-*p*-dioxin
CAS Registry No: 39227-26-8
Molecular Formula: Cl$_2$C$_6$H$_2$OC$_6$Cl$_4$
Molecular Weight: 391.0

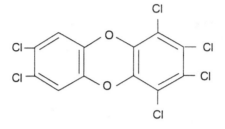

Melting Point (°C):

    273-275 (Pohland & Yang 1972; Rordorf 1987,1989)

    273   (quoted, Shiu et al. 1988)

    259-261 (quoted, Friesen et al. 1985)

Boiling Point (°C):

    487.7   (calculated, Rordorf 1987,1989)

Density (g/cm^3 at 20°C):

Molar Volume (cm^3/mol):

    317.4   (LeBas method, Shiu et al. 1988)

    155.38 (calculated as per Pearlman 1986, De Voogt et al. 1990)

    151.44 (calculated as per Govers & De Voogt 1989, De Voogt et al. 1990)

    1.686   (intrinsic V$_I$/100, Hawker 1990)

    206.96 (calculated-liquid density, Govers et al. 1990)

Molecular Volume (Å^3): 239 (Friesen & Webster 1990)

Total Surface Area, TSA (Å^2):

    321   (Friesen & Webster 1990)

Heat of Fusion, ΔH_{fus}, kcal/mol:

    7.22   (Friesen & Webster 1990)

Entropy of Fusion, ΔS_{fus}, cal/mol K (e.u.):

    21.03   (Rordorf 1987,1989)

Fugacity Ratio at 25 °C (assuming ΔS = 13.5 e.u.), F:

    0.00352 (Shiu et al. 1988)

Water Solubility (g/m^3 or mg/L at 25°C):

    0.00000442; 0.0000101 (20°C, ^{14}C-labeled, gen. col.-HPLC/LSC; calculated, Friesen
        et al. 1985; quoted, Shiu et al. 1988)

    0.0000057 (20°C, ^{14}C-labeled, gen. col.-HPLC/LSC, Webster et al. 1986b)

0.00000442 (selected, Shiu et al. 1988)

0.0000040; 0.00000644 (21°C; 26 °C, gen. col.-HPLC/LSC, Friesen & Webster 1990)

0.0000044 (20°C, quoted, De Voogt et al. 1990)

0.000044 (quoted, Paterson et al. 1990)

Vapor Pressure (Pa at 25°C):

5.10×10^{-9} (calculated, Rordorf 1985a,b, 1987,1989; quoted, Shiu et al. 1988)

3.20×10^{-7} (gas saturation, estimated from vapor pressure vs. temperature plot, Rordorf et al. 1986)

3.96×10^{-6} (subcooled liquid value, GC-RT, Eitzer & Hites 1988)

7.85×10^{-6} (subcooled liquid, GC/Ms, Eitzer & Hites 1988)

5.10×10^{-9} (quoted, Paterson et al. 1990)

1.00×10^{-8} (gas saturation, estimated from extrapolated vapor pressure vs. halogen substitution no. plot, Rordorf et al. 1990)

Henry's Law Constant (Pa m³/mol):

4.52 (calculated-P/C, Shiu et al. 1988)

Octanol/Water Partition Coefficient, log K_{ow}:

10.36, 10.39, 10.89; 10.22, 10.44, 10.89 (HPLC-RT, linear; quadratic regressions, Sarna et al. 1984)

10.40, 9.65; 10.22, 9.19 (HPLC-RT, linear; quadratic regressions, Webster et al. 1985; quoted, Sijm et al. 1989a)

10.44, 7.79 (HPLC-RT, Burkhard & Kuehl 1986)

7.79 (quoted, De Voogt et al. 1990)

7.80 (selected, Shiu et al. 1988; quoted, Gobas et al. 1988; Sijm et al. 1989a; Gobas & Schrap 1990; Hawker 1990; Paterson et al. 1990)

8.00 (quoted, Hawker 1990)

7.30 (calculated, Broman et al. 1991)

9.53 (quoted, HPLC-RT, Chessells et al. 1991)

9.53, 9.13 (mean lit. value, calculated-f const., Chessells et al. 1991)

Bioconcentration Factor, log BCF:

3.36 (rainbow trout, steady-state, wet weight, Muir et al. 1985)

3.63 (fathead minnows, steady-state, wet weight, Muir et al. 1985, quoted, Adams et al. 1986)

3.36, 3.63 (rainbow trout, minnows, quoted, De Voogt et al. 1990)

4.00, 5.00 (fathead minnows, wet wt. based, lipid based, quoted, Gobas & Schrap 1990)

3.36, 3.63 (quoted, Hawker 1990)

3.41, 3.76 (fathead minnows, quoted, Opperhuizen & Sijm 1990)
3.45, 3.23 (rainbow trout, quoted, Opperhuizen & Sijm 1990)

Sorption Partition Coefficient, log K_{OC}:
    5.41, 6.02, 6.22 (20°C, fulvic acid, humic acid, Aldrich humic acid, Webster et al. 1986b)
    5.02 (DOC, De Voogt et al. 1990)
    7.10 (calculated, Broman et al. 1991)

Half-Lives in the Environment:
    Air: estimated reaction rate constant, 0.005 hour^{-1} (Paterson et al. 1990); photodegradation in a rotary photoreactor adsorbed to clean silica gel by filtered < 290 nm of light, 140 hours (Koester & Hites 1992).
    Surface water: under conditions of variable sunlight intensity at 40°N latitude: 7.57 days in spring with rate constant of 1.06×10^{-6} sec^{-1}, 6.27 days in summer with rate constant of 1.280×10^{-6} sec^{-1}, 11.87 days in fall with rate constant of 0.676×10^{-6} sec^{-1}, 21.57 days in winter with rate constant of $0.37s \times 10^{-6}$ sec^{-1}, and 76.82 days averaged over full year with rate constant of 0.252×10^{-6} sec^{-1} (Choudhry Webster 1985a, 1986); photolysis half-life of 24.5 hours in water-acetonitrile (2:3, v/v) at 313 nm and the calculated midday, midseason direct phototransformation half-lives near water bodies at 40°N latitude: 7.6 days in spring, 6.3 days in summer, 12.0 days in fall and 22.0 days in winter (Choudhry & Webster 1989); 81 days in sunlit surface water and 2.5 days in surface water of actual pond (Friesen et al. 1990); estimated reaction rate constant, 0.002 hour^{-1} (Paterson et al. 1990).
    Groundwater:
    Sediment: estimated reaction rate constant to be 4.0×10^{-6} hour^{-1} (Paterson et al. 1990).
    Soil: estimated reaction rate constant to be 2.8×10^{-6} hour^{-1} (Paterson et al. 1990).
    Biota: 43 days in whole body of rainbow trout (Muir et al. 1988; quoted, Muir et al. 1990)

Environmental Fate Rate Constants or Half-Lives:
    Volatilization:
    Photolysis: calculated rate constants under conditions of variable sunlight intensity at 40°N latitude in aqueous acetonitrile (4:6 v/v) solution: 1.06×10^{-6} sec^{-1} for spring with half-life of 7.57 days, 1.280×10^{-6} sec^{-1} for summer with half-life of 6.27 days, 0.676×10^{-6} sec^{-1} for fall with half-life of 11.87 days, $0.37s \times 10^{-6}$ sec^{-1} for winter with half-life of 21.57 days and 0.252×10^{-6} sec^{-1} averaged over full year with half-life of 31.85 days (Choudhry & Webster 1985a, 1985c, 1986); 7.86×10^{-6} sec^{-6} in water-acetonitrile (2:3, v/v) under direct sunlight (Choudhry & Webster

1985b); photolysis rate constant of 7.86×10^{-6} sec^{-1} with a half-life of 24.5 hours in water-acetonitrile (2:3, v/v) at 313 nm and the calculated midday, midseason direct sunlight photolysis first-order rate constant in aquatic bodies for various seasons: 9.16×10^2 day^{-1} for spring, 11.06×10^2 day^{-1} for summer, 5.84×10^2 day^{-1} for fall, 3.21×10^2 day^{-1} for winter and with half-lives: 7.6 days in spring, 6.3 days in summer, 12.0 days in fall and 22.0 days in winter (Choudhry & Webster 1989); 0.28 day^{-1} in natural water and 0.019 day^{-1} in distilled water-acetonitrile (Friesen et al. 1990); photolytic half-life in the fly ash extract in tetradecane calculated to be 38 hours (Tysklind & Rappe 1991); photodegradation half-life in a rotary photoreactor absorbed to silica gel by filtered < 290 nm of light, 140 hours (Koester & Hites 1992).

Hydrolysis:

Oxidation: oxidative degradation rate constant of water dissolved PCDD is 5.02×10^4 liter gram^{-1} minute^{-1} under alkaline condition at pH 10 and 20°C (Palauschek & Scholz 1987).

Biodegradation:

Biotransformation:

Bioconcentration, Uptake (k_1) and Elimination (k_2) Rate Constants:

k_1: 102 day^{-1} (rainbow trout, Muir et al. 1985; quoted, Opperhuizen & Sijm 1990)

k_2: 0.046 day^{-1} (rainbow trout, Muir et al. 1985; quoted, Opperhuizen & Sijm 1990)

k_1: 112 day^{-1} (fathead minnows, Muir et al. 1985; quoted, Opperhuizen & Sijm 1990)

k_2: 0.030 day^{-1} (fathead minnows, Muir et al. 1985; quoted, Opperhuizen & Sijm 1990)

k_2: 0.015 day^{-1} (rainbow trout, quoted, Opperhuizen & Sijm 1990)

k_2: 0.0066 day^{-1} (fathead minnows, quoted, Opperhuizen & Sijm 1990)

Common Name: 1,2,3,4,6,7,8-Heptachloro-dibenzo-*p*-dioxin
Synonym: 1,2,3,4,6,7,8-H₇CDD
Chemical Name: 1,2,3,4,6,7,8-heptachloro-dibenzo-*p*-dioxin
CAS Registry No: 35822-46-9
Molecular Formula: $Cl_3C_6HOC_6Cl_4$
Molecular Weight: 425.2

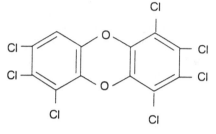

Melting Point (°C):
    265   (quoted, Shiu et al. 1988)
    264-265 (quoted, Friesen et al. 1985; Rordorf 1987,1989)
Boiling Point (°C):
    507.2   (Rordorf 1987,1989)
Density (g/cm³ at 20°C):
Molar Volume (cm³/mol):
    338.3   (LeBas method, Shiu et al. 1988)
    164.52   (calculated as per Pearlman 1986, De Voogt et al. 1990)
    159.96   (calculated as per Govers & De Voogt 1989, De Voogt et al. 1990)
    1.778   (intrinsic V_I/100, Hawker 1990)
    218.32   (calculated-liq. density, Govers et al. 1990)
Molecular Volume (Å³): 253 (Friesen & Webster 1990)
Total Surface Area, TSA (Å²): 338 (Friesen & Webster 1990)
Heat of Fusion, ΔH_{fus}, kcal/mol: 7.27 (Friesen & Webster 1990)
Entropy of Fusion, ΔS_{fus}, cal/mol K (e.u.):
    23.90   (Rordorf 1987,1989)
Fugacity Ratio at 25 °C (assuming ΔS = 13.5 e.u.), F:
    0.00423 (Shiu et al. 1988)

Water Solubility (g/m³ or mg/L at 25°C):
    0.0000024, 0.00000242 (20°C, ¹⁴C-labeled, gen. col.-HPLC/LSC; calculated, Friesen
            et al. 1985; quoted, Shiu et al. 1988)
    0.0000024 (selected, Shiu et al. 1988)
    0.000561, 0.000848 (quoted, calculated-χ , Nirmalakhanden & Speece 1989)
    0.0000023; 0.00000256 (21°C; 26 °C, gen. col.-HPLC/LSC, Friesen & Webster 1990)
    0.0000024 (20°C, quoted, De Voogt et al. 1990)
    0.000002, 0.00000203 (quoted, calculated-QSAR, Fiedler & Schramm 1990)

Vapor Pressure (Pa at 25°C):

7.5x10^{-10} (calculated, Rordorf 1985a,b, 1987,1989; quoted, Shiu et al. 1988)

3.2x10^{-8} (gas saturation, estimated from vapor pressure vs. temperature plot, Rordorf et al. 1986)

1.024x10^{-6} (subcooled liquid value, GC-RT, Eitzer & Hites 1988)

1.93x10^{-6} (subcooled liquid, GC/MS, Eitzer & Hites 1988)

3.2x10^{-9} (gas saturation, estimated from extrapolated vapor pressure vs. halogen substitution no. plot, Rordorf et al. 1990)

Henry's Law Constant (Pa m^3/mol):

0.133 (calculated-P/C, Shiu et al. 1988)

Octanol/Water Partition Coefficient, log K_{OW}:

11.29, 11.42, 11.90; 11.03, 11.50, 11.98 (HPLC-RT, linear; quadratic regressions, Sarna et al. 1984)

11.05, 11.50 (HPLC-RT, Sarna et al. 1984; quoted, Shiu et al. 1988)

11.38, 10.55; 11.05, 9.69 (HPLC-RT, linear; quadratic regressions, Webster et al. 1985; quoted, Sijm et al. 1989a)

8.20 (HPLC-RT, Burkhard & Kuehl 1986; quoted, Shiu et al. 1988)

8.20 (quoted, De Voogt et al. 1990)

8.00 (selected, Shiu et al. 1988; quoted, Gobas et al. 1988; Sijm et al. 1989a; Gobas & Schrap 1990; Hawker 1990; Loonen et al. 1991)

7.92 (slow stirring-GC/MSD, from extract of fly ash, Sijm et al. 1989a)

9.86, 8.00 (calculated-QSAR, quoted, Fiedler & Schramm 1990)

8.00 (quoted, Hawker 1990)

8.00 (calculated, Broman et al. 1991)

10.52 (quoted, HPLC-RT, Chessells et al. 1991)

10.32, 9.85 (mean lit. value, calculated-f const., Chessells et al. 1991)

Bioconcentration Factor, log BCF:

2.71 (fathead minnows, steady-state, wet weight, Muir et al. 1985, quoted, Adams et al. 1986)

3.15 (rainbow trout, steady-state, wet weight, Muir et al. 1985)

2.71, 3.15 (fathead minnows, rainbow trout, quoted, De Voogt et al. 1990)

3.32, 4.32 (fathead minnows, wet wt. based, lipid based, quoted, Gobas & Schrap 1990)

2.71, 3.15 (quoted, Hawker 1990)

2.70, 2.70 (fathead minnows, quoted, Opperhuizen & Sijm 1990)

3.23, 3.04 (rainbow trout, quoted, Opperhuizen & Sijm 1990)

Sorption Partition Coefficient, log K$_{OC}$:

    6.69   (organic carbon, calculated-QSAR, Fiedler & Schramm 1990)

    5.47   (DOC, De Voogt et al. 1990)

    7.80   (calculated, Broman et al. 1991)

Half-Lives in the Environment:

    Air:

    Surface water: 11 hours in n-hexane solution to natural sunlight as well as to fluorescent black light (Dobbs & Grant 1979; quoted, Choudhry & Webster 1982); solution photolysis half-life at 1.0 m from a GE Model sunlamp, 30 hours (Nestrick et al. 1980); photolysis half-life of 30 hours in n-hexadecane solution (Mamantov 1984); direct sunlight photolysis half-lives in aquatic bodies at latitude 40°N for various seasons: spring, 56.46 days; summer, 47.33 days; fall, 87.86 days; winter, 155.79 days and averaged over full year, 2393 days (Choudhry & Webster 1985b, 1986); photolysis half-life of 190.97 hours in water-acetonitrile (2:3, v/v) at 313 nm and the calculated midday, midseason direct phototransformation half-lives near water bodies at 40°N latitude: 57 days in spring, 47 days in summer, 88 days in fall and 156 days in winter (Choudhry & Webster 1989); half-life of 81 days in sunlit filtered and sterilized surface water and 2.5 days in surface water of actual pond at 50°N latitude (Friesen et al. 1990); 53.4 and 32.6 hours in native and ^{13}C-labelled congeners respectively in extract from fly ash (Tysklind & Rappe 1991).

    Groundwater:

    Sediment:

    Soil:

    Biota: 17.2 days in fathead minnows (Adams et al. 1986); >336 days in carp (Kuehl et al. 1987); 39 days in whole body of rainbow trout (Muir et al. 1988; quoted, Muir et al. 1990); 27.2 days in lactating cows (Olling et al. 1991).

Environmental Fate Rate Constants or Half-Lives:

    Volatilization:

    Photolysis: 11 hours in n-hexane solution to natural sunlight as well as to fluorescent black light (Dobbs & Grant 1979; quoted, Choudhry & Webster 1982); solution photolysis half-life at 1.0 m from a GE Model sunlamp, 30 hours and surface photolysis half-life on clean soft glass surface under the same conditions, 52.3 hours (Nestrick et al. 1980); 30 hours in n-hexadecane solution (Mamantov 1984); first order rate constant of 1.02×10^{-6} sec^{-1} in water-acetonitrile (2:3, v/v) at 313 nm and calculated direct sunlight photolysis half-lives in aquatic bodies at latitude 40°N for various seasons: spring, 56.46 days; summer, 47.33 days; fall, 87.86 days; winter, 155.79 days and averaged over full year, 2392.68 days (Choudhry & Webster 1985b, 1986); photolysis rate constant of 1.02×10^{-6} sec^{-1}

with a half-life of 190.97 hours in water-acetonitrile solution (2:3, v/v) at 313 nm and the calculated midday, midseason direct sunlight photolysis first-order rate constant in aquatic bodies for various seasons: 1.24×10^2 day^{-1} for spring, 1.48×10^2 day^{-1} for summer, 0.80×10^2 day^{-1} for fall, 0.45×10^2 day^{-1} for winter with half-lives: 57 days in spring, 47 days in summer, 88 days in fall and 156 days in winter (Choudhry & Webster 1989); sunlight photolysis rate constant of 0.28 day^{-1} in filtered and sterilized natural water and 0.019 day^{-1} in (2:32, v/v) distilled water-acetonitrile solution at 50°N (Friesen et al. 1990); half-lives of 53.4 and 32.6 hours in native and ^{13}C-labelled congeners respectively in extract from fly ash (Tysklind & Rappe 1991).

Oxidation: oxidative degradation rate constant of water dissolved PCDD by ozone is 5.46×10^4 liter gram^{-1} minute^{-1} under alkaline condition at pH 10 and 20°C (Palauschek & Scholz 1987).

Hydrolysis:

Biodegradation:

Biotransformation:

Bioconcentration, Uptake (k_1) and Elimination (k_2) Rate Constants:

k_1: 56 day^{-1} (rainbow trout, Muir et al. 1985; quoted, Opperhuizen & Sijm 1990)

k_2: 0.042 day^{-1} (rainbow trout, Muir et al. 1985; quoted, Opperhuizen & Sijm 1990)

k_1: 19 day^{-1} (fathead minnows, Muir et al. 1985; quoted, Adams et al. 1986; Opperhuizen & Sijm 1990)

k_2: 0.040 day^{-1} (fathead minnows, Muir et al. 1985; quoted, Adams et al. 1986; Opperhuizen & Sijm 1990)

k_2: 0.0092 day^{-1} (fathead minnows, quoted, Opperhuizen & Sijm 1990)

k_2: 0.0110 day^{-1} (rainbow trout, quoted, Opperhuizen & Sijm 1990)

Common Name: Octachloro-dibenzo-*p*-dioxin
Synonym: O$_8$CDD, OCDD
Chemical Name: octachloro-dibenzo-*p*-dioxin
CAS Registry No: 3268-87-9
Molecular Formula: Cl$_4$C$_6$OC$_6$Cl$_4$
Molecular Weight: 460.0

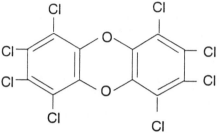

Melting Point (°C):

    330   (Pohland & Yang 1972; quoted, Rordorf 1986,1987,1989)

    318-326 (quoted lit., Pohland & Yang 1972)

    332 (quoted, Shiu et al. 1988)

    325-326 (quoted, Friesen et al. 1985; Shiu et al. 1987)

Boiling Point (°C):

    510   (Rordorf 1987, 1989)

Density (g/cm^3 at 20°C):

Molar Volume (cm^3/mol):

    359.2   (LeBas method, Shiu et al. 1987,1988)

    173.66 (calculated as per Pearlman 1986, De Voogt et al. 1990)

    168.48 (calculated as per Govers & De Voogt 1989, De Voogt et al. 1990)

    1.868   (intrinsic V$_I$/100, Hawker 1990)

    229.6, 229.11 (calculated-liq. density, crystalline volume, Govers et al. 1990)

Molecular Volume (Å^3):

Total Surface Area, TSA (Å^2):

    314.90 (Doucette & Andren 1987, 1988b)

Entropy of Fusion, ΔS_{fus}, cal/mol K (e.u.):

    24.37   (Rordorf 1986,1987,1989)

Fugacity Ratio at 25 °C (assuming ΔS = 13.5 e.u.), F:

    0.00107 (Shiu et al. 1987)

    0.00115 (Shiu et al. 1988)

Water Solubility (g/m^3 or mg/L at 25°C):

    2.0x10^{-6} (Barrie et al. 1983)

    4.0x10^{-7} (^{14}C-labelled, gen. col.-HPLC/LSC, Webster et al. 1983; quoted, Shiu et al.
         1988)

4.0x10^{-7}, 3.17x10^{-7} (20°C, ^{14}C-labeled, gen. col.-HPLC/LSC, calculated, Friesen et al. 1985)

4.0x10^{-7} (20°C, ^{14}C-labelled, gen. col.-HPLC/LSC, Webster et al. 1985)

1.8x10^{-4} (Opperhuizen 1986)

7.36x10^{-6} (gen. col.-HPLC/UV, Shiu et al. 1987)

1.0x10^{-7} (^{14}C-labeled-LSC, Srinivasan & Fogler 1987)

7.4x10^{-8} (gen. col.-GC/ECD, Shiu et al. 1988)

7.4x10^{-8} (extrapolated, gen. col.-GC/ECD, Doucette & Andren 1988a)

4.0x10^{-7} (20°C, quoted, De Voogt et al. 1990)

7.36x10^{-7} (quoted, Paterson et al. 1990)

4.01x10^{-7}-7.46x10^{-8}; 2.90x10^{-7} (quoted; calculated-QSAR, Fiedler & Schramm 1990)

7.0x10^{-8} (quoted, Gobas & Schrap 1990)

6.44x10^{-5} (quoted, Servos et al. 1992a)

Vapor Pressure (Pa at 25°C):

2.40x10^{-5}, 1.8x10^{-5} (quoted, calculated-volatilization rate, Dobbs & Cull 1982)

2.40x10^{-5} (quoted, Cull et al. 1983)

8.70x10^{-6} (20°C, gas saturation, Webster et al. 1985; quoted, Shiu et al. 1988)

1.10x10^{-10} (calculated, Rordorf 1985a,b, 1986a,b, 1987,1989; quoted, Shiu et al. 1987,1988)

2.51x10^{-10} (gas saturation, estimated from vapor pressure vs. temperature plot, Rordorf 1986)

2.77x10^{-7} (subcooled liquid value, GC-RT, Eitzer & Hites 1988)

1.10x10^{-10} (quoted, Paterson et al. 1990)

2.72x10^{-7} (subcooled liquid value, GC-RT, Eitzer & Hites 1989)

1.60x10^{-10} (gas saturation, estimated from extrapolated vapor pressure vs. halogen substituted no. plot, Rordorf et al. 1990)

1.2x10^{-7} (quoted, Servos et al. 1992a)

Henry's Law Constant (Pa m^3/mol):

0.683   (calculated-P/C, Shiu et al. 1987,1988)

Octanol/Water Partition Coefficient, log K_{ow}:

8.31 (Garten & Trablka 1983)

8.50 (Bruggeman et al. 1984; quoted, Opperhuizen 1986)

12.21, 12.60, 12.97; 11.82, 12.72, 13.08 (HPLC-RT, linear; quadratic regressions, Sarna et al. 1984)

10.56, 7.53 (calculated, Doucette 1985; quoted, Shiu et al. 1988)

11.16, 12.72 (HPLC-RT, Sarna et al. 1984; quoted, Shiu et al. 1988)

12.26, 11.35; 11.76, 10.07 (HPLC-RT, linear; quadratic regressions, Webster et al. 1985; quoted, Sijm et al. 1989a)

8.60 (HPLC-RT, Burkhard & Kuehl 1986; quoted, Geyer et al. 1987; Shiu et al. 1988)

7.59, 10.56, 7.33 (gen. col.-GC/ECD, calculated-f-const., TSA, Doucette & Andren 1987)

7.59 (quoted, Geyer et al. 1987)

8.40 (quoted, Gobas et al., 1987)

7.59-8.60 (selected, Shiu et al. 1987)

7.59, 10.56, 10.09, 7.83, 7.65, 7.53, 7.46 (selected exptl. value, calculated-π const, f-const, HPLC-RT, MW, -χ , TSA, Doucette and Andren 1988b)

8.20 (selected, Shiu et al. 1988; quoted, Gobas et al. 1988; Sijm et al. 1989a; Friedler & Schramm 1990; Gobas et al., 1990; Gobas & Schrap 1990; Hawker 1990; Paterson et al. 1990; Loonen et al. 1991; Servos et al. 1992a,b)

7.59 (gen. col.-GC, Shiu et al. 1988)

8.60 (quoted, De Voogt et al. 1990)

10.56 (calculated-QSAR, Fiedler & Schramm 1990)

8.20 (quoted, Hawker 1990)

8.20 (calculated, Broman et al. 1991)

7.59, 10.96 (quoted values: gen. col. method, HPLC-RT, Chessells et al. 1991)

10.28, 10.56 (mean lit. value, calculated-f const., Chessells et al. 1991)

Bioconcentration Factor, log BCF:

-1.10 (rodents, Garten & Trabalka 1983)

3.35 (fathead minnows, steady-state, wet weight, Muir et al. 1985,1986; quoted, Adams et al. 1986)

1.93 (rainbow trout, steady-state, wet weight, Muir et al. 1985,1986)

1.93, 3.35 (trout fry, fathead minnows, Muir et al. 1986)

1.00, 2.05 (human fat, calculated-lipid base, Geyer et al. 1987)

0.903, 1.93 (human fat, calculated-wet wt. based, Geyer et al. 1987)

1.90, 3.10 (guppy: in whole fish, in lipid, Gobas et al. 1987)

6.33 (plant parts, calculated-vapor pressure, Reischl et al. 1989)

5.85 (plant parts, calculated-vapor pressure & HLC, Reischl et al. 1989)

3.35, 1.93 (fathead minnows, rainbow trout, quoted, De Voogt et al. 1990)

3.35, 1.93 (quoted, Hawker 1990)

2.85. 3.97 (guppy: wet weight base, lipid weight base, Gobas & Schrap 1990)

4.35, 5.35 (fathead minnows, quoted, Gobas & Schrap 1990)

3.93, 4.93 (rainbow trout, quoted, Gobas & Schrap 1990)

2.15, 1.49 (rainbow trout, quoted, Opperhuizen & Sijm 1990)

3.34 (fathead minnows, quoted, Opperhuizen & Sijm 1990)

2.85 (predicted for biota held in lake enclosures exposed to water concentration of 0.1 ng/g for 0-10 days, Servos et al. 1992b)

2.15　(caged invertebrates exposed to water concn. 2.0 ng/L for 0-10 days, Servos et al. 1992b)

2.32　(caged unionid clams exposed to water concn. 2.9 ng/L for 0-10 days, Servos et al. 1992b)

2.32　(caged white suckers gill exposed to water concn. 2.9 ng/L for 0-10 days, Servos et al. 1992b)

1.89　(caged white suckers carcass exposed to water concn. 1.1 ng/L for 0-10 days, Servos et al. 1992b)

2.42　(caged invertebrates exposed to water concn. 0.6 ng/L for 14-24 days, Servos et al. 1992b)

2.34　(caged unionid clams exposed to water concn. 0.5 ng/L for 14-24 days, Servos et al. 1992b)

2.75　(caged white suckers gill exposed to water concn. 1.3 ng/L for 14-24 days, Servos et al. 1992b)

2.24　(caged white suckers carcass exposed to water concn. 0.4 ng/L for 14-24 days, Servos et al. 1992b)

3.97　(caged white suckers gill exposed to water concn. 2.8 ng/L for 0-104 days, Servos et al. 1992b)

3.95　(caged white suckers carcass exposed to water concn. 2.7 ng/L ffor 0-104 days, Servos et al. 1992b)

Sorption Partition Coefficient, log K_{OC}:
　　7.08　(organic carbon, calculated-QSAR, Fiedler & Schramm 1990)
　　5.92　(DOC, De Voogt et al. 1990)
　　7.90　(calculated, Broman et al. 1991)

Half-Lives in the Environment:
　　Air: estimated reaction rate constant, 0.0015 hour[-1] (Paterson et al. 1990); photodegradation half-life in a rotary photo-reactor adsorbed to clean silica gels by filtered < 290 nm of light, 270 hours (Koester & Hites 1992).
　　Surface water: solution photolysis half-life at 1.0 m from a GE Model RS sunlamp, 24.3 hours (Nestrick et la. 1980); direct sunlight photolysis half-lives in aquatic bodies at latitude 40°N for various seasons: spring, 20.53 days; summer, 17.85 days; fall, 31.26 days; winter, 50.45 days and averaged over full year, 863.22 days (Choudhry & Webster 1986); rate constant of 1.064×10^6 sec[-1] in water-acetonitrile (2:3, v/v) under direct sunlight (Choudhry & Webster 1985a,1986); photolysis half-life of 183.95 hours in water-acetonitrile solution (2:3, v/v) at 313 nm and the calculated midday, midseason direct photolysis half-lives near water bodies at 40°N latitude: 21 days in spring, 18 days in summer, 31 days in fall and 50 days in winter (Choudhry & Webster 1989); half-life of 4.0 days in the water column of an experimental lake in northwestern Ontario (Servos et

al. 1989); calculated transformation rate constant in simulated lake enclosure, 1.6×10^{-3} hour^{-1} (Servos et al. 1992a).

Groundwater:

Sediment: estimated reaction rate constant, 5×10^{-3} to 1.0×10^{-6} hour^{-1} (Paterson et al. 1990); calculated half-life, 10 years with a transformation rate constant of 7.9×10^{-6} hour^{-1} (Servos et al. 1992a).

Soil: undergoes photoreduction on soil surfaces to lower chlorinated congeners (Kieatiwong et al. 1990); estimated reaction rate constant, 7.0×10^{7} hour^{-1} (Paterson et al. 1990).

Biota: half-life of elimination from rat, 21 days (Norback et al. 1975; quoted, Birbaum 1985); elimination half-life of 13.9 days for fathead minnows (Muir et al. 1985, Adams et al. 1986); elimination half-life of 5-13 days for both rainbow trout and fathead minnows (Muir et al. 1986); mean biological half-life in rainbow trout, about 15 days (Niimi 1986); 15 days in rainbow trout (Niimi & Oliver 1986; quoted, Muir et al. 1986,1990).

Environmental Fate Rate Constants or Half-Lives:

Volatilization:

Photolysis: photolytic half-life by both natural sunlight and fluorescent black light in hexane solution, 16 hours (Dobbs & Grant 1979; quoted, Choudry & Webster 1982); solution photolysis half-life at 1.0 m from a GE Model RS sunlamp, 24.3 hours and surface photolysis on clean soft glass surface under the same condition is 815 hours (Nestrick et al. 1980); half-life photolysis in hexadecane solution, 24.3 hours (Mamantov 1984); first-order rate constnat of photolysis in water acetonenitrile (2:3, v/v) at 313 nm was 1.06×10^{-6} sec^{-1} and calculated direct sunlight photolysis half-lives in aquatic bodies at latitude 40°N for various seasons: spring, 20.53 days; summer, 17.85 days; fall, 31.26 days; winter, 50.45 days; and averaged over full year, 863.22 days (Choudhry & Webster 1986; quoted, Muto et al. 1991); photolysis rate constant of 1.06×10^{-6} sec^{-1} with a half-life of 183.95 hours in water-acetonitrile solution (2:3, v/v) at 313 nm and the calculated midday, midseason direct sunlight photolysis first-order rate constant in aquatic bodies for various seasons: 3.45×10^{2} day^{-1} for spring, 3.97×10^{2} day^{-1} for summer, 2.27×10^{2} day^{-1} for fall, 1.40×10^{2} day^{-1} for winter with half-lives: 21 days in spring, 18 days in summer, 31 days in fall and 50 days in winter (Choudhry & Webster 1989); calculated photolysis decay rate constants when irradiated with UV light at 254 nm for fly ash suspensions in distilled water, 1.77×10^{5} sec^{-1} and 4.62×10^{5} sec^{-1} for fly ash suspensions in water-acetonitrile solution (2:3, v/v) is 1.42×10^{5} sec^{-1} and for fly ash suspensions in water-acetonitrile solution (2:3, v/v) with ozone is 2.74×10^{5} sec^{-1} (Muto et al. 1991); photolytic half-lives in extract of fly ash and in tetradecane solution: 37.3 hours-native congener, 29.6 hours for ^{13}C-labelled congener (Tysklind & Rappe 1991).

Oxidation: oxidative degradation rate of water dissolved OCDD by ozone is 1.51×10^4 liter gram^{-1} minute^{-1} under alkaline condition at pH 10 and 20°C (Palauschek & Scholz 1987).

Hydrolysis:

Biodegradation:

Biotransformation:

Bioconcentration, Uptake (k_1) and Elimination (k_2) Rate Constants:

k_1: 142 day^{-1} (fathead minnows, Muir et al. 1985; quoted, Adams et al. 1986)

k_2: 0.05 day^{-1} (fathead minnows, Muir et al. 1985, quoted, Adams et al. 1986)

k_1: 11.0 day^{-1} (rainbow trout, Muir et al. 1985)

k_2: 0.12 day^{-1} (rainbow trout, Muir et al. 1985)

k_1: 5.0-17.0 day^{-1} (rainbow trout, Muir et al. 1986; quoted, Opperhuizen & Sijm 1990)

k_2: 0.103-0.142 day^{-1} (rainbow trout, Muir et al. 1986; quoted, Opperhuizen & Sijm 1990)

k_1: 142 day^{-1} (fathead minnows, Muir et al. 1986; quoted, Opperhuizen & Sijm 1990)

k_2: 0.053 day^{-1} (fathead minnows, Muir et al. 1986; quoted, Opperhuizen & Sijm 1990)

k_1: 984 day^{-1} (guppy, Gobas & Schrap 1990)

k_2: 1.4 day^{-1} (guppy, Gobas & Schrap 1990)

k_2: 0.046 day^{-1} (rainbow trout, quoted, Opperhuizen & Sijm 1990)

k_1: 60 day^{-1} (filter-feeder, Servos et al. 1992b)

k_2: 0.12 day^{-1} (filter-feeder, Servos et al. 1992b)

k_1: 30 day^{-1} (small fish, estimated from Muir et al. 1985 and Servos et at. 1989, Servos et al. 1992b)

k_2: 0.08 day^{-1} (small fish, estimated from Muir et al. 1985 and Servos et al. 1989, Servos et al. 1992b)

3.2 Summary Tables and QSPR Plots

Table 3.1 Physical-chemical properties of some chlorinated dioxins at 25 °C

Compounds	CAS no.	formula	chlorine no. n	MW g/mol	M.P. °C	B.P. °C	fugacity ratio, F	$V_M^{[a]}$ cm³/mol (LeBas)	$V^{[b]}$ cm³/mol from ρ	$V_I/100^{[c]}$ cm³/mol intrinsic	TSA[d] Å²
Dibenzo-p-dioxin	262-23-4	$C_{12}H_8O_2$	0	184.0	123	283.5	0.1073	192.0	143.82		186.5
1-Chloro-	39227-53-7	$C_{12}H_7O_2Cl$	1	218.5	105.5	315.5	0.160	212.9	157.41		
2-Chloro-	39227-54-8	$C_{12}H_7O_2Cl$	1	218.5	89	316	0.233	212.9	154.85		203.6
2,3-Dichloro-	29446-15-9	$C_{12}H_6O_2Cl_2$	2	253.0	164	358	0.0422	233.8	164.07		
2,7-Dichloro-	33857-26-0	$C_{12}H_6O_2Cl_2$	2	253.0	210	373.5	0.0148	233.8	165.88		
2,8-Dichloro-	38964-22-6	$C_{12}H_6O_2Cl_2$	2	253.0	151		0.0567	233.8	165.88		
1,2,4-Trichloro-	39227-58-2	$C_{12}H_5O_2Cl_3$	3	287.5	129	375	0.0936	254.7	179.66		
1,2,3,4-Tetrachloro-	30756-58-8	$C_{12}H_4O_2Cl_4$	4	322.0	190	419	0.0233	275.6	186.71		250.3
1,2,3,7-Tetrachloro-	67028-18-6	$C_{12}H_4O_2Cl_4$	4	322.0	172	438.3	0.0352	275.6	186.52		
1,3,6,8-Tetrachloro-	30746-58-8	$C_{12}H_4O_2Cl_4$	4	322.0	219	438.3	0.0121	275.6	192.34	1.508	
2,3,7,8-Tetrachloro-	1746-01-6	$C_{12}H_4O_2Cl_4$	4	322.0	305	446.5	0.00170	275.6	184.32		
1,2,3,4,7-Pentachloro-	39227-61-7	$C_{12}H_3O_2Cl_5$	5	356.4	195	464.7	0.0208	296.5	197.74		
1,2,3,4,7,8-Hexachloro-	39227-26-8	$C_{12}H_2O_2Cl_6$	6	391.0	273	487.7	0.00352	317.4	206.96	1.688	
1,2,3,4,6,7,8-Heptachloro-	35822-46-9	$C_{12}HO_2Cl_7$	7	425.2	265	507.2	0.00422	338.3	218.3	1.778	
Octachloro-	3268-87-9	$C_{12}O_2Cl_8$	8	460.0	322	510	0.00115	359.2	229.6	1.868	315

Note:
(a) Shiu et al. 1988
(b) Govers et al. 1990
(c) Hawker 1990
(d) Doucette & Andren 1987

429

Table 3.2   Selected physical-chemical properties of some chlorinated dioxins at 25 °C

| COMPOUND | Selected properties | | | | | log K_{ow} | Henry's law const., H |
| | vapor pressure | | solubility | | | | |
	P^s Pa	P_L Pa (liquid)	S mg/m^3	C^s mmol/m^3	C_L mmol/m^3 (liquid)		Pa m^3/mol
Dibenzo-p-dioxin	0.055	0.512	865	4.701	43.80	4.30	11.70
1-CDD	0.012	0.075	417	1.908	11.94	4.75	6.288
2-CDD	0.017	0.0730	295	1.350	5.799	5.00	12.59
2,3-DCDD	0.00039	0.00924	14.9	0.059	1.396	5.60	6.622
2,7-DCDD	0.00012	0.00811	3.75	0.0148	1.002	5.75	8.096
2,8-DCDD	0.00014	0.00247	16.7	0.0660	1.164	5.60	2.121
1,2,4-T$_3$CDD	0.0001	0.00107	8.41	0.0293	0.312	6.35	3.419
1,2,3,4-TCDD	6.40×10^{-6}	2.75×10^{-4}	0.55	0.0017	0.0732	6.60	3.747
1,2,3,7-TCDD	1.00×10^{-6}	2.84×10^{-5}	0.42	0.0013	0.0371	6.90	0.767
1,3,6,8-TCDD	7.00×10^{-7}	5.81×10^{-5}	0.32	0.000993	0.0824	7.10	0.704
2,3,7,8-TCDD	2.00×10^{-7}	1.18×10^{-4}	0.0193	0.00006	0.0352	6.80	3.337
1,2,3,4,7-PCDD	8.80×10^{-8}	4.23×10^{-6}	0.118	0.000331	0.0159	7.40	0.266
1,2,3,4,7,8-H$_6$CDD	5.10×10^{-9}	1.45×10^{-6}	0.00442	1.13×10^{-5}	0.00321	7.80	1.084
1,2,3,4,6,7,8-H$_7$CDD	7.50×10^{-10}	1.77×10^{-7}	0.0024	5.64×10^{-6}	0.00133	8.00	1.273
OCDD	1.10×10^{-10}	9.53×10^{-7}	0.000074	1.61×10^{-7}	0.00014	8.20	0.684

Table 3.3 Suggested half-life classes of polychlorinated dibenzo-p-dioxins
in various environmental compartments

Compounds	Air class	Water class	Soil class	Sediment class
	3	3	6	7
2-Chloro-	4	4	7	8
2,7-Dichloro-	4	4	7	8
2,8-Dichloro-	4	4	7	8
1,2,4-Trichloro-	4	4	7	8
1,2,3,4-TCDD	4	5	8	9
2,3,7,8-TCDD	4	5	8	9
1,2,3,4,7-PCDD	5	5	8	9
1,2,3,4,7,8-H_6CDD	5	6	9	9
1,2,3,4,6,7,8-H_7CDD	5	6	9	9
OCDD	5	7	9	9

where,

Class	Mean half-life (hours)	Range (hours)
1	5	< 10
2	17 ($\sim$ 1 day)	10-30
3	55 ($\sim$ 2 days)	30-100
4	170 ($\sim$ 1 week)	100-300
5	550 ($\sim$ 3 weeks)	300-1,000
6	1700 ($\sim$ 2 months)	1,000-3,000
7	5500 ($\sim$ 8 months)	3,000-10,000
8	17000 ($\sim$ 2 years)	10,000-30,000
9	55888 ($\sim$ 6 years)	> 30,000

431

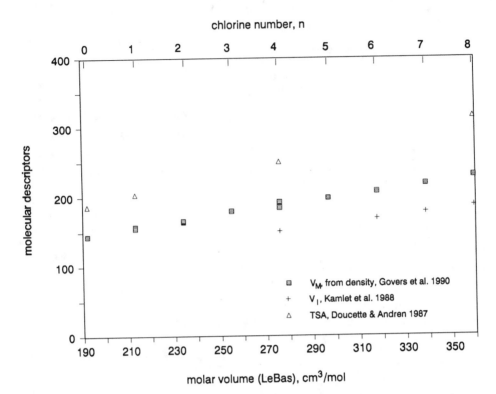

Figure 3.1 Molecular descriptors versus LeBas molar volume and chlorine number.

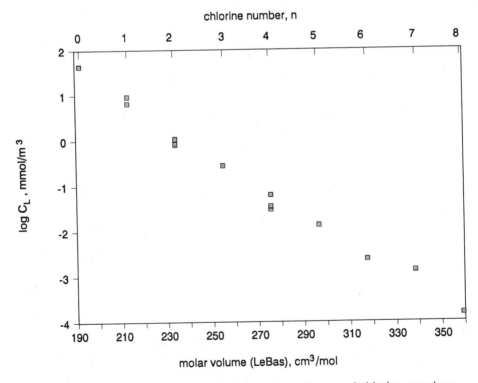

Figure 3.2 Plot of log C_L (liquid solubility) vs molar volume and chlorine number.

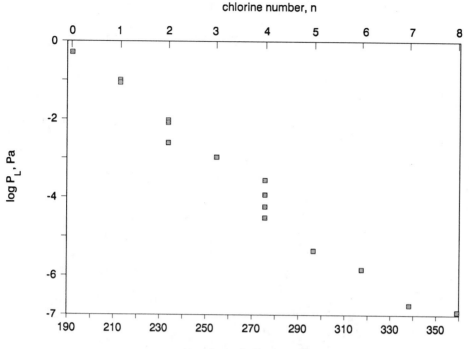

Figure 3.3   Plot of log P_L (liquid vapor pressure) versus molar volume   and chlorine number.

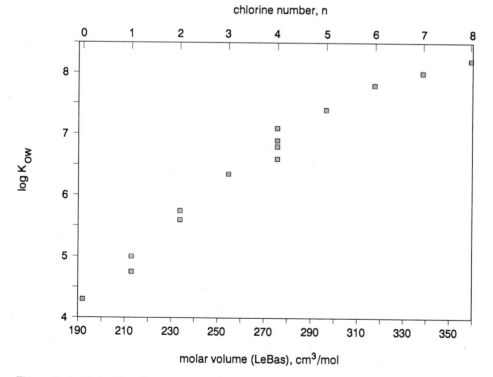

Figure 3.4   Plot of log K_{OW} versus LeBas molar volume and chlorine number.

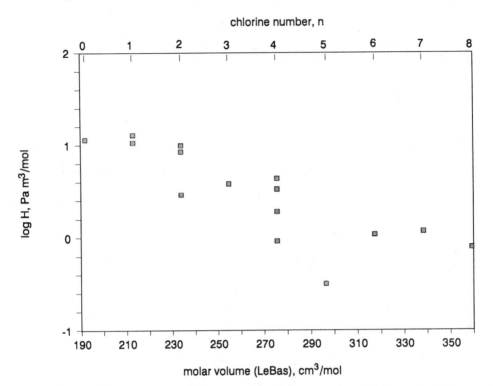

Figure 3.5   Plot of log H (Henry's law constant) versus molar volume   and chlorine number.

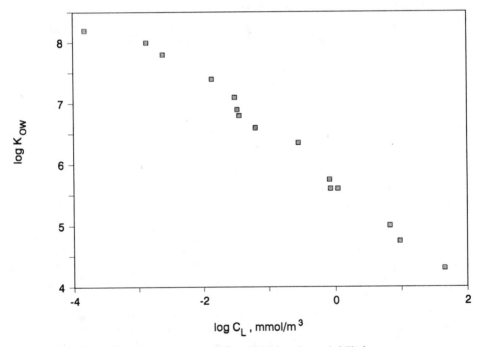

Figure 3.6   Plot of log K_{OW} versus Log C_L (liquid molar solubility).

3.3 Illustrative Fugacity Calculations: Level I, II and III

Chemical name: Dibenzo-p-dioxin

Level I calculation: (six compartment model)

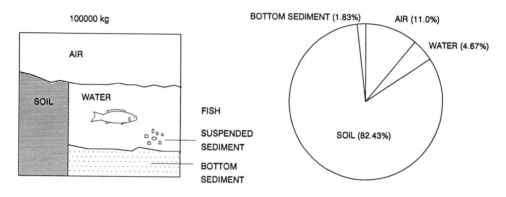

Distribution of mass

physical-chemical properties:

MW: 184 g/mol

M.P.: 123 °C

Fugacity ratio: 0.1073

vapor pressure: 0.055 Pa

solubility: 0.865 g/m^3

log K_{OW}: 4.30

Compartment	Z	Concentration			Amount	Amount
	mol/m3 Pa	mol/m3	mg/L (or g/m3)	ug/g	kg	%
Air	4.034E-04	5.983E-10	1.101E-07	9.287E-05	11009	11.009
Water	8.547E-02	1.268E-07	2.333E-05	2.333E-05	4665.08	4.6651
Soil	3.356E+01	4.978E-05	9.159E-03	3.816E-03	82432	82.432
Biota (fish)	8.527E+01	1.265E-04	2.327E-02	2.327E-02	4.6540	4.65E-03
Suspended sediment	2.098E+02	3.111E-04	5.724E-02	3.816E-02	57.245	5.72E-02
Bottom sediment	6.713E+01	9.956E-05	1.832E-02	7.633E-03	1831.83	1.8318
Total					100000	100

f = 1.483E-06 Pa

Chemical name: Dibenzo-p-dioxin

Level II calculation: (six compartment model)

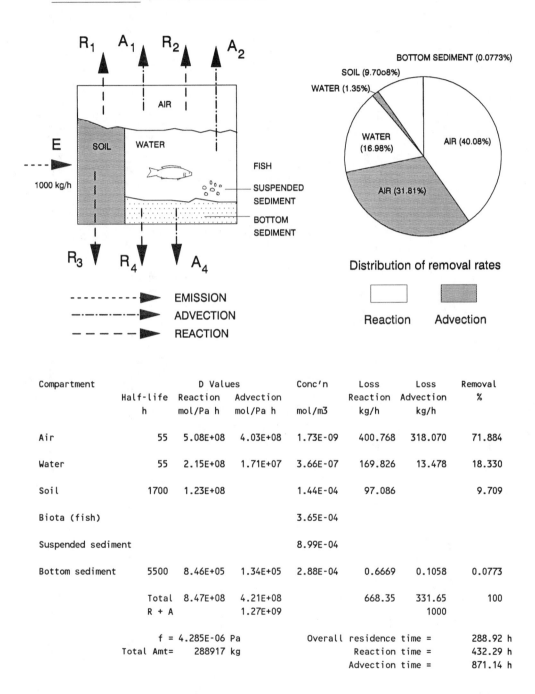

Distribution of removal rates

Reaction   Advection

Compartment	Half-life h	D Values Reaction mol/Pa h	Advection mol/Pa h	Conc'n mol/m3	Loss Reaction kg/h	Loss Advection kg/h	Removal %
Air	55	5.08E+08	4.03E+08	1.73E-09	400.768	318.070	71.884
Water	55	2.15E+08	1.71E+07	3.66E-07	169.826	13.478	18.330
Soil	1700	1.23E+08		1.44E-04	97.086		9.709
Biota (fish)				3.65E-04			
Suspended sediment				8.99E-04			
Bottom sediment	5500	8.46E+05	1.34E+05	2.88E-04	0.6669	0.1058	0.0773
Total		8.47E+08	4.21E+08		668.35	331.65	100
R + A			1.27E+09			1000	

f = 4.285E-06 Pa

Total Amt=   288917 kg

Overall residence time =   288.92 h
Reaction time =   432.29 h
Advection time =   871.14 h

437

Fugacity Level III calculations: (four compartment model)

Chemical name: Dibenzo-p-dioxin

Four-compartment fugacity diagram showing AIR (1), WATER (2), SOIL (3), and SEDIMENT (4) with emission (E), reaction (R), advection (A) and transfer D-value arrows (E_1, E_2, E_3, R_1, R_2, R_3, R_4, A_1, A_2, A_4; D_{12}, D_{21}, D_{13}, D_{31}, D_{24}, D_{42}, D_{32}).

Transfer D values shown: 1.370E07, 7.995E05, 1.824E06, 1.382E07, 4.148E05, 2.197E05, 1.134E06.

Phase Properties and Rates:

Compartment	Bulk Z mol/m3 Pa	Half-life h	D Values Reaction mol/Pa h	Advection mol/Pa h
Air (1)	4.035E-04	55	5.08E+08	4.04E+08
Water (2)	8.661E-02	55	2.18E+08	1.73E+07
Soil (3)	1.681E+01	1700	1.23E+08	
Sediment (4)	1.349E+01	5500	8.50E+05	1.35E+05

E(1,2,3): 4.04E+08, 1.73E+07, 1.35E+05

	E(1)=1000	E(2)=1000	E(3)=1000	E(1,2,3)
Overall residence time =	49.73	97.34	2429.83	302.02 h
Reaction time =	88.24	107.40	2437.42	425.54 h
Advection time =	113.95	1039.54	780365.85	1040.55 h

EMISSION (E)
REACTION (R)
ADVECTION (A)
TRANSFER D VALUE mol/Pa h

Phase Properties, Compositions, Transport and Transformation Rates:

Emission, kg/h | Fugacity, Pa | Concentration, g/m3 | Amounts, kg | Total Amount, kg

E(1)	E(2)	E(3)	f(1)	f(2)	f(3)	f(4)	C(1)	C(2)	C(3)	C(4)	m(1)	m(2)	m(3)	m(4)	Total Amount, kg
1000	0	0	5.864E-06	3.240E-07	8.588E-08	3.050E-08	4.354E-07	5.163E-06	2.656E-04	7.573E-04	4.354E+04	1.033E+04	4.780E+03	3.787E+04	4.973E+04
0	1000	0	3.212E-07	2.174E-05	4.703E-09	2.047E-05	2.384E-08	3.464E-04	1.455E-05	5.082E-04	2.384E+03	6.929E+04	2.618E+02	2.541E+04	9.734E+04
0	0	1000	3.872E-08	7.449E-08	4.364E-05	7.014E-08	2.875E-09	1.187E-06	1.350E-01	1.741E-04	2.875E+02	2.374E+02	2.429E+06	8.707E+01	2.430E+06
600	300	100	3.619E-06	6.724E-06	6.724E-06	6.330E-06	2.687E-07	1.071E-04	1.366E-02	1.572E-02	2.687E+04	2.143E+04	2.459E+05	7.859E+03	3.020E+05

Emission, kg/h | Loss, Reaction, kg/h | Loss, Advection, kg/h | Intermedia Rate of Transport, kg/h

E(1)	E(2)	E(3)	R(1)	R(2)	R(3)	R(4)	A(1)	A(2)	A(4)	T12 air-water	T13 air-soil	T21 water-air	T31 soil-air	T32 soil-water	T24 water-sed	T42 sed-water
1000	0	0	5.486E+02	1.301E+01	1.95E+00	4.771E-02	4.354E+02	1.033E+00	7.573E-03	1.491E+01	1.968E+00	8.169E+01	1.263E-02	6.555E-03	6.762E-02	1.233E-02
0	1000	0	3.004E+01	8.730E+02	1.07E-01	3.202E+00	2.384E+01	6.929E+01	5.082E+00	8.165E-01	1.078E-01	5.481E+01	6.919E-04	3.590E-04	4.537E+00	8.275E-01
0	0	1000	3.622E+00	2.992E+00	9.90E+02	1.097E-02	2.875E+00	2.374E+00	1.741E-03	9.843E-02	1.299E+00	1.878E-01	6.420E+00	3.331E+00	1.555E+00	2.836E-03
600	300	100	3.385E+02	2.700E+02	1.00E+02	9.902E-01	2.687E+02	2.143E+01	1.572E-01	9.200E+00	1.214E+01	1.695E+01	6.498E-01	3.371E-01	1.403E+00	2.559E-01

Level III Distribution

Chemical name: Dibenzo-p-dioxin

Distribution of mass

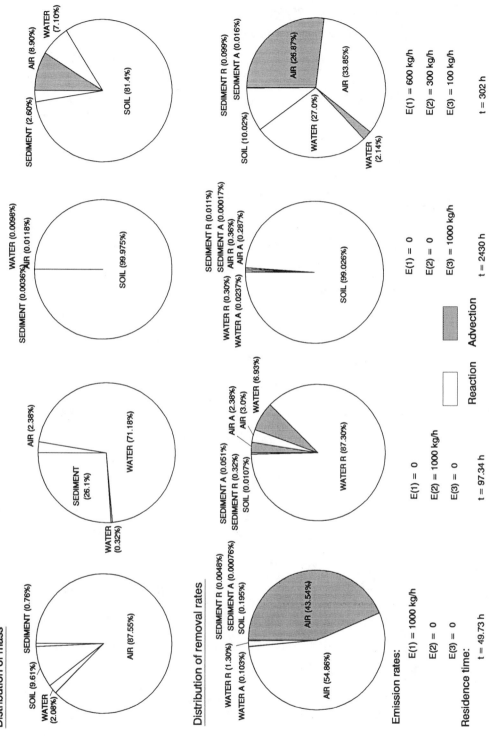

Distribution of removal rates

Emission rates:

E(1) = 1000 kg/h	E(1) = 0	E(1) = 0	E(1) = 600 kg/h
E(2) = 0	E(2) = 1000 kg/h	E(2) = 0	E(2) = 300 kg/h
E(3) = 0	E(3) = 0	E(3) = 1000 kg/h	E(3) = 100 kg/h

Residence time:

| t = 49.73 h | t = 97.34 h | t = 2430 h | t = 302 h |

439

Chemical name: 2-Chlorodibenzo-p-dioxin

Level I calculation: (six compartment model)

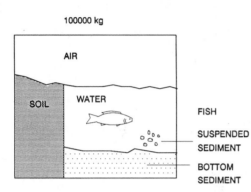

100000 kg

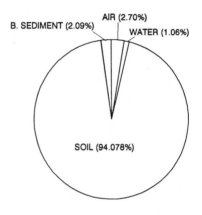

AIR (2.70%)
B. SEDIMENT (2.09%)
WATER (1.06%)
SOIL (94.078%)

Distribution of mass

physical-chemical properties:

MW: 218.5 g/mol

M.P.: 89.0°C

Fugacity ratio: 0.233

vapor pressure: 0.017 Pa

solubility: 0.295 g/m^3

log K_{ow}: 5.0

Compartment	Z	Concentration			Amount	Amount
	mol/m3 Pa	mol/m3	mg/L (or g/m3)	ug/g	kg	%
Air	4.034E-04	1.235E-10	2.698E-08	2.276E-05	2698.08	2.698
Water	7.942E-02	2.431E-08	5.312E-06	5.312E-06	1062.31	1.0623
Soil	1.563E+02	4.784E-05	1.045E-02	4.355E-03	94078	94.078
Biota (fish)	3.971E+02	1.215E-04	2.656E-02	2.656E-02	5.3116	5.31E-03
Suspended sediment	9.768E+02	2.990E-04	6.533E-02	4.355E-02	65.332	6.53E-02
Bottom sediment	3.126E+02	9.568E-05	2.091E-02	8.711E-03	2090.63	2.0906
Total					100000	100

f = 3.061E-07 Pa

Chemical name: 2-Chlorodibenzo-p-dioxin

Level II calculation: (six compartment model)

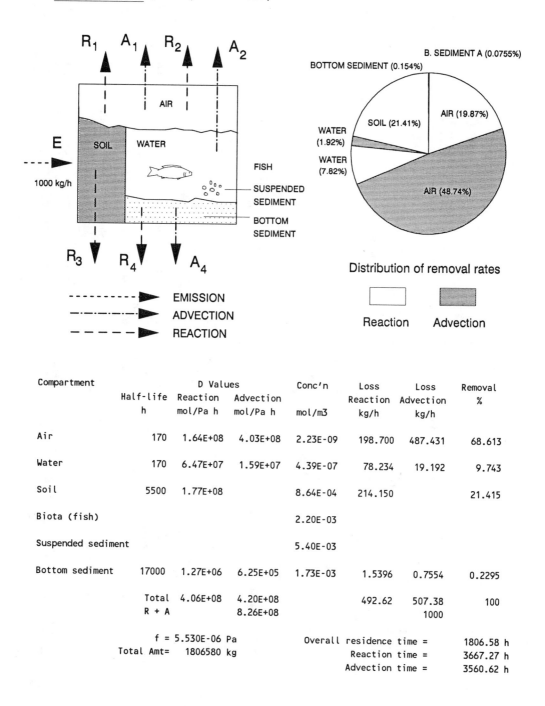

Distribution of removal rates

Reaction   Advection

Compartment	Half-life h	D Values Reaction mol/Pa h	Advection mol/Pa h	Conc'n mol/m3	Loss Reaction kg/h	Loss Advection kg/h	Removal %
Air	170	1.64E+08	4.03E+08	2.23E-09	198.700	487.431	68.613
Water	170	6.47E+07	1.59E+07	4.39E-07	78.234	19.192	9.743
Soil	5500	1.77E+08		8.64E-04	214.150		21.415
Biota (fish)				2.20E-03			
Suspended sediment				5.40E-03			
Bottom sediment	17000	1.27E+06	6.25E+05	1.73E-03	1.5396	0.7554	0.2295
	Total	4.06E+08	4.20E+08		492.62	507.38	100
	R + A		8.26E+08			1000	

f = 5.530E-06 Pa
Total Amt= 1806580 kg

Overall residence time =   1806.58 h
Reaction time =   3667.27 h
Advection time =   3560.62 h

Fugacity Level III calculations: (four compartment model)

Chemical name: 2-Chlorodibenzo-p-dioxin

Phase Properties and Rates:

Compartment	Bulk Z mol/m3 Pa	Half-life h	D Values Reaction mol/Pa h	Advection mol/Pa h
Air (1)	4.041E-04	170	1.65E+08	4.04E+08 h
Water (2)	8.470E-02	170	6.91E+07	1.69E+07 h
Soil (3)	7.817E+01	5500	1.77E+08	
Sediment (4)	6.258E+01	17000	1.28E+06	6.26E+05 h

	$E(1)=1000$	$E(2)=1000$	$E(3)=1000$	$E(1,2,3)$
Overall residence time =	131.09	760.18	7881.78	1094.89 h
Reaction time =	431.34	1036.31	7912.16	2180.91 h
Advection time =	188.33	2852.97	2053385	2198.73 h

EMISSION (E) ----
REACTION (R) ---·
ADVECTION (A) ----
TRANSFER D VALUE mol/Pa h

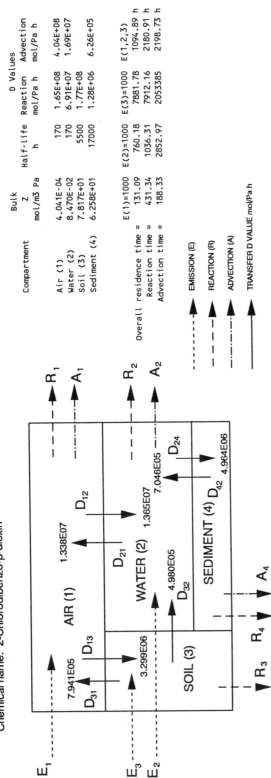

Amounts, kg:

$m(1)$	$m(2)$	$m(3)$	$m(4)$	Total Amount, kg
6.920E+04	3.848E+03	4.451E+04	1.354E+04	1.311E+05
8.987E+03	1.650E+05	5.781E+03	1.679E+03	7.602E+05
3.328E+03	4.772E+02	7.879E+06	1.679E+03	7.882E+06
4.425E+04	5.185E+04	8.164E+05	1.824E+05	1.095E+06

Intermedia Rate of Transport, kg/h:

T12 air-water	T21 water-air	T13 air-soil	T31 soil-air	T32 soil-water	T24 water-sed	T42 sed-water
2.338E+01	3.039E+00	3.037E+00	2.512E-02	1.575E-02	1.128E+00	3.049E-01
3.037E+01	1.303E+02	7.337E-01	3.263E-01	2.046E-03	4.834E+01	1.307E+01
1.124E-01	3.768E-01	2.717E-02	4.447E+00	2.789E+00	1.398E+00	3.781E-02
1.495E+02	4.094E+01	3.612E-01	4.607E-01	2.890E-01	1.519E+01	4.108E+00

Phase Properties, Compositions, Transport and Transformation Rates:

Emission, kg/h — Fugacity, Pa & Concentration, g/m3:

E(1)	E(2)	E(3)	f(1)	f(2)	f(3)	f(4)	C(1)	C(2)	C(3)	C(4)
1000	0	0	7.837E-06	1.040E-06	1.448E-07	1.980E-06	6.920E-07	1.924E-05	2.473E-03	2.708E-02
0	1000	0	1.018E-06	4.457E-05	1.880E-08	8.490E-05	8.987E-08	8.249E-04	3.212E-04	1.161E+00
0	0	1000	3.769E-08	1.289E-07	2.563E-05	2.456E-07	3.328E-09	2.386E-06	4.377E-01	3.358E-03
600	300	100	5.011E-06	1.401E-05	2.655E-06	2.668E-05	4.425E-07	2.592E-04	4.535E-02	3.648E-01

Emission, kg/h — Loss, Reaction, kg/h & Loss, Advection, kg/h:

E(1)	E(2)	E(3)	R(1)	R(2)	R(3)	R(4)	A(1)	A(2)	A(4)
1000	0	0	2.821E+02	1.569E+01	5.61E+01	5.520E-01	6.920E+02	3.848E+00	2.708E-01
0	1000	0	3.663E+01	6.725E+02	7.28E-01	2.366E+01	8.987E+01	1.650E+02	1.161E+01
0	0	1000	1.357E+00	1.945E+00	9.93E+02	6.844E-02	3.328E+00	4.772E-01	3.358E-02
600	300	100	1.804E+02	2.114E+02	1.03E+02	7.436E+00	4.425E+02	5.185E+01	3.648E+00

Level III Distribution

Chemical name: 2-Chlorodibenzo-p-dioxin

Distribution of mass

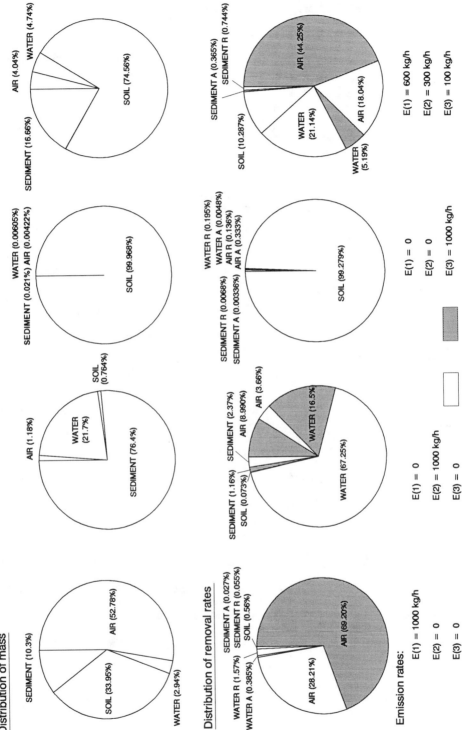

Distribution of removal rates

Reaction | Advection

Emission rates:

E(1) = 1000 kg/h
E(2) = 0
E(3) = 0

Residence time:

t = 131 h

E(1) = 0
E(2) = 1000 kg/h
E(3) = 0

t = 760 h

E(1) = 0
E(2) = 0
E(3) = 1000 kg/h

t = 7881 h

E(1) = 600 kg/h
E(2) = 300 kg/h
E(3) = 100 kg/h

t = 1095 h

Chemical name: 2,7-Dichlorodibenzo-p-dioxin

Level I calculation: (six compartment model)

100000 kg

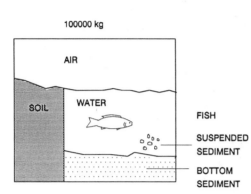

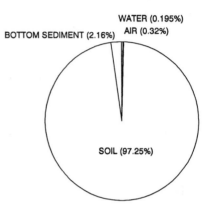

Distribution of mass

physical-chemical properties:

MW: 253 g/mol

M.P.: 210 $^\circ$C

Fugacity ratio: 0.0148

vapor pressure: 0.00012 Pa

solubility: 0.00375 g/m^3

log K_{OW}: 5.75

Compartment	Z	Concentration			Amount	Amount
	mol/m3 Pa	mol/m3	mg/L (or g/m3)	ug/g	kg	%
Air	4.034E-04	1.260E-11	3.189E-09	2.690E-06	318.90	0.319
Water	1.235E-01	3.859E-09	9.764E-07	9.764E-07	195.28	0.1953
Soil	1.367E+03	4.271E-05	1.081E-02	4.502E-03	97252	97.252
Biota (fish)	3.473E+03	1.085E-04	2.745E-02	2.745E-02	5.4907	5.49E-03
Suspended sediment	8.543E+03	2.669E-04	6.754E-02	4.502E-02	67.536	6.75E-02
Bottom sediment	2.734E+03	8.542E-05	2.161E-02	9.005E-03	2161.15	2.1611
	Total				100000	100

f = 3.124E-08 Pa

Chemical name: 2,7-Dichlorodibenzo-p-dioxin

Level II calculation: (six compartment model)

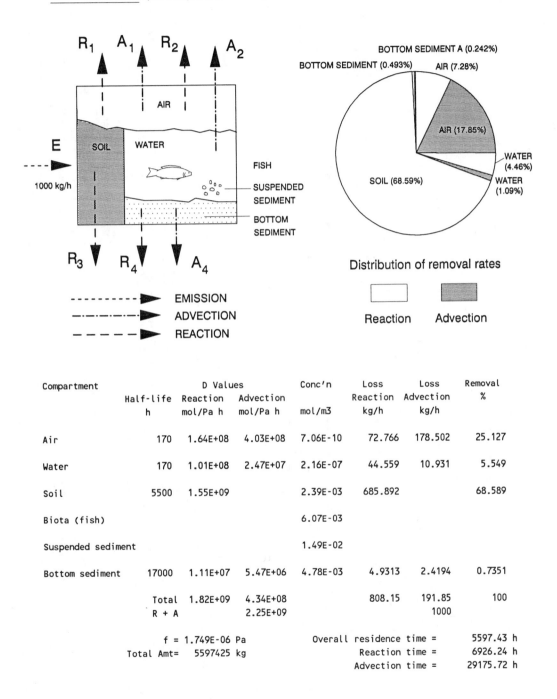

Distribution of removal rates

Reaction   Advection

Compartment	Half-life h	D Values Reaction mol/Pa h	D Values Advection mol/Pa h	Conc'n mol/m3	Loss Reaction kg/h	Loss Advection kg/h	Removal %
Air	170	1.64E+08	4.03E+08	7.06E-10	72.766	178.502	25.127
Water	170	1.01E+08	2.47E+07	2.16E-07	44.559	10.931	5.549
Soil	5500	1.55E+09		2.39E-03	685.892		68.589
Biota (fish)				6.07E-03			
Suspended sediment				1.49E-02			
Bottom sediment	17000	1.11E+07	5.47E+06	4.78E-03	4.9313	2.4194	0.7351
Total		1.82E+09	4.34E+08		808.15	191.85	100
R + A			2.25E+09			1000	

f = 1.749E-06 Pa

Total Amt= 5597425 kg

Overall residence time = 5597.43 h
Reaction time = 6926.24 h
Advection time = 29175.72 h

445

Fugacity Level III calculations: (four compartment model)

Chemical name: 2,7-Dichlorodibenzo-p-dioxin

Phase Properties and Rates:

Compartment	Bulk Z mol/m3 Pa	Half-life h	D Values Reaction mol/Pa h	Advection mol/Pa h
Air (1)	4.094E-04	170	1.67E+08	4.09E+08
Water (2)	1.697E-01	170	1.38E+08	3.39E+07
Soil (3)	6.835E+02	5500	1.55E+09	
Sediment (4)	5.469E+02	17000	1.11E+07	5.47E+06

	E(1)=1000	E(2)=1000	E(3)=1000	E(1,2,3)
Overall residence time =	373.45	2583.02	7926.29	1791.60 h
Reaction time =	1154.52	3441.67	7931.45	3450.90 h
Advection time =	552.01	10353.23	12192602	3726.06 h

Legend:
- EMISSION (E)
- REACTION (R)
- ADVECTION (A)
- TRANSFER D VALUE mol/Pa h

Diagram (four compartment model):

AIR (1) — E_1, R_1, A_1
WATER (2) — E_2, R_2, A_2
SOIL (3) — E_3, R_3
SEDIMENT (4) — A_4, R_4

D values:
D_{13} 8.335E05, D_{31} 1.807E07, D_{12} 1.520E07, D_{21} 1.712E07, D_{24} 5.591E06, D_{32} 1.786E06, D_{42} 4.284E07

Phase Properties, Compositions, Transport and Transformation Rates:

Emission, kg/h

E(1)	E(2)	E(3)
1000	0	0
0	1000	0
0	0	1000
600	300	100

Fugacity, Pa

f(1)	f(2)	f(3)	f(4)
6.477E-06	5.056E-07	7.535E-08	9.754E-07
4.485E-07	1.804E-05	5.218E-09	3.480E-05
3.992E-09	2.102E-08	2.545E-06	4.055E-08
4.021E-06	5.717E-06	3.013E-07	1.103E-05

Concentration, g/m3

C(1)	C(2)	C(3)	C(4)
6.708E-07	2.171E-05	1.303E-02	1.350E-01
4.646E-08	7.744E-04	9.024E-04	4.814E+00
4.135E-10	9.024E-07	4.402E-01	5.610E-01
4.165E-07	2.454E-04	5.211E-02	1.526E+00

Amounts, kg

m(1)	m(2)	m(3)	m(4)	Total Amount, kg
6.708E+04	4.342E+03	2.345E+05	6.748E+04	3.735E+05
4.646E+03	1.549E+05	1.624E+04	2.407E+06	2.583E+06
4.135E+01	1.805E+02	7.923E+06	2.805E+03	7.926E+06
4.165E+01	4.909E+04	9.379E+05	7.629E+05	1.792E+06

Emission, kg/h

E(1)	E(2)	E(3)
1000	0	0
0	1000	0
0	0	1000
600	300	100

Loss, Reaction, kg/h

R(1)	R(2)	R(3)	R(4)
2.735E+02	1.770E+01	2.96E+01	2.751E+00
1.894E+01	6.314E+02	2.05E+00	9.813E+01
1.686E-01	7.357E+00	9.98E+02	1.143E-01
1.698E+02	2.001E+02	1.18E+02	3.110E+01

Loss, Advection, kg/h

A(1)	A(2)	A(4)
6.708E+02	4.342E+00	1.350E+00
4.646E+01	1.549E+02	4.814E+01
4.135E-01	1.805E-01	5.610E-01
4.165E+02	4.909E+02	1.526E+01

Intermedia Rate of Transport, kg/h

T12 air-water	T13 air-soil	T21 water-air	T31 soil-air	T32 soil-water	T24 water-sed	T42 sed-water
2.805E+01	2.960E+01	1.945E+00	1.589E-02	3.405E-02	5.480E+00	1.380E+00
1.943E+00	2.050E+01	6.939E+01	1.100E-03	2.358E-03	1.955E+02	4.922E+01
1.729E-02	1.825E-02	8.085E-02	5.368E-01	1.150E+00	2.278E-01	5.736E-02
1.742E+01	1.838E+01	2.199E+01	6.354E-01	1.362E-01	6.196E+01	1.560E+01

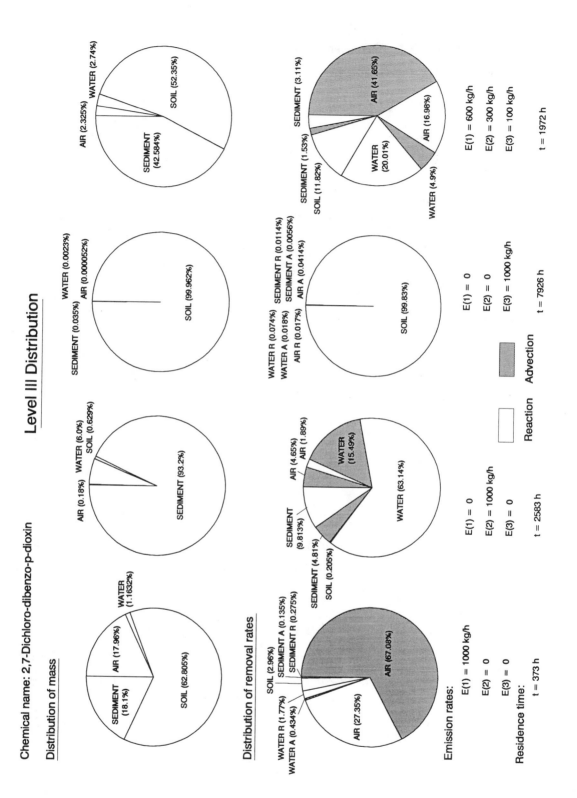

Chemical name: 2,7-Dichloro-dibenzo-p-dioxin

Level III Distribution

Distribution of mass

Distribution of removal rates

Emission rates:

E(1) = 1000 kg/h	E(1) = 0
E(2) = 0	E(2) = 1000 kg/h
E(3) = 0	E(3) = 0

Residence time:

t = 373 h   t = 2583 h

E(1) = 0
E(2) = 0
E(3) = 1000 kg/h

t = 7926 h

E(1) = 600 kg/h
E(2) = 300 kg/h
E(3) = 100 kg/h

t = 1972 h

Reaction   Advection

447

Chemical name: 2,8-Dichlorodibenzo-p-dioxin

Level I calculation: (six compartment model)

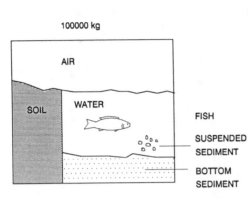

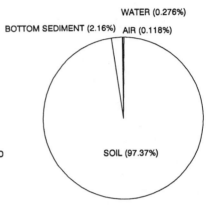

WATER (0.276%)

BOTTOM SEDIMENT (2.16%)   AIR (0.118%)

SOIL (97.37%)

Distribution of mass

physical-chemical properties:

MW: 253 g/mol

M.P.: 151 $^\circ$C

Fugacity ratio: 0.0567

vapor pressure: 1.40E-4 Pa

solubility: 0.0167 g/m^3

log K_{OW}: 5.60

Compartment	Z	Concentration			Amount	Amount
	mol/m3 Pa	mol/m3	mg/L (or g/m3)	ug/g	kg	%
Air	4.034E-04	4.670E-12	1.182E-09	9.967E-07	118.15	0.1182
Water	4.715E-01	5.458E-09	1.381E-06	1.381E-06	276.17	0.2762
Soil	3.694E+03	4.276E-05	1.082E-02	4.508E-03	97369	97.369
Biota (fish)	9.385E+03	1.086E-04	2.749E-02	2.749E-02	5.4973	5.50E-03
Suspended sediment	2.309E+04	2.673E-04	6.762E-02	4.508E-02	67.617	6.76E-02
Bottom sediment	7.388E+03	8.552E-05	2.164E-02	9.016E-03	2163.75	2.1638
Total					100000	100

f = 1.158E-08 Pa

Chemical name: 2,8-Dichlorodibenzo-p-dioxin

Level II calculation: (six compartment model)

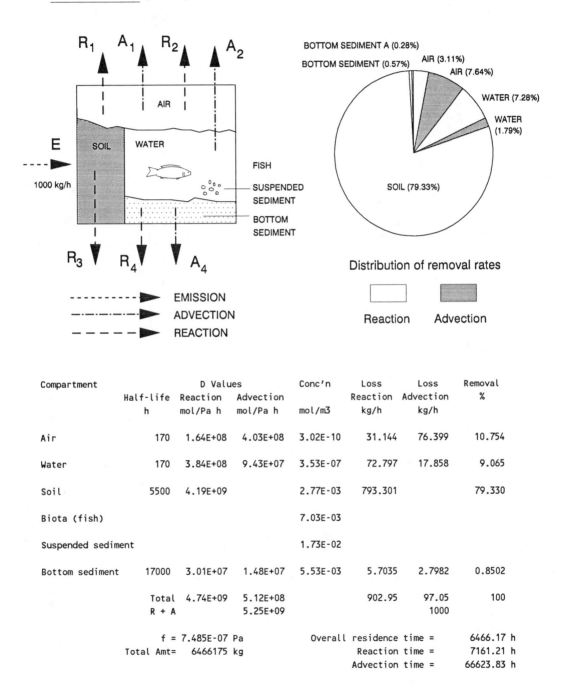

Distribution of removal rates

Reaction   Advection

Compartment	Half-life h	D Values Reaction mol/Pa h	Advection mol/Pa h	Conc'n mol/m3	Loss Reaction kg/h	Loss Advection kg/h	Removal %
Air	170	1.64E+08	4.03E+08	3.02E-10	31.144	76.399	10.754
Water	170	3.84E+08	9.43E+07	3.53E-07	72.797	17.858	9.065
Soil	5500	4.19E+09		2.77E-03	793.301		79.330
Biota (fish)				7.03E-03			
Suspended sediment				1.73E-02			
Bottom sediment	17000	3.01E+07	1.48E+07	5.53E-03	5.7035	2.7982	0.8502
Total		4.74E+09	5.12E+08		902.95	97.05	100
R + A			5.25E+09			1000	

f = 7.485E-07 Pa

Total Amt= 6466175 kg

Overall residence time = 6466.17 h
Reaction time = 7161.21 h
Advection time = 66623.83 h

Fugacity Level III calculations: (four compartment model)
Chemical name: 2,8-Dichlorodibenzo-p-dioxin

Phase Properties and Rates:

Compartment	Bulk Z mol/m3 Pa	Half-life h	Reaction mol/Pa h	Advection mol/Pa h
Air (1)	4.230E-04	170	1.72E+08	4.23E+08
Water (2)	5.963E-01	170	4.86E+08	1.19E+08
Soil (3)	1.847E+03	5500	4.19E+09	
Sediment (4)	1.478E+03	17000	3.01E+07	1.48E+07

	E(1)=1000	E(2)=1000	E(3)=1000	E(1,2,3)
Overall residence time =	824.35	2193.49	7927.12	1945.37 h
Reaction time =	2237.18	2828.28	7930.79	3513.35 h
Advection time =	1305.33	9773.00	17106453	4358.94 h

EMISSION (E) - - - -
REACTION (R) - - - -
ADVECTION (A) -·-·-
TRANSFER D VALUE mol/Pa h

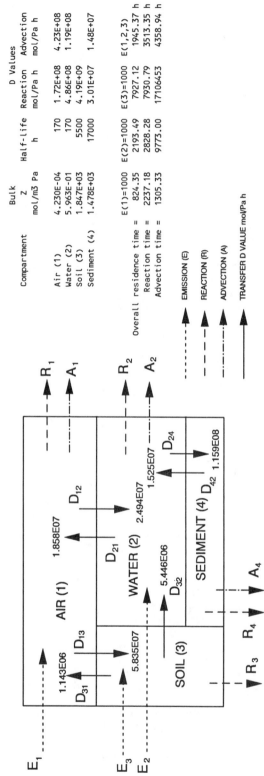

450

Phase Properties, Compositions, Transport and Transformation Rates:

Emission, kg/h

E(1)	E(2)	E(3)
1000	0	0
0	1000	0
0	0	1000
600	300	100

Fugacity, Pa

f(1)	f(2)	f(3)	f(4)
5.829E-06	2.052E-07	8.106E-08	3.954E-07
1.524E-07	5.568E-06	2.120E-09	1.073E-05
1.786E-09	7.283E-09	9.420E-07	1.403E-08
3.543E-06	1.794E-06	1.435E-07	3.457E-06

Concentration, g/m3

C(1)	C(2)	C(3)	C(4)
6.239E-07	3.096E-05	3.788E-02	1.478E-01
1.631E-08	8.400E-04	9.906E-04	4.012E+00
1.912E-10	1.099E-06	4.402E-01	5.248E-03
3.792E-07	2.707E-04	6.705E-02	1.293E-02

Amounts, kg

m(1)	m(2)	m(3)	m(4)	Total Amount, kg
6.239E+04	6.191E+03	6.818E+05	7.392E+04	8.243E+05
1.631E+03	1.680E+05	1.783E+04	2.006E+06	2.193E+06
1.912E+01	2.198E+02	7.924E+06	2.624E+03	7.927E+06
3.792E+04	5.414E+04	1.207E+06	6.464E+05	1.945E+06

Emission, kg/h

E(1)	E(2)	E(3)
1000	0	0
0	1000	0
0	0	1000
600	300	100

Loss, Reaction, kg/h

R(1)	R(2)	R(3)	R(4)
2.543E+02	2.524E+01	8.59E+01	3.014E+00
6.650E+00	6.849E+02	2.25E+00	8.177E+01
7.792E-02	8.959E-01	9.98E+02	1.070E-01
1.546E+02	2.207E+02	1.52E+02	2.635E+01

Loss, Advection, kg/h

A(1)	A(2)	A(4)
6.239E+02	6.191E+00	1.478E+00
1.631E+01	1.680E+02	4.012E+00
1.912E-01	2.198E-01	5.248E-02
3.792E+02	5.414E+01	1.293E+01

Intermedia Rate of Transport, kg/h

T12 air-water	T21 water-air	T13 air-soil	T31 soil-air	T32 soil-water	T24 water-sed	T42 sed-water
3.678E+01	9.646E-01	8.605E+01	2.345E-02	1.117E-01	6.017E-01	1.525E+00
9.617E-01	2.618E+01	2.250E+00	6.131E-04	2.921E-03	1.633E+02	4.139E+01
1.127E-02	3.424E-02	2.637E-02	2.725E-01	1.298E+00	2.136E-01	5.414E-02
2.235E+01	8.435E+01	5.231E+01	4.150E-02	1.977E-01	5.262E+01	1.334E+01

Level III Distribution

Chemical name: 2,8-Dichlorodibenzo-p-dioxin

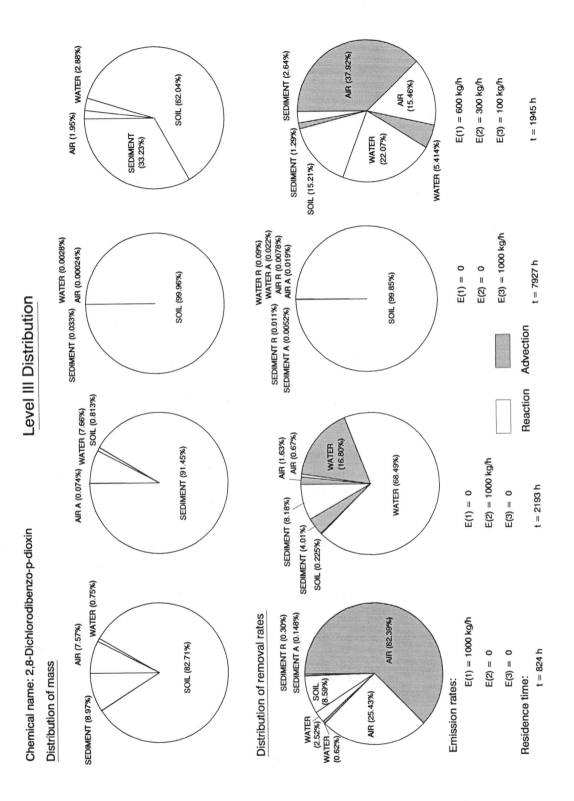

Chemical name: 1,2,4-Trichlorodibenzo-p-dioxin

Level I calculation: (six compartment model)

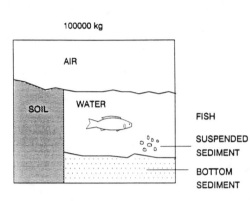

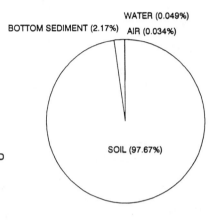

Distribution of mass

physical-chemical properties:

MW: 287.5 g/mol

M.P.: 129 °C

Fugacity ratio: 0.0936

vapor pressure: 0.0001 Pa

solubility: 0.00841 g/m^3

log K_{OW}: 6.35

Compartment	Z mol/m3 Pa	Concentration mol/m3	mg/L (or g/m3)	ug/g	Amount kg	Amount %
Air	4.034E-04	1.182E-12	3.397E-10	2.866E-07	33.971	0.0340
Water	2.925E-01	8.568E-10	2.463E-07	2.463E-07	49.265	0.0493
Soil	1.289E+04	3.775E-05	1.085E-02	4.522E-03	97673	97.673
Biota (fish)	3.274E+04	9.590E-05	2.757E-02	2.757E-02	5.5145	5.51E-03
Suspended sediment	8.055E+04	2.359E-04	6.783E-02	4.522E-02	67.828	6.78E-02
Bottom sediment	2.578E+04	7.550E-05	2.171E-02	9.044E-03	2170.51	2.1705
Total					100000	100

f = 2.929E-09 Pa

Chemical name: 1,2,4-Trichlorodibenzo-p-dioxin

Level II calculation: (six compartment model)

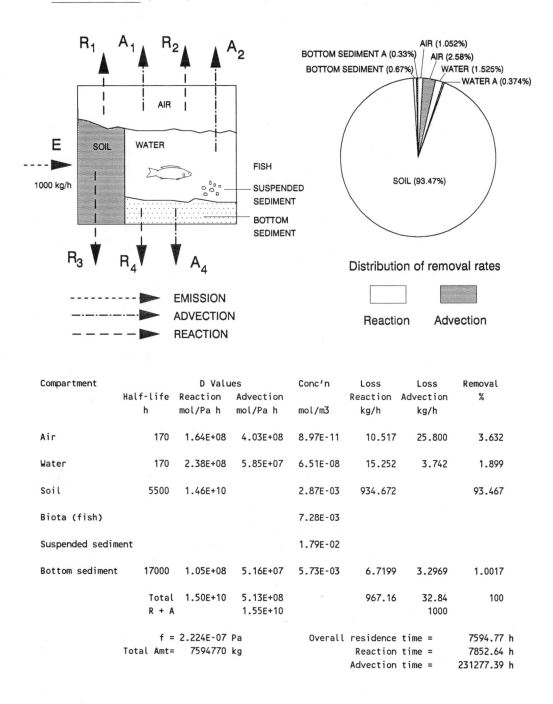

Distribution of removal rates

Reaction   Advection

Compartment	Half-life h	D Values Reaction mol/Pa h	Advection mol/Pa h	Conc'n mol/m3	Loss Reaction kg/h	Loss Advection kg/h	Removal %
Air	170	1.64E+08	4.03E+08	8.97E-11	10.517	25.800	3.632
Water	170	2.38E+08	5.85E+07	6.51E-08	15.252	3.742	1.899
Soil	5500	1.46E+10		2.87E-03	934.672		93.467
Biota (fish)				7.28E-03			
Suspended sediment				1.79E-02			
Bottom sediment	17000	1.05E+08	5.16E+07	5.73E-03	6.7199	3.2969	1.0017
	Total	1.50E+10	5.13E+08		967.16	32.84	100
	R + A		1.55E+10			1000	

f = 2.224E-07 Pa
Total Amt= 7594770 kg

Overall residence time = 7594.77 h
Reaction time = 7852.64 h
Advection time = 231277.39 h

453

Fugacity Level III calculations: (four compartment model)
Chemical name: 1,2,4-Trichlorodibenzo-p-dioxin

Phase Properties and Rates:

Compartment	Bulk Z mol/m3 Pa	Half-life h	D Values Reaction mol/Pa h	Advection mol/Pa h
Air (1)	4.487E-04	170	1.83E+08	4.49E+08
Water (2)	7.280E-01	170	5.94E+08	1.46E+08
Soil (3)	6.444E+03	5500	1.46E+10	
Sediment (4)	5.155E+03	17000	1.05E+08	5.16E+07

	E(1)=1000	E(2)=1000	E(3)=1000	E(1,2,3)
Overall residence time =	1518.13	4865.47	7933.36	3163.86 h
Reaction time =	3599.58	6411.45	7935.36	5448.46 h
Advection time =	2625.40	20177.96	31508929	7545.35 h

EMISSION (E)
REACTION (R)
ADVECTION (A)
TRANSFER D VALUE mol/Pa h

Diagram labels: E_1, R_1, A_1, AIR (1), D_{12}, D_{21}, WATER (2), R_2, A_2, D_{13}, D_{31}, 9.841E05, 1.260E08, 1.773E07, 3.162E07, D_{24} 5.184E07, D_{42} 4.030E08, SEDIMENT (4), 1.292E07, D_{32}, SOIL (3), E_3, E_2, R_3, R_4, A_4

454

Phase Properties, Compositions, Transport and Transformation Rates:

Emission, kg/h

E(1)	E(2)	E(3)
1000	0	0
0	1000	0
0	0	1000
600	300	100

Fugacity, Pa

f(1)	f(2)	f(3)	f(4)
4.410E-06	1.320E-07	3.798E-08	2.553E-07
7.377E-08	3.285E-06	6.353E-10	6.350E-06
3.618E-10	2.909E-09	2.378E-07	5.623E-09
2.668E-06	1.065E-06	4.676E-08	2.059E-06

Concentration, g/m3

C(1)	C(2)	C(3)	C(4)
5.689E-07	2.764E-05	7.036E-02	3.783E-01
9.517E-08	6.875E-04	1.177E-03	9.412E+00
4.667E-11	6.088E-07	4.405E-01	8.335E-03
3.442E-07	2.229E-04	8.662E-02	3.051E+00

Amounts, kg

m(1)	m(2)	m(3)	m(4)	Total Amount, kg
5.689E+04	5.527E+03	1.267E+06	1.892E+05	1.518E+06
9.517E+02	1.375E+05	2.119E+04	4.706E+06	4.865E+06
4.667E+00	1.218E+02	7.929E+06	4.167E+03	7.933E+06
3.442E+04	4.458E+04	1.559E+06	1.526E+06	3.164E+06

Loss, Reaction, kg/h

R(1)	R(2)	R(3)	R(4)
2.319E+02	2.253E+01	1.60E+02	7.711E+00
3.880E+00	5.605E+02	2.67E+00	1.918E-02
1.903E-02	4.964E-01	9.99E+02	1.699E-01
1.403E+02	1.817E+02	1.96E+02	6.219E+01

Loss, Advection, kg/h

A(1)	A(2)	A(4)
5.689E+02	5.527E+00	3.783E+00
9.517E+02	1.375E+00	9.412E+00
4.667E-02	1.218E-02	8.335E-03
3.442E+02	4.458E+01	3.051E+01

Intermedia Rate of Transport, kg/h

T12 air-water	T13 air-soil	T21 water-air	T24 water-sed	T31 soil-air	T32 soil-water	T42 sed-water
4.008E+01	1.597E+02	6.729E-01	1.530E+01	1.075E-02	1.410E-01	3.805E+00
6.705E-01	2.672E+00	1.674E+01	3.806E+02	1.797E-04	2.359E-03	9.465E+01
3.288E-03	1.310E-02	1.482E-02	3.370E-01	6.727E-02	8.829E-01	8.382E-02
2.425E+01	9.664E+01	5.427E+00	1.234E+02	1.323E-02	1.736E-01	3.069E+01

Level III Distribution

Chemical name: 1,2,4-Trichloro-dibenzo-p-dioxin

Distribution of mass

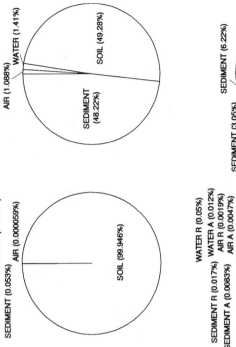

Distribution of removal rates

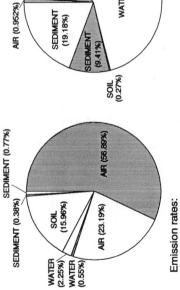

Advection Reaction

Chemical name: 1,2,3,4-Tetrachlorodibenzo-p-dioxin

Level I calculation: (six compartment model)

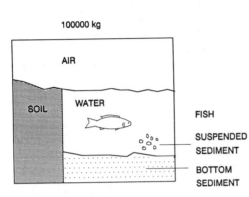

100000 kg

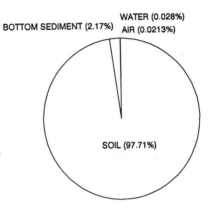

WATER (0.028%)
BOTTOM SEDIMENT (2.17%) AIR (0.0213%)

SOIL (97.71%)

Distribution of mass

physical-chemical properties:

MW: 322 g/mol

M.P.: 190 $^{\circ}$C

Fugacity ratio: 0.0233

vapor pressure: 6.50×10^{-6} Pa

solubility: 0.00055 g/m^3

log K$_{OW}$: 6.60

Compartment	Z mol/m3 Pa	Concentration mol/m3	Concentration mg/L (or g/m3)	Concentration ug/g	Amount kg	Amount %
Air	4.034E-04	6.606E-13	2.127E-10	1.795E-07	21.272	0.0213
Water	2.628E-01	4.303E-10	1.386E-07	1.386E-07	27.713	0.0277
Soil	2.059E+04	3.372E-05	1.086E-02	4.523E-03	97706	97.706
Biota (fish)	5.231E+04	8.566E-05	2.758E-02	2.758E-02	5.5164	5.52E-03
Suspended sediment	1.287E+05	2.107E-04	6.785E-02	4.523E-02	67.852	6.79E-02
Bottom sediment	4.118E+04	6.743E-05	2.171E-02	9.047E-03	2171.25	2.1713
Total					100000	100

f = 1.638E-09 Pa

456

Chemical name: 1,2,3,4-Tetrachlorodibenzo-p-dioxin

Level II calculation: (six compartment model)

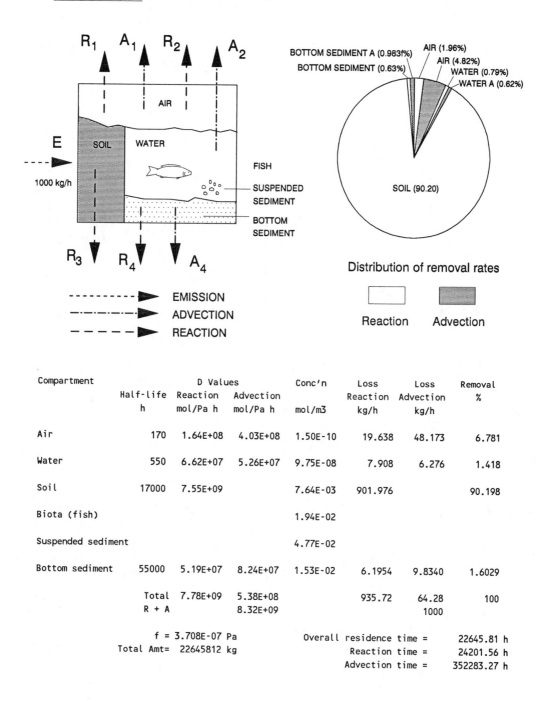

Distribution of removal rates

Reaction   Advection

Compartment	Half-life h	D Values Reaction mol/Pa h	Advection mol/Pa h	Conc'n mol/m3	Loss Reaction kg/h	Loss Advection kg/h	Removal %
Air	170	1.64E+08	4.03E+08	1.50E-10	19.638	48.173	6.781
Water	550	6.62E+07	5.26E+07	9.75E-08	7.908	6.276	1.418
Soil	17000	7.55E+09		7.64E-03	901.976		90.198
Biota (fish)				1.94E-02			
Suspended sediment				4.77E-02			
Bottom sediment	55000	5.19E+07	8.24E+07	1.53E-02	6.1954	9.8340	1.6029
Total R + A		7.78E+09	5.38E+08 8.32E+09		935.72	64.28 1000	100

f = 3.708E-07 Pa
Total Amt= 22645812 kg

Overall residence time = 22645.81 h
Reaction time = 24201.56 h
Advection time = 352283.27 h

457

Fugacity Level III calculations: (four compartment model)

Chemical name: 1,2,3,4-Tetrachlorodibenzo-p-dioxin

Phase Properties and Rates:

Compartment	Bulk Z mol/m3 Pa	Half-life h	D Values Reaction mol/Pa h	Advection mol/Pa h
Air (1)	5.772E-04	170	2.35E+08	5.77E+08
Water (2)	9.585E-01	550	2.42E+08	1.92E+08
Soil (3)	1.029E+04	17000	7.55E+09	
Sediment (4)	8.235E+03	55000	5.19E+07	8.24E+07

	E(1)=1000	E(2)=1000	E(3)=1000	E(1,2,3)	
Overall residence time =	9353	14811	24504	12506	h
Reaction time =	17113	31048	24539	21905	h
Advection time =	20626	28323	17276774	29144	h

- R_1, A_1, R_2, A_2, R_3, R_4, A_4, E_1, E_2, E_3
- - - - EMISSION (E)
- – – – REACTION (R)
- – · – ADVECTION (A)
- ⟶ TRANSFER D VALUE mol/Pa h

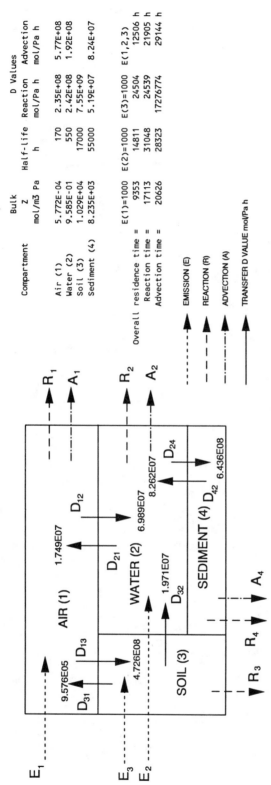

458

Phase Properties, Compositions, Transport and Transformation Rates:

Emission, kg/h E(1)	E(2)	E(3)	Fugacity, Pa f(1)	f(2)	f(3)	f(4)	Concentration, g/m3 C(1)	C(2)	C(3)	C(4)	Amounts, kg m(1)	m(2)	m(3)	m(4)	Total Amount, kg
1000	0	0	2.294E-06	1.922E-07	1.432E-07	5.704E-07	4.265E-07	5.931E-05	4.746E-01	1.513E+00	4.265E+00	1.186E+04	8.542E+06	7.563E+05	9.353E+06
0	1000	0	4.725E-08	3.661E-06	2.948E-09	1.087E-06	8.782E-09	1.130E-03	9.773E-01	2.882E+01	8.782E+02	2.260E+05	1.759E+05	1.441E+07	1.481E+07
0	0	1000	4.130E-10	9.553E-09	4.100E-07	2.835E-08	7.677E-11	2.948E-06	1.359E+00	7.519E-02	7.677E+00	5.896E+02	2.447E+07	3.759E+04	2.450E+07
600	300	100	1.391E-06	1.215E-06	1.278E-07	3.605E-06	2.585E-07	3.749E-04	4.236E-01	9.560E-01	2.585E+04	7.497E+04	7.625E+06	4.780E+06	1.251E+07

Emission, kg/h E(1)	E(2)	E(3)	Loss, Reaction, kg/h R(1)	R(2)	R(3)	R(4)	Loss, Advection, kg/h A(1)	A(2)	A(4)
1000	0	0	1.738E+02	1.495E+01	3.48E+02	9.530E+00	4.265E+02	1.186E+01	1.513E+01
0	1000	0	3.580E+00	2.848E+02	7.17E+02	1.815E+02	8.782E+00	2.260E+02	2.882E+02
0	0	1000	3.130E-02	7.430E+01	9.97E+02	4.737E-01	7.677E-02	5.896E-01	7.519E-02
600	300	100	1.054E+02	9.447E+01	3.11E+02	6.023E+01	2.585E+02	7.497E+01	9.560E+01

Intermedia Rate of Transport, kg/h

	T12 air-water	T21 water-air	T13 air-soil	T31 soil-air	T32 soil-water	T24 water-sed	T42 sed-water
	5.164E+01	1.082E+00	3.492E+02	4.415E-02	9.087E-02	3.983E+01	1.517E+01
	1.063E+00	2.062E+01	7.191E+00	9.091E-04	1.871E-02	7.588E+02	2.891E+02
	9.296E-03	5.379E-02	6.286E-02	1.264E-01	2.603E+00	1.980E+00	7.543E-01
	3.130E+01	6.839E+00	2.117E+02	3.940E-02	8.111E-01	2.517E+02	9.591E+01

Chemical name: 1,2,3,4-Tetrachloro-dibenzo-p-dioxin   Level III Distribution

Distribution of mass

Distribution of removal rates

Emission rates:

E(1) = 1000 kg/h

E(2) = 0

E(3) = 0

Residence time:

t = 9353 h

E(1) = 0

E(2) = 1000 kg/h

E(3) = 0

t = 14811 h

E(1) = 0

E(2) = 0

E(3) = 1000 kg/h

t = 24504 h

E(1) = 600 kg/h

E(2) = 300 kg/h

E(3) = 100 kg/h

t = 12506 h

Reaction   Advection

Chemical name: 2,3,7,8-Tetrachlorodibenzo-p-dioxin

Level I calculation: (six compartment model)

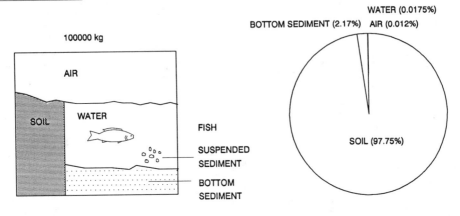

Distribution of mass

physical-chemical properties:

MW: 322 g/mol

M.P.: 305 °C

Fugacity ratio: 0.0017

vapor pressure: 0.0000002 Pa

solubility: 0.0000193 g/m^3

log K_{OW}: 6.80

Compartment	Z mol/m3 Pa	Concentration mol/m3	mg/L (or g/m3)	ug/g	Amount kg	Amount %
Air	4.034E-04	3.656E-13	1.177E-10	9.930E-08	11.771	0.0118
Water	2.997E-01	2.716E-10	8.745E-08	8.745E-08	17.489	0.0175
Soil	3.721E+04	3.372E-05	1.086E-02	4.524E-03	97726	97.726
Biota (fish)	9.455E+04	8.568E-05	2.759E-02	2.759E-02	5.5175	5.52E-03
Suspended sediment	2.326E+05	2.108E-04	6.787E-02	4.524E-02	67.865	6.79E-02
Bottom sediment	7.443E+04	6.744E-05	2.172E-02	9.049E-03	2171.68	2.1717
Total					100000	100

f = 9.062E-10 Pa

Chemical name: 2,3,7,8-Tetrachlorodibenzo-p-dioxin

Level II calculation: (six compartment model)

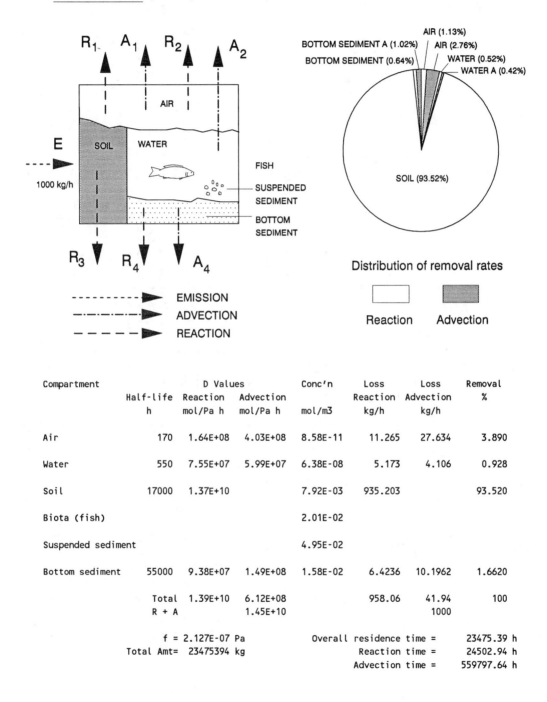

Distribution of removal rates

Reaction   Advection

Compartment	Half-life h	D Values Reaction mol/Pa h	Advection mol/Pa h	Conc'n mol/m3	Loss Reaction kg/h	Loss Advection kg/h	Removal %
Air	170	1.64E+08	4.03E+08	8.58E-11	11.265	27.634	3.890
Water	550	7.55E+07	5.99E+07	6.38E-08	5.173	4.106	0.928
Soil	17000	1.37E+10		7.92E-03	935.203		93.520
Biota (fish)				2.01E-02			
Suspended sediment				4.95E-02			
Bottom sediment	55000	9.38E+07	1.49E+08	1.58E-02	6.4236	10.1962	1.6620
Total		1.39E+10	6.12E+08		958.06	41.94	100
R + A			1.45E+10			1000	

f = 2.127E-07 Pa
Total Amt= 23475394 kg

Overall residence time =   23475.39 h
            Reaction time =   24502.94 h
           Advection time =   559797.64 h

461

Fugacity Level III calculations: (four compartment model)

Chemical name: 2,3,7,8-Tetrachlorodibenzo-p-dioxin

Phase Properties and Rates:

Compartment	Bulk Z mol/m3 Pa	Half-life h	D Values Reaction mol/Pa h	D Values Advection mol/Pa h
Air (1)	8.151E-04	170	3.32E+08	8.15E+08
Water (2)	1.557E+00	550	3.92E+08	3.11E+08
Soil (3)	1.861E+04	17000	1.37E+10	
Sediment (4)	1.489E+04	55000	9.38E+07	1.49E+08

	$E(1)=1000$	$E(2)=1000$	$E(3)=1000$	$E(1,2,3)$
Overall residence time =	12326	15692	24508	14554 h
Reaction time =	19588	33178	24541	23500 h
Advection time =	33251	29775	17912170	38233 h

EMISSION (E)

REACTION (R)

ADVECTION (A)

TRANSFER D VALUE mol/Pa h

AIR (1) 1.778E07

9.904E05 D_{13}

D_{31}

D_{12} D_{21} 1.416E08 1.492E08 D_{24}

WATER (2) 1.115E09

3.484E07 D_{32}

SOIL (3)

SEDIMENT (4) D_{42} 1.163E09

E_1 R_1 A_1

E_3 R_2 A_2

E_2

R_3 R_4 A_4

Phase Properties, Compositions, Transport and Transformation Rates:

Emission, kg/h

E(1)	E(2)	E(3)
1000	0	0
0	1000	0
0	0	1000
600	300	100

Fugacity, Pa

f(1)	f(2)	f(3)	f(4)
1.293E-06	1.295E-07	1.053E-07	3.844E-07
1.594E-08	2.155E-06	1.298E-09	6.399E-06
1.341E-10	5.495E-09	2.269E-07	1.631E-08
7.804E-07	7.248E-07	8.627E-08	2.152E-06

Concentration, g/m3

C(1)	C(2)	C(3)	C(4)
3.393E-07	6.492E-05	6.310E-01	1.843E+00
4.183E-09	1.081E-03	7.780E-03	3.067E-02
3.520E-11	2.755E-06	1.359E+00	7.820E-02
2.048E-07	3.634E-04	5.169E-01	1.031E-01

Amounts, kg

m(1)	m(2)	m(3)	m(4)	Total Amount, kg
3.393E+02	1.298E+04	1.136E+07	9.213E+05	1.233E+07
4.183E+00	2.161E+05	1.400E+05	1.534E+07	1.569E+07
3.520E-02	5.510E+02	2.447E+07	3.910E+04	2.451E+07
2.048E+02	7.269E+04	9.304E+06	5.157E+06	1.455E+07

Emission, kg/h

E(1)	E(2)	E(3)
1000	0	0
0	1000	0
0	0	1000
600	300	100

Loss, Reaction, kg/h

R(1)	R(2)	R(3)	R(4)
1.383E+02	1.636E+01	4.63E+02	1.161E+01
1.705E+00	2.723E+02	5.71E+00	1.932E+02
1.435E-02	6.943E-01	9.97E+02	4.926E-01
8.350E+01	9.158E+01	3.79E+02	6.498E+01

Loss, Advection, kg/h

A(1)	A(2)	A(4)
3.393E+02	1.298E+01	1.843E+01
4.183E+00	2.161E+00	3.067E+02
3.520E-02	5.510E-02	7.820E-01
2.048E+02	7.269E+01	1.031E+02

Intermedia Rate of Transport, kg/h

T12 air-water	T13 air-soil	T21 water-air	T31 soil-air	T32 soil-water	T24 water-sed	T42 sed-water
5.894E+01	4.642E+01	7.412E-01	3.359E+02	1.182E+00	4.849E+01	1.846E+01
7.266E-01	5.723E+00	1.234E+01	4.141E-04	1.457E-02	8.073E+02	3.073E+02
6.114E-03	4.816E-02	3.146E-02	7.236E-02	2.545E+00	2.058E+00	7.835E-01
3.558E+01	2.803E+02	4.149E+00	2.751E-02	9.678E-01	2.715E-02	1.034E+02

Chemical name: 2,3,7,8-Tetrachlorodibenzo-p-dioxin   Level III Distribution

Distribution of mass

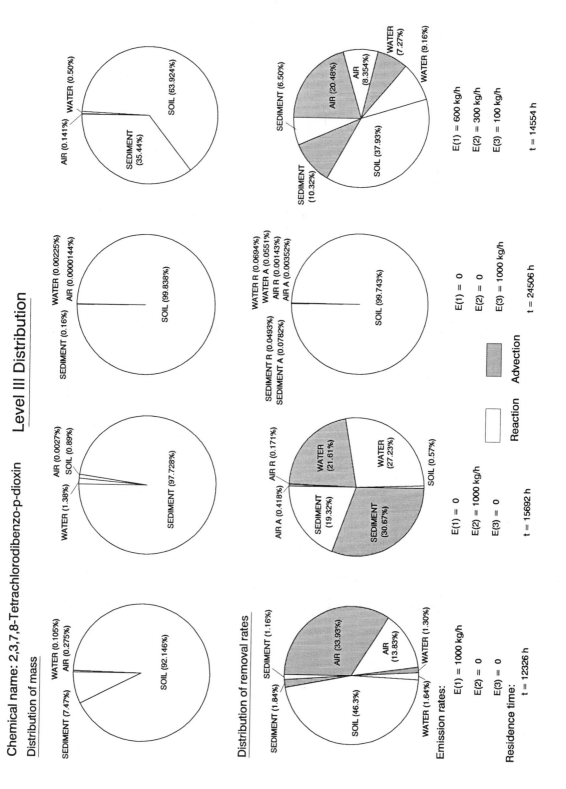

Distribution of removal rates

Emission rates:

E(1) = 1000 kg/h   E(1) = 0   E(1) = 0   E(1) = 600 kg/h
E(2) = 0   E(2) = 1000 kg/h   E(2) = 0   E(2) = 300 kg/h
E(3) = 0   E(3) = 0   E(3) = 1000 kg/h   E(3) = 100 kg/h

Residence time:

t = 12326 h   t = 15692 h   t = 24506 h   t = 14554 h

463

Chemical name: 1,2,3,4,7-Pentachlorodibenzo-p-dioxin

Level I calculation: (six compartment model)

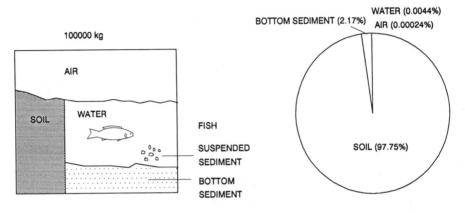

Distribution of mass

physical-chemical properties:

MW: 356.4 g/mol

M.P.: 195 °C

Fugacity ratio: 0.0208

vapor pressure: 8.80×10^{-8} Pa

solubility: 0.000118 g/m^3

log K_{OW} : 7.40

Compartment	Z	Concentration			Amount	Amount
	mol/m3 Pa	mol/m3	mg/L (or g/m3)	ug/g	kg	%
Air	4.034E-04	6.610E-15	2.356E-12	1.987E-09	0.236	2.36E-04
Water	3.762E+00	6.165E-11	2.197E-08	2.197E-08	4.394	0.00439
Soil	1.860E+06	3.047E-05	1.086E-02	4.525E-03	97750	97.750
Biota (fish)	4.725E+06	7.742E-05	2.759E-02	2.759E-02	5.5188	5.52E-03
Suspended sediment	1.162E+07	1.905E-04	6.788E-02	4.525E-02	67.882	6.79E-02
Bottom sediment	3.720E+06	6.095E-05	2.172E-02	9.051E-03	2172.22	2.1722
Total					100000	100

f = 1.639E-11 Pa

Chemical name: 1,2,3,4,7-Pentachlorodibenzo-p-dioxin

Level II calculation: (six compartment model)

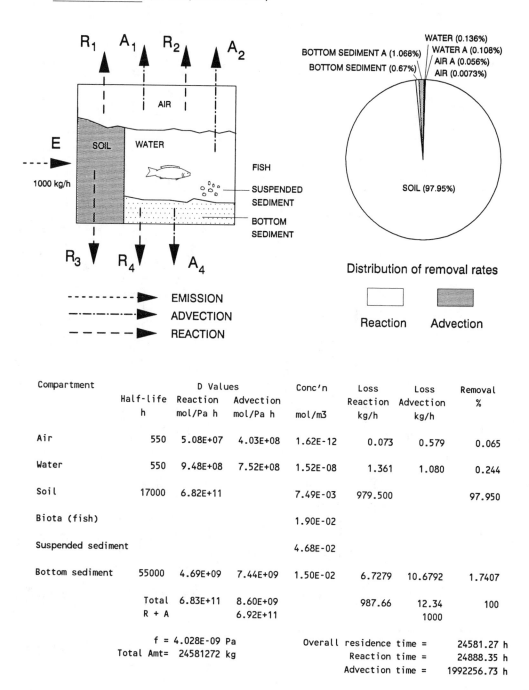

Distribution of removal rates

Reaction   Advection

Compartment	Half-life h	D Values Reaction mol/Pa h	D Values Advection mol/Pa h	Conc'n mol/m3	Loss Reaction kg/h	Loss Advection kg/h	Removal %
Air	550	5.08E+07	4.03E+08	1.62E-12	0.073	0.579	0.065
Water	550	9.48E+08	7.52E+08	1.52E-08	1.361	1.080	0.244
Soil	17000	6.82E+11		7.49E-03	979.500		97.950
Biota (fish)				1.90E-02			
Suspended sediment				4.68E-02			
Bottom sediment	55000	4.69E+09	7.44E+09	1.50E-02	6.7279	10.6792	1.7407
	Total	6.83E+11	8.60E+09		987.66	12.34	100
	R + A		6.92E+11			1000	

f = 4.028E-09 Pa
Total Amt= 24581272 kg

Overall residence time =   24581.27 h
Reaction time =   24888.35 h
Advection time =   1992256.73 h

465

Fugacity Level III calculations: (four compartment model)

Chemical name: 1,2,3,4,7-Pentachlorodibenzo-p-dioxin

Diagram (four-compartment model):

E_1 --→ R_1, A_1

AIR (1) 1.996E07
D_{31} 4.021E06 D_{13}
D_{12} D_{21}

E_3, E_2 --→ R_2, A_2

WATER (2) 3.461E09   7.443E09
1.691E09 D_{32} 3.097E10

SEDIMENT (4) D_{42} 5.813E10 D_{24}

SOIL (3)

R_3, R_4, A_4

Compartment	Bulk Z mol/m3 Pa	Half-life h	D Values Reaction mol/Pa h	Advection mol/Pa h
Air (1)	1.186E-02	550	1.49E+09	1.19E+10
Water (2)	6.661E+01	550	1.68E+10	1.33E+10
Soil (3)	9.299E+05	17000	6.82E+11	
Sediment (4)	7.440E+05	55000	4.69E+09	7.44E+09

	E(1)=1000	E(2)=1000	E(3)=1000	E(1,2,3)
Overall residence time =	17136	16910	24512	17806 h
Reaction time =	24062	36409	24545	26715 h
Advection time =	59534	31574	18493271	53391 h

Legend:
- ---- EMISSION (E)
- ---- REACTION (R)
- -·-· ADVECTION (A)
- → TRANSFER D VALUE mol/Pa h

Phase Properties, Compositions, Transport and Transformation Rates:

Emission, kg/h

E(1)	E(2)	E(3)
1000	0	0
0	1000	0
0	0	1000
600	300	100

Fugacity, Pa

f(1)	f(2)	f(3)	f(4)
5.872E-08	3.140E-09	2.658E-09	9.327E-09
1.772E-11	4.242E-08	8.021E-13	1.260E-10
3.890E-13	1.049E-10	4.102E-09	3.115E-10
3.524E-08	1.462E-08	2.006E-09	4.343E-08

Concentration, g/m3

C(1)	C(2)	C(3)	C(4)
2.482E-07	7.454E-05	8.811E-01	2.473E+00
7.489E-11	1.007E-03	2.658E-04	3.341E+01
1.644E-12	2.490E-06	1.359E+00	8.259E-02
1.489E-07	3.471E-04	6.647E-01	1.151E+01

Amounts, kg

m(1)	m(2)	m(3)	m(4)	Total Amount, kg
2.482E+04	1.491E+04	1.586E+07	1.236E+06	1.714E+07
3.341E+01	2.014E+05	4.785E+03	1.670E+07	1.691E+07
8.259E-02	4.979E+02	2.447E+07	4.129E+04	2.451E+07
1.151E+01	6.942E+04	1.196E+07	5.757E+06	1.781E+07

Loss, Reaction, kg/h

R(1)	R(2)	R(3)	R(4)
3.127E+01	1.879E+01	6.47E+02	1.558E+01
9.436E-03	2.538E+02	1.95E-01	2.105E+02
2.072E-04	6.274E+01	9.98E+02	5.203E-01
1.877E+01	8.747E+01	4.88E+02	7.254E+01

Loss, Advection, kg/h

A(1)	A(2)	A(4)
2.482E+02	1.491E+01	2.473E+01
7.489E-02	2.014E+02	3.341E+02
1.644E-03	4.979E-01	8.259E-02
1.489E+02	6.942E+01	1.151E+02

Intermedia Rate of Transport, kg/h

T12 air-water	T13 air-soil	T21 water-air	T24 water-sed	T31 soil-air	T32 soil-water	T42 sed-water
7.242E+01	6.481E+02	2.233E-02	6.505E+01	3.810E-03	1.602E+00	2.474E+01
2.185E-02	1.956E-01	3.017E-01	8.788E+02	1.150E-06	4.834E-04	3.342E-02
4.797E-04	4.293E-03	7.459E-04	2.173E+00	5.878E-03	2.472E-03	8.263E-01
4.346E-01	3.889E+02	1.040E-01	3.029E+02	2.874E-03	1.209E+00	1.152E+02

Chemical name: 1,2,3,4,7-Pentachloro-dibenzo-p-dioxin   Level III Distribution

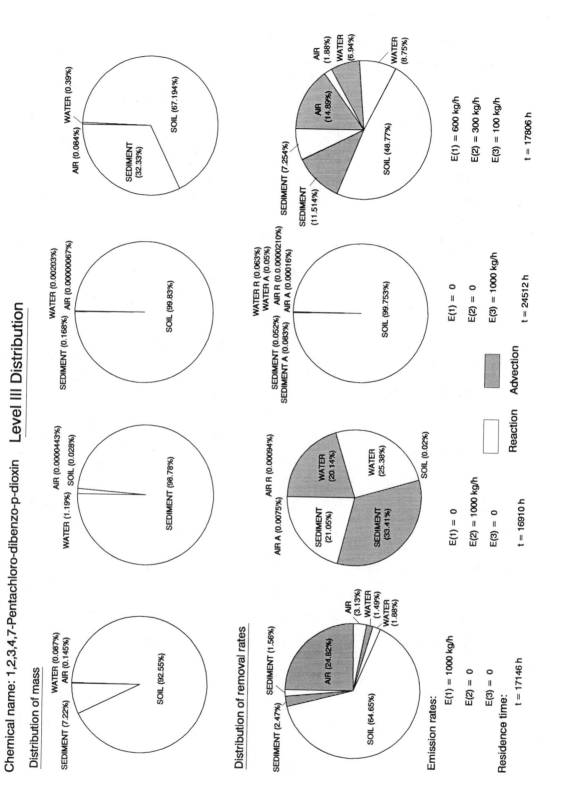

Distribution of mass

Pie 1:
WATER (0.087%)
AIR (0.145%)
SEDIMENT (7.22%)
SOIL (92.55%)

Pie 2:
AIR (0.0000443%)
SOIL (0.028%)
WATER (1.19%)
SEDIMENT (98.78%)

Pie 3:
WATER (0.00203%)
AIR (0.00000677%)
SEDIMENT (0.168%)
SOIL (99.83%)

Pie 4:
AIR (1.88%)
WATER (6.94%)
WATER (8.75%)
AIR (14.89%)
SEDIMENT (7.254%)
SEDIMENT (11.514%)
SOIL (48.77%)

Distribution of removal rates

Pie 1:
AIR (3.13%)
WATER (1.49%)
WATER (1.88%)
SEDIMENT (1.56%)
SEDIMENT (2.47%)
AIR (24.82%)
SOIL (64.65%)

Pie 2:
AIR R (0.00094%)
AIR A (0.0075%)
WATER (20.14%)
WATER (25.38%)
SOIL (0.02%)
SEDIMENT (21.05%)
SEDIMENT (33.41%)

Pie 3:
WATER R (0.063%)
WATER A (0.05%)
AIR R (0.0000210%)
AIR A (0.00016%)
SEDIMENT (0.052%)
SEDIMENT A (0.083%)
SOIL (99.753%)

Reaction   Advection

Emission rates:

E(1) = 1000 kg/h   E(1) = 0   E(1) = 0   E(1) = 600 kg/h
E(2) = 0   E(2) = 1000 kg/h E(2) = 0   E(2) = 300 kg/h
E(3) = 0   E(3) = 0   E(3) = 1000 kg/h E(3) = 100 kg/h

Residence time:

t = 17146 h   t = 16910 h   t = 24512 h   t = 17806 h

467

Chemical name: 1,2,3,4,7,8-Hexachlorodibenzo-p-dioxin

Level I calculation: (six compartment model)

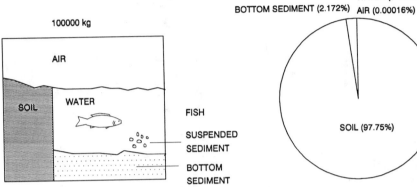

100000 kg

AIR

SOIL   WATER

FISH

SUSPENDED
SEDIMENT

BOTTOM
SEDIMENT

WATER (0.00175%)
BOTTOM SEDIMENT (2.172%)   AIR (0.00016%)

SOIL (97.75%)

Distribution of mass

physical-chemical properties:

MW: 391 g/mol

M.P.: 273 °C

Fugacity ratio: 0.00352

vapor pressure: 5.10×10^{-9} Pa

solubility: 4.42×10^{-6} g/m^3

log K_{OW} : 7.80

Compartment	Z mol/m3 Pa	Concentration			Amount kg	Amount %
		mol/m3	mg/L (or g/m3)	ug/g		
Air	4.034E-04	4.072E-15	1.592E-12	1.343E-09	0.1592	1.59E-04
Water	2.217E+00	2.237E-11	8.747E-09	8.747E-09	1.7494	0.00175
Soil	2.752E+06	2.778E-05	1.086E-02	4.526E-03	97752	97.752
Biota (fish)	6.993E+06	7.058E-05	2.759E-02	2.759E-02	5.5190	5.52E-03
Suspended sediment	1.720E+07	1.736E-04	6.788E-02	4.526E-02	67.884	6.79E-02
Bottom sediment	5.505E+06	5.556E-05	2.172E-02	9.051E-03	2172.28	2.1723
Total					100000	100

f = 1.009E-11 Pa

Chemical name: 1,2,3,4,7,8-Hexachlorodibenzo-p-dioxin

Level II calculation: (six compartment model)

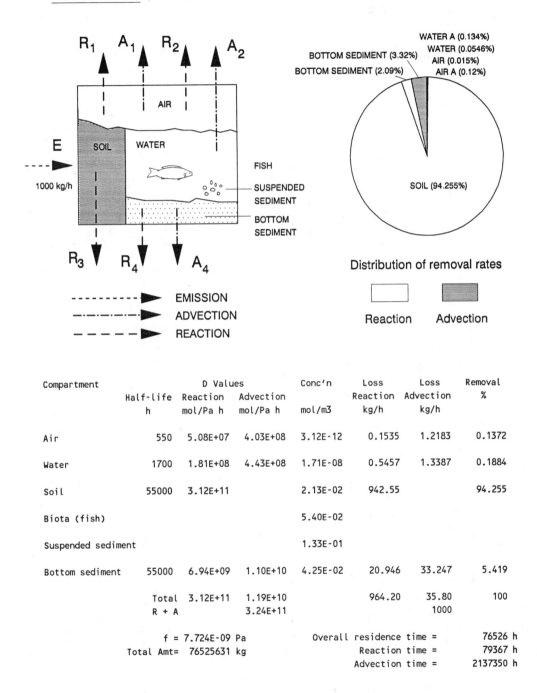

Compartment	Half-life h	D Values Reaction mol/Pa h	Advection mol/Pa h	Conc'n mol/m3	Loss Reaction kg/h	Loss Advection kg/h	Removal %
Air	550	5.08E+07	4.03E+08	3.12E-12	0.1535	1.2183	0.1372
Water	1700	1.81E+08	4.43E+08	1.71E-08	0.5457	1.3387	0.1884
Soil	55000	3.12E+11		2.13E-02	942.55		94.255
Biota (fish)				5.40E-02			
Suspended sediment				1.33E-01			
Bottom sediment	55000	6.94E+09	1.10E+10	4.25E-02	20.946	33.247	5.419
Total		3.12E+11	1.19E+10		964.20	35.80	100
R + A			3.24E+11			1000	

f = 7.724E-09 Pa
Total Amt= 76525631 kg

Overall residence time =   76526 h
Reaction time =   79367 h
Advection time = 2137350 h

Fugacity Level III calculations: (four compartment model)

Chemical name: 1,2,3,4,7,8-Hexachloro-dibenzo-p-dioxin

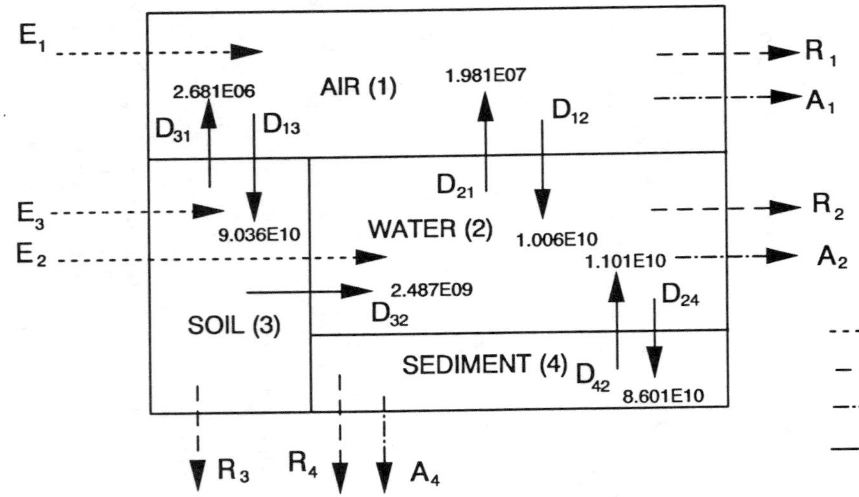

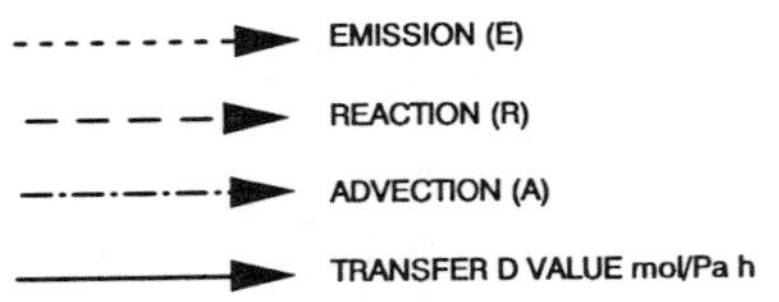

Phase Properties and Rates:

Compartment	Bulk Z mol/m3 Pa	Half-life h	D Values Reaction mol/Pa h	Advection mol/Pa h
Air (1)	3.386E-02	550	4.27E+09	3.39E+10
Water (2)	9.522E+01	1700	7.76E+09	1.90E+10
Soil (3)	1.376E+06	55000	3.12E+11	
Sediment (4)	1.101E+06	55000	6.94E+09	1.10E+10

	E(1)=1000	E(2)=1000	E(3)=1000	E(1,2,3)
Overall residence time =	52983	20656	78901	45877 h
Reaction time =	75113	58324	79306	72940 h
Advection time =	179829	31983	15446147	123643 h

Legend:
- EMISSION (E) ----
- REACTION (R) - - -
- ADVECTION (A) -·-·-
- TRANSFER D VALUE mol/Pa h ——

Diagram labels (four compartment model):
E_1, R_1, A_1, R_2, A_2, R_3, R_4, A_4, E_2, E_3
- AIR (1): 1.981E07
- WATER (2)
- SOIL (3)
- SEDIMENT (4)
- D_{12} = 1.006E10 ; D_{21}
- D_{13} = 9.036E10 ; D_{31} = 2.681E06
- D_{24} = 1.101E10 ; D_{42} = 8.601E10
- D_{32} = 2.487E09

Concentration, g/m3

C(1)	C(2)	C(3)	C(4)
2.444E-07	9.241E-05	2.853E+00	3.174E+00
6.042E-11	1.188E-03	7.053E-04	4.081E+00
2.560E-12	9.395E-06	4.374E+00	3.227E+00
1.467E-07	4.129E-04	2.149E+00	1.418E+00

Amounts, kg

m(1)	m(2)	m(3)	m(4)	Total Amount, kg
2.444E+04	1.848E+04	5.135E+07	1.587E+06	5.298E+07
6.042E+00	2.377E+05	1.270E+04	2.041E+07	2.066E+07
2.560E-01	1.879E+03	7.874E+07	1.613E+05	7.890E+07
1.467E+04	8.258E+04	3.869E+07	7.090E+06	4.588E+07

Intermedia Rate of Transport, kg/h

T12 air-water	T13 air-soil	T21 water-air	T31 soil-air	T32 soil-water	T24 water-sed	T42 sed-water
7.261E+01	6.522E+02	1.923E-02	5.558E-03	5.156E+00	8.347E+01	3.174E+01
1.795E-02	1.612E-01	2.472E-01	1.374E-06	1.275E-03	1.073E+03	4.082E+02
7.606E-04	6.832E-03	1.955E-03	8.521E-03	7.905E-03	8.487E+00	3.227E+00
4.357E+02	3.914E+02	8.590E-02	4.187E-02	3.885E+00	3.730E+02	1.418E+02

Phase Properties, Compositions, Transport and Transformation Rates:

Emission, kg/h

E(1)	E(2)	E(3)
1000	0	0
0	1000	0
0	0	1000
600	300	100

Fugacity, Pa

f(1)	f(2)	f(3)	f(4)
1.846E-08	2.482E-09	5.302E-09	7.373E-09
4.564E-12	3.192E-08	1.311E-12	9.481E-08
1.934E-13	2.523E-10	8.129E-09	7.496E-10
1.108E-08	1.109E-08	3.995E-09	3.294E-08

Loss, Reaction, kg/h

R(1)	R(2)	R(3)	R(4)
3.080E+01	7.534E+00	6.47E+02	1.999E+01
7.613E-03	9.688E+01	1.60E-01	2.571E+02
3.226E-04	7.660E-01	9.92E+02	2.033E+01
1.848E+01	3.366E+01	4.87E+02	8.933E+01

Emission, kg/h

E(1)	E(2)	E(3)
1000	0	0
0	1000	0
0	0	1000
600	300	100

Loss, Advection, kg/h

A(1)	A(2)	A(3)	A(4)
2.444E+02	1.848E+01	2.560E-01	3.174E+01
6.042E-02	2.377E+02	6.042E-02	4.081E+02
2.560E-03	1.879E+00	2.560E-03	3.227E+00
1.467E+02	8.258E+01	1.467E+02	1.418E+02

Chemical name: 1,2,3,4,7,8-Hexachlorodibenzo-p-dioxin   Level III Distribution

Distribution of mass

Distribution of removal rates

Emission rates:

E(1) = 1000 kg/h	E(1) = 0	E(1) = 0	E(1) = 600 kg/h
E(2) = 0	E(2) = 1000 kg/h	E(2) = 0	E(2) = 300 kg/h
E(3) = 0	E(3) = 0	E(3) = 1000 kg/h	E(3) = 100 kg/h

Residence time:

| t = 52983 h | t = 20656 h | t = 78910 h | t = 45877 h |

Reaction   Advection

471

Chemical name: 1,2,3,4,6,7,8-Heptachlorodibenzo-p-dioxin

Level I calculation: (six compartment model)

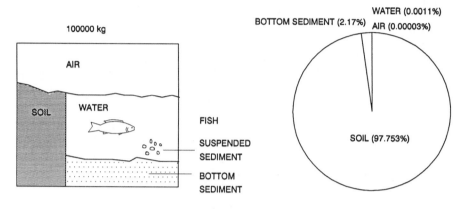

Distribution of mass

physical-chemical properties:

MW: 425.2 g/mol

M.P.: 265 °C

Fugacity ratio: 0.0042

vapor pressure: 7.50×10^{-10} Pa

solubility: 2.40×10^{-6} g/m^3

log K_{OW}: 8.0

Compartment	Z	Concentration			Amount	Amount
	mol/m3 Pa	mol/m3	mg/L (or g/m3)	ug/g	kg	%
Air	4.034E-04	6.958E-16	2.958E-13	2.496E-10	0.0296	2.96E-05
Water	7.526E+00	1.298E-11	5.519E-09	5.519E-09	1.104	0.00110
Soil	1.481E+07	2.554E-05	1.086E-02	4.526E-03	97753	97.753
Biota (fish)	3.763E+07	6.490E-05	2.760E-02	2.760E-02	5.5190	5.52E-03
Suspended sediment	9.257E+07	1.597E-04	6.788E-02	4.526E-02	67.884	6.79E-02
Bottom sediment	2.962E+07	5.109E-05	2.172E-02	9.051E-03	2172.29	2.1723
	Total				100000	100

f = 1.725E-12 Pa

Chemical name: 1,2,3,4,6,7,8-Heptachlorodibenzo-p-dioxin

Level II calculation: (six compartment model)

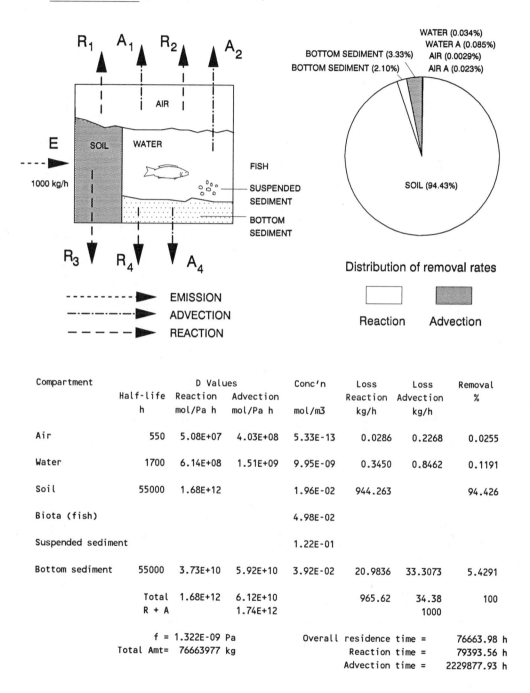

Distribution of removal rates

Reaction   Advection

Compartment	Half-life h	D Values Reaction mol/Pa h	Advection mol/Pa h	Conc'n mol/m3	Loss Reaction kg/h	Loss Advection kg/h	Removal %
Air	550	5.08E+07	4.03E+08	5.33E-13	0.0286	0.2268	0.0255
Water	1700	6.14E+08	1.51E+09	9.95E-09	0.3450	0.8462	0.1191
Soil	55000	1.68E+12		1.96E-02	944.263		94.426
Biota (fish)				4.98E-02			
Suspended sediment				1.22E-01			
Bottom sediment	55000	3.73E+10	5.92E+10	3.92E-02	20.9836	33.3073	5.4291
Total		1.68E+12	6.12E+10		965.62	34.38	100
R + A			1.74E+12			1000	

f = 1.322E-09 Pa
Total Amt= 76663977 kg

Overall residence time =   76663.98 h
Reaction time =   79393.56 h
Advection time = 2229877.93 h

473

Fugacity Level III calculations: (four compartment model)

Chemical name: 1,2,3,4,6,7,8-Heptachlorodibenzo-p-dioxin

Phase Properties and Rates:

Compartment	Bulk Z mol/m3 Pa	Half-life h	D Values Reaction mol/Pa h	Advection mol/Pa h
Air (1)	2.734E-01	550	3.44E+10	2.73E+11
Water (2)	5.080E+02	1700	4.14E+10	1.02E+11
Soil (3)	7.405E+06	55000	1.68E+12	
Sediment (4)	5.924E+06	55000	3.73E+10	5.92E+10

	E(1)=1000	E(2)=1000	E(3)=1000	E(1,2,3)
Overall residence time =	53142	20707	78902	45987 h
Reaction time =	75153	58449	79306	72991 h
Advection time =	181444	32068	15475628	124305 h

Legend: EMISSION (E) · REACTION (R) · ADVECTION (A) · TRANSFER D VALUE mol/Pa h

Diagram (four compartment model):

- Compartments: AIR (1), WATER (2), SOIL (3), SEDIMENT (4)
- Emissions: E_1, E_2, E_3
- Reaction/Advection losses: R_1, A_1, R_2, A_2, R_3, R_4, A_4
- Transfer D values:
  - D_{12} = 8.193E10
  - D_{21} = 5.925E10
  - D_{13} = 7.372E11
  - D_{31} = 7.202E06
  - D_{24} = 2.006E07
  - D_{32} = 1.336E10
  - D_{42} = 4.828E11

Phase Properties, Compositions, Transport and Transformation Rates:

Emission, kg/h

E(1)	E(2)	E(3)
1000	0	0
0	1000	0
0	0	1000
600	300	100

Fugacity, Pa

f(1)	f(2)	f(3)	f(4)
2.087E-09	4.260E-10	9.087E-10	1.265E-09
9.740E-14	5.471E-09	4.241E-14	1.625E-08
9.647E-15	4.319E-11	1.389E-09	1.283E-10
1.252E-09	1.901E-09	6.842E-10	5.647E-09

Amounts, kg

m(1)	m(2)	m(3)	m(4)	Total Amount, kg
2.426E+04	1.840E+04	5.151E+07	1.594E+06	5.314E+07
1.132E+00	2.363E+05	2.404E+03	2.047E+07	2.071E+07
1.121E-01	1.866E+03	7.874E+07	1.616E+05	7.890E+07
1.456E+04	8.213E+04	3.878E+07	7.113E+06	4.599E+07

Concentration, g/m3

C(1)	C(2)	C(3)	C(4)
2.426E-07	9.201E-05	2.861E+00	3.187E+00
1.132E-11	1.182E-03	1.335E-04	4.094E+02
1.121E-12	9.329E-06	4.374E+00	3.232E+00
1.456E-07	4.107E-04	2.154E+00	1.423E+00

Loss, Reaction, kg/h

R(1)	R(2)	R(3)	R(4)
3.057E+01	7.502E+00	6.49E+02	2.008E+01
1.427E-03	9.634E+01	3.03E-02	2.579E+02
1.413E-04	7.606E-01	9.92E+02	2.036E+00
1.834E+01	3.348E+01	4.89E+02	8.962E+01

Loss, Advection, kg/h

A(1)	A(2)	A(4)
2.426E+02	1.840E+01	3.187E+01
1.132E-02	2.363E+02	4.094E+02
1.121E-03	1.866E+00	3.232E+00
1.456E+02	8.213E+01	1.423E+02

Intermedia Rate of Transport, kg/h

T12 air-water	T21 water-air	T13 air-soil	T31 soil-air	T32 soil-water	T24 water-sed	T42 sed-water
7.270E+01	3.634E+03	6.541E+02	2.783E-03	5.164E+00	8.383E+01	3.188E+02
3.393E-03	4.667E-02	3.053E-02	1.299E-07	2.410E-04	1.077E+03	4.094E+02
3.361E-04	3.684E-04	3.024E-03	4.254E-03	7.894E+00	8.500E+00	3.232E+00
4.362E+01	1.456E+02	1.622E-02	2.095E-03	3.888E+00	3.742E+02	1.423E+02

Chemical name: 1,2,3,4,6,7,8-Heptachlorodibenzo-p-dioxin   Level III Distribution

Distribution of mass

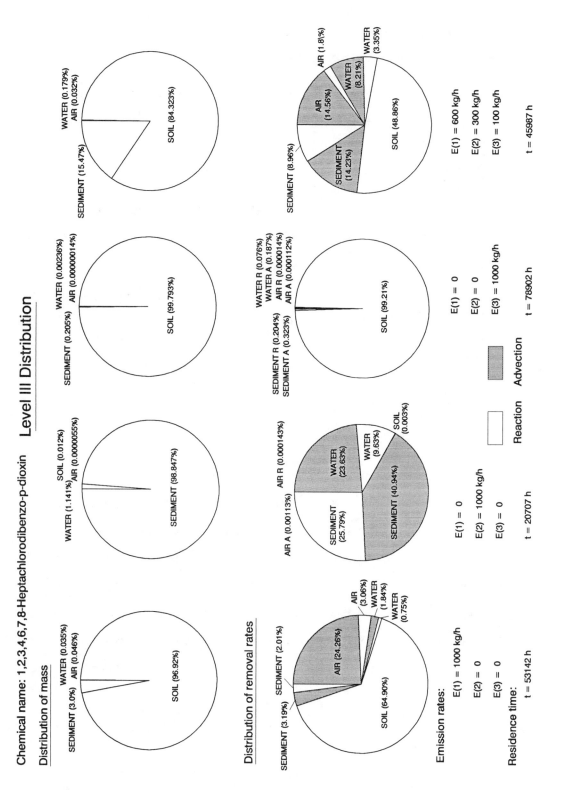

Emission rates:

E(1) = 1000 kg/h   E(1) = 0   E(1) = 0   E(1) = 600 kg/h

E(2) = 0   E(2) = 1000 kg/h E(2) = 0   E(2) = 300 kg/h

E(3) = 0   E(3) = 0   E(3) = 1000 kg/h E(3) = 100 kg/h

Residence time:

t = 53142 h   t = 20707 h   t = 78902 h   t = 45987 h

Chemical name: Octachlorodibenzo-p-dioxin

Level I calculation: (six compartment model)

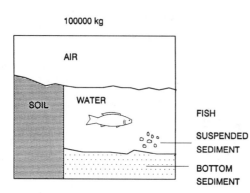

100000 kg

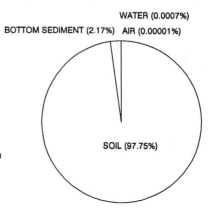

WATER (0.0007%)
BOTTOM SEDIMENT (2.17%) AIR (0.00001%)

SOIL (97.75%)

Distribution of mass

physical-chemical properties:

MW: 460 g/mol

M.P.: 322 $^\circ$C

Fugacity ratio: 0.00115

vapor pressure: 1.10×10^{-10} Pa

solubility: 7.4×10^{-8} g/m^3

log K_{OW}: 8.20

Compartment	Z	Concentration			Amount	Amount
	mol/m3 Pa	mol/m3	mg/L (or g/m3)	ug/g	kg	%
Air	4.034E-04	2.088E-16	9.606E-14	8.103E-11	0.0096	9.61E-06
Water	1.462E+01	7.570E-12	3.482E-09	3.482E-09	0.696	0.00070
Soil	4.561E+07	2.361E-05	1.086E-02	4.526E-03	97754	97.754
Biota (fish)	1.159E+08	5.999E-05	2.760E-02	2.760E-02	5.5191	5.52E-03
Suspended sediment	2.851E+08	1.476E-04	6.788E-02	4.526E-02	67.884	6.79E-02
Bottom sediment	9.123E+07	4.722E-05	2.172E-02	9.051E-03	2172.30	2.1723
	Total				100000	100

f = 5.176E-13 Pa

Chemical name: Octachloro-dibenzo-p-dioxin

Level II calculation: (six compartment model)

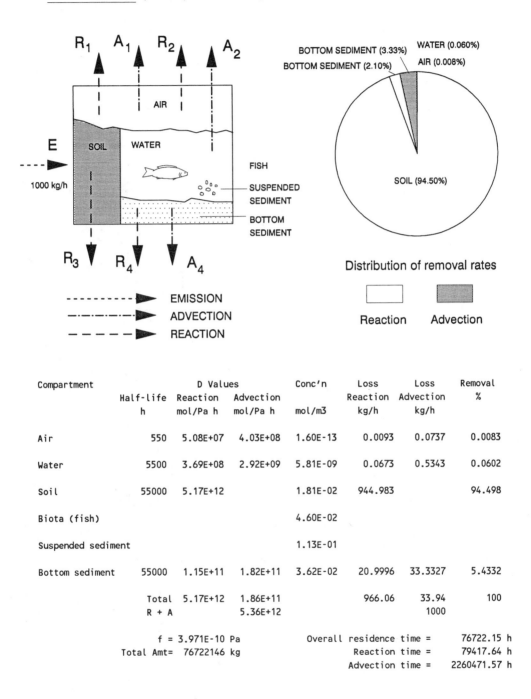

Distribution of removal rates

Reaction   Advection

Compartment	Half-life h	D Values Reaction mol/Pa h	Advection mol/Pa h	Conc'n mol/m3	Loss Reaction kg/h	Loss Advection kg/h	Removal %
Air	550	5.08E+07	4.03E+08	1.60E-13	0.0093	0.0737	0.0083
Water	5500	3.69E+08	2.92E+09	5.81E-09	0.0673	0.5343	0.0602
Soil	55000	5.17E+12		1.81E-02	944.983		94.498
Biota (fish)				4.60E-02			
Suspended sediment				1.13E-01			
Bottom sediment	55000	1.15E+11	1.82E+11	3.62E-02	20.9996	33.3327	5.4332
	Total	5.17E+12	1.86E+11		966.06	33.94	100
	R + A		5.36E+12			1000	

f = 3.971E-10 Pa
Total Amt= 76722146 kg

Overall residence time =   76722.15 h
Reaction time =   79417.64 h
Advection time = 2260471.57 h

477

Fugacity Level III calculations: (four compartment model)

Chemical name: Octachlorodibenzo-p-dioxin

Phase Properties and Rates:

Compartment	Bulk Z mol/m3 Pa	Half-life h	D Values Reaction mol/Pa h	Advection mol/Pa h
Air (1)	5.086E-01	550	6.41E+10	5.09E+11
Water (2)	1.556E+03	5500	3.92E+10	3.11E+11
Soil (3)	2.281E+07	55000	5.17E+12	
Sediment (4)	1.825E+07	55000	1.15E+11	1.82E+11

	E(1)=1000	E(2)=1000	E(3)=1000	E(1,2,3)
Overall residence time =	53270	75702	78914	46518 h
Reaction time =	75702	72008	79347	75736 h
Advection time =	179767	32129	14468882	120581 h

EMISSION (E)

REACTION (R)

ADVECTION (A)

TRANSFER D VALUE mol/Pa h

Phase Properties, Compositions, Transport and Transformation Rates:

Emission, kg/h

E(1)	E(2)	E(3)	f(1)	f(2)	f(3)	f(4)
1000	0	0	1.036E-09	1.372E-10	2.728E-10	4.075E-10
0	1000	0	1.690E-14	1.762E-09	4.447E-15	5.234E-09
0	0	1000	2.698E-15	1.390E-11	4.169E-10	4.128E-11
600	300	100	6.218E-10	6.123E-10	2.054E-10	1.819E-09

Fugacity, Pa

C(1)	C(2)	C(3)	C(4)
2.425E-07	9.819E-05	2.862E+00	3.420E+00
3.953E-12	1.261E-03	4.666E-05	4.393E+01
6.312E-13	9.946E-06	4.374E+00	3.464E-01
1.455E-07	4.382E-04	2.155E+00	1.526E+01

Concentration, g/m3

m(1)	m(2)	m(3)	m(4)
2.425E+04	1.964E+04	5.152E+07	1.710E+06
3.953E-01	2.522E+05	8.398E+02	2.196E+07
6.312E-02	1.989E+03	7.874E+07	1.732E+05
1.455E+04	8.765E+04	3.878E+07	7.632E+06

Amounts, kg

Total Amount, kg

| 5.327E+07 |
| 2.222E+07 |
| 7.891E+07 |
| 4.652E+07 |

Emission, kg/h

E(1)	E(2)	E(3)	R(1)	R(2)	R(3)	R(4)
1000	0	0	3.055E+01	2.474E+00	6.49E+02	2.155E+01
0	1000	0	4.981E-04	3.178E+01	1.06E-02	2.767E+02
0	0	1000	7.953E-05	2.506E-01	9.92E+02	2.183E+00
600	300	100	1.833E+01	1.104E+01	4.89E+02	9.617E+01

Loss, Reaction, kg/h

A(1)	A(2)	A(3)	A(4)
2.425E+02	1.964E+01	3.953E+02	3.420E+01
3.953E-03	2.522E+02	1.630E-02	4.393E+02
6.312E-04	1.989E+00	2.183E+00	3.464E+00
1.455E+02	8.765E+01	9.617E+01	1.526E+02

Loss, Advection, kg/h

Intermedia Rate of Transport, kg/h

	T12	T13	T21	T31	T32	T24	T42
	air-water	air-soil	water-air	soil-air	soil-water	water-sed	sed-water
	7.270E+01	6.543E+02	1.269E-03	1.619E-03	5.160E+00	8.996E+01	3.421E+01
	1.185E-03	1.067E-02	1.630E-01	2.639E-08	8.412E-05	1.155E+03	4.393E+02
	1.892E-04	1.703E-03	1.286E-04	2.474E-03	7.887E+00	9.111E+00	3.465E+00
	4.362E+01	5.665E-03	3.926E+01	1.219E-03	3.885E+00	4.015E+02	1.527E+02

Chemical name: Octachloro-dibenzo-p-dioxin   Level III Distribution

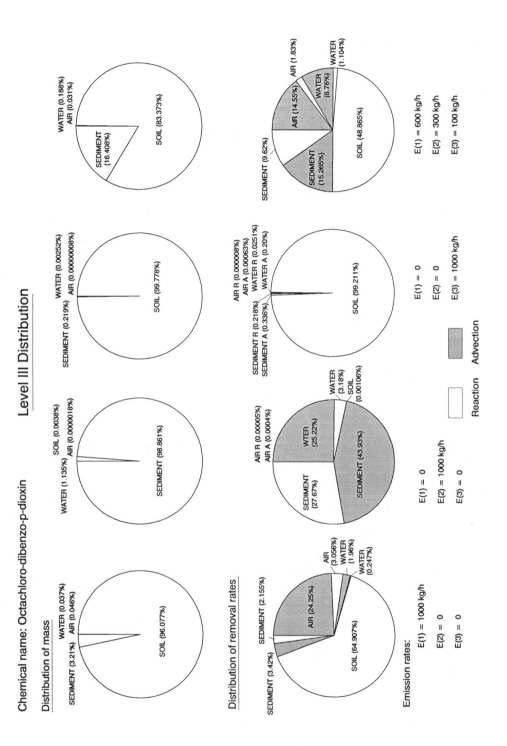

Distribution of mass

WATER (0.037%)
AIR (0.046%)
SEDIMENT (3.21%)
SOIL (96.077%)

WATER (1.135%)
SOIL (0.00038%)
AIR (0.0000018%)
SEDIMENT (98.861%)

WATER (0.00252%)
AIR (0.0000008%)
SEDIMENT (0.219%)
SOIL (99.778%)

WATER (0.188%)
AIR (0.031%)
SEDIMENT (16.408%)
SOIL (83.373%)

Distribution of removal rates

SEDIMENT (3.42%)
SEDIMENT (2.155%)
AIR (24.25%)
AIR (3.056%)
WATER (1.96%)
WATER (0.247%)
SOIL (64.907%)

AIR R (0.00005%)
AIR A (0.0004%)
WTER (25.22%)
WATER (3.18%)
SOIL (0.00106%)
SEDIMENT (27.67%)
SEDIMENT (43.93%)

AIR R (0.000008%)
AIR A (0.00063%)
WATER R (0.0251%)
WATER A (0.20%)
SEDIMENT R (0.218%)
SEDIMENT A (0.336%)
SOIL (99.211%)

AIR (1.83%)
WATER (1.104%)
AIR (14.55%)
WATER (8.76%)
SEDIMENT (9.62%)
SEDIMENT (15.265%)
SOIL (48.865%)

Emission rates:

E(1) = 1000 kg/h
E(2) = 0
E(3) = 0

E(1) = 0
E(2) = 1000 kg/h
E(3) = 0

E(1) = 0
E(2) = 0
E(3) = 1000 kg/h

E(1) = 600 kg/h
E(2) = 300 kg/h
E(3) = 100 kg/h

Reaction   Advection

3.4   COMMENTARY ON THE PHYSICAL-CHEMICAL PROPERTIES AND ENVIRONMENTAL FATE

QSPR Plots

The QSPR plots of the "dioxins" in Figure 3.1 indicate that total surface area (TSA) and molar volume (V_M) are well correlated, although it is probable that TSA is fundamentally more accurate as a descriptor. There may be loss of correlation by about half a log-unit by using the simpler V_M as descriptor. Little significance should be placed on the absolute values of the LeBas molar volume.

Figure 3.2 shows the expected steady drop in subcooled liquid solubility as a function of V_M with a slope of about 0.035 log-units per cm^3/mol, or about a factor of 5.0 in solubility (0.70 log units) for every 20 cm^3/mol added. This is about the increase in V_M for each chlorine added. It must be emphasized that some of the properties of the more chlorinated dioxins with very low solubilities and high melting points are subject to considerable error, especially in the fugacity ratio. It seems likely that these solubilities are predictable within a factor of about 4. There seems to be a tendency for the slope of the solubility-V_M curve to be reduced for the larger congeners. This may, or may not, be real.

The vapor pressure QSPR plot in Figure 3.3 is similar but with slightly more scatter in the data, and with a steeper slope of about 0.047 log-units per cm^3/mol, corresponding to a factor of 8.7 per 20 cm^3/mol.

The log K_{OW} plot in Figure 3.4 does not contain the somewhat suspect fugacity ratio correction, and more data have been determined, thus the relationship with V_M is better established. The slope for up to a chlorine number of 6 is about 0.03 log-unit per cm^3/mol. The implication is that a 20 cm^3/mol increase in V_M causes K_{OW} to increase by 0.60 log units or a factor of 4.0 while water solubility falls by a factor of 5. There appears to be a "levelling-off" at high chlorine numbers. The near-inverse K_{OW}-solubility relationship is shown in the plot of log K_{OW} versus solubility in Figure 3.6 in which the slope is about 0.8.

The log H plot in Figure 3.5 shows considerable scatter but the scale is expanded. There is a definite decrease in H with molar volume, i.e., the more chlorinated dioxins have lower air-water partition coefficients. The slope of about 0.012 units per cm^3/mol, or a factor of about 1.7 per 20 cm^3/mol, is attributable to the fact that increasing molar volumes causes a more rapid decrease in vapor pressure than in solubility.

In summary, a 20 cm^3/mol increase in LeBas molar volume (corresponding to an additional chlorine atom) causes approximately
  (1) a decrease in log solubility of 0.70 units (factor of 5.0).
  (2) a decrease in log vapor pressure of 0.94 units (factor of 8.7).
  (3) a decrease in log H of 0.24 units (factor of 1.7)
  (4) an increase in log K_{OW} of 0.60 units (factor of 4.0).

The net result is a consistent increase in hydrophobicity with molecular mass and also a steadily increased tendency, but of reduced magnitude, to partition from air into water.

Selected Values

The physical-chemical properties listed for each chemical were inspected, examined in the light of the QSPR plots and values selected as reported in Tables 3.1 and 3.2. The sparse reaction rate data were examined and half-life classes suggested in Table 3.3. The air half-lives range from 55 h to 550 h for pentachlorodibenzo-*p*-dioxin (penta-CDD) and higher congeners. Water phase half-lives are generally longer, with the soil half-lives believed to be about a factor of 30 or 3 classes longer still. Sediment half-lives are believed to be longer still. The higher congeners are thus well-preserved in sediments. Most interest has focussed on the remarkably toxic 2,3,7,8-TCDD which appears to be quite persistent when in sorbed state in soils and sediments. The half-lives assigned to the air phase are sufficiently long that most loss from a local atmosphere is by advection rather than reaction, thus for local estimations of exposure all that is needed is the knowledge that the half-life exceeds the local atmospheric residence time. The rates in water are regarded as quite speculative and depend on the local conditions of microbial activity and sunlight intensity.

These values should be used only for approximate general assessments. For more detailed site-specific assessments estimates should be made of partitioning and reactivity under local conditions, preferably by experiment.

Evaluative Calculations

Level I, II and III calculations are shown for eleven selected "dioxins", the aim being to cover the range of properties from dibenzo-*p*-dioxin (DD) to octachlorodibenzo-*p*-dioxin (OCDD). It is possible to infer the likely environmental behavior of other congeners from those of similar structure and hence properties.

The Level I calculation shows the strong tendency of the relatively involatile and hydrophobic dioxins to partition into soil. About 15% of the DD partitions into air and water, but this figure drops to about 5% for a monochloro-dibenzo-*p*-dioxin (mono-CDD), less than 1% for a dichloro-dibenzo-*p*-dioxin (di-CDD), less than 0.1% for a trichloro-dibenzo-*p*-dioxin (tri-CDD) and even less for the more highly chlorinated congeners. Clearly these substances "seek" solid organic carbon phases.

The Level II calculation suggests that reaction and advection in air are dominant for dioxin and mono-DDs. For the di-CDDs, soil reaction processes become more important because of the reduced tendency to partition into air. The tri-CDDs and higher congeners are almost entirely removed by reaction in soil despite the relatively long half-lives. The strong partitioning to organic carbon leaves little opportunity for removal into other media.

The Level III calculations are regarded as most faithful to reality. If discharged to air the DD and, to a lesser extent, the mono-CDDs are found in air they are primarily removed from that medium by advection and reaction. The di-CDD is mainly found in soil and this partitioning is even more pronounced for the higher congeners. Less than 0.1% of the OCDD remains in the air. This is the result of the decreasing volatility, the increasing hydrophobicity and the increasing tendency to partition to aerosols and thus become subject to deposition from the atmosphere.

If discharged to water there is a tendency to partition, i.e., deposit to sediments which increases with chlorination, with some 90% of the di-CDD being associated with sediment, the corresponding figure being 97% for the tri-CDD, 98% fo the tetra-TCDD and ultimately, 99% for the octa-CDD. The primary removal mechanism is reaction in water, with advection from water and reaction in sediment also becoming important for the higher congeners.

If discharged to soil, all congeners display a strong tendency to remain there, subject to relatively slow degrading reactions.

The overall environmental persistence of these chemicals increases considerably with chlorine number. For a ratio of 60:30:10 discharge to air, water and soil, the persistence or residence time increases from 12 days for DD to 70 days for di-CDD to 600 days for tetra-CDD and ultimately to over 5 years for octa-CDD.

In summary, this series of chemicals shows a wide variation in properties and persistence. Progressive addition of chlorines causes reduced volatility and increased hydrophobicity which is displayed as a strong tencency to partition into soils and sediments. The important transport processes are deposition from air to water in association with aerosols, and from water to sediment with particles in the water column. As this tendency to partition into soil phase increases, the rate of reaction in these media become very important determinants of environmental persistence, and thus of concentration.

3.5 REFERENCES

Adams, W.J., Blaine, K.A. (1986)   A water solubility determination of 2,3,7,8-TCDD. *Chemosphere* 15, 1397-1400.

Adams, W.J., DeGraeve, G.M., Sabourin, T.D., Cooney, J.D., Mosher, G.M. (1986) Toxicity and bioconcentration of 2,3,7,8-TCDD to fathead minnows (*pimephales promelas*). *Chemosphere* 15, 1503-1511.

Andren, A.W. (1986) personal communication.

Atkinson, R. (1987a) Estimation of OH radical reaction rate constants and atmospheric lifetimes for polychlorobiphenyls, dibenzo-p-dioxins, and dibenzofurans. *Environ. Sci. Technol.* 21, 305-307.

Atkinson, R. (1987b) A structure-activity relationship for the estimation of rate constants for the gas-phase reactions of OH radicals with organic compounds. *Int'l. J. Chem. Kinet.* 19, 799-828.

Atkinson, R., Arey, J., Zielinska, B., Aschmann, S.M. (1987) Kinetics and products of the gas phase reactions of OH radicals and N_2O_5 with naphthalene and biphenyl. *Environ. Sci. Technol.* 21, 1014-1022.

Atkinson, R., Arey, J., Zielinska, B., Aschmann, S. M. (1990) *Int'l. J. Chem. Kinet.* 22, 999.

Atkinson, R., Aschmann, S.M. (1987) Kinetics of the gas-phase reactions of alkylnaphthalenes with O_3, N_2O_5 and OH radicals at 298 $\pm$ 2 K. *Atmos. Environ.* 21, 2323-2326.

Atkinson, R., Aschmann, S.M. (1988) *Int'l. J. Chem. Kinet.* 20, 513.

Atkinson, R., Lloyd, A.C., Wings, L. (1982)   An updated chemical mechanism for hydrocarbon/NO_X/SO_2 photooxidations. *Atmos. Environ.* 16, 1341.

Atkinson, R., Tuazon, E.C., Arey, J. (1990) *Int'l. J. Chem. Kinet.* 22, 1071.

Banerjee, S., Yalkowsky, S.H., Valvani, S.C. (1980) Water solubility and octanol/water partition coefficient of organics. Limitations of solubility-partition coefficient correlation. *Environ. Sci. Technol.* 14:1227-1229.

Barbeni, M., Pramauro, E., Pelizzetti, E., Borgarello, E., Serpone, N., Jamieson, M.A. (1986) Photochemical degradation of chlorinated dioxins, biphenyls, phenols and benzene on semiconductor dispersion. *Chemosphere* 15, 1913-1916.

Barrie, W.G.K., Sarna, L.P., Muir, D.C.G. (1983) K_{ow} of 1,3,6,8-T_4CDD and OCDD by reverse phase HPLC. Abstract, *ACS Symposium*, 186th National Meeting, Div. of Environmental Chemistry 23(2), 316.

Bidleman, T.F. (1984). Estimation of vapor pressures for nonpolar organic compounds by capillary gas chromatography. Anal. Chem. 56:2490-2496.

Bidleman, T.F., Foreman, W.T. (1987). Vapor-particle partitioning of semivolatile organic compounds. In: *Sources and Fates of Aquatic Pollutants*. Hite, R.A., Eisenreich, S.J., Editors, *Advances in Chemistry Series* 216, American Chemical Society, Washington, D.C.

Birnbaum, L.S. (1985)   The role of structure in the disposition of halogenated aromatic xenobiotics. *Environ. Health Perspectives* 61, 11-20.

Birnbaum, L.S., Decad, G.M., Matthews, H.B. (1980) Disposition and excretion of 2,3,7,8-tetrachlorodibenzofuran in the rat. *Toxicol. Appl. Pharmacol.* 55, 342-352.

Birnbaum, L.S., Decad, G.M., Matthews, H.B., McConnell, E.E. (1981) Fate of 2,3,7,8-tetrachlorodibenzofuran in the monkey. *Toxicol. Appl. Pharmacol.* 57, 189-196.

Boddington, M.J., Gilman, A.P., Newhook, R.C., Braune, B.M., Hay, D.J., Shantova, V. (1990) *Polychlorinated Dioxins and Polychlorinated Dibenzofurans.* Priority Substances List Assessment Report no.1, Canadian Environmental Protection Act, Environment Canada, Ottawa.

Boublik, T., Fried, V., Hala, E. (1973) *The Vapour Pressure of Pure Substances.* Elsevier, Amsterdam.

Boublik, T., Fried, V., Hala, E. (1984) *The Vapour Pressures of Pure Substances.* (second revised edition), Elsevier, Amsterdam.

Branson, D.R., Takahashi, I.T., Parker, W.M., Blau, G.E. (1983) Bioconcentration kinetics of 2,3,7,8-tetrachlorodibenzo-p-dioxin (TCDD) in rainbow trout. *Dioxin Symposium*, September of 1983, Michigan State University, E. Lansing, Michigan.

Branson, D.R., Takahashi, I.T., Parker, W.M., Blau, G.E. (1985) Bioconcentration kinetics of 2,3,7,8-tetrachlorodibenzo-p-dioxin in rainbow trout. *Environ. Toxicol. & Chem.* 4, 779-788.

Broman, D., Näf, C., Rolff, C., Zebühr, Y. (1991) Occurence and dynamics of polychlorinated dibenzo-p-dioxins and dibenzofurans and polycyclic aromatic hydrocarbons in mixed surface layer of remote coastal and offshore waters of the Baltic. *Environ. Sci. Technol.* 25(11), 1850-1864.

Burkhard, L.P., Kuehl, D.W. (1986). n-Octanol/water partition coefficients by reverse phase liquid chromatography/mass spectrometry for eight tetrachlorinated planar molecules. Chemosphere 15:163-167.

Burkhard, L.P. (1985). Evaluation of reverse phase LC/MS for estimation of n-octanol/water partition coefficients of organic chemicals. *Chemosphere* 14:1551-1560.

Buser, H.R. (1988) Rapid photolytic decomposition of brominated and brominated/chlorinated dibenzodioxins and dibenzofurans. *Chemosphere* 17, 889-903.

Callahan, M.A., Slimak, M.W., Gabel, N.W., May, I.P., Fowler, C.F., Freed, J.R., Jennings, P., Durfee, R.L., Whitmore, F.C., Maestri, B., Mabey, W.R., Holt, B.R., Gould, C. (1979). *Water-Related Environmental Fate of 129 Priority Pollutants.* Vol. I, EPA Report No. 440/4-79-029a. Versar, Inc., Springfield, Virginia.

Chessells, M., Hawker, D.W., Connel, D.W. (1991) Critical evaluation of the measurement of 1-octanol/water partition coefficient of hydrophobic compounds. *Chemosphere* 12, 1175-1190.

Chiou, C.T., Porter, P.E., Schmedding, D.W. (1983). Partition equilibria of nonionic organic compounds between soil organic matter and water. *Environ. Sci. Technol.* 17:227-231.

Chiou, C.T., Freed, V.H., Schmedding, D.W., Kohnert, R.L. (1977). Partition coefficient and bioaccumulation of selected organic chemicals. *Environ. Sci. Technol.* 11:5:475-478.

Chiou, C.T., Peters, L.J., Freed, V.H. (1979). A physical concept of soil-water equilibria for nonionic organic compounds. Science 206:831-832.

484

Choudhary, G.G., Hutzinger, O. (1982) Photochemical formation and degradation of polychlorinated dibenzofurans and dibenzo-p-dioxins. *Residue Rev.* 84, 113-161.

Choudhary, G.G., Webster, G.R.B. (1985a) Environmantal photochemistry of PCDDs. Part I. Kinetics and quantum yields of the photodegradation of 1,2,3,4,7-penta- and 1,2,3,4,7,8-hexachlorodibenzo-p-dioxin in aqueous acetonitrile. *Chemosphere* 14, 9-26.

Choudhary, G.G., Webster, G.R.B. (1985b) Quantum yields for the photodecomposition of polychlorinated dibenzo-p-dioxins (PCDDs) in water-acetonitrile solution. *Chemosphere* 14, 893-896.

Choudhary, G.G., Webster, G.R.B. (1985c) Protocol guidelines for the investigations of photochemical fate of pesticides in water, air and soil. *Residue Rev.* 96, 80-136.

Choudhary, G.G., Webster, G.R.B. (1986) Photochemical quantum yields and sunlight half-lives of polychlorodibenzo-p-dioxins in aquatic systems. *Chemosphere* 15, 1935-1940.

Choudhry, G.G., Webster, G.R.B. (1989) Environmental photochemistry of OCDDs. 2. Quantum yields of the direct phototransformation of 1,2,3,7-tetra-, 1,3,6,8-tetra-, 1,2,3,4,6,7,8-hepta-, and 1,2,3,4,6,7,8,9-octachlorodibenzo-p-dioxin in aqueous acetonitrile and their sunlight half-lives. *J. Agric. Food Chem.* 37, 254-251.

Clark, K.E., Mackay, D. (1991) Dietary uptake and biomagnification of four chlorinated hydrocarbons by guppies. *Environ. Toxicol. Chem.* 10, 1205-1217.

Connett, P., Webster, T. (1987) An estimation of the relative human exposure to 2,3,7,8-TCDD emissions via inhalation and ingestion of cow's milk. *Chemosphere* 16, 2079-2084.

Corbet, R.L., Muir, D.C.G., Webster, G.R.B. (1983) Fate of 1,3,6,8-T₄CDD in an outdoor-aquatic system. *Chemosphere* 12(4/5), 523-527.

Corbet, R.L., Webster, G.R.B., Muir, D.C.G. (1988) Fate of 1,3,6,8-tetrachlorodibenzo-p-dioxin in an outdoor-aquatic system. *Environ. Toxicol. & Chem.* 7, 167-180.

Crosby, D.G. (1985) The degradation and disposal of chlorinated dioxins. In: *Dioxins in the Environment*. Kamrin, M.A., Rogers, P.W., Editors, Hamisphere Publication Corporation, Washington. pp.195-204.

Crosby, D.G., Moilanen, K.W. (1973) Photodecomposition of chlorinated biphenyls and dibenzofurans. *Bull. Environ. Contam. Toxicol.* 6, 372.

Crosby, D.G., Wong, A.S. (1977) Environmental degradation 2,3,7,8-tetrachlorodibenzo-p-dioxin (TCDD). *Science* 195, 1337-1778.

Crosby, D.G., Wong, A.S., Plimmer, J.R., Woolson, E.A. (1971) Photodecomposition of chlorinated dibenzo-p-dioxins. *Science* 173, 748.

Crummett, W.B., Stehl, R.H. (1973) Determination of chlorinated dibenzo-p-dioxins and dibenzofurans in various materials. *Environ. Health Perspectives* 5, 15-25.

Cull, M.R., Dobbs, A.J., Williams, N. (1983) Polychlorodibenzo-p-dioxins (PCDDs) in commercial pentachlorophenol (PCP) used in wood preservation. *Chemosphere* 12(4/5), 483-485.

Davenport, J., Gu, C-L., Hendrey, D.G., Mill, T. (1984) Estimation of rate constants for reaction of hydroxyl radical with organic compounds.

Davies, R.P., Dobbs, A.J. (1984). The prediction of bioconcentration in fish. Water Res. 18:1253-1262.

Dean, J.D., Ed. (1979) *Lange's Handbook of Chemistry.* 12th ed., McGraw-Hill, Inc., New York.

Dean, J.D., Ed. (1985) *Lange's Handbook of Chemistry.* 13th ed. McGraw-Hill, Inc., New York.

Decad, G.M., Birnbaum, L.S., Matthews, H.B. (1981a) 2,3,7,8-tetrachlorodibenzofuran tissue distribution and excretion in guinea pigs. *Toxicol. Appl. Pharmacol.* 57, 231-240.

Decad, G.M., Birnbaum, L.S., Matthews, H.B. (1981b) Distribution and excretion of 2,3,7,8-tetrachlorodibenzofuran in C57BL/6J and DBA/2J mice. *Toxicol. Appl. Pharmacol.* 59, 564-573.

De Voogt, P., Muir, D.C.G., Webster, G.R.B., Govers, H. (1990) Quantitative structure-activity relationships for the bioconcentration in fish of seven polychlorinated dibenzodioxins. *Chemosphere* 21, 1385-1396.

Di Domenico, A., Silano, V., Viviano, G., Zapponi, G. (1980) *Ecotoxicol. Environ. Safety* 4, 339.

Dobbs, A.J., Grant, C. (1979) Photolysis of highly chlorinated dibenzo-p-dioxins by sunlight. *Nature (London)* 278, 163-165.

Dobbs, A.J., Cull, M.R. (1982) Volatilization of chemicals-relative loss rates and the estimation of vapor pressures. *Environ. Pollut.* (series B) 3, 289-298.

Doucette, W.J. (1985) *Measurement and Estimation of Octanol/Water Partition Coefficients and Aqueous Solubilities for Halogenated Aromatic Hydrocarbons.* Ph.D. thesis, University of Wisconsin-Madison.

Doucette, W.J., Andren, A.W. (1987) Correlation of octanol/water partition coefficients and total molecular surface area for highly hydrophobic aromatic compounds. *Environ. Sci. Technol.* 21, 821-824.

Doucette, W.J., Andren, A.W. (1988a) Aqueous solubility of selected biphenyl, furan and dioxin congeners. *Chemosphere* 17, 243-252.

Doucette, W.J., Andren, A.W. (1988b) Estimation of octanol/water partition coefficients: Evaluation of six methods for highly hydrophobic aromatic hydrocarbons. *Chemosphere* 17, 345-359.

Dougherty, E.J., McPeters, A.L., Overcach, M.R. (1991) Kinetics of photodegradation of 2,3,7,8-tetrachlorodibenzo-p-dioxin: theoretical maximum rate of soil decontamination. *Chemosphere* 23, 589-600.

Dulin, D., Drossman, H., Mill, T. (1986) Products and quantum yields for photolysis of chloroaromatics in water. *Environ. Sci. Technol.* 20, 72-77.

Dunn III, W.J., Koehler, M., Stalling, D.L., Schwartz, T.R. (1986) Relationship between gas chromatographic retention of polychlorinated dibenzofurans and calculated molecular surface area. *Anal. Chem.* 58, 1835-1838.

Eduljee, G. (1987) Volatility of TCDD and PCB from soil. *Chemosphere* 16, 907-920.

Eichler, W., Editor (1965) *Handbuch der insektizidkunde.* Veb. Verlag Volk. Gesundheit, Berlin, pp. 1-84.

Eitzer, B.D., Hites, R.A. (1988) Vapor pressures of chlorinated dioxins and dibenzofurans. *Environ. Sci. Technol.* 22, 1362-1364.

Eitzer, B.D., Hites, R.A. (1989) Polychlorinated dibenzo-p-dioxins and dibenzofurans in the ambient atmosphere of Bloomington, Indiana. *Environ. Sci. Technol.* 23, 1389-1395.

Esposito, M.P., Tiernan, T.O., Dryden, F.E. (1980) *Dioxins.* EPA 600/2-80-197, pp.257-270, Cincinnati, Ohio.

Exner, J.H., Johnson, J.D., Ivins, O.D., Wass, M.N., Miller, R.A. (1982) Process for destroying tetrachlorodibenzo-p-dioxin in a hazardous waste. In: *Detoxication of Hazardous Waste.* Exner, J.H., Editor, pp.269-287, Ann Arbor Science Publishers, Ann Arbor, Michigan.

Facchetti, S., Balasso, C., Fichtner, G., Frare, A., Leoni, A., Mauri, C., Vasconi, M. (1986) Studies on the absorption of TCDD by some plant species. *Chemosphere* 15(9-12), 1387-1388.

Facchetti, S., Balasso, C., Fichtner, G., Frare, A., Leoni, A., Mauri, C., Vasconi, M. (1986) Studies on the absorption of TCDD by plant species. In: *Chlorinated Dioxins and Dibenzofurans in Perspective.* Rappe, C., Choudhary, G., Keith, L.H., Editors, Lewis Publishers, Inc., Chelsea, Michigan. pp.225-235.

Fiedler, H., Schramm, K.-W. (1990) QSAR generated octanol-water partition coefficients of selected mixed halogenated dibenzodioxins and dibenzofurans. *Chemosphere* 20, 1597-1602.

Friesen, K.J., Sarna, L.P., Webster, G.R.B. (1985) Aqueous solubility of polychlorinated dibenzo-p-dioxins determined by high pressure chromatography. *Chemosphere* 14, 1267-1274.

Friesen, K.J., Webster, G.R.B. (1990) Temperature dependence of the aqueous solubilities of highly chlorinated dibenzo-p-dioxins. *Environ. Sci. Technol.* 24, 97-101.

Friesen, K.J., Muir, D.C.G., Webster, G.R.B. (1990) Evidence of sensitized photolysis of polychlorinated dibenzo-p-dioxins in natural waters under sunlight conditions. *Environ. Scil Technol.* 24, 1739-1744.

Friesen, K.J., Vilk, J., Muir, D.C.G. (1990) Aqueous solubilities of selected 2,3,7,8-substituted polychlorinated dibenzofurans (PCDFs). *Chemosphere* 20, 27-32.

Fujita, T., Iwasa, J., Hansch, C. (1964). A new substituent constant, "pi" derived from partition coefficients. *J. Am. Chem. Soc.* 86, 5175-5180.

Garten, Jr., C.T., Trabalka, J.R. (1983) Evaluation of models for predicting terrestrial food chain behavior of xenobiotics. *Environ. Sci. Technol.* 17, 590-595.

Gasiewicz, T.A., Geiger, L.E., Rucci, G., Neal, R.A. (1983) Distribution, excretion and metabolism of 2,3,7,8-tetrachlorodibenzo-p-dioxin in C57BZL/6J, DBA/2J and B6D3F/J mice. *Drug Metab. Disp.* 11, 397-403.

Geyer, H., Kraus, A.G., Klein, W., Richter, E., Korte, F. (1980) Relationship between water solubility and bioaccumulation potential of organic chemicals in rats. *Chemosphere* 9, 277-291.

Geyer, H., Politzki, G., Freitag, D. (1984) Prediction of ecotoxicological behaviour of chemicals: relationship between n-octanol/water partition coefficient and bioaccumulation of organic chemicals by Alga Chlorella. *Chemosphere* 13, 269-184.

Geyer, H.J., Scheunert, I., Korte, F. (1986) Bioconcentration potential (BCP) of 2,3,7,8-tetrachlorodibenzo-p-dioxin (2,3,7,8-TCDD) in terrestrial organisms including humans. *Chemophere* 15, 1495-1502.

Geyer, H.J., Scheunert, I., Korte, F. (1987) Correlation between the bioconcentration potential of organic environmental chemicals in humans and their n-octanol/water partition coefficients. *Chemosphere* 16, 239-252.

Gilman, H., Dietrich, J.J. (1957) Halogen derivatives of dibenzo-p-dioxin. *J. Am. Chem. Soc.* 79, 1439-1441.

Gobas, F.A.P.C., Shiu, W.Y., Mackay, D. (1987) Factors determining partitioning of hydrophobic organic chemicals in aquatic organisms. In: *QSAR in Environmental Toxicology II*. Kaiser, K.L.E., (Ed.), D. Reidel Publ. Co., Dordrecht, Holland.

Gobas, F.A.P.C., Muir, D.C.G., Mackay, D. (1988) Dynamics of dietary bioaccumulation and faecal elimination of hydrophobic organic chemicals in fish. *Chemosphere* 17, 493-962.

Gobas, F.A.P.C, Lahittete, J.M., Garofalo, G., Shiu, W.Y., Mackay, D. (1988) A novel method for measuring membrane-water partition coefficients of hydrophobic organic chemicals: comparison with 1-octanol-water partitioning. *J. Pharm. Sci.* 77, 265-272.

Gobas, F.A.P.C., Schrap, S.M. (1990) Bioconcentration fo some polychlorinated dibenzo-p-dioxins and octachlorodibenzofuran in the guppy (*Poecilia reticulata*). *Chemosphere* 20, 495-512.

Govers, H., De Voogt, P. (1989) *Quant. Struct.-Act. Relat.* 8, 11-16.

Govers, H.A.J., Luijk, R., R., Evers, E.H.G. (1990) Calculation of heat of vaporization, molar volume and solubility parameter of polychlorodibenzo-p-dioxins. *Chemosphere* 20, 287-294.

Gray, A.P., Cepa, S.P., Solomon, I.J., Anilin, O. (1976) *J. Org. Chem.* 41, 2435.

Hansch, C., Leo, A. (1985) *Medchem Project* Issue No. 26. Pomona College, Claremont, California.

Hansch, C., Leo, A. (1979) *Substituent Constants for Correlation Analysis in Chemistry and Biology*. Wiley, New York.

Hawker, D. (1990) Description of fish bioconcentration factors in terms of solvatochromic parameters. *Chemosphere* 20, 467-477.

Howard, P.H., Boethling, R.S., Jarvis, W.F., Meylan, W.M., Michalenko, E.M. (1991) *Handbook of Environmental Degradation Rates*. Lewis Publishers, Inc., Chelsea, Michigan.

Ioannou, Y.M., Birnbaum, L.S., Matthews, H.B. (1983) Toxicity and distribution of 2,3,7,8-tetrachlorodibenzofuran in male guinea pigs. *J. Toxicol. Environ. Health 12, 541-553.*

Isensee, A.R., Jones, G.E. (1975) Distribution of 2,3,7,8-tetrachlorodibenzo-*p*-dioxin (TCDD) in aquatic model ecosystem. *Environ. Sci. Technol.* 9, 668-672.

Jaber, H.M., Podoll, T. (1983) Stanford Research Institute (SRI), unpublished results.

Jaber, H.M., Smith, J.H., Cwirla, A.N. (1982) *Evaluation of Gas Saturation Methods to Measure Vapor Pressure*. (EPA Contract No. 68-01-5117) SRI International, Menlo Park, CA.

Jackson, D.R., Roulier, M.H., Grotta, H.M., Rust, S.W., Warner, J.S. (1986) Solubility of 2,3,7,8-TCDD in contaminated soils. In: *Chlorinated Dioxins and Dibenzofurans in Perspective*. Rappe, C., Choudhry, G.G., Leith, L.H., Eds., pp. 185-200, Lewis Publishers, Inc., Chelsea, MI.

Jackson, D.R., Roulier, M.H., Grotta, H.M., Rust, S.W., Warner, J.S., Arthur, M.F., DeRoos, F.L. (1985) Leaching potential of 2,3,7,8--TCDD in contaminated soils. In: *Land Disposal of Hazardous Waste-Proceedings of the Eleventh Annual Research Symposium*. Cincinnati, Ohio. U.S. EPA Report-600/9-85-013. pp.153-168.

Johnson, H. (1982) In: *Aquatic Fate Process Data for Organic Priority Pollutants*. Mabey, W.R., Ed., EPA Final Report on Contract 68-01-3867; U.S. Government Printing Office, Washington, D.C.

Jury, W.A., Russo, D., Streile, G., El Abd, H. (1990) Evaluation of volatilization by organic chemicals residing below the soil surface. *Water Resources Res.* 26, 13-26.

Kaiser, K.L.E. (1983) A non-linear function for the calculation of partition coefficients of aromatic compounds with multiple chlorine substitution. *Chemosphere* 12, 1159-1165.

Kearney, P.C., Isensee, A.R., Helling, C.S., Woolsen, E.A., Plimmer, J.R. (1971) Environmental significance of chlorodioxins. In: *Chlorodioxins-Origin and Fate. Adv. Chem. Ser.* 120, 105-111.

Kearney, P.C., Woolson, E.A., Ellington, Jr., C.P. (1972) Persistence and metabolism of chlorodioxins in soils. *Environ. Sci. Technol.* 6, 1017-1019.

Kearney, P.C., Woolson, E.A., Isensee, A.R., Helling, C.S. (1973) Tetrachlorodibenzodioxin in the environment: Sources, fate and decontamination. *Envirn. Health Perspectives* (No. 5), 273-277.

Kenaga, E.E., Goring, C.A.I. (1980) In: *Aquatic Toxicology*. Eaton, J.G., Parrish, P.R., Hendrick, A.C. (eds.). Am. Soc. for Testing and Materials, STP 707, pp 78-115.

Kenaga, E.E. (1980a) Predicted bioconcentration factors and soil sorption coefficients of pesticides and other chemicals. *Ecotoxicol. Environ. Safety* 4, 26-38.

Kenaga, E.E. (1980b) Correlation of bioconcentration factors of chemicals in aquatic and terrestrial organisms with their physical and chemical properties. *Environ. Sci. Technol.* 14, 553-556.

Kimbrough, R.D., Falk, H., Stehr, P., Fries, G. (1984) Health implications of 2,3,7,8-tetrachlorodibezodioxin contamination of residential soil. *J. Toxicol. & Environ. Health* 14, 47-94.

Kieatiwong, S., Nguyen, L.V., Hebert, V.R., Hackett, M., Miller, G.C., Millie, M.J., Mitzel R. (1990) Photolysis chlorinated dioxins in organic solvent and on soils. *Environ. Sci. Technol.* 24, 1575-1580.

Kissel, J., Robarge, G. (1988) Assessing the elimination of 2,3,7,8-TCDD from humans with a physiologically based pharmacokinetic model. *Chemosphere* 17, 2017-2027.

Kleeman, J.M., Olson, J.R., Chen, S.M., Peterson, R.E. (1986) Metabolism and disposition of 2,3,7,8-tetrachloro-p-dioxin in rainbow trout. *Toxicol. Appl. Pharmacol.* 83, 391-401.

Koester, C.J., Hites, R.A. (1992) Photodegradation of polychlorinated dioxins and dibenzofurans adsorbed to fly ash. *Environ. Sci. Technol.* 26, 502-507.

Kuehl, D.W., Cook, P.M., Batterman, A.R. (1986) Uptake and depuration studies of PCDDs and PCDFs in freshwater fish. *Chemosphere* 15(9-12), 2023-2026.

Kuehl, D.W., Cook, P.M., Batterman, A.R., Lothenbach, D., Butterworth, B.C. (1987) Bioavailability of polychlorinated dibenzo-p-dioxins and dibenzofurans from contaminated Wisconsin river sediment to carp. *Chemosphere* 16(4), 667-679.

Kuroki, H., Haraguchi, K., Masuda, Y. (1984) Synthesis of polychlorinated dibenzofuran isomers and their gas chromatographic profiles. *Chemosphere* 13(4), 561-573.

Lee, M.D., Wilson, J.T., Ward, C.H. (1984) Microbial degradation of selected aromatics in a hazardous waste site. *Devel. Indust. Microbiol.* 25, 557-565.

Leo, A. (1985). Medchem. Project. Issue No.26, Pomona College, Claremont, CA.

Leo, A., Hansch, C., Elkins, D. (1971) Partition coefficients and their uses. *Chemical Reviews* 71, 525-616.

Lodge, K.B. (1989) Solubility studies using a generator column for 2,3,7,8-tetrachlorodibenzo-p-dioxin. *Chemosphere* 18, 933-940.

Lodge, K.B., Cook, P.M. (1989) Partition Studies of dioxin between sediment and water: The measurement of K_{oc} for Lake Ontario sediment. *Chemosphere* 19(1-6), pp. 439-444.

LOGP and Related Computerized Data Base, Pomona College Med-Chem. Project, Pomona College, Claremont, CA.. Technical Data Base (TDS) Inc.

Loonen, H., Parsons, J.R., Govers, H.A.J. (1991) Dietary accumulation of PCDDs and PCDFs in guppies. *Chemosphere* 23, 1349-1357.

Lyman, W.J., Reehl, W.F., Rosenblatt, D.H. (1982) *Handbook on Chemical Property Estimation Methods, Environmental Behavior of Organic Compounds*. McGraw-Hill, New York.

Mabey, W., Mill, T. (1978). Critical review of hydrolysis of organic compounds in water under environmental conditions. *J. Phys. Chem. Ref Data* 7, 383-415.

Mabey, W., Smith, , J.H., Podoll, R.T., Johnson, H.L., Mill, T., Chou, T.W., Gates, J., Waight-Partridge, I., Vanderberg, D. (1982). *Aquatic Fate Process for Organic Priority Pollutants*. EPA Report No. 440/4-81-014.

Mackay, D., Paterson, S., Cheung, B. (1985) Evaluation the environmental fate of chemicals. The fugacity-level III approach as applied to 2,3,7,8-TCDD. *Chemosphere* 15, 1397-1400.

Mamantov, A. (1984) Linear correlation between photolysis rates and toxicity of polychlorinated dibenzo-p-dioxins. *Environ. Sci. Technol.* 18, 808-810.

Marple, L., Berridge, B., Throop, L. (1986a) Measurement of the water-octanol partition coefficient of 2,3,7,8-tetrachlorodibenzo-p-dioxin. *Environ. Sci. Technol.* 20, 397-399.

Marple, L., Brunck, R., Throop, L. (1986b) Water solubility of 2,3,7,8-tetrachlorodibenzo-p-dioxin. *Environ. Sci. Technol.* 20, 180-182.

Matsumura, F., Benezet, H.J. (1973) Studies on the bioaccumulation and microbial degradation of 2,3,7,8-tetrachlorodibenzo-p-dioxin. *Environ. Health Perspet.* 5, 253-258.

McKone, T., Ryan, P. (1989) Human exposures to chemicals through food chains: An uncertainty analysis. *Environ. Sci. Technol.* 23, 1154-1163.

Mehrle, P.M., Buckler, D.R., Little, E.E., Smith, L.M., Petty, J.D., Peterman, P.H., Stalling, D.L., De Graeve, G.M.,Coyle, J.J., Adams, W.J. (1988) Toxicity and bioconcentration of 2,3,7,8-tetrachlorodibenzodioxin and 2,3,7,8-tetrachloro-dibenzofuran in rainbow trout. *Environ. Toxicol. & Chem.* 7, 47-62.

Michaels, R. (1989) *Health Risk Assessment for the Planned St. Lawrence County, New York.* New York Resource Recovery Facility. RAMTRAC, Long Island City, New York.

Mill, T. (1985) Prediction of the environmental fate of tetrachlorodibenzodioxin. In: *Dioxins in the Environment.* Kamrin, M.A., Rogers, P.W., Editors, Hemisphere Publication Corp., Washington. pp. 173-193.

Mill, T., Mabey, W.R., Bomberger, D.C., Chou, T.W., Hendry, D.G., Smith, J.H. (1982) *Laboratory Protocols for Evaluating the Fate of Organic Chemicals in Air and Water.* EPA Final Report, U.S. EPA 600/3-82-022.

Mills, W.B., Dean, J.D., Porcella, D.B., Gherini, S.A., Hudson, R.J.M., Frick, W.E., Rupp, G.L., Bowie, G.L. (1982). *Water Quality Assessment: A Screening Procedure for Toxic and Conventional Pollutants.* Part 1, EPA-600/6-82-004a.

Muir, D.C.G., Townsend, B.E., Webster, G.R.B. (1984) Bioavailability of ^{14}C-1,3,6,8-tetrachlodibenzodioxin and ^{14}C-octachlorodibenzodioxin to aquatic insects in sediment and water. In: *Chlorinated Dioxins and Dibenzofurans in the Total Environment, II.* Keith, L.H., Rappe, C., Choudhary, G., Eds., pp. 89-102, Butterworth, Boston, MA.

Muir, D.C.G., Marshall, W.K., Webster, G.R.B. (1985) Bioconcentration of PCDDs by fish: effects of molecular structure and water chemistry. *Chemosphere* 14, 829-833.

Muir, D.C.G., Yarechewski, A.L. (1988) Dietary accumulation of four chlorinated dioxin congeners by rainbow trout and fathead minnows. *Environ. Toxicol. Chem.* 7, 227-236.

Muir, D.C.G., Yarechewski, A.L., Knoll, A., Webster, G.R.B. (1986) Bioconcentration and disposition of 1,3,6,8-tetrachlorodibenzo-p-dioxin and octachlorodibenzo-p-dioxin by rainbow trout and fathead minnows. *Environ. Toxicol. Chem.* 5, 261-272.

Muir, D.C.G., Yarechewski, A.L., Metner, D.A., Lockhart, W.L., Webster, G.R.B., Friesen, K.J. (1990) Dietary accumulation and sustained hepatic mixed function oxidase enzyme induction by 2,3,7,8-pentachlorodibenofuran in rainbow trout. *Environ. Toxicol. & Chem.* 9, 1463-1472.

Muto, H., Shinada, M., Takizawa, Y. (1991) Heterogeneous photolysis of polychlorinated dibenzo-p-dioxins on fly ash in water-acetonitrile solution in relation to the reaction with ozone. *Environ. Sci. Technol.* 25, 316-322.

Nash, R.G., Beall, M.L. (1980) Distribution of Silvex, 2,4-D and TCDD applied to turf in chambers and field plots. *J. Agric. Food Chem.* 28, 614-623.

Nauman, C.H., Schaum., J.L. (1987) Human exposure estimation for 2,3,7,8-TCDD. *Chemosphere* 16, 1851-1856.

Neely, W.B. (1976) Predicting the flux of organics across the air/water interface. In: National Conference on Control of Hazardous Material Spills, New Orlands.

Neely, W.B. (1979) Estimation rate constants for the uptake and clearance of chemicals by fish. *Environ. Sci. Technol.* 13, 1506-1510.

Neely, W.B., Branson, D.R., Blau, G.E. (1974) Partition coefficient to measure bioconcentration potential of organic chemicals in fish. *Environ. Sci. Technol.* 8, 1113-1115.

Nestrick, T.J., Lamparski, L.L., Townsend D.I. (1980) Identification of tetrachlorodibenzo-p-dioxin isomers at the 1-ng level by photolytic degradation and pattern recognition techniques. *Anal. Chem.* 52, 1865-1874.

Niimi, A.J., Oliver, B.G. (1983) Biological half-lives of polychlorinated biphenyl (PCB) congeners in whole fish and muscle of rainbow trout (*Salmo gairdnberi*). *Can. J. Fish. Aquat. Sci.* 40, 1388-1394.

Niimi, A.J., Oliver, B.G. (1986) Biological half-lives of chlorinated dibenzo-p-dioxins and dibenzofurans in rainbow trout (*Salmo gairdneri*). *Envrion. Toxicol. Chem.* 5, 49-53.

Nirmalakhandan, N.N., Speece, R.E. (1989) Prediction of aqueous solubility of organic chemicals based on molecular structure. 2. Application to PNAs, PCBs, PCDDs, etc. *Environ. Sci. Technol.* 23, 708-713.

Norback, D.H., Engblom, J.F., Allen, J.R. (1975) Tissue distribution and excretion of octachlorodibenzo-p-dioxin in rat. *Toxicol. Appl. Pharmacol.* 32, 330-338.

OECD (1981) *OECD Gudielines for Testing of Chemicals.* Organization for Economic Co-operation and Development. OECD, Paris.

O'Keefe, P.W., Hilker, D.R., Smith, R.M., Aldous, K.M., Donnelly, R.J., Long, D., Pope, D.H. (1986) Nonaccumulation of chlorinated dioxins and furans by goldfish exposed to contaminated sediment and flyash. *Bull. Environ. Contam. Toxicol.* 36, 452-459.

Olling, M., Derks, H.J.G.M., Berende, P.L.M., Liem, A.K.D., de Jong, A.P.J.M. (1991) Toxicokinetics of eight ^{13}C-labelled polychlorinated dibenzo-p-dioxins and -furans in lactating cows. *Chemosphere* 23, 1377-1385.

Olsen, J.R., Gasiewicz, T.A., Neal, R.A. (1980) Tissue distribution, excretion and metabolism of 2,3,7,8-tetrachlorodibenzo-p-dioxin (TCDD) in the golden Syrian hamster. *Toxico. Appl. Pharmacol.* 56, 78-85.

Opperhuizen, A. (1986) Bioconcentration of hydrophobic chemicals in fish. In: *Aquatic Toxicology and Environmental Fate.* 9th Vol., Poston, T.M., Purdy, R., Editors, ASTM STP 921, pp.304-315., American Society for Testing and Materials, Philadephia.

Opperhuizen, A., Vander Velde, E.W., Gobas, F.A.P.C., Liem, D.A.K., Van der Steen, J.M., Hutzinger, O. (1985) Relationship between bioconcentration in fish and steric factors of hydrophobic chemicals. *Chemosphere* 14, 1871-1896.

Opperhuizen, A., Wagenaar, W.J., Van der Wielen, F.W.M., Van den Berg, M., Olie, K., O. Hutzinger, O., Gobas, F.A.P.C. (1986) Uptake and elimation of PCDD/PCDF congeners by fish after aqueous exposure to a fly ash extract from a municipal incinerator. *Chemosphere* 15, 2049-2053.

Opperhuizen, A., Sijm, D.T.H.M. (1990) Bioaccumulation and biotransformation of polychlorinated dibenzo-p-dioxins and dibenzofurans in fish. *Environ. Toxicol. & Chem.* 9, 175-186.

Orth, R.G., Ritchie, C., Hileman, F. (1989) Measurement of the photoinduced loss of vapor phase TCDD. *Chemosphere* 18, 1275-1282.

Palauschek, N., Scholz, S. (1987) Destruction of polychlorinated dibenzo-p-dioxins and dibenzofurans in contaminated water samples using ozone. *Chemosphere* 16, 1857-1863.

Paterson, S., Shiu, W.Y., Mackay, D., Phyper, J.D. (1990) Dioxins from combustion processes: environmental fate and exposure. In: *Emissions from Combustion Processes: Origin, Measurement, Control.* Clement, R., Kagel, R., Eds., pp. 405-423, Lewis Publishers, Inc., Chelsea, MI.

Pearlman, R.S. (1986) In: *Partition Coefficient.* Dunn III, W.J., Editor, Pergamon Press, Oxford, U.K., pp. 3-20.

Pelizzetti, E., Borgarello, M., Minero, C., Pramauro, E., Serpone, N. (1988) Photolytic degradation of polychlorinated dioxins and polychlorinated biphenyl in aqueous suspensions of semiconductors irradiated with simulated solar light. *Chemosphere* 17, 499-510.

Perkow, J., Eschewroeder, A., Goyer, M., Stevens, J., Wechsler, A. (1980) *An Exposure and Risk Assessment for 2,3,7,8-Tetrachlorodibenzo-p-Dioxin.* Office of Water Regulations and Standards, U.S. EPA, Washington D.C..

Peterson, J.T. (1976) *Calculated Actinic Fluxes (290-700 nm) for Air Pollution Photochemistry Applications.* EPA Final Report 600/4-76-025.

Pirkle, J., Wolfe, W., Patterson, D., Needham, L., Michalek, J., Miner, J., Peterson, M., Phillips, D. (1989) Estimates of the half-life of 2,3,7,8-tetrachlorodibenzodioxin in Vietnam veterans of operation ranch hand. *J. Toxicol. Environ. Health* 27, 165-171.

Plimmer, J.R. (1978) Photolysis of TCDD and trifluralin on silica and soil. *Bull. Environ. Contam. Toxicol.* 20, 82-92.

Plimmer, J.R., Klingbiel, U.I., Crosby, D.G., Wong, A.S. (1973) Photochemistry of dibenzo-p-dioxins. *Adv. Chem. Ser.* 120, 44-54.

Podall, R.T., Jaber, H.M., Mill, T. (1986) Tetrachlorodibenzodioxin: rates of volatilization and photolysis in the environment. *Environ. Sci. Technol.* 20 (5), 490-492.

Pohland, A.E., Yang, G.C. (1972) Preparation and characterization of chlorinated dibenzo-p-dioxins. *J. Agric. & Food Chem.* 20(6), 1093-1099.

Poiger, H., Schlatler, C. (1986) Pharmacokinetics of 2,3,7,8-TCDD in man. *Chemosphere* 15, 1489-1494.

Puri, R.K., Clevenger, T.E., Kapila, S., Yanders, A.F., Malhotra, R.K. (1989) Studies of parameters affecting translocation of tetrachlorodibenzo-p-dioxin in soil. *Chemosphere* 18, 1291-1296.

Quensen III, J.F., Matsumura, F. (1983) Oxidative degradation of 2,3,7,8-tetrachlorodibenzo-p-dioxin by microorganisms. *Environ. Toxicol. & Chem.* 2, 261-268.

Reischl, A., Reissinger, M., Thoma, H., Hutzinger, O. (1989) Uptake and accumulation of PCDD/F in terrestrial plants: basic considerations. *Chemosphere* 19, 467-474.

Rekker, R.F. (1977) *The Hydrophobic Fragmental Constants. Its Derivation and Application, a Means of Characterizing Membrane Systems.* Elsevier Sci. Publ. Co., Oxford, England.

Rordorf, B.F. (1985a) Thermodynamic properties of polychlorinated compounds: The vapor pressures and enthalpies of sublimation of ten dibenzo-para-dioxins. *Thermochimca Acta* 85, 435-438.

Rordorf, B.F. (1985b) Thermodynamic and thermal properties of polychlorinated compounds: The vapor pressures and flow tube kinetics of ten dibenzo-para-dioxins. *Chemosphere* 14, 885-892.

Rordorf, B.F. (1986a) Private communication, Ciba Geigy Ltd., Switzerland.

Rordorf, B.F. (1986b) Thermal properties of dioxins, furans and related compounds. *Chemosphere* 15, 1325-1332.

Rordorf, B.F. (1987) Prediction of vapor pressures, boiling points and enthalpies of fusion for twenty-nine halogenated dibenzo-p-dioxins. *Therochimica Acta* 112, 117-122.

Rordorf, B.F. (1989) Prediction of vapor pressures, boiling points and enthalpies of fusion for twenty-nine halogenated dibenzo-p-dioxins and fifty-five dibenzofurans by a vapor pressure correlation method. *Chemosphere* 18, 783-788.

Rordorf, B.F., Freeman, R.A., Schroy, J.M., Glasgow, D.G. (1986a) Mobility of HCX at Times Beach, Missouri. *Chemosphere* 15, 2069-2072.

Rordorf, B.F., Sarna, L.P., Webster, G.R.B. (1986b) Vapor pressure determination for several polychlorodioxins by two gas saturation methods. *Chemosphere* 15, 2073-2076.

Rordorf, B.F., Sarna, L.P., Webster, G.R.B., Safe, S.H., Safe, L.M., Lenoir, D., Schwind, K.H., Hutzinger, O. (1990) Vapor pressure measurements on halogenated dibenzo-p-dioxins and dibenzofurans. An extended data set for a correlation method. *Chemosphere* 20, 1603-1609.

Rose, J.Q., Ramsey, J.C., Wentzler, T.H., Hummel, R.H., Gehring, P.J. (1976) The fate of 2,3,7,8-tetrachlorodibenzo-p-dioxin following single and repeated doses to the rat. *Toxicol. Appl. Pharmacol.* 36, 209-226.

Sarna, L.P., Hodge, P.E., Webster, G.R.B. (1984) Octanol-water partition coefficients of chlorinated dioxins and dibenzofurans by reversed-phase HPLC using several C_{18} columns. *Chemosphere* 13, 975-983.

Schecter, A., Ryan, J.J. (1991) Brominated and chlorinated dioxin blood levels in a chemist 34 years after exposure to 2,3,7,8-tetrachlorodibenzodioxin and 2,3,7,8-tetrabromodibenzodioxin. *Chemosphere* 23, 1921-1924.

Schroy, J.M., Hileman, F.D., Cheng, S.C. (1985a) Physical/chemical properties of 2,3,7,8-tetrachlorodibenzo-p-dioxin. *ASTM STP* 891, pp 409-421.

Schroy, J.M., Hileman, F.D., Cheng, S.C. (1985b) Physical/chemical of 2,3,7,8-TCDD. *Chemosphere* 14, 873-886.

Servos, M.R., Muir, D.C.G. (1989a) Effect of dissolved organic matter from Canadian shield lakes on the bioavailability of 1,3,6,8-tetrachlorodibenzo-p-dioxin to the amphipod *Crangonyx laurentianus*. *Environ. Toxicol. Chem.* 8, 141-150.

Servos, M.R., Muir, D.C.G. (1989b) Effect of suspended sediment concentration on the sediment to water partition coefficient for 1,3,6,8-tetrachlorodibenzo-p-dioxin. *Environ. Sci. Technol.* 23, 1302-1306.

Servos, M.R., Muir, D.C.G., Webster, G.R.B. (1989) The effect of dissolved organic matter on the bioavailability of polychlorinated dibenzo-p-dioxins. *Aquat. Toxicol.* 14, 169-184.

Servos, M.R., Muir, D.C.G., Whittle, D.M., Sergeant, D.B., Webster, G.R.B. (1989) Bioavailability of octachlorodibenzo-p-dioxin in aquatic ecosystems. *Chemosphere* 19, 969-972.

Shaub, W.M., Tsang, W. (1983) Dioxin formation in incinerators. *Environ. Sci. Technol.* 17, 721-730.

Shaub, W.M., Tsang, W. (1983) Physical and chemical properties of dioxins in relation to their disposal. In: *Human and Environmental Risks of Chlorinated Dioxins and Related Compounds.* Tucker, R.E., Young, A.L., Gray, A.P., Editors, pp.731-747, Plenum Press, New York.

Shiu, W.Y., Mackay, D. (1992) Temperature dependence of aqueous solubility and octanol-water partition coefficient of dibenzofuran. (to be submitted to *J. Chem. Eng. Data*)

Shiu, W.Y., Gobas, F.A.P.C., Mackay, D. (1987) Physical-chemical properties of three congeneric series of chlorinated aromatic hydrocarbons. In: *QSAR in Environmental Toxicology - II.* Kaiser, K.L.E., Ed., pp. 347-362, D.Reidel Publishing Co., Dordrect, Netherlands.

Shiu, W.Y., Doucette, W., Gobas, F.A.P.C., Mackay, D., Andren, A.W. (1988) Physical-chemical properties of chlorinated dibenzo-p-dioxins. *Environ. Sci. Technol.* 22, 651-658.

Sijm, T.H.M., Opperhuezen, A. (1988) Biotransformation, bioaccumulation and lethality of 2,8-dichlorodibenzo-p-dioxin: a proposal to explain the biotic fate and toxicity of PCDD's and PCDF's. *Chemosphere* 17, 83-99.

Sijm, D.T.H.M., Wever, H., de Vries, P.J., Opperhuizen, A. (1989a) Octan-1-ol/water partition coefficients of polychlorinated dibenzo-p-dioxins and dibenzofurans: experimental values determined with a stirring method. *Chemosphere* 19, 263-266.

Sijm, D.T.H.M., Wever, H., Opperhuizen, A. (1989b) Influence of biotransformation on the accumulation of PCDDs from fly-ash in fish. *Chemosphere* 19, 475-480.

Sijm, D.T.H.M., Yarechewski, A.L., Muir, D.C.G., Webster, G.R.B., Seinen, W., Opperhuizen, A. (1990) Biotransformation and tissue distribution of 1,2,3,7-tetrachlorodibenzo-p-dioxin, 1,2,3,7-pentachlorodibenzo-p-dioxin and 2,3,4,7,8-pentachlorodibenzofuran in rainbow trout. *Chemosphere* 21, 845-866.

Singh, H.B. (1977) Preliminary estimation of average tropospheric HO concentrations in the northern and southern hemispheres. *Geophys. Res. Lett.* 4, 453.

Smith, J.H., Mabey, W.R., Bahonos, N., Holt, B.R., Lee, S.S., Chou, T.W., Bomberger, D.C., Mill, T. (1978) *Environmental Pathways of Selected Chemicals in Freshwater Systems: Part II. Laboratory Studies.* Interagency Energy-environment Research and Development Program Report. EPA-600/7-78-074. Environmental Research Laboratory Office of Research and Development. U.S. Environmental Protection Agency, Athens, Georgia 30605, p. 304.

Speece, R.E. (1990) Comment on "Prediction of aqueous solubility of organic chemicals based on molecular structure. 2. Application to PNAs, PCBs, PCDDs, etc." *Environ. Sci. Technol.* 24(6), 929-930.

Srinivasan, K.R., Fogler, H.S. (1987) Binding of OCDD, 2,3,7,8-TCDD and HCB to clay-based sorbents. In: *Chlorinated Dioxins and Dibenzofurans in Perspective.* Rappe, C., Choudhary, G., Keith, L.H., Eds., pp. 531-539. Lewis Publishers, Inc., Chelsea, MI.

Stephenson, R.M., Malanowski, A. (1987) *Handbook of the Thermodynamics of Organic Compounds.* Elsevier, New York.

The Merck Index (1989) *An Encyclopedia of Chemicals, Drugs and Biologicals.* Budavari, S., Editor, Merck & Co., Inc., Rahway, N.J., 11th edition.

The Merck Index (1983) *An Encyclopedia of Chemicals, Drugs and Biologicals.* Windholz, M. Editor, Merck and Co., Inc., Rahway, N.J., U.S.A., 10th edition.

Thibodeaux, L.J. (1979) *Chemodynamics.* John Wiley & Sons, New York.

Thibodeauz, L.J., Lipsky, D. (1985) *Haz. Waste & Haz. Mat.* 2, 225.

Travis, C.C., Hattemer-Frey, H. (1987) Human exposure to 2,3,7,8-TCDD. *Chemosphere* 16, 2331-2342.

Tysklind, M., Rappe, C. (1991) Photolytic transformation of polychlorinated dioxins and dibenzofurans in fly ash. *Chemosphere* 23, 1365-1375.

Van den Berg, M., Olie, K. (1985) Polychlorinated dibenzofurans (PCDFs). Environmental occurence and physical, chemical and biological properties. *Toxicol. & Environ. Chem.* 9, 171-217.

Van den Berg, M., De Jongh, J., Eckhart, P., Van der Wielen, F.W.M. (1989) The elimination and absence of pharmacokinetic interaction of some polychlorinated dibenzofurans (PCDFs) in the liver of the rat. *Chemosphere* 18(1-6), 665-675.

Veith, G.D., Austin, N.M., Morris, R.T. (1979a) A rapid method for estimation log P for organic chemicals. *Water Res.* 13, 43-47.

Veith, G.D., Defor, D.L. Bergstedt, B.V. (1979b) Measuring and estimating the bioconcentration factor of chemicals in fish. *J. Fish Res. Board Can.* 26, 1040-1048.

Veith, G.D., Call, D.J., Brooke, L.T. (1983). Structure toxicity relationships for the fathead minnow, Pimephales promelas, narcotic industrial chemicals. *Can. J. Fish Aquat. Sci.* 40, 743-748.

Veith, G.D., Macek, K.J., Petrocelli, S.R., Carroll, J. (1980) An evaluation of using partition coefficients and water solubility to estimate bioconcentration factors for organic chemicals in fish. In: *Aquatic Toxicology. ASTM STP* 707, Eaton, J.G., Parrish, P.R., Hendricks, A.C., Eds., pp 116-129, Amer. Soc. for Testing and Materials, Philadelphia.

Verschueren, K. (1977) *Handbook of Environmental Data on Organic Chemicals.* van Nostrands Reinhold, New York.

Verschueren, K. (1983) *Handbook of Environmental Data on Organic Chemicals*, 2nd edition, Van Nostrand Reinhold, New York.

Walker, M.K., Spitsbergen, J.M., Olson, J.R., Peterson, R.E. (1991) 2,3,7,8-tetrachlorodibenzo-p-dioxin (TCDD) toxicity during early life stage development of lake trout (*salvelinus namaycush*). *Can. J. Fish Aquatic Sci.* 48, 875-883.

Walters, R.W., Guiseppi-Elie, A. (1988) Sorption of 2,3,7,8-tetrachlorodibenzo-p-dioxin to soils for water/methanol mixtures. *Environ. Sci. Technol.* 22, 819-825.

Walters, R.W., Ostazeski, S.A., Guiseppi-Elie, A. (1989) Sorption of 2,3,7,8-tetrachlorodibenzo-p-dioxin from water by surface soils. *Environ. Sci. Technol.* 23(4), 480-484.

Ward, C.T., Matsumura, F. (1978) Fate of 2,3,7,8-tetrachloro-p-dioxin (TCDD) in a model aquatic environment. *Arch. Environ. Contam. Toxicol.* 7, 349-357.

Ward, C.H., Tomson, M.B., Bedient, P.B., Lee, M.D. (1986) Transport and fate processes in the subsurface. *Water Res. Symp.* 13, 19-39.

Warner, H. et al. (1980) *Determination of Henry's Law Constants of Selected Priority Pollutants.* MERL, Cincinnati.

Wasik, S.P., Miller, M.M., Teware, Y.B., May, W.E., Sonnefeld, W.J., DeVoe, H., Zoller, W.H. (1983) Determination of the vapor pressure, aqueous solubility, and octanol/water partition coefficient of hydrophobic substances by coupled generator column/liquid chromatographic methods. *Residue Rev.* 85, 29-42.

Watts, C.D., Moore, K. (1987) Fate and transport of organic compounds in river. In: *Organic Micropollutants in the Aquatic Environment.* Angelletti, G., Bjorseth, A., Eds., Proceedings of the 5th European Symposium, Rome, Italy, pp.154-169, Kluwer Academic Publ.

Weast, R. (1984) *Handbook of Chemistry and Physics.* 64th ed., CRC Press, Boca Raton, Florida.

Weast, R. (1972-73) *Handbook of Chemistry and Physics.* 53rd ed., CRC Press, Cleveland.

Weast, R. (1977) *Handbook of Chemistry and Physics.* 57th ed., CRC Press, Boca Raton, Florida.

Weast, R. (1982-83) *Handbook of Chemistry and Physics.* 63rd ed., CRC Press, Boca Raton, Florida.

Weast, R.C., Ed. (1983-84). *Handbook of Chemistry and Physics.* 64th ed., CRC Press, Boca Raton, Florida.

Webster, G.R.B., Sarna, L.P., Muir, D.C.G. (1983) Presented at the American Chemical Society National Meeting in Washington D.C..

Webster, G.R.B., Friesen, K.J., Sarna, L.P., Muir, D.C.G. (1985) Environmental fate modelling of chlorodioxins: Determination of physical constants. *Chemosphere* 14, 609-622.

Webster, G.R.B., Muldrew, D.H., Graham, J.J., Sarna, L.P., Muir, D.C.G. (1986) Dissolved organic matter mediated aquatic transport of chlorinated dioxins. *Chemosphere* 15, 9-12.

Webster, T., Connett, P. (1990) The use of bioconcentration factors in estimating the 2,3,7,8-TCDD content of cow's milk. *Chemosphere* 20(7-9), 779-786.

Webster, T., Connett, P. (1991) Estimating bioconcentration factors and half-lives in humans using physiologically based pharmacokinetic modelling: 2,3,7,8-TCDD. *Chemosphere* 23, 1763-1768.

Yalkowsky, S.H., Mishra, D.S. (1990) Comment on "Prediction of aqueous solubility of organic chemicals based on molecular structure. 2. Application to PNAs, PCBs, PCDDs, etc.". *Environ. Sci. Technol.* 24(6), 927-929.

Zepp, R.G. (1980) Assessing the photochemistry of pollutants in aquatic environments. In: Dynamics, Exposure and Hazard Assessment of Toxic Chemicals. Haque, R., Ed., pp.69-110, Ann Arbor Publ. Inc., Ann Arbor, Michigan.

Zeep, R.G. (1991) Photochemical fate of agrochemicals in natural waters. In: *Pesticide Chemistry-Advances in International Research, Development, and Legislation.* Frehse, H., editor., pp.329-345. VCH, Weinheim, Federal Republic of Germany.

Zitko, V., Choi, P.M.K. (1973) Oral toxicity of chlorinated dibenzofurans to juvenile Atlantic salmon. *Bull. Environ. Contam. Toxicol.* 10, 120-122.

4. Chlorinated Dibenzofurans

4.1   List of Chemicals and Data Compilations

Common Name: Dibenzofuran
Synonym: diphenylene oxide
Chemical Name: dibenzofuran
CAS Registry No: 132-64-9
Molecular Formula: $C_6H_4OC_6H_4$
Molecular Weight: 168.21

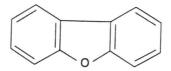

Melting Point (°C):
    86.7   (Weast 1982-83)
    82   (Banerjee et al. 1980; Pearlman et al. 1984)
    84-92 (Stephenson & Malnowski 1987)
    86.5   (Rordorf 1986,1989)
Boiling Point (°C): 287
    287   (quoted, Weast 1982-83; Stephenson & Malanowski 1987)
    273   (calculated, Rordorf 1986,1989)
Density (g/cm³ at 20°C):
    1.0886 (99°C, Weast 1982-83)
Molar Volume (cm³/mol):
    226.4 (LeBas method)
    154.4 (calculated-density, Stephenson & Malanowski 1987)
Molecular Volume (Å³):
    152.438 (Pearlman et al. 1984)
Total Surface Area, TSA (Å²):
    183.75 (Pearlman et al. 1984)
    176.3 (Doucette & Andren 1987)
Heat of Fusion, ΔH_{fus}, kcal/mol:
    4.6845 (Rordorf 1989)
Entropy of Fusion, ΔS_{fus}, cal/mol K (e.u.):
    12.90 (Rordorf 1986,1989)
Fugacity Ratio at 25 °C (assuming ΔS = 13.5 e.u.):
    0.246 (at M.P. = 86.5°C)

Water Solubility (g/m³ or mg/L at 25°C):
    3.11   (Lu et al. 1978; quoted, Pearlman et al. 1984)
    10.03 (shake flask-HPLC/UV, Banerjee et al. 1980)

6.56　(selected average, Pearlman et al. 1984)
4.22　(gen. col.-GC, Doucette & Andren 1988a, quoted, Friesen et al. 1990)
1.00　(selected, Isnard & Lambert 1988 1989)
4.75　(gen. col.-HPLC/UV, Shiu & Mackay 1992)

Vapor Pressure (Pa at 25°C):
2.026　(extrapolated liquid value, Antoine eqn., Stephenson & Malanowski 1987)
0.35　(gas saturation, Rordorf 1986,1989)
0.40　(gas saturation, estimated from extrapolated vapor pressure vs. halogen substitution no. plot, Rordorf et al. 1990)

Henry's Law Constant (Pa m³/mol):

Octanol/Water Partition Coefficient, log K_{OW}:
4.12　(Hansch & Leo 1979; quoted, Birnbaum 1985)
4.17　(shake flask-HPLC/UV, Banerjee et al. 1980)
4.12　(Veith et al. 1979; quoted, Mackay 1982)
3.91, 4.12, 4.18; 3.96, 4.10, 4.17 (HPLC-RT, linear regressions; quadratic regressions, Sarna et al. 1984)
3.92　(re-evaluated HPLC-RT data, Burkhard & Kuehl 1986)
4.31　(gen. col.-GC/ECD, Doucette & Andren 1987)
4.12, 4.04 (calculated-f const., TSA, K_{OW}, Doucette & Andren 1987)
4.31, 4.12, 4.04, 3.86, 4.11, 4.57, 4.13 (selected exptl., calculated-π-const, f-const., HPLC-RT, MW, χ, TSA, Doucette & Andren 1988b)
4.12　(selected, Isnard & Lambert 1988, 1989)
4.33　(gen. col.-HPLC/UV, Shiu & Mackay 1992)

Bioconcentration Factor, log BCF:
3.13　(fathead minnow, 28 days exposure, Veith et al. 1979)
3.13, 2.80 (quoted exptl., calculated-K_{OW}, Mackay 1982)
3.13, 3.13 (quoted exptl., calculated-χ, Sabljic 1987)
3.13　(calculated-K_{OW}, Isnard & Lambert 1989)

Sorption Partition Coefficient, log K_{OC}:

502

Half-Lives in the Environment:

    Air: 1.9-19 hours, based on estimated rate constants for reaction with hydroxyl radicals in air (Atkinson 1987b; Howard et al. 1991).

    Surface water: 168-672 hours, based on aerobic acclimated and unacclimated groundwater die-away test data (Lee et al. 1984; quoted; Ward et al. 1986; Howard et al. 1991).

    Groundwater: 205-835 hours, based on aerobic acclimated and unacclimated groundwater die-away teat data (Lee et al. 1984; quoted, Ward et al. 1986; Howard et al. 1991).

    Sediment:

    Soil: 168-672 hours, based on aerobic acclimated and unacclimated groundwater die-away test data (Lee et al. 1984; quoted, Ward et al. 1986; Howard et al. 1991).

    Biota:

Environmental Fate Rate Constants or Half-Lives:

    Volatilization:

    Photolysis:

    Oxidation: gas phase reaction rate constant with OH radicals was calculated to be 3.4×10^{-11} cm^3 molecule^{-1} sec^{-1} corresponding to an atmospheric lifetime of about 8 hours (Atkinson 1987a) and 3.3×10^{-11} cm^3 molecule^{-1} sec^{-1} (Atkinson 1987b); photooxidation half-life was estimated to be between 1.9-19 hours, based on estimated rate constant with OH radicals in air (Atkinson 1987b; quoted, Howard et al. 1991).

    Hydrolysis:

    Biodegradation: biodegradation half-life was estimated to be 168-672 hours and anaerobic half-life of 672-2688 hours, based on aerobic acclimated and unacclimated groundwater die-away test data (Lee et al. 1984; quoted, Ward et al. 1986; Howard et al. 1991); nonautoclaved groundwater samples at hazardous waste site with a concentration of approximate 0.09 mg/liter are degraded by microbes at rates about 30% per week while the levels of the controls decreased only about half that rate (Lee et al. 1984).

    Biotransformation:

    Bioconcentration, Uptake (k_1) and Elimination (k_2) Rate Constants:

Common Name: 2,8-Dichlorodibenzofuran
Synonym: 2,8-DCDF
Chemical Name: 2,8-dichlorodibenzofuran
CAS Registry No: 5409-83-6
Molecular Formula: $C_6H_4OC_6H_2Cl_2$
Molecular Weight: 237.1

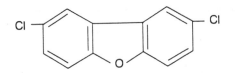

Melting Point (°C):
    184-185   (Kuroki et al. 1984; quoted, Rordorf 1989)
    185   (Van den Berg et al. 1985)
Boiling Point (°C):
    375   (calculated, Rordorf 1989)
Density (g/cm³ at 20°C):
Molar Volume (cm³/mol):
    226.4   (LeBas method)
Molecular Volume (Å³):
Total Surface Area, TSA (Å²):
    212.8   (Doucette & Andren 1987)
    216.56   (van der Waals surface area, Dunn III et al. 1986)
Heat of Fusion, ΔH_{fus}, kcal/mol:
    5.60   (Rordorf 1989)
Entropy of Fusion, ΔS_{fus}, cal/mol K, (e.u.):
    13.145   (Rordorf 1989)
Fugacity Ratio at 25 °C (assuming ΔS = 13.5 e.u.), F:
    0.0268

Water Solubility (g/m³ or mg/L at 25°C):
    0.0145   (gen. col.-GC, Doucette & Andren 1988a, quoted, Friesen et al. 1990)

Vapor Pressure (Pa at 25°C):
    3.9x10⁻⁴ (calculated-B.P. and Δ_{fus} , Rordorf 1989)

Henry's Law Constant (Pa m³/mol):

Octanol/Water Partition Coefficient, log K_{OW}:
    5.56, 5.97, 6.16; 5.65, 5.95, 6.15 (HPLC-RT, linear regressions; quadratic regressions, Sarna et al. 1984)
    5.30   (re-evaluated HPLC-RT data, Burkhard & Kuehl 1986)
    5.44   (gen. col.-GC/ECD, Doucette & Andren 1987)
    5.65, 4.91 (calculated-f const., calculated-TSA-K_{OW}, Doucette & Andren 1987)
    5.44, 5.65, 5.53, 5.12, 4.92, 5.42, 5.01 (selected exptl., calculated-π-const, f-const., HPLC-RT, MW, χ, TSA, Doucette & Andren 1988b)

Bioconcentration Factor, log BCF:

Sorption Partition Coefficient, log K_{OC}:

Half-Lives in the Environment:
    Air:
    Surface water:
    Groundwater:
    Sediment:
    Soil:
    Biota: mean biological half-life in rainbow trout, about 12 days (Niimi 1986).

Environmental Fate Rate Constants or Half-Lives:
    Volatilization:
    Photolysis: when 5 mg/liter in methanol was irradiated by sunlight simulator more than 95% disappears within 48 hours while a similar experiment with highly purified methanol solution reveals only very slow photolysis within the same period of irradiation and results were same with 10 mg/liter in methanol solution (Crosby & Moilanen 1973; quoted, Choudhry & Hutzinger 1982).
    Hydrolysis:
    Oxidation:
    Biodegradation:
    Biotransformation:
    Bioconcentration, Uptake (k_1) and Elimination (k_2) Rate Constants:

Common Name: 2,3,7,8-Tetrachlorodibenzofuran
Synonym: 2,3,7,8-TCDF
Chemical Name: 2,2,7,8-Tetrachlorodibenzofuran
CAS Registry No: 51207-31-9
Molecular Formula: $C_6H_2Cl_2OC_6H_2Cl_2$
Molecular Weight: 306

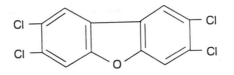

Melting Point (°C):
      227-228 (Gray et al. 1976)
      219-221 (Kuroki et al. 1984)
      227-228 (quoted, Van den Berg et al. 1985)
      227, 228 (quoted, Rordorf 1989)
Boiling Point (°C):
      438.3 (calculated, Rordorf 1989)
Density (g/cm³ at 20°C):
Molar Volume (cm³/mol):
      268.2 (LeBas method)
Molecular Volume (Å³):
Total Surface Area, TSA (Å²):
      247.73 (van der Waals surface area, Dunn III et al. 1986)
      212.8 (Doucette & Andren 1987)
Heat of Fusion, ΔH_{fus}, kcal/mol:
      8.748 (Rordorf 1989)
Entropy of Fusion, ΔS_{fus}, cal/mol K (e.u.):
      17.447 (Rordorf 1989)
Fugacity Ratio at 25 °C (assuming ΔS = 13.5 e.u.), F:
      0.0000251

Water Solubility (g/m³ or mg/L at 25°C):
      0.000419 (22.7 °C, gen. col.-HPLC/LSC, Friesen et al. 1990; quoted, Mackay 1991)
      0.00040 (quoted, Ma et al. 1990)
      0.00351 (calculated-QSAR, Fiedler & Schramm 1990)

Vapor Pressure (Pa at 25°C):
     2.0×10^{-6} (quoted, Van den Berg et al. 1985)
     0.000002 (correlated, Rordorf 1989)
     0.0001228 (GC-RT, subcooled liquid value, Eitzer & Hites 1988)

Henry's Law Constant (Pa m^3/mol):
     1.50   (calculated for tetrachloro-PCDFs, Eitzer & Hites 1989)

Octanol/Water Partition Coefficient, log K_{ow}:
     5.82   (HPLC-RT, Burkhard & Kuehl 1986)
     6.53   (slow stirring, GC/MS, from mixture of fly ash extract, Sijm et al. 1989a)
     6.19   (correlated, Ma et al. 1990)
     7.10   (calculated-QSAR, Fiedler & Schramm 1990)
     7.70   (calculated, Broman et al. 1991)

Bioconcentration Factor, log BCF:
     4.82   (guppy, Opperhuizen et al. 1986)
     3.78, 3.39 (rainbow trout, exposed to: 0.41 ng/L, 3.93 ng/L for 28 days, Mehrle et al. 1988)
     3.64, 3.41 (rainbow trout, quoted, Opperhuizen & Sijm 1990)
     3.82   (guppy, quoted, Opperhuizen & Sijm 1990)

Sorption Partition Coefficient, log K_{oc}:
     5.20   (organic carbon, calculated-QSAR, Fiedler & Schramm 1990)
     7.50   (calculated, Broman et al. 1991)

Half-Lives in the Environment:
     Air:
     Surface water: 90 minutes in isooctane solution in summer sunlight (Palauschek & Scholz 1987).
     Groundwater:
     Sediment:
     Soil:
     Biota: half-lives of elimination: from rat, 2 days (Birnbaum et al. 1980; quoted, Birnbaum 1985); 8 days from monkey (Birnbaum et al. 1981; quoted, Birnbaum 1985); 2-4 days from mouse (Decad et al. 1981b; quoted, Birnbaum 1985); 40 days from guinea pig (Decad et al. 1981a; Ioannou et al. 1983; quoted, Birnbaum 1985); elimination half-lives: in guinea pigs, 20 days; rats, < 2 days;

moneys, 8 days; and mice, 2-4 days (quoted, Van den Berg et al. 1985); half-life in carp, <336 days (Kuehl et al. 1987); elimination half-life of 3.0 days, 0.27 day (rainbow trout, exposed to 0.41 ng/L, 3.93 ng/L for 28 days, Mehrle et al. 1988); elimination half-life of 0.8 day for lactating cows (Olling et al. 1991).

Environmental Fate Rate Constants or Half-Lives:

Volatilization:

Photolysis: sunlight induced photolysis half-life, 220 minutes in isooctane solution, and solid phase photolysis half-life, 120 hours with PCDF dispersed as solid films (Buser 1988); photolytic half-life in extract of fly ash and in tetradecane solution: 9.8 hours for native congener and 3.0 hours for ^{13}C-labelled congener (Tysklind & Rappe 1991).

Hydrolysis:

Oxidation: room temperature OH radical reaction rate constant calculated to be 2.3×10^{-12} cm^3 molecule^{-1} sec^{-1} (Atkinson 1987b); oxidative degradation rate constant for water dissolved PCDF by ozone is 1.32×10^5 liter gram^{-1} minute^{-1} under alkaline conditions at pH 10 and 20°C (Palauschek & Scholz 1987).

Biodegradation:

Biotransformation:

Bioconcentration, Uptake (k_1) and Elimination (k_2) Rate Constants:

k_1: 400 day^{-1} (guppy, exposed to fly ash extract, Opperhuizen et al. 1986; quoted, Opperhuizen & Sijm 1990)

k_2: 0.062 day^{-1} (guppy, exposed to fly ash extract, Opperhuizen et al. 1986; quoted, Opperhuizen & Sijm 1990)

k_1: 1228, 6853 day^{-1} (rainbow trout, exposed to: 0.41 ng/L, 3.93 ng/L for 28 days, Mehrle et al. 1988; quoted, Opperhuizen & Sijm 1990)

k_2: 0.28, 2.60 day^{-1} (rainbow trout, exposed to: 0.41 ng/L, 3.93 ng/L for 28 days, Mehrle et al. 1988; quoted, Opperhuizen & Sijm 1990)

Common Name: 2,3,4,7,8-Pentachlorodibenzofuran
Synonym: 2,3,4,7,8-PCDF
Chemical Name: 2,3,4,7,8-pentachlorodibenzofuran
CAS Registry No: 51207-31-4
Molecular Formula: $C_6H_2Cl_2OC_6HCl_3$
Molecular Weight: 340.42

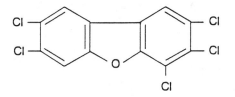

Melting Point (°C):
     196-196.5 (Kuroki et al. 1984)
     196, 196.5 (Rordorf 1989)
Boiling Point (°C):
     464.7   (calculated, Rordorf 1989)
Density (g/cm³ at 20°C):
Molar Volume (cm³/mol):
     289.1   (LeBas method)
Molecular Volume (Å³):
Total Surface Area, TSA (Å²):
     262.71   (van der Waals surface area, Dunn III et al. 1988)
Heat of Fusion, ΔH_{fus}, kcal/mol:
     75.05   (Rordorf 1989)
Entropy of Fusion, ΔS_{fus}, cal/mol K (e.u.):
     21.51   (Rordorf 1989)
Fugacity Ratio at 25 °C (assuming ΔS = 13.5 e.u.):
     0.001125

Water Solubility (g/m³ or mg/L at 25°C):
     0.000515   (calculated-QSAR, Fiedler & Schramm 1990)
     0.000236   (22.7°C, gen. col.-GC/MS, Friesen et al. 1990)

Vapor Pressure (Pa at 25°C):
     3.50×10^{-7} (calculated-B.P and Δ_{fus}, Rordorf 1989)
     2.17×10^{-5} (subcooled liquid value, Eitzer & Hites 1988)

Henry's Law Constant (Pa m^3/mol):

Octanol/Water Partition Coefficient, log K$_{ow}$:
6.92   (slow stirring-GC/MSD, from fly-ash extract, Sijm et al. 1989a; quoted, Loonen et al. 1991)
7.82   (calculated-QSAR, Fiedler & Schramm 1990)
7.60   (calculated, Broman et al. 1991)

Bioconcentration Factor, log BCF:
3.70   (guppy, Opperhuizen et al. 1986; quoted, Opperhuizen & Sijm 1990)

Sorption Partition Coefficient, log K$_{oc}$:
5.59   (organic carbon, calculated-QSAR, Fiedler & Schramm 1990)
7.40   (calculated, Broman et al. 1991)

Half-Lives in the Environment:
Air:
Surface water:
Groundwater:
Sediment:
Soil:
Biota: elimination half-life of 108 days in the liver of female rats (Van den Berg et al. 1989); 61, 69 days in rainbow trout at the 0.82 ng/g exposure concentration from 0-140 days data, at 9.0 ng/g exposure concentration from 0-180 days data (Muir et al. 1990); 65 days in whole body rainbow trout with 31 days dietary exposure (Muir et al. 1990); elimination half-life of 48.5 days for lactating cows (Olling et al. 1991); photolytic half-lives in the extract from fly ash in tetradecane solution, calculated to be 3.5 hours for native congener and 3.1 hours for ^{13}C labelled congener (Tysklind & Rappe 1991).

Environmental Fate Rate Constants or Half-Lives:
Volatilization:
Photolysis: photolytic degradation half-lives of PCDD in extract of fly ash in tetradecane solution, 3.5 hours for native congener and 3.1 hours for ^{13}C-labelled congener (Tysklind & Rappe 1991).
Hydrolysis:
Oxidation:

Biodegradation:

Biotransformation:

Bioconcentration, Uptake (k_1) and Elimination (k_2) Rate Constants:

k_2: 0.0027 day^{-1} (carp, calculated from data reported by Kuehl et al. 1987, Sijm et al. 1990)

k_1:   400 mL gram^{-1} day^{-1} (guppy, Opperhuizen et al. 1986; quoted, Opperhuizen & Sijm 1990; Sijm et al. 1990)

k_2: 0.079 day^{-1} (guppy, Opperhuizen et al. 1986; quoted, Opperhuizen & Sijm 1990; Sijm et al. 1990)

k_2: 0.0064 day^{-1} (liver of female rat, Van den Berg et al. 1989)

k_2: -2.1x10^{-2}, 1.5x10^{-2} day^{-1} (rainbow trout, 2 to 21 days exposure: metabolic inhibitor PBO-treated, control, Sijm et al. 1990)

k_2: 10.1x10^{-3} day^{-1} (rainbow trout, 0-140 days exposure at 0.82 ng/g PCDF concn., Muir et al. 1990; quoted, Sijm et al. 1990)

k_2: 8.10x10^{-3} day^{-1} (rainbow trout, 0-180 days exposure at 0.82 ng/g PCDF concn., Muir et al. 1990)

k_2: 12.6x10^{-3} day^{-1} (rainbow trout, 0-140 days exposure at 9.01 ng/g PCDF concn., Muir et al. 1990);

k_2: 11.4x10^{-3} day^{-1} (rainbow trout, 0-180 days exposure at 9.01 ng/g PCDF concn., Muir et al. 1990)

Common Name: 1,2,3,4,7,8-Hexachlorodibenzofuran
Synonym: 1,2,3,4,7,8-HCDF
Chemical Name: 1,2,3,4,7,8-Hexachlorodibenzofuran
CAS Registry No: 70658-26-9
Molecular Formula: $C_6H_2Cl_2OC_6Cl_4$
Molecular Weight: 374.87

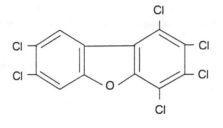

Melting Point (°C):
     225.5-226.5   (Kuroki et al. 1984)
     225.5, 226.5 (quoted, Rordorf 1989)
Boiling Point (°C):
     487.7 (calculated, Rordorf 1989)
Density (g/cm³ at 20°C):
Molar Volume (cm³/mol):
     310   (LeBas method)
Molecular Volume (Å³):
Total Surface Area, TSA (Å²):
     274.71 (van der Waals surface area, Dunn III et al. 1988)
Heat of Fusion, ΔH_{fus}, kcal/mol:
     75.76 (Rordorf 1989)
Entropy of Fusion, ΔS_{fus}, cal/mol K (e.u.):
     22.94 (Rordorf 1989)
Fugacity Ratio at 25 °C (assuming ΔS = 13.5 e.u.):
     0.0103 (at M.P. = 226°C)

Water Solubility (g/m³ or mg/L at 25°C):
     8.25×10^{-6} (22.7°C, gen. col.-GC/MS, Friesen et al. 1990)

Vapor Pressure (Pa at 25°C):
     3.20×10^{-8} (calculated, Rordorf 1989)
     8.093×10^{-6} (subcooled liquid value, GC-RT, Eitzer & Hites 1988)
     8.90×10^{-8} (gas saturation, estimated from extrapolated vapor pressure vs. halogen substitution no. plot, Rordorf et al. 1990)

Henry's Law Constant (Pa m³/mol):

Octanol/Water Partition Coefficient, log K_{OW}:
       7.70   (calculated, Broman et al. 1991)

Bioconcentration Factor, log BCF:

Sorption Partition Coefficient, log K_{OC}:
       7.40   (calculated, Broman et al. 1991)

Half-Lives in the Environment:
      Air:
      Surface water:
      Groundwater:
      Sediment:
      Soil:
      Biota: about 336 days in carp (Kuehl et al. 1987); elimination half-life of 48.5 days from
            lactating cows (Olling et al. 1991).

Environmental Fate Rate Constants or Half-Lives:
      Volatilization:
      Photolysis: photolytic half-life for the PCDD in extract of fly ash in tetradecane solution,
            5.5 hours for native congener (Tysklind & Rappe 1991).
      Hydrolysis:
      Oxidation: oxidative degradation rate constant of water dissolved PDCF by ozone is 7.28
            $x10^4$ liter gram^{-1} minute^{-1} under alkaline conditions at pH 10 and 20°C
            (Palauschek & Scholz 1987).
      Biodegradation:
      Biotransformation:
      Bioconcentration, Uptake (k_1) and Elimination (k_2) Rate Constants:

Common Name: 1,2,3,4,6,7,8-Heptachlorodibenzofuran
Synonym: 1,2,3,4,6,7,8-HCDF
Chemical Name: 1,2,3,4,6,7,8-Heptachlorodibenzofuran
CAS Registry No: 67462-39-4
Molecular Formula: $C_6HCl_3OC_6Cl_4$
Molecular Weight: 409.31

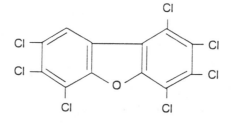

Melting Point (°C):
    236-237   (Kuroki et al. 1984)
    236, 237 (quoted, Rordorf 1989)
Boiling Point (°C):
    507.2   (calculated, Rordorf 1989)
Density (g/cm³ at 20°C):
Molar Volume (cm³/mol):
    330.9   (LeBas method)
Molecular Volume (Å³):
Total Surface Area, TSA (Å²):
    289.57 (van der Waals surface area, Dunn III et al. 1988)
Heat of Fusion, ΔH_{fus}, kcal/mol:
    75.76   (Rordorf 1989)
Entropy of Fusion, ΔS_{fus}, cal/mol K (e.u.):
    25.33   (Rorfdorf 1989)
Fugacity Ratio at 25°C (assuming ΔS = 13.5 e.u.):
    0.00819

Water Solubility (g/m³ or mg/L at 25°C):
    1.35×10^{-6} (22.7°C, gen. col.-GC/MS, Friesen et al. 1990)
    1.08×10^{-5} (calculated-QSAR, Fiedler & Schramm 1990)

Vapor Pressure (Pa at 25°C):
    4.70×10^{-9} (calculated, Rordorf 1989)
    2.24×10^{-6} (subcooled liquid value, GC-RT, Eitzer & Hites 1988)
    1.93×10^{-6} (subcooled liquid, Eitzer & Hites 1989)

5.10x10^{-8} (gas saturation, estimated from extrapolated vapor pressure vs. halogen
           substitution no. plot, Rordorf et al. 1990)

Henry's Law Constant (Pa m^3/mol):

Octanol/Water Partition Coefficient, log K_{OW}:
    7.92   (slow stirring-GC/MSD, from fly-ash extract, Sijm et al. 1989a)
    9.25   (calculated-QSAR, Friedler & Schramm 1990)
    7.90   (quoted, Loonen et al. 1991)
    8.10   (calculated, Broman et al. 1991)

Bioconcentration Factor, log BCF:

Sorption Partition Coefficient, log K_{OC}:
    6.37   (organic carbon, calculated-QSAR, Friedler & Schramm 1990)
    6.00   (organic carbon, calculated, Broman et al. 1991)
    7.90   (calculated, Broman et al. 1991)

Half-Lives in the Environment:
    Air:
    Surface water:
    Groundwater:
    Sediment:
    Soil:
    Biota: depuration half-life, > 336 days for carp in Lake Superior water (Keuhl et al.
        1987); elimination half-life from lactating cows, 33.9 days (Olling et al. 1991).

Environmental Fate Rate Constants or Half-Lives:
    Volatilization:
    Photolysis: photolytic degradation half-lives with extract of fly ash and in tetradecane
        solution, 9.8 hours for native congener and 3.7 hours for ^{13}C-labelled congener
        (Tysklind & Rappe 1991).
    Hydrolysis:

515

Oxidation: oxidative degradation rate constant of water dissolved PCDF by ozone is 1.08 $\times 10^5$ liter gram^{-1} minute^{-1} under alkaline conditions at pH 10 and 20°C (Palauschek & Scholz 1987).

Biodegradation:

Biotransformation:

Bioconcentration, Uptake (k_1) and Elimination (k_2) Rate Constants:

Common Name: 1,2,3,4,7,8,9-Heptachlorodibenzofuran
Synonym: 1,2,3,4,7,8,9-HCDF
Chemical Name: 1,2,3,4,7,8,9-Heptachlorodibenzofuran
CAS Registry No: 55673-89-7
Molecular Formula: $C_6HCl_3OC_6Cl_4$
Molecular Weight: 409.31

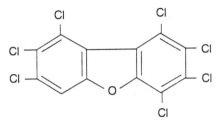

Melting Point (°C):
  221-223   (Kuroki et al. 1984)
  221, 223   (quoted, Rordorf 1989)
Boiling Point (°C):
  507.2   (calculated, Rordorf 1989)
Density (g/cm³ at 20°C):
Molar Volume (cm³/mol):
  330.9   (LeBas method)
Molecular Volume (Å³):
Total Surface Area, TSA (Å²):
  285.99   (van der Waals surface area, Dunn III et al. 1988)
Heat of Fusion, kcal/mol:
  78.87   (Rordorf 1989)
Entropy of Fusion, cal/mol K (e.u.):
  26.05   (Rordorf 1989)
Fugacity Ratio at 25 °C (assuming ΔS = 13.5 e.u.), F:

Water Solubility (g/m³ or mg/L at 25°C):

Vapor Pressure (Pa at 25°C):
  6.200×10^{-9} (calculated, Rordorf 1989)
  1.305×10^{-6} (subcooled liquid value, GC-RT, Eitzer & Hites 1988)
  1.011×10^{-6} (subcooled liquid, Eitzer & Hites 1989)

Henry's Law Constant (Pa m³/mol):

Octanol/Water Partition Coefficient, log K_{OW}:
    6.90   (calculated, Broman et al. 1991)

Bioconcentration Factor, log BCF:

Sorption Partition Coefficient, log K_{OC}:
    5.00   (organic carbon, calculated, Broman et al. 1991)
    6.70   (calculated, Broman et al. 1991)

Half-Lives in the Environment:
    Air:
    Surface water:
    Groundwater:
    Sediment:
    Soil:

Environmental Fate Rate Constants or Half-Lives:
    Volatilization:
    Photolysis: photolysis half-life in the extract from fly ash in tetradecane, calculated to be
            3.3 hours for ^{13}C labelled congener (Tysklind & Rappe 1991).
    Hydrolysis:
    Oxidation:
    Biodegradation:
    Biotransformation:
    Bioconcentration, Uptake (k_1) and Elimination (k_2) Rate Constants:

Common Name: Octachlorodibenzofuran
Synonym: OCDF
Chemical Name: octachlorodibenzofuran
CAS Registry No: 39001-02-0
Molecular Formula: $C_6Cl_4OC_6Cl_4$
Molecular Weight: 443.76

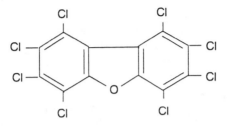

Melting Point (°C):
    330   (Doucette 1985)
    258-260 (quoted, Rordorf 1986,1989)
Boiling Point (°C):
    510   (calculated, Rordorf 1986)
    537   (calculated, Rorforf 1989)
Density (g/cm³ at 20°C):
Molar Volume (cm³/mol):
    351.8 (LeBas method)
Molecular Volume (Å³):
Total Surface Area, TSA (Å²):
    298.0 (Doucette 1985, Doucette & Andren 1987)
    300.44 (van der Waals surface area, Dunn III et al. 1986)
Heat of Fusion, ΔH_{fus}, kcal/mol:
    77.22 (Rordorf 1989)
Entropy of Fusion, ΔS_{fus}, cal/mol K (e.u.):
    25.813 (Rordorf 1986,1989)
Fugacity Ratio at 25 °C (assuming ΔS_{fus} = 13.5 e.u.):
    0.00496

Water Solubility (g/m³ or mg/L at 25°C):
    1.16×10^{-6} (gen. col.-GC, Doucette & Andren 1988a, quoted, Friesen et al. 1990b)
    1.54×10^{-6} (calculated-QSAR, Fiedler & Schramm 1990)
    4.00×10^{-8} (quoted, Gobas & Schrap 1990)

Vapor Pressure (Pa at 25°C):

    1.90×10^{-7} (quoted, Van den Berg et al. 1985)

    5.00×10^{-10} (gas saturation, Rordorf 1989)

    2.60×10^{-7} (subcooled liquid, GC/MS, Eitzer & Hites 1989)

    5.00×10^{-10} (gas saturation, estimated from extrapolated vapor pressure vs. halogen substitution no. plot, Rordorf et al. 1990)

Henry's Law Constant (Pa m³/mol):

    0.10   (estimated, Clark & Mackay 1991)

Octanol/Water Partition Coefficient, log K_{OW}:

    13.06, 13.22, 13.78; 12.54, 13.37, 13.93 (HPLC-RT, linear regressions; quadratic regressions, Sarna et al. 1984)

    8.78   (re-evaluated HPLC-RT data, Burkhard & Kuehl 1986; quoted, Geyer et al. 1987)

    7.97   (gen. col.-GC/ECD, Doucette & Andren 1987,1988; quoted, Geyer et al. 1987; Sijm et al. 1989a)

    9.96, 6.94 (calculated-f const., TSA, K_{OW}, Doucette & Andren 1987)

    7.97, 9.96, 9.99, 7.90, 7.51, 7.32, 7.05 (selected exptl., calculated-π-const, f-const., HPLC-RT, MW, χ, TSA, Doucette & Andren 1988b)

    9.96   (calculated-QSAR, Fiedler & Schramm 1990)

    8.20   (quoted, Gobas & Schrap 1990)

    7.60   (calculated, Broman et al. 1991)

    7.97   (quoted, Clark & Mackay 1991)

Bioconcentration Factor, log BCF:

    1.613, 0.70 (human fat, calculated-different K_{OW}s, wet weight basis, Geyer et al. 1987)

    1.71, 0.778 (human fat, calculated-different K_{OW}s, lipid basis, Geyer et al. 1987)

    2.77, 3.89 (guppy, wet weight based, lipid weight based, Gobas & Schrap 1990)

Sorption Partition Coefficient, log K_{OC}:

    6.75   (calculated-QSAR, Fiedler & Schramm 1990)

    6.00   (organic carbon, calculated, Broman et al. 1991)

    7.40   (calculated, Broman et al. 1991)

Half-Lives in the Environment:

    Air:

    Surface water:

    Groundwater:

Sediment:

Soil:

Biota: mean biological half-life in rainbow trout, about 7 days (Niimi 1986); 7 to 12
          days in rainbow trout (Niimi & Oliver 1986; quoted, Muir et al. 1990; Clark
          & Mackay 1991).

Environmental Fate Rate Constants or Half-Lives:

Volatilization:

Photolysis: photolytic half-life from extract of fly ash in tetradecane solution, 2.1 hours
          for native congener (Tysklind & Rappe 1991).

Hydrolysis:

Oxidation: oxidative degradation rate for water dissolved OCDF by ozone is 1.05×10^4
          liter gram^{-1} minute^{-1} under alkaline condition at pH 10 and 20°C (Palauschek &
          Scholz 1987).

Biodegradation:

Biotransformation: metabolism half-time 10^7 hours (guppy, Clark & Mackay 1991).

Bioconcentration, Uptake (k_1) and Elimination (k_2) Rate Constants:

k_2:   0.012 day^{-1} (rainbow trout, Nimii & Oliver 1986; quoted, Opperhuizen
         & Sijm 1990)

k_1:   824 day^{-1} (guppy, Gobas & Schrap 1990)

k_2:   1.4 day^{-1} (guppy, Gobas & Schrap 1990)

4.2   Summary Tables and QSPR Plots

Table 4.1 Physical-chemical properties of some chlorinated furans at 25 °C

COMPOUND	CAS no.	formula	Cl no. n	MW g/mol	M.P. °C	B.P. °C	fugacity ratio, F at 25 °C	V_M[a] cm³/mol (LeBas)	TSA[b] Å²	TSA[c] Å²
Dibenzofuran	132-64-9	$C_{12}H_8O$	0	168.2	86.5	287	0.2465	184.6	176.3	
2,8-Dichloro-	5409-83-6	$C_{12}H_7OCl$	2	237.1	184	375	0.0268	226.4	212.8	216.6
2,3,7,8-Tetra-	51207-31-9	$C_{12}H_4OCl_4$	4	306.0	227	438.3	0.010	268.2		247.7
2,3,4,7,8-Penta-	57117-31-4	$C_{12}H_3OCl_5$	5	340.42	196	464.7	0.0204	289.1		262.71
1,2,3,4,7,8-Hexa-	70648-26-9	$C_{12}H_2OCl_6$	6	374.87	225.5	487.7	0.0104	310		276.37
1,2,3,6,7,8-Hexa-	57117-44-9	$C_{12}H_2OCl_6$	6	374.87	232	487.7	0.00897	310		274.87
1,2,3,4,6,7,8-Hepta-	67562-39-4	$C_{12}HOCl_7$	7	409.31	236	507.2	0.00819	330.9		289.57
1,2,3,4,7,8,9-Hepta-	55673-89-7	$C_{12}HOCl_7$	7	409.31	221	507.2	0.0115	330.9		285.99
Octachloro-	39001-02-0	$C_{12}OCl_8$	8	443.8	258	537	0.00496	351.8	298	300.4

Note:
(a) calculated for this work
(b) Doucette & Andren 1987
(c) Dunn et al. 1986

Table 4.2 Selected physical-chemical properties of some chlorinated furans at 25 °C

Compounds	Selected properties						Henry's law const., H
	vapour pressure		solubility				
	P^S	P_L	S	C^S	C_L	log K_{OW}	calcd.
	Pa	Pa	mg/m^3	mmol/m^3	mmol/m^3		Pa m^3/mol
		(liquid)			(liquid)		
Dibenzofuran	0.30	1.217	4750	28.240	114.59	4.31	10.62
2,8-CDF	0.00039	1.46×10^{-2}	14.5	0.0612	2.286	5.44	6.377
2,3,7,8-TCDF	2.00×10^{-6}	1.99×10^{-4}	0.419	1.37×10^{-3}	0.136	6.1	1.461
2,3,4,7,8-PCDF	3.50×10^{-7}	1.72×10^{-5}	0.236	6.93×10^{-4}	0.0341	6.5	0.505
1,2,3,4,7,8-H$_6$CDF	3.20×10^{-8}	3.08×10^{-6}	0.00825	2.20×10^{-5}	2.12×10^{-3}	7.0	1.454
1,2,3,6,7,8-H$_6$CDF	3.50×10^{-8}	3.61×10^{-6}	0.0177	4.72×10^{-5}	5.27×10^{-3}		0.741
1,2,3,4,6,7,8-H$_7$CDF	4.70×10^{-9}	5.74×10^{-7}	0.00135	3.30×10^{-6}	4.03×10^{-4}	7.4	1.425
1,2,3,4,7,8,9-H$_7$CDF	6.20×10^{-9}	5.39×10^{-7}					
OCDF	5.0×10^{-10}	1.01×10^{-7}	0.00116	2.61×10^{-6}	5.27×10^{-5}	8.0	0.191

Table 4.3 Suggested half-life classes of polychlorinated dibenzofurans in various environmental compartments

Compounds	Air class	Water class	Soil class	Sediment class
Dibenzofuran	3	4	6	7
2,8-Dichloro-	4	5	7	8
2,3,7,8-TCDF	4	5	8	9
2,3,4,7,8-Pentachloro-	5	5	8	9
1,2,3,4,6,7,8-Heptachloro-	5	6	8	9
Octachloro-	5	7	9	9

where,

Class	Mean half-life (hours)	Range (hours)
1	5	< 10
2	17 (~ 1 day)	10-30
3	55 (~ 2 days)	30-100
4	170 (~ 1 week)	100-300
5	550 (~ 3 weeks)	300-1,000
6	1700 (~ 2 months)	1,000-3,000
7	5500 (~ 8 months)	3,000-10,000
8	17000 (~ 2 years)	10,000-30,000
9	55000 (~ 6 years)	> 30,000

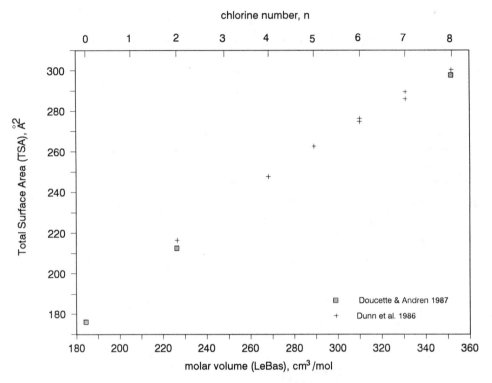

Figure 4.1   Plot of TSA versus LeBas molar volume and chlorine number.

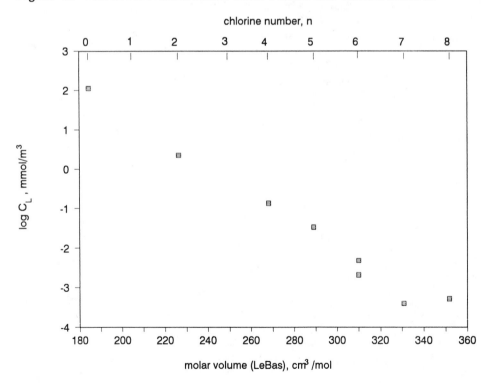

Figure 4.2 Plot of log C_L (liquid solubility) versus molar volume and chlorine number.

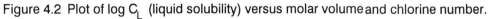

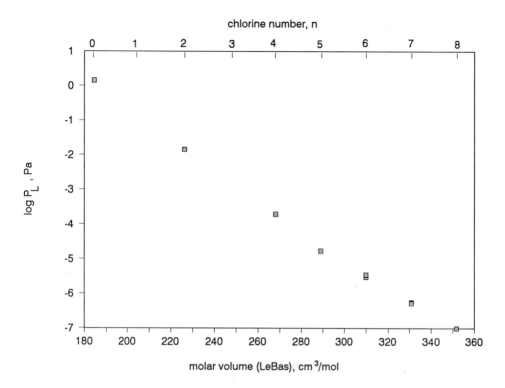

Figure 4.3 Plot of log P_L (liquid vapor pressure) versus molar volume and chlorine number.

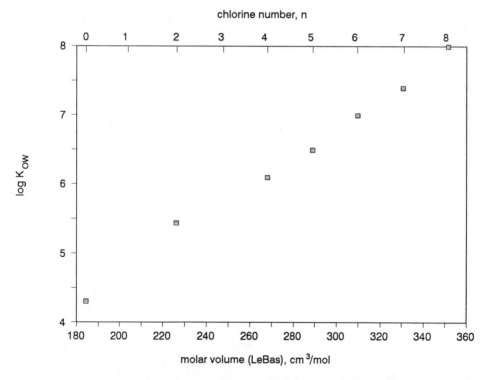

Figure 4.4 Plot of log K_{OW} versus LeBas molar volume and chlorine number.

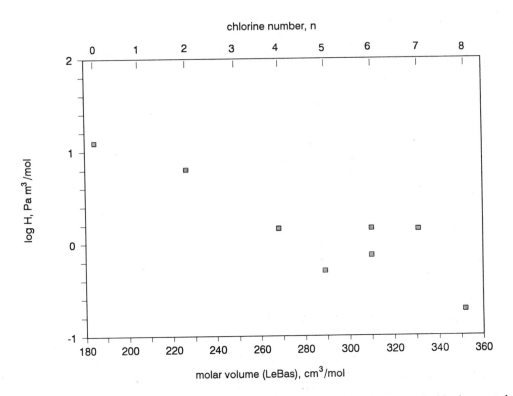

Figure 4.5　Plot of log H (Henry's law constants) vs molar volume and chlorine number.

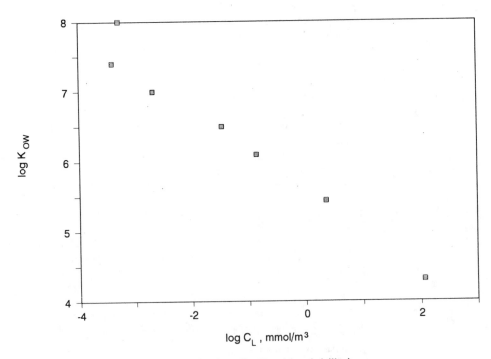

Figure 4.6　Plot of log K_{OW} versus log C_L (liquid solubility).

4.3 Illustrative Fugacity Calculations: Level I, II and III

Chemical name: Dibenzofuran

Level I calculation: (six compartment model)

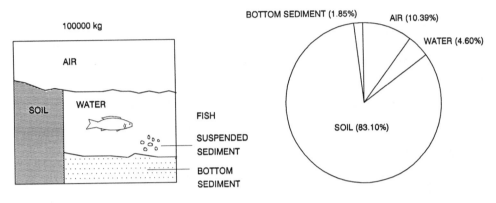

Distribution of mass

physical-chemical properties:

MW: 168.2 g/mol

M.P.: 86.5 °C

Fugacity ratio: 0.247

vapor pressure: 0.30 Pa

solubility: 4.50 g/m^3

log K_{OW}: 4.31

Compartment	Z mol/m3 Pa	Concentration			Amount kg	Amount %
		mol/m3	mg/L (or g/m3)	ug/g		
Air	4.034E-04	6.180E-10	1.039E-07	8.769E-05	10395	10.395
Water	8.918E-02	1.366E-07	2.298E-05	2.298E-05	4595.83	4.596
Soil	3.583E+01	5.490E-05	9.233E-03	3.847E-03	83100	83.100
Biota (fish)	9.104E+01	1.395E-04	2.346E-02	2.346E-02	4.6917	4.69E-03
Suspended sediment	2.240E+02	3.431E-04	5.771E-02	3.847E-02	57.708	5.77E-02
Bottom sediment	7.167E+01	1.098E-04	1.847E-02	7.694E-03	1846.67	1.8467
Total					100000	100

f = 1.532E-06 Pa

Chemical name: Dibenzofuran

Level II calculation: (six compartment model)

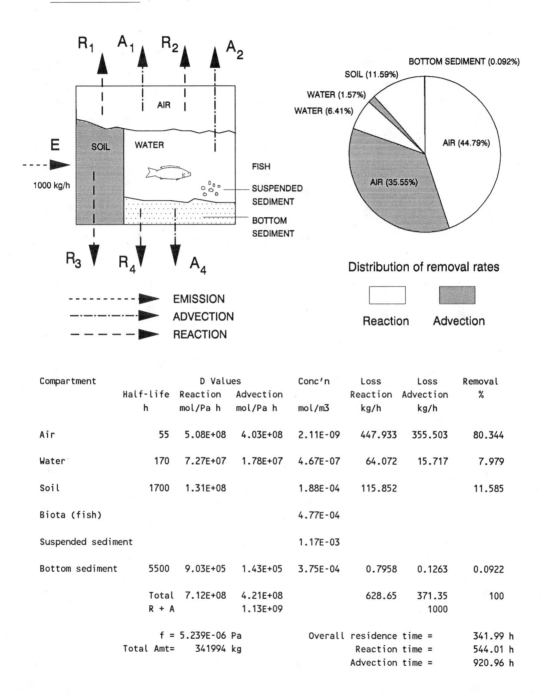

Distribution of removal rates

Reaction   Advection

Compartment	Half-life h	D Values Reaction mol/Pa h	D Values Advection mol/Pa h	Conc'n mol/m3	Loss Reaction kg/h	Loss Advection kg/h	Removal %
Air	55	5.08E+08	4.03E+08	2.11E-09	447.933	355.503	80.344
Water	170	7.27E+07	1.78E+07	4.67E-07	64.072	15.717	7.979
Soil	1700	1.31E+08		1.88E-04	115.852		11.585
Biota (fish)				4.77E-04			
Suspended sediment				1.17E-03			
Bottom sediment	5500	9.03E+05	1.43E+05	3.75E-04	0.7958	0.1263	0.0922
	Total	7.12E+08	4.21E+08		628.65	371.35	100
	R + A		1.13E+09			1000	

f = 5.239E-06 Pa

Total Amt= 341994 kg

Overall residence time = 341.99 h
Reaction time = 544.01 h
Advection time = 920.96 h

Fugacity Level III calculations: (four compartment model)

Chemical name: Dibenzofuran

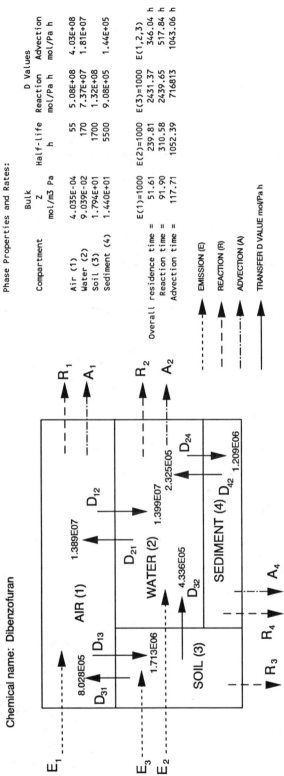

Phase Properties and Rates:

Compartment	Bulk Z mol/m3 Pa	Half-life h	D Values Reaction mol/Pa h	D Values Advection mol/Pa h
Air (1)	4.035E-04	55	5.08E+08	4.03E+08
Water (2)	9.039E-02	170	7.37E+07	1.81E+07
Soil (3)	1.794E+01	1700	1.32E+08	
Sediment (4)	1.440E+01	5500	9.08E+05	1.44E+05

	E(1)=1000	E(2)=1000	E(3)=1000	E(1,2,3)
Overall residence time =	51.61	239.81	2431.37	346.04 h
Reaction time =	91.90	310.58	2439.65	517.84 h
Advection time =	117.71	1052.39	716813	1043.06 h

- - - - EMISSION (E)
- - - - REACTION (R)
- · - · - ADVECTION (A)
→ TRANSFER D VALUE mol/Pa h

Phase Properties, Compositions, Transport and Transformation Rates:

Emission, kg/h E(1)	E(2)	E(3)	Fugacity, Pa f(1)	f(2)	f(3)	f(4)	Concentration, g/m3 C(1)	C(2)	C(3)	C(4)
1000	0	0	6.423E-06	8.428E-07	8.278E-08	7.935E-07	4.358E-07	1.281E-05	2.498E-04	1.923E-03
0	1000	0	8.364E-07	5.586E-05	1.078E-06	5.259E-05	5.676E-08	8.492E-04	3.253E-01	1.274E-01
0	0	1000	4.153E-08	1.873E-07	4.474E-05	1.764E-07	2.818E-09	2.848E-06	1.350E-01	4.273E-04
600	300	100	4.109E-06	1.728E-05	4.526E-06	1.627E-05	2.788E-07	2.627E-04	1.366E-02	3.942E-02

Emission, kg/h E(1)	E(2)	E(3)	Loss, Reaction, kg/h R(1)	R(2)	R(3)	R(4)	Loss, Advection, kg/h A(1)	A(2)	A(3)	A(4)
1000	0	0	5.492E+02	1.045E+01	1.83E+00	1.211E-01	4.358E+02	2.563E+00		1.923E-02
0	1000	0	7.151E+01	6.923E+02	2.39E-01	8.027E+00	5.676E+01	1.698E+02		1.274E+00
0	0	1000	3.551E+00	2.322E+00	9.91E+02	2.692E-02	2.818E+00	5.695E-01		4.273E-03
600	300	100	3.513E+02	2.142E+02	1.00E+02	2.484E+00	2.788E+02	5.255E+01		3.942E-01

Amounts, kg m(1)	m(2)	m(3)	m(4)	Total Amount, kg
4.358E+04	2.563E+03	4.497E+03	9.613E+02	5.161E+04
5.676E+03	1.698E+05	5.856E+02	6.371E+04	2.398E+05
2.818E+02	5.695E+02	2.430E+06	2.136E+02	2.431E+06
2.788E+04	5.255E+04	2.459E+05	1.971E+04	3.460E+05

Intermedia Rate of Transport, kg/h

T12 air-water	T21 water-air	T13 air-soil	T31 soil-air	T24 water-sed	T32 soil-water	T42 sed-water
1.511E+01	1.969E+00	1.850E+00	1.118E-02	1.714E-01	6.036E-03	3.103E-02
1.968E+00	1.305E+02	2.410E-01	1.456E-03	1.136E+01	7.861E-04	2.057E+00
9.771E-02	4.375E-01	1.196E-02	6.041E+00	3.809E+00	3.262E+00	6.897E-03
9.668E+00	4.037E+01	1.184E+00	6.112E-01	3.514E+00	3.301E-01	6.363E-01

Level III Distribution

Chemical name: Dibenzofuran

Distribution of mass

SEDIMENT (1.86%)
SOIL (8.71%)
WATER (4.97%)
AIR (84.46%)

AIR (2.367%)
SEDIMENT (26.6%)
WATER (70.82%)
SOIL (0.244%)

SEDIMENT (0.00088%)
WATER (0.023%)
AIR (0.012%)
SOIL (99.956%)

SEDIMENT (5.70%)
AIR (8.06%)
WATER (15.18%)
SOIL (71.1%)

Distribution of removal rates

SOIL (0.183%)
SEDIMENT (0.0121%)
SEDIMENT A (0.00192%)
WATER (1.04%)
WATER A (0.256%)
AIR (43.58%)
AIR (54.92%)

SEDIMENT (0.803%)
SOIL (0.024%)
SEDIMENT A (0.127%)
AIR (5.676%)
AIR (7.151%)
WATER (16.98%)
WATER (69.235%)

SEDIMENT A (0.000427%)
SEDIMENT R (0.00269%)
WATER A (0.057%)
AIR (0.355%)
WATER (0.232%)
AIR A (0.282%)
SOIL (99.071%)

SEDIMENT (0.039%)
SEDIMENT (0.248%)
AIR (27.9%)
AIR (35.1%)
SOIL (10.0%)
WATER (21.4%)
WATER (5.25%)

Reaction   Advection

Emission rates:

E(1) = 1000 kg/h
E(2) = 0
E(3) = 0

E(1) = 0
E(2) = 1000 kg/h
E(3) = 0

E(1) = 0
E(2) = 0
E(3) = 1000 kg/h

E(1) = 600 kg/h
E(2) = 300 kg/h
E(3) = 100 kg/h

Residence time:

t = 51.61 h

t = 240 h

t = 2431 h

t = 346 h

Chemical name: 2,8-Dichlorodibenzofuran

Level I calculation: (six compartment model)

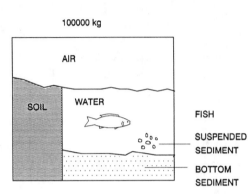

100000 kg

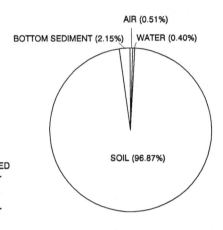

AIR (0.51%)

BOTTOM SEDIMENT (2.15%) | WATER (0.40%)

SOIL (96.87%)

Distribution of mass

physical-chemical properties:

MW: 237.1 g/mol

M.P.: 184 °C

Fugacity ratio: 0.0268

vapor pressure: 0.00039 Pa

solubility: 0.0145 g/m^3

log K_{OW} : 5.44

Compartment	Z mol/m3 Pa	Concentration			Amount kg	Amount %
		mol/m3	mg/L (or g/m3)	ug/g		
Air	4.034E-04	2.155E-11	5.108E-09	4.309E-06	510.85	0.511
Water	1.568E-01	8.375E-09	1.986E-06	1.986E-06	397.13	0.397
Soil	8.500E+02	4.539E-05	1.076E-02	4.485E-03	96867	96.867
Biota (fish)	2.159E+03	1.153E-04	2.734E-02	2.734E-02	5.4690	5.47E-03
Suspended sediment	5.312E+03	2.837E-04	6.727E-02	4.485E-02	67.269	6.73E-02
Bottom sediment	1.700E+03	9.079E-05	2.153E-02	8.969E-03	2152.59	2.1526
Total					100000	100

f = 5.341E-08 Pa

Chemical name: 2,8-Dichlorodibenzofuran

Level II calculation: (six compartment model)

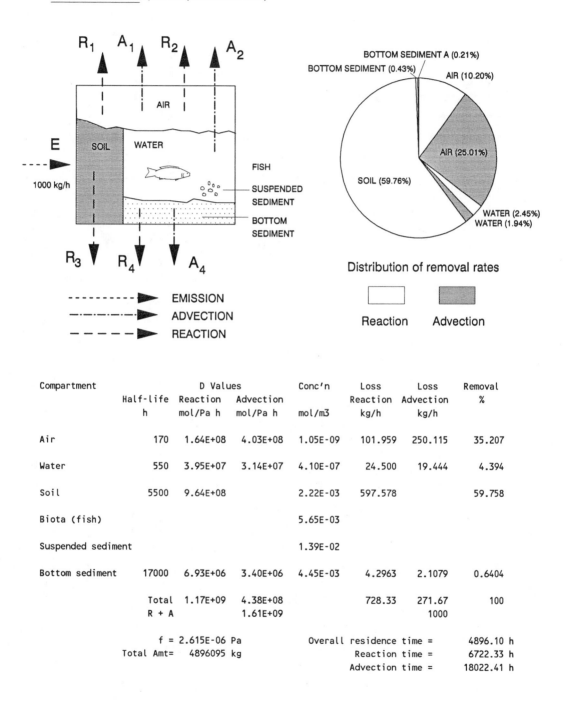

Distribution of removal rates

Reaction   Advection

Compartment	Half-life h	D Values Reaction mol/Pa h	Advection mol/Pa h	Conc'n mol/m3	Loss Reaction kg/h	Loss Advection kg/h	Removal %
Air	170	1.64E+08	4.03E+08	1.05E-09	101.959	250.115	35.207
Water	550	3.95E+07	3.14E+07	4.10E-07	24.500	19.444	4.394
Soil	5500	9.64E+08		2.22E-03	597.578		59.758
Biota (fish)				5.65E-03			
Suspended sediment				1.39E-02			
Bottom sediment	17000	6.93E+06	3.40E+06	4.45E-03	4.2963	2.1079	0.6404
Total		1.17E+09	4.38E+08		728.33	271.67	100
R + A			1.61E+09			1000	

f = 2.615E-06 Pa
Total Amt= 4896095 kg

Overall residence time =   4896.10 h
Reaction time =   6722.33 h
Advection time =   18022.41 h

535

Fugacity Level III calculations: (four compartment model)
Chemical name: 2,8-Dichloro-dibenzofuran

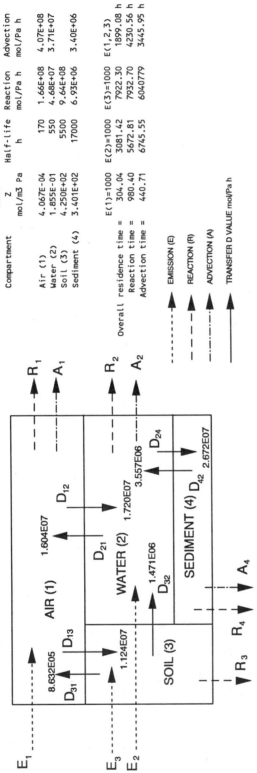

Phase Properties and Rates:

Compartment	Bulk Z mol/m3 Pa	Half-life h	D Values Reaction mol/Pa h	Advection mol/Pa h	E(1,2,3)
Air (1)	4.067E-04	170	1.66E+08	4.07E+08	1899.08 h
Water (2)	1.855E-01	550	4.68E+07	3.71E+07	4230.56 h
Soil (3)	4.250E+02	5500	9.64E+08	3.40E+06	3445.95 h
Sediment (4)	3.401E+02	17000	6.93E+06		

	E(1)=1000	E(2)=1000	E(3)=1000
Overall residence time =	304.04	3081.42	7922.30
Reaction time =	980.40	5672.81	7932.70
Advection time =	440.71	6745.55	6040779

EMISSION (E)
REACTION (R)
ADVECTION (A)
TRANSFER D VALUE mol/Pa h

Phase Properties, Compositions, Transport and Transformation Rates:

Emission, kg/h

E(1)	E(2)	E(3)
1000	0	0
0	1000	0
0	0	1000
600	300	100

Fugacity, Pa

f(1)	f(2)	f(3)	f(4)
7.045E-06	1.012E-06	8.196E-08	1.948E-06
9.436E-07	3.535E-05	1.098E-08	6.799E-05
7.729E-09	5.470E-08	4.365E-06	1.052E-07
4.511E-06	1.122E-05	4.890E-07	2.158E-05

Concentration, g/m3

C(1)	C(2)	C(3)	C(4)
6.794E-07	4.454E-05	8.259E-03	1.570E-01
9.100E-08	1.555E-03	1.106E-03	5.483E+00
7.454E-10	2.406E-06	4.399E-01	8.485E-03
4.350E-07	4.934E-04	4.927E-02	1.740E+00

Loss, Reaction, kg/h

R(1)	R(2)	R(3)	R(4)
2.770E+02	1.122E+01	1.87E+01	3.201E+00
3.710E+00	3.918E+02	2.51E+00	1.118E+00
3.039E-01	6.064E-01	9.98E+02	1.729E-01
1.773E+02	1.243E+02	1.12E+02	3.546E+01

Loss, Advection, kg/h

A(1)	A(2)	A(4)
6.794E+02	8.907E+00	1.570E+00
9.100E+00	3.110E+02	5.483E+01
7.454E-01	4.812E-01	8.485E-02
4.350E+02	9.869E+01	1.740E+01

Amounts, kg

m(1)	m(2)	m(3)	m(4)	Total Amount, kg
6.794E+04	8.907E+03	1.487E+05	7.852E+04	3.040E+05
9.100E+03	3.110E+05	1.991E+04	2.741E+06	3.081E+06
7.454E+01	4.812E+02	7.918E+06	4.242E+03	7.922E+06
4.350E+04	9.869E+04	8.869E+05	8.700E+05	1.899E+06

Intermedia Rate of Transport, kg/h

T12 air-water	T21 water-air	T13 air-soil	T31 soil-air	T32 soil-water	T24 water-sed	T42 sed-water
2.872E+01	3.851E+00	1.878E+01	1.677E-02	2.858E-02	6.414E+00	1.642E+00
3.847E+00	1.345E+02	2.515E+00	2.247E-03	3.828E-03	2.239E+02	5.734E+01
3.151E-02	2.081E-01	2.060E-01	8.933E-01	1.522E+00	3.465E+02	8.873E-02
1.839E+01	4.267E+01	1.202E+01	1.001E-01	1.705E-01	7.106E+01	1.820E+01

Level III Distribution

Chemical name: 2,8-Dichloro-dibenzofuran

Distribution of mass

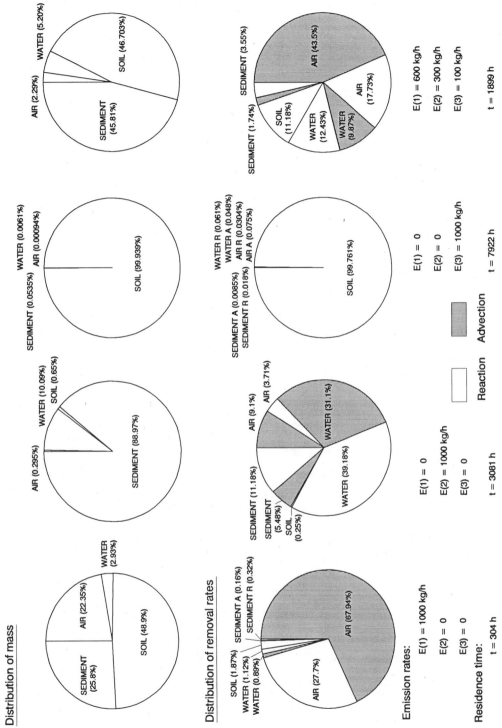

Distribution of removal rates

Emission rates:

E(1) = 1000 kg/h	E(1) = 0	E(1) = 0	E(1) = 600 kg/h
E(2) = 0	E(2) = 1000 kg/h	E(2) = 0	E(2) = 300 kg/h
E(3) = 0	E(3) = 0	E(3) = 1000 kg/h	E(3) = 100 kg/h

Residence time:

t = 304 h   t = 3081 h   t = 7922 h   t = 1899 h

Reaction   Advection

Chemical name: 2,3,7,8-Tetrachloro-dibenzofuran

Level I calculation: (six compartment model)

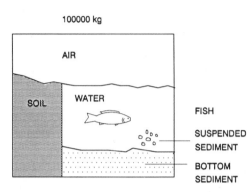

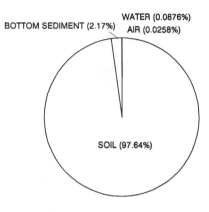

Distribution of mass

Physical-chemical properties:

MW: 306 g/mol

M.P.: 227 °C

Fugacity ratio: 0.010

vapor pressure: 0.000002 Pa

solubility: 0.000419 mg/m^3

log K_{OW}: 6.10

Compartment	Z mol/m3 Pa	Concentration mol/m3	mg/L (or g/m3)	ug/g	Amount kg	Amount %
Air	4.034E-04	8.432E-13	2.580E-10	2.177E-07	25.803	0.0258
Water	6.846E-01	1.431E-09	4.379E-07	4.379E-07	87.580	0.0876
Soil	1.696E+04	3.546E-05	1.085E-02	4.521E-03	97643	97.643
Biota (fish)	4.310E+04	9.008E-05	2.756E-02	2.756E-02	5.5128	5.51E-03
Suspended sediment	1.060E+05	2.216E-04	6.781E-02	4.521E-02	67.808	6.78E-02
Bottom sediment	3.392E+04	7.091E-05	2.170E-02	9.041E-03	2169.85	2.1699
Total					100000	100

f = 2.090E-09 Pa

Chemical name: 2,3,7,8-Tetrachloro-dibenzofuran

Level II calculation: (six compartment model)

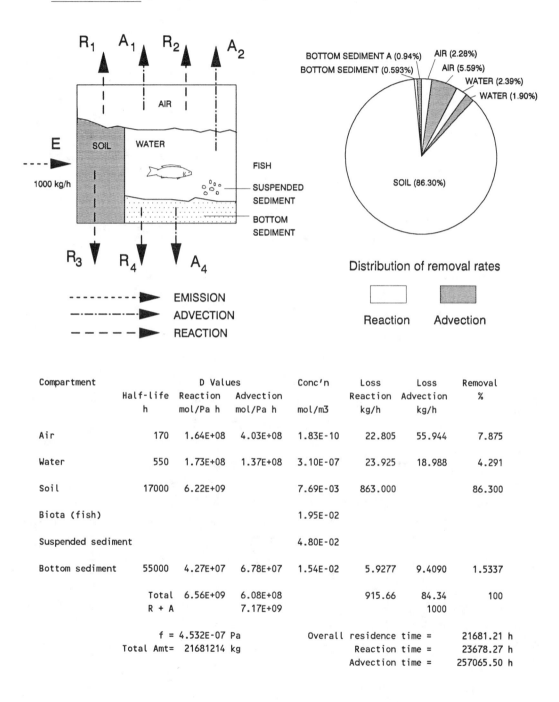

Distribution of removal rates

Reaction   Advection

Compartment	Half-life h	D Values Reaction mol/Pa h	Advection mol/Pa h	Conc'n mol/m3	Loss Reaction kg/h	Loss Advection kg/h	Removal %
Air	170	1.64E+08	4.03E+08	1.83E-10	22.805	55.944	7.875
Water	550	1.73E+08	1.37E+08	3.10E-07	23.925	18.988	4.291
Soil	17000	6.22E+09		7.69E-03	863.000		86.300
Biota (fish)				1.95E-02			
Suspended sediment				4.80E-02			
Bottom sediment	55000	4.27E+07	6.78E+07	1.54E-02	5.9277	9.4090	1.5337
Total		6.56E+09	6.08E+08		915.66	84.34	100
R + A			7.17E+09			1000	

f = 4.532E-07 Pa
Total Amt= 21681214 kg

Overall residence time = 21681.21 h
Reaction time = 23678.27 h
Advection time = 257065.50 h

539

Fugacity Level III calculations: (four compartment model)

Chemical name: 2,3,7,8-Tetrachloro-dibenzofuran

540

Diagram labels: E_1, R_1, A_1, E_3, E_2, R_2, A_2, AIR (1), WATER (2), SOIL (3), SEDIMENT (4), R_3, R_4, A_4, D_{12}, D_{21}, D_{13}, D_{31}, D_{32}, D_{24}, D_{42}
1.905E07, 1.332E06, 6.624E08, 9.270E07, 1.835E07, 6.853E07, 5.308E08

Phase Properties and Rates:

Compartment	Bulk Z mol/m3 Pa	Half-life h	D Values Reaction mol/Pa h	Advection mol/Pa h
Air (1)	6.466E-04	170	2.64E+08	6.47E+08
Water (2)	1.258E+00	550	3.17E+08	2.52E+08
Soil (3)	8.481E+03	17000	6.22E+09	
Sediment (4)	6.786E+03	55000	4.27E+07	6.79E+07

	E(1)=1000	E(2)=1000	E(3)=1000	E(1,2,3)
Overall residence time =	10434	11475	24490	12152 h
Reaction time =	17881	23096	24528	20286 h
Advection time =	25053	22805	15622394	30304 h

Legend:
- EMISSION (E)
- REACTION (R)
- ADVECTION (A)
- TRANSFER D VALUE mol/Pa h

Phase Properties, Compositions, Transport and Transformation Rates:

Emission, kg/h E(1)	E(2)	E(3)	Fugacity, Pa f(1)	f(2)	f(3)	f(4)	Concentration, g/m3 C(1)	C(2)	C(3)	C(4)	Amounts, kg m(1)	m(2)	m(3)	m(4)	Total Amount, kg
1000	0	0	1.963E-06	2.030E-07	2.088E-07	6.014E-07	3.884E-07	7.812E-05	5.419E-01	1.249E+00	3.884E+04	1.562E+04	9.755E+06	6.243E+05	1.043E+07
0	1000	0	4.085E-08	3.575E-06	4.346E-09	1.059E-05	8.082E-09	1.376E-03	1.128E-02	2.199E-01	8.082E+02	2.752E+05	2.030E+05	1.100E+07	1.147E+07
0	0	1000	5.389E-10	1.055E-08	5.235E-07	3.125E-08	1.066E-10	4.060E-06	1.359E+00	6.489E-02	1.066E+01	8.120E+02	2.446E+07	3.245E+04	2.449E+07
600	300	100	1.190E-06	1.195E-06	1.789E-07	3.541E-06	2.355E-07	4.600E-04	4.644E-01	7.355E-01	2.355E+04	9.201E+04	8.360E+06	3.677E+06	1.215E+07

Emission, kg/h E(1)	E(2)	E(3)	Loss, Reaction, kg/h R(1)	R(2)	R(3)	R(4)	Loss, Advection, kg/h A(1)	A(2)	A(4)
1000	0	0	1.583E+02	1.969E+01	3.98E+02	7.867E+00	3.884E+02	1.562E+01	1.249E+01
0	1000	0	3.295E+00	3.467E+02	8.28E+00	1.385E+02	8.082E+00	2.752E+02	2.199E+02
0	0	1000	4.347E-02	1.023E+00	9.97E+02	4.088E-01	1.066E-01	8.120E-01	6.489E-02
600	300	100	9.598E-01	1.159E+02	3.41E+02	4.632E+01	2.355E+02	9.201E+01	7.353E+01

Intermedia Rate of Transport, kg/h

T12 air-water	T13 air-soil	T21 water-air	T31 soil-air	T32 soil-water	T24 water-sed	T42 sed-water
5.568E+01	3.989E+02	1.183E+00	8.514E-02	1.172E+00	3.297E+01	1.261E+01
1.159E+00	8.302E+01	2.084E+01	1.772E-03	2.440E-02	5.806E+02	2.221E+02
1.529E-02	1.095E-01	6.149E-02	2.135E-01	2.939E+00	1.713E+00	6.554E-01
3.375E+01	2.419E+02	6.967E+01	7.296E-02	1.005E+00	1.941E+02	7.427E+01

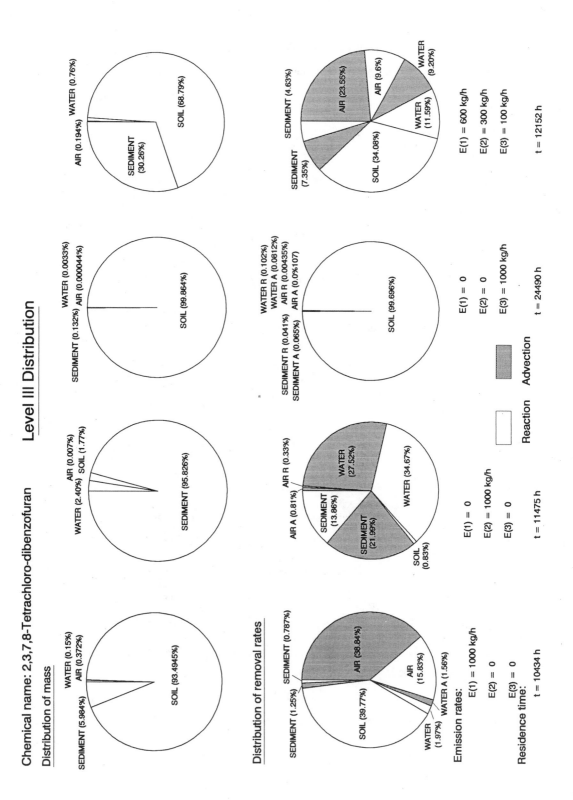

Chemical name: 2,3,7,8-Tetrachloro-dibenzofuran   Level III Distribution

541

Chemical name: 2,3,4,7,8-Pentachloro-dibenzofuran

Level I calculation: (six compartment model)

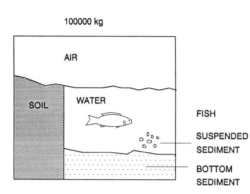

100000 kg

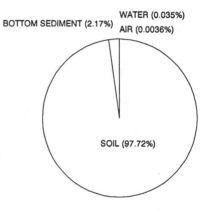

Distribution of mass

physical-chemical properties:

MW: 340.42 g/mol

M.P.: 196 °C

fugacity ratio: 0.0204

vapor pressure: 3.50×10^{-7} Pa

solubility: 2.36×10^{-4} g/m^3

log K_{OW}: 6.50

Compartment	Z mol/m3 Pa	Concentration mol/m3	mg/L (or g/m3)	ug/g	Amount kg	Amount %
Air	4.034E-04	1.044E-13	3.553E-11	2.997E-08	3.553	0.00355
Water	1.981E+00	5.125E-10	1.745E-07	1.745E-07	34.892	0.0349
Soil	1.233E+05	3.189E-05	1.086E-02	4.524E-03	97717	97.717
Biota (fish)	3.132E+05	8.103E-05	2.758E-02	2.758E-02	5.5170	5.52E-03
Suspended sediment	7.704E+05	1.993E-04	6.786E-02	4.524E-02	67.859	6.79E-02
Bottom sediment	2.465E+05	6.379E-05	2.171E-02	9.048E-03	2171.48	2.1715
Total					100000	100

f = 2.587E-10 Pa

Chemical name: 2,3,4,7,8-Pentachloro-dibenzofuran

Level II calculation: (six compartment model)

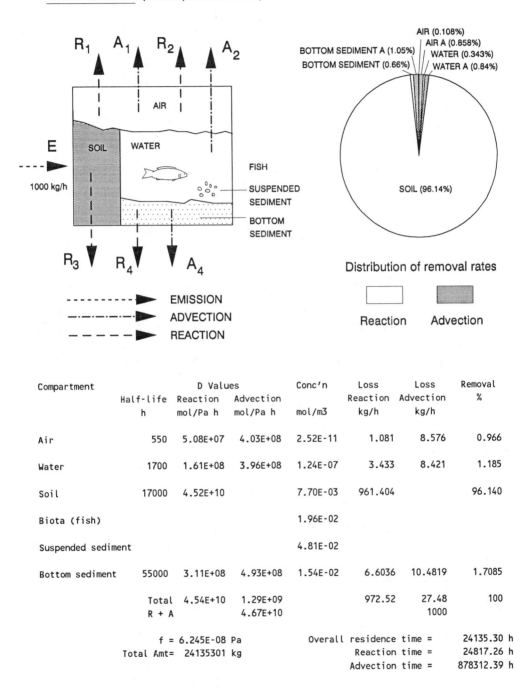

Distribution of removal rates

Reaction   Advection

Compartment	Half-life h	D Values Reaction mol/Pa h	D Values Advection mol/Pa h	Conc'n mol/m3	Loss Reaction kg/h	Loss Advection kg/h	Removal %
Air	550	5.08E+07	4.03E+08	2.52E-11	1.081	8.576	0.966
Water	1700	1.61E+08	3.96E+08	1.24E-07	3.433	8.421	1.185
Soil	17000	4.52E+10		7.70E-03	961.404		96.140
Biota (fish)				1.96E-02			
Suspended sediment				4.81E-02			
Bottom sediment	55000	3.11E+08	4.93E+08	1.54E-02	6.6036	10.4819	1.7085
Total		4.54E+10	1.29E+09		972.52	27.48	100
R + A			4.67E+10			1000	

f = 6.245E-08 Pa
Total Amt= 24135301 kg

Overall residence time =   24135.30 h
Reaction time =   24817.26 h
Advection time =   878312.39 h

543

Fugacity Level III calculations: (four compartment model)

Chemical name: 2,3,4,7,8-Pentachloro-dibenzofuran

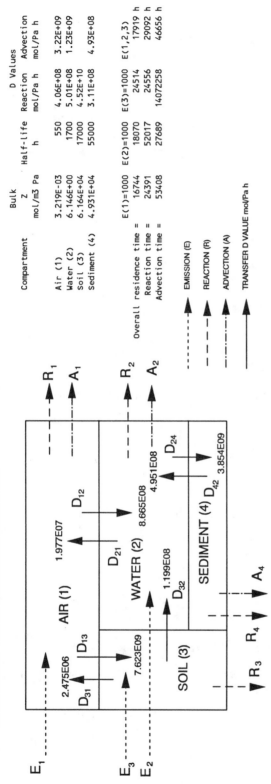

Phase Properties and Rates:

Compartment	Bulk Z mol/m3 Pa	Half-life h	D Values Reaction mol/Pa h	D Values Advection mol/Pa h
Air (1)	3.219E-03	550	4.06E+08	3.22E+09
Water (2)	6.146E+00	1700	5.01E+08	1.23E+09
Soil (3)	6.164E+04	17000	4.52E+10	
Sediment (4)	4.931E+04	55000	3.11E+08	4.93E+08

	E(1)=1000	E(2)=1000	E(3)=1000	E(1,2,3)
Overall residence time =	16744	18070	24514	17919 h
Reaction time =	24391	52017	24556	29092 h
Advection time =	53408	27689	14072258	46656 h

EMISSION (E) ——▶
REACTION (R) – – –▶
ADVECTION (A) –·–·▶
TRANSFER D VALUE mol/Pa h ——▶

Phase Properties, Compositions, Transport and Transformation Rates:

Emission, kg/h | Fugacity, Pa | Concentration, g/m3

E(1)	E(2)	E(3)	f(1)	f(2)	f(3)	f(4)	C(1)	C(2)	C(3)	C(4)
1000	0	0	2.426E-07	5.201E-08	4.078E-08	1.543E-07	2.658E-07	1.088E-04	8.556E-01	2.591E+00
0	1000	0	1.160E-09	7.106E-07	1.949E-10	2.109E-06	1.271E-09	1.487E-03	4.090E-03	3.540E-02
0	0	1000	1.630E-11	1.881E-09	6.478E-08	5.582E-09	1.787E-11	3.936E-06	1.359E+00	9.370E-01
600	300	100	1.459E-07	2.446E-07	3.100E-08	7.258E-07	1.599E-07	5.117E-04	6.505E-01	1.218E+00

Emission, kg/h | Loss, Reaction, kg/h | Loss, Advection, kg/h

E(1)	E(2)	E(3)	R(1)	R(2)	R(3)	R(4)	A(1)	A(2)	A(3)	A(4)
1000	0	0	3.350E+01	8.872E+00	6.28E+02	1.632E+00	2.658E+02	2.176E+01		2.591E+01
0	1000	0	1.601E-01	1.212E+02	3.00E+00	2.230E+02	1.271E+00	2.974E+02		3.540E+02
0	0	1000	2.251E-03	3.209E-01	9.97E+02	5.903E-01	1.787E-02	7.871E+01		9.370E+01
600	300	100	2.015E+01	4.172E+01	4.77E+02	7.675E+01	1.599E+02	1.023E+02		1.218E+02

Amounts, kg | Total Amount, kg

m(1)	m(2)	m(3)	m(4)	Total Amount, kg
2.658E+04	2.176E+07	1.540E+07	1.295E+06	1.674E+07
1.271E+02	2.974E+05	7.362E+06	1.770E+07	1.807E+07
1.787E+00	7.871E+02	2.447E+07	4.685E+04	2.451E+07
1.599E+04	1.023E+05	1.171E+07	6.092E+06	1.792E+07

Intermedia Rate of Transport, kg/h

T12 air-water	T13 air-soil	T21 water-air	T24 water-sed	T31 soil-air	T32 soil-water	T42 sed-water
7.155E+01	6.295E+02	3.500E-01	6.824E+01	3.435E-02	1.664E+00	2.601E+01
3.421E-01	3.009E+00	4.782E+00	9.324E+02	1.642E-04	7.954E-02	3.554E+02
4.809E-03	4.231E-02	1.266E-02	2.468E+00	5.457E+02	2.643E+00	9.407E-01
4.304E+01	3.786E+02	1.646E+00	3.209E+02	2.612E-02	1.265E+00	1.223E+02

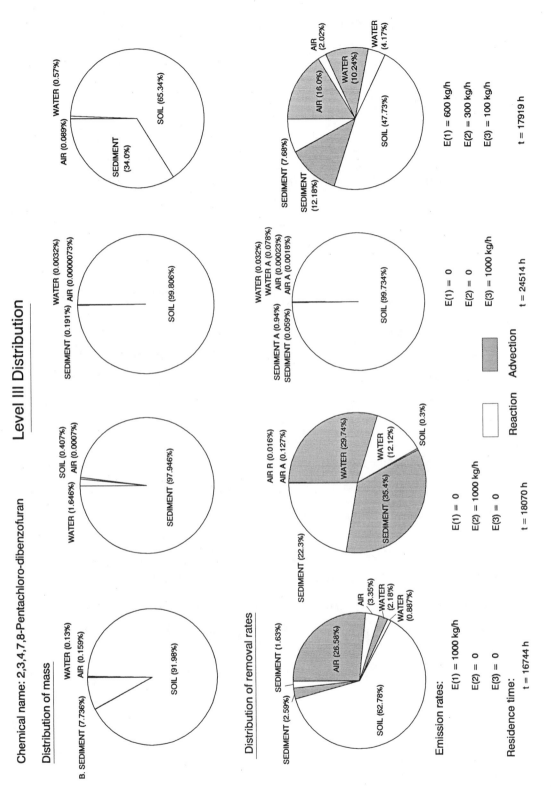

Chemical name: 2,3,4,7,8-Pentachloro-dibenzofuran

Level III Distribution

Distribution of mass

Distribution of removal rates

545

Chemical name: 1,2,3,4,7,8-Hexachloro-dibenzofuran

Level I calculation: (six compartment model)

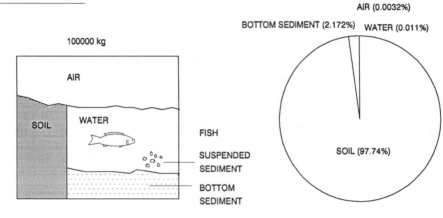

Distribution of mass

physical-chemical properties:

MW: 374.87 g/mol

M.P.: 225.5°C

Fugacity ratio: 0.0104

vapor pressure: 3.20×10^{-8} Pa

solubility: 8.25×10^{-6} g/m^3

log K_{OW}: 7.0

Compartment	Z mol/m3 Pa	Concentration			Amount kg	Amount %
		mol/m3	mg/L (or g/m3)	ug/g		
Air	4.034E-04	8.635E-14	3.237E-11	2.731E-08	3.237	0.00324
Water	6.877E-01	1.472E-10	5.518E-08	5.518E-08	11.037	0.0110
Soil	1.353E+05	2.897E-05	1.086E-02	4.525E-03	97740	97.740
Biota (fish)	3.439E+05	7.360E-05	2.759E-02	2.759E-02	5.5183	5.52E-03
Suspended sediment	8.459E+05	1.811E-04	6.788E-02	4.525E-02	67.875	6.79E-02
Bottom sediment	2.707E+05	5.794E-05	2.172E-02	9.050E-03	2172.01	2.1720
Total					100000	100

f = 2.140E-10 Pa

Chemical name: 1,2,3,4,7,8-Hexachlorodibenzofuran

Level II calculation: (six compartment model)

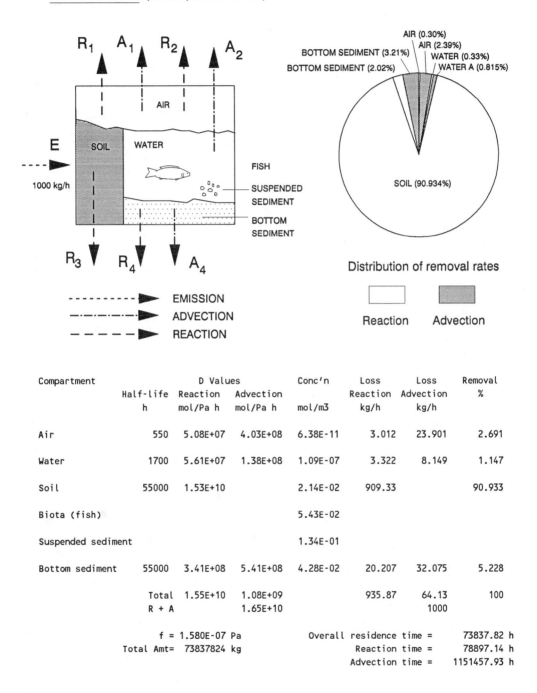

Distribution of removal rates

Reaction   Advection

Compartment	Half-life h	D Values Reaction mol/Pa h	D Values Advection mol/Pa h	Conc'n mol/m3	Loss Reaction kg/h	Loss Advection kg/h	Removal %
Air	550	5.08E+07	4.03E+08	6.38E-11	3.012	23.901	2.691
Water	1700	5.61E+07	1.38E+08	1.09E-07	3.322	8.149	1.147
Soil	55000	1.53E+10		2.14E-02	909.33		90.933
Biota (fish)				5.43E-02			
Suspended sediment				1.34E-01			
Bottom sediment	55000	3.41E+08	5.41E+08	4.28E-02	20.207	32.075	5.228
Total		1.55E+10	1.08E+09		935.87	64.13	100
R + A			1.65E+10			1000	

f = 1.580E-07 Pa
Total Amt= 73837824 kg

Overall residence time =   73837.82 h
Reaction time =   78897.14 h
Advection time = 1151457.93 h

Fugacity Level III calculations: (four compartment model)
Chemical name: 1,2,3,4,7,8-Hexachloro-dibenzofuran

Phase Properties and Rates:

Compartment	Bulk Z mol/m3 Pa	Half-life h	D Values Reaction mol/Pa h	Advection mol/Pa h
Air (1)	1.613E-02	550	2.03E+09	1.61E+10
Water (2)	5.261E+00	1700	4.29E+08	1.05E+09
Soil (3)	6.767E+04	55000	1.53E+10	
Sediment (4)	5.414E+04	55000	3.41E+08	5.41E+08

	E(1)=1000	E(2)=1000	E(3)=1000	E(1,2,3)
Overall residence time =	52724	20006	78884	45525 h
Reaction time =	75003	56654	79300	72585 h
Advection time =	177497	30928	15034152	122112 h

- - - - - EMISSION (E)
– · – · – REACTION (R)
───── ADVECTION (A)
TRANSFER D VALUE mol/Pa h

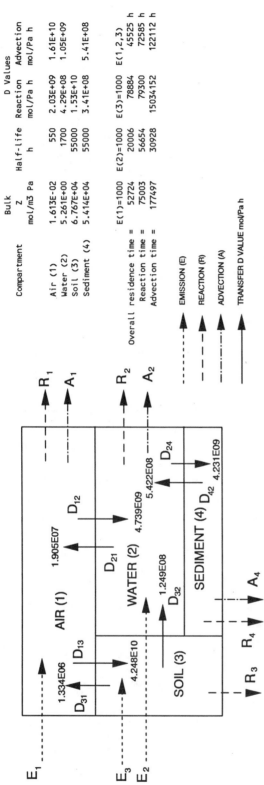

548

Phase Properties, Compositions, Transport and Transformation Rates:

Emission, kg/h

E(1)	E(2)	E(3)
1000	0	0
0	1000	0
0	0	1000
600	300	100

Fugacity, Pa

f(1)	f(2)	f(3)	f(4)
4.082E-08	5.033E-08	1.120E-07	1.495E-07
1.887E-10	6.476E-07	1.277E-06	1.923E-06
5.045E-12	5.231E-09	1.724E-07	1.554E-08
2.455E-08	2.250E-07	8.462E-08	6.681E-07

Concentration, g/m3

C(1)	C(2)	C(3)	C(4)
2.469E-07	9.927E-05	2.842E+00	3.033E+00
1.141E-09	1.277E-03	1.314E-02	3.903E+01
3.051E-11	1.032E-05	4.374E+00	3.153E-01
1.485E-07	4.438E-04	2.147E-01	1.356E+01

Amounts, kg

m(1)	m(2)	m(3)	m(4)	Total Amount, kg
2.469E+04	1.985E+04	5.116E+07	1.517E+06	5.272E+07
1.141E+02	2.554E+05	2.366E+05	1.951E+07	2.001E+07
3.051E+00	2.064E+03	7.872E+07	1.576E+05	7.888E+07
1.485E+04	8.875E+04	3.864E+07	6.780E+06	4.552E+07

Intermedia Rate of Transport, kg/h

T12 air-water	T21 water-air	T13 air-soil	T31 soil-air	T32 soil-water	T24 water-sed	T42 sed-water
7.251E+01	3.595E-01	6.500E+02	5.608E-02	5.246E+00	7.982E+01	3.037E+01
3.352E-01	4.625E+00	3.005E+00	2.593E-04	2.426E-02	1.027E+03	3.908E+02
8.962E-03	3.737E-02	8.034E-02	8.629E-02	8.072E+00	8.296E+00	3.157E+00
4.361E+01	1.607E+00	3.909E+02	4.236E-02	3.962E+00	3.568E+02	1.358E+02

Emission, kg/h

E(1)	E(2)	E(3)
1000	0	0
0	1000	0
0	0	1000
600	300	100

Loss, Reaction, kg/h

R(1)	R(2)	R(3)	R(4)
3.110E+01	8.093E+00	6.45E+02	1.911E+01
1.438E-01	1.041E+02	2.98E+00	2.459E+02
3.845E-03	8.412E-01	9.92E+02	1.986E+00
1.871E+01	3.618E+01	4.87E+02	8.543E+01

Loss, Advection, kg/h

A(1)	A(2)	A(4)
2.469E+02	1.985E+01	3.033E+01
1.141E+00	2.554E+02	3.903E+02
3.051E-02	2.064E+00	3.153E+02
1.485E+02	8.875E+02	1.356E+02

Chemical name: 1,2,3,4,7,8-Hexachloro-dibenzofuran   Level III Distribution

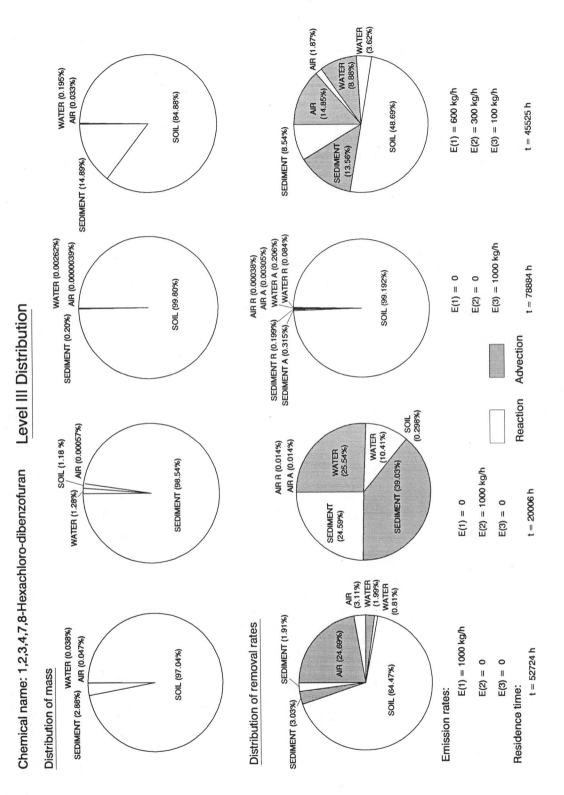

Chemical name: 1,2,3,4,6,7,8-Heptachloro-dibenzofuran

Level I calculation: (six compartment model)

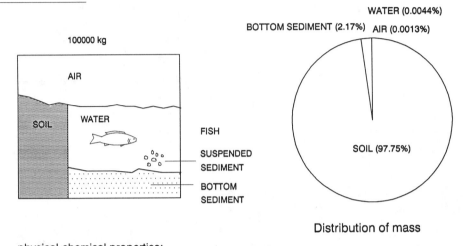

Distribution of mass

physical-chemical properties:

MW: 409.3 g/mol

M.P.: 236 °C

Fugacity ratio: 0.0082

vapor pressure: 4.70 x 10^{-9} Pa

solubility: 1.35 x10^{-6} g/m^3

log K_{ow}: 7.40

Compartment	Half-life h	D Values Reaction mol/Pa h	Advection mol/Pa h	Conc'n mol/m3	Loss Reaction kg/h	Loss Advection kg/h	Removal %
Air	550	5.08E+07	4.03E+08	2.33E-11	1.203	9.548	1.075
Water	1700	5.72E+07	1.40E+08	4.06E-08	1.354	3.322	0.468
Soil	55000	3.93E+10		2.01E-02	931.04		93.104
Biota (fish)				5.10E-02			
Suspended sediment				1.25E-01			
Bottom sediment	55000	8.74E+08	1.39E+09	4.01E-02	20.690	32.841	5.353
	Total	3.94E+10	1.93E+09		954.29	45.71	100
	R + A		4.14E+10			1000	

f = 5.782E-08 Pa
Total Amt= 75594099 kg

Overall residence time = 75594.10 h
Reaction time = 79215.04 h
Advection time = 1653765.59 h

550

Chemical name: 1,2,3,4,6,7,8-Heptachlorodibenzofuran

Level II calculation: (six compartment model)

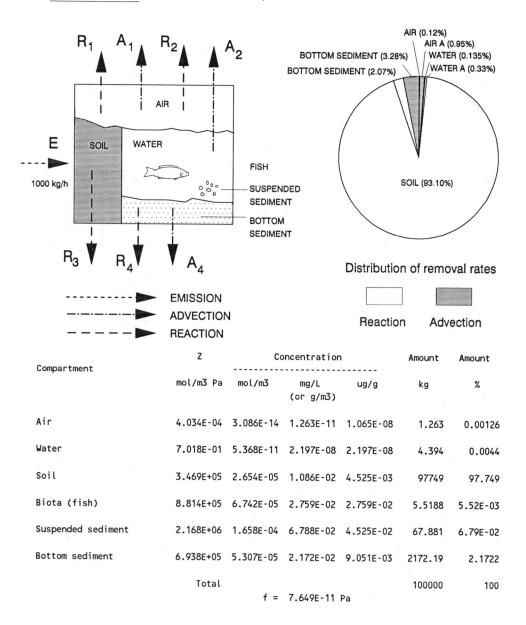

Distribution of removal rates

Reaction   Advection

Compartment	Z	Concentration			Amount	Amount
	mol/m3 Pa	mol/m3	mg/L (or g/m3)	ug/g	kg	%
Air	4.034E-04	3.086E-14	1.263E-11	1.065E-08	1.263	0.00126
Water	7.018E-01	5.368E-11	2.197E-08	2.197E-08	4.394	0.0044
Soil	3.469E+05	2.654E-05	1.086E-02	4.525E-03	97749	97.749
Biota (fish)	8.814E+05	6.742E-05	2.759E-02	2.759E-02	5.5188	5.52E-03
Suspended sediment	2.168E+06	1.658E-04	6.788E-02	4.525E-02	67.881	6.79E-02
Bottom sediment	6.938E+05	5.307E-05	2.172E-02	9.051E-03	2172.19	2.1722
Total					100000	100

f = 7.649E-11 Pa

Fugacity Level III calculations: (four compartment model)

Chemical name: 1,2,3,4,6,7,8-Heptachloro-dibenzofuran

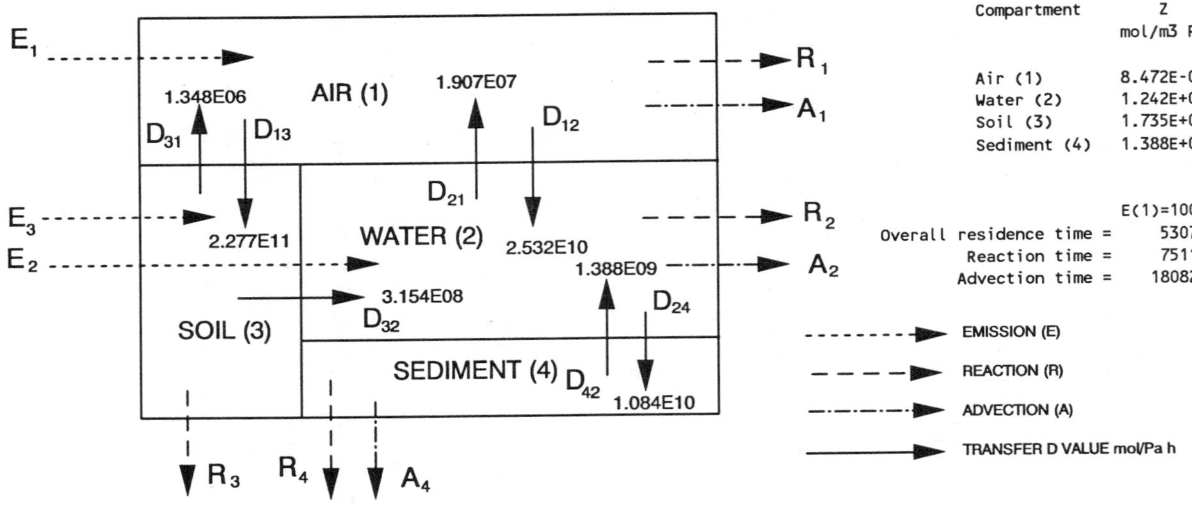

Diagram labels:

E_1, R_1, A_1 — AIR (1), 1.907E07, D_{12}, D_{21}, D_{13}, 1.348E06, D_{31}, 2.277E11

E_3, R_2, A_2 — WATER (2) 2.532E10, 3.154E08, D_{32}, D_{42} 1.084E10, D_{24}, 1.388E09

E_2, SOIL (3), SEDIMENT (4), R_3, R_4, A_4, A_2

Legend:
- EMISSION (E) — dashed
- REACTION (R) — dashed
- ADVECTION (A) — dash-dot
- TRANSFER D VALUE mol/Pa h — solid arrow

Phase Properties and Rates:

Compartment	Bulk Z mol/m3 Pa	Half-life h	D Values Reaction mol/Pa h	Advection mol/Pa h
Air (1)	8.472E-02	550	1.07E+10	8.47E+10
Water (2)	1.242E+01	1700	1.01E+09	2.48E+09
Soil (3)	1.735E+05	55000	3.93E+10	
Sediment (4)	1.388E+05	55000	8.74E+08	1.39E+09

	E(1)=1000	E(2)=1000	E(3)=1000	E(1,2,3)
Overall residence time =	53071	20476	78896	45875 h
Reaction time =	75118	57846	79304	72866 h
Advection time =	180824	15326739	123847 h	

Phase Properties, Compositions, Transport and Transformation Rates:

Emission, kg/h | Fugacity, Pa | Concentration, g/m3

E(1)	E(2)	E(3)	f(1)	f(2)	f(3)	f(4)	c(1)	c(2)	c(3)	c(4)
1000	0	0	7.014E-09	1.859E-08	4.027E-08	5.521E-08	2.432E-07	9.453E-05	2.859E+00	3.136E+00
0	1000	0	1.307E-11	2.387E-07	7.505E-11	7.091E-07	4.533E-10	1.214E-03	5.328E-03	4.028E+01
0	0	1000	3.423E-13	1.899E-09	6.161E-08	5.641E-09	1.187E-11	9.658E-06	4.374E+00	3.204E+00
600	300	100	4.212E-09	8.297E-08	3.034E-08	2.464E-07	1.461E-07	4.219E-04	2.154E+00	1.400E+00

Emission, kg/h | Loss, Reaction, kg/h | Loss, Advection, kg/h

E(1)	E(2)	E(3)	R(1)	R(2)	R(3)	R(4)	A(1)	A(2)	A(4)
1000	0	0	3.065E+01	7.707E+00	6.48E+00	1.976E+01	2.432E+02	1.891E+01	3.136E+01
0	1000	0	5.712E-02	9.898E+01	1.21E+00	2.537E+02	4.533E-01	2.428E+02	4.028E+02
0	0	1000	1.496E-03	7.874E-01	9.92E+02	2.019E+00	1.187E-02	1.932E+00	3.204E+00
600	300	100	1.841E+01	3.440E+01	4.89E+02	8.818E+01	1.461E+02	8.438E+01	1.400E+02

Amounts, kg

m(1)	m(2)	m(3)	m(4)	Total Amount, kg
2.432E+04	1.891E+04	5.146E+07	1.568E+06	5.307E+07
4.533E+01	2.428E+05	9.590E+04	2.014E+07	2.048E+07
1.187E+00	1.932E+03	7.873E+07	1.602E+05	7.890E+07
1.461E+04	8.438E+04	3.878E+07	6.998E+06	4.588E+07

Intermedia Rate of Transport, kg/h

T12 air-water	T21 water-air	T13 air-soil	T31 soil-air	T32 soil-water	T24 water-sed	T42 sed-water
7.268E+01	1.451E+01	6.536E+02	2.221E-02	5.198E+00	8.249E+01	3.138E+01
1.354E-01	1.864E+00	1.218E+00	4.140E-05	9.687E-03	1.059E+03	4.030E+02
3.547E-03	1.483E-02	3.190E-02	3.398E-02	7.953E-02	8.428E+00	3.206E+00
4.365E+01	6.477E-01	3.925E+02	1.674E-02	3.917E-01	3.682E+02	1.400E+02

Chemical name: 1,2,3,4,6,7,8-Heptachloro-dibenzofuran   Level III Distribution

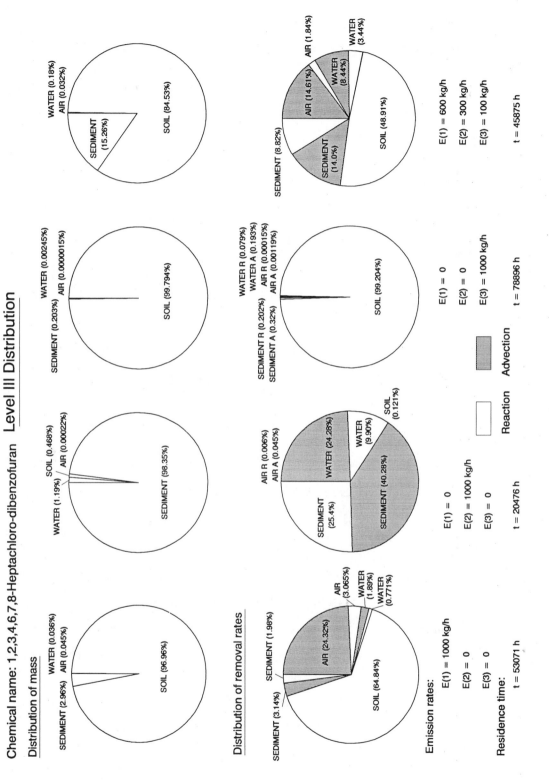

Distribution of mass

Pie chart 1:
- WATER (0.036%)
- AIR (0.045%)
- SEDIMENT (2.96%)
- SOIL (96.96%)

Pie chart 2:
- SOIL (0.468%)
- AIR (0.00022%)
- WATER (1.19%)
- SEDIMENT (98.35%)

Pie chart 3:
- WATER (0.00245%)
- AIR (0.0000015%)
- SEDIMENT (0.203%)
- SOIL (99.794%)

Pie chart 4:
- WATER (0.18%)
- AIR (0.032%)
- SEDIMENT (15.26%)
- SOIL (84.53%)

Distribution of removal rates

Pie chart 5:
- AIR (3.065%)
- WATER (1.89%)
- WATER (0.771%)
- SEDIMENT (1.98%)
- AIR (24.32%)
- SEDIMENT (3.14%)
- SOIL (64.84%)

Pie chart 6:
- AIR R (0.006%)
- AIR A (0.045%)
- WATER (24.28%)
- WATER (9.90%)
- SOIL (0.121%)
- SEDIMENT (40.28%)
- SEDIMENT (25.4%)

Pie chart 7:
- WATER R (0.079%)
- WATER A (0.193%)
- AIR R (0.00015%)
- AIR A (0.00119%)
- SEDIMENT R (0.202%)
- SEDIMENT A (0.32%)
- SOIL (99.204%)

Pie chart 8:
- AIR (1.84%)
- WATER (3.44%)
- WATER (8.44%)
- AIR (14.61%)
- SEDIMENT (14.0%)
- SEDIMENT (8.82%)
- SOIL (48.91%)

Reaction | Advection

Emission rates:

	Chart 1	Chart 2	Chart 3	Chart 4
E(1)	1000 kg/h	0	0	600 kg/h
E(2)	0	1000 kg/h	0	300 kg/h
E(3)	0	0	1000 kg/h	100 kg/h

Residence time:

Chart 1	Chart 2	Chart 3	Chart 4
t = 53071 h	t = 20476 h	t = 78896 h	t = 45875 h

553

Chemical name: Octachlorodibenzofuran

Level I calculation: (six compartment model)

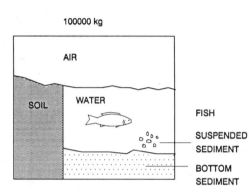

100000 kg

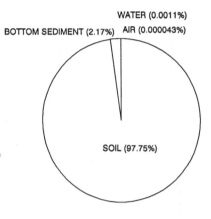

WATER (0.0011%)

BOTTOM SEDIMENT (2.17%) AIR (0.000043%)

SOIL (97.75%)

Distribution of mass

physical-chemical properties:

MW: 443.8 g/mol

M.P.: 258 °C

Fugacity ratio: 0.0050

vapor pressure: 5.0×10^{-10} Pa

solubility: 1.16×10^{-6} g/m^3

log K_{OW}: 8.0

Compartment	Z	Concentration			Amount	Amount
	mol/m3 Pa	mol/m3	mg/L (or g/m3)	ug/g	kg	%
Air	4.034E-04	9.597E-16	4.259E-13	3.593E-10	0.0426	4.26E-05
Water	5.228E+00	1.244E-11	5.519E-09	5.519E-09	1.1038	0.0011
Soil	1.029E+07	2.447E-05	1.086E-02	4.526E-03	97753	97.753
Biota (fish)	2.614E+07	6.218E-05	2.760E-02	2.760E-02	5.5190	5.52E-03
Suspended sediment	6.430E+07	1.530E-04	6.788E-02	4.526E-02	67.884	6.79E-02
Bottom sediment	2.058E+07	4.895E-05	2.172E-02	9.051E-03	2172.29	2.1723
	Total				100000	100

f = 2.379E-12 Pa

Chemical name: Octachlorodibenzofuran

Level II calculation: (six compartment model)

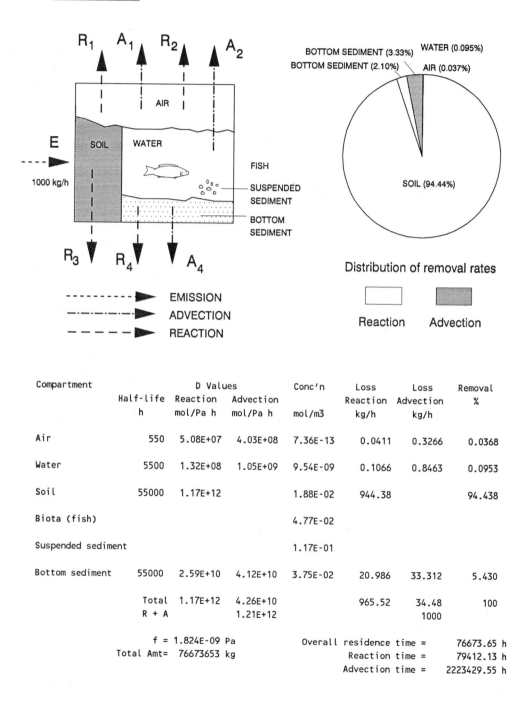

Distribution of removal rates

Reaction   Advection

Compartment	Half-life h	D Values Reaction mol/Pa h	D Values Advection mol/Pa h	Conc'n mol/m3	Loss Reaction kg/h	Loss Advection kg/h	Removal %
Air	550	5.08E+07	4.03E+08	7.36E-13	0.0411	0.3266	0.0368
Water	5500	1.32E+08	1.05E+09	9.54E-09	0.1066	0.8463	0.0953
Soil	55000	1.17E+12		1.88E-02	944.38		94.438
Biota (fish)				4.77E-02			
Suspended sediment				1.17E-01			
Bottom sediment	55000	2.59E+10	4.12E+10	3.75E-02	20.986	33.312	5.430
Total		1.17E+12	4.26E+10		965.52	34.48	100
R + A			1.21E+12			1000	

f = 1.824E-09 Pa
Total Amt= 76673653 kg

Overall residence time = 76673.65 h
Reaction time = 79412.13 h
Advection time = 2223429.55 h

555

Fugacity Level III calculations: (four compartment model)

Chemical name: Octachloro-dibenzofuran

Phase Properties and Rates:

Compartment	Bulk Z mol/m3 Pa	Half-life h	D Values Reaction mol/Pa h	D Values Advection mol/Pa h
Air (1)	4.807E-01	550	6.06E+10	4.81E+11
Water (2)	3.529E+02	5500	8.89E+09	7.06E+10
Soil (3)	5.144E+06	55000	1.17E+12	4.12E+10
Sediment (4)	4.115E+06	55000	2.59E+10	

	E(1)=1000	E(2)=1000	E(3)=1000	E(1,2,3)
Overall residence time =	53265	75701	78914	46506 h
Reaction time =	75701	75701	79347	75730 h
Advection time =	179726	32070	14447823	120514 h

EMISSION (E)
REACTION (R)
ADVECTION (A)
TRANSFER D VALUE mol/Pa h

E$_1$ R$_1$ A$_1$

AIR (1)   2.002E07

D$_{12}$ D$_{21}$

5.273E06   D$_{13}$ D$_{31}$ 1.297E12

WATER (2)   1.441E11   R$_2$ A$_2$

E$_3$ E$_2$

9.283E09   D$_{32}$ 4.116E10   D$_{24}$

SOIL (3)   SEDIMENT (4)   D$_{42}$ 3.215E11

R$_3$ R$_4$ A$_4$

Phase Properties, Compositions, Transport and Transformation Rates:

Emission, kg/h

E(1)	E(2)	E(3)
1000	0	0
0	1000	0
0	0	1000
600	300	100

Fugacity, Pa

f(1)	f(2)	f(3)	f(4)
1.137E-09	6.295E-10	1.254E-09	1.870E-09
8.164E-14	8.084E-09	9.003E-14	2.401E-08
5.742E-15	6.382E-11	1.916E-09	1.896E-10
6.821E-10	2.809E-09	9.438E-10	8.345E-09

Concentration, g/m3

C(1)	C(2)	C(3)	C(4)
2.425E-07	9.858E-05	2.862E+00	3.415E+00
1.741E-11	1.266E-03	2.055E-04	4.386E+00
1.225E-12	9.994E-06	4.374E+00	3.462E+00
1.455E-07	4.399E-04	2.155E+00	1.524E+00

Amounts, kg

m(1)	m(2)	m(3)	m(4)	Total Amount, kg
2.425E+04	1.972E+04	5.151E+07	1.707E+06	5.327E+07
1.741E+00	2.532E+05	3.699E+03	2.193E+07	2.218E+07
1.225E-01	1.999E+03	7.874E+07	1.731E+05	7.891E+07
1.455E+04	8.799E+04	3.878E+07	7.620E+06	4.651E+07

Intermedia Rate of Transport, kg/h

	T12	T13	T21	T24	T31	T32	T42
	air-water	air-soil	water-air	water-sed	soil-air	soil-water	sed-water
	7.270E+01	6.542E+02	5.592E-03	8.982E+01	2.934E+03	5.164E+01	3.415E+01
	5.221E-03	4.698E-02	7.181E-02	1.153E+03	2.107E-07	3.709E-04	4.386E+02
	3.672E-04	3.305E-03	5.669E-04	9.106E+00	4.484E-03	7.894E-03	3.462E+00
	4.362E-02	3.926E+02	2.496E-02	4.008E+02	2.209E-03	3.888E+00	1.524E+02

Emission, kg/h

E(1)	E(2)	E(3)
1000	0	0
0	1000	0
0	0	1000
600	300	100

Loss, Reaction, kg/h

R(1)	R(2)	R(3)	R(4)
3.056E+01	2.484E+00	6.49E+02	2.151E+01
2.194E-03	3.190E+01	4.66E+02	2.763E+02
1.543E-04	2.518E-01	9.92E+02	2.181E+00
1.833E+01	1.109E+01	4.89E+02	9.601E+01

Loss, Advection, kg/h

A(1)	A(2)	A(4)
2.425E+02	1.972E+01	3.415E+01
1.741E-02	2.532E+02	4.386E+02
1.225E-03	1.999E+00	3.462E+00
1.455E+02	8.799E+01	1.524E+02

Level III Distribution

Chemical name: Octachloro-dibenzofuran

Distribution of mass

Distribution of removal rates

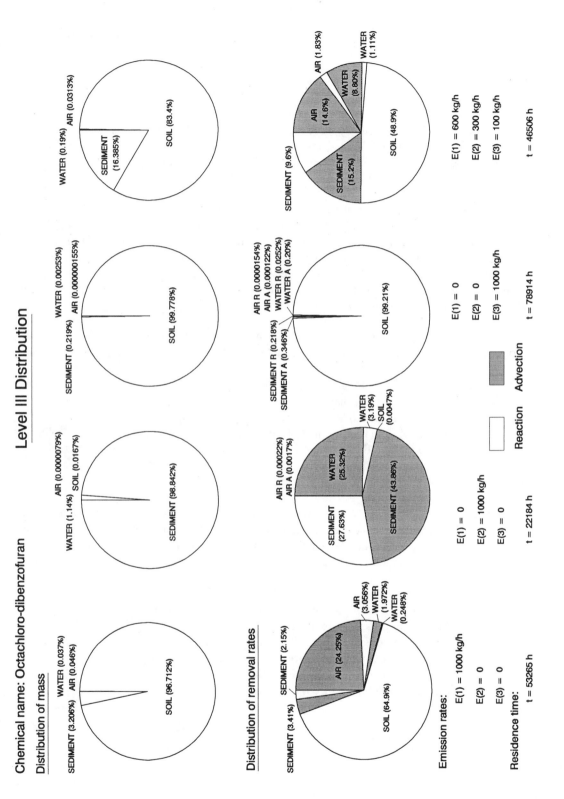

557

4.4 COMMENTARY ON THE PHYSICAL-CHEMICAL PROPERTIES AND ENVIRONMENTAL FATE

QSPR Plots

The QSPR plot of the polychlorinated dibenzofurans (PCDFs) in Figure 4.1 indicates that total surface area (TSA) and molar volume (V_M) are well correlated as with the dioxins. It is probable that TSA is fundamentally more accurate as a descriptor. There may be some loss of correlation using the simpler V_M as descriptor. Little significance should be placed on the absolute values of the LeBas molar volume, especially for the higher congeners. The more easily calculated V_M is regarded as adequate for the present proposes.

Figure 4.2 shows the expected steady drop in subcooled liquid solubility as a function of V_M with a slope of about 0.036 log-units per cm³/mol, or about a factor of 5.2 in solubility (0.72 log-units) for every 20 cm³/mol added. This is about the increase in V_M for each chlorine added. It should be noted that some of the properties of the more chlorinated PCDFs with very low solubilities and high melting points are subject to considerable error, especially in the fugacity ratio. It seems likely that these solubilities are predictable within a factor of about 5. This behavior is very similar to that of the "dioxins".

The vapor pressure QSPR plot in Figure 4.3 is similar but with a steeper slope of about 0.043 log-units per cm³/mol, corresponding to a factor of 7.2 (i.e., 0.86 log-unit) per 20 cm³/mol.

The log K_{OW} plot in Figure 4.4 does not contain the somewhat suspect fugacity ratio correction, and shows very linear behavior. The dependence on V_M is better established than for solubility. The slope is about 0.022 log-unit per cm³/mol. The implication is that a 20 cm³/mol increase in V_M causes K_{OW} to increase by 0.44 log-unit or a factor of 2.75 while water solubility falls by a factor of 5. There does not appear to be a "levelling-off" at high chlorine numbers as occurs with the "dioxins". The plot of log K_{OW} versus solubility in Figure 4.6 has a slope of about 0.6 compared with a slope of 0.8 for the "dioxins".

The log H plot in Figure 4.5 shows more scatter but there appears to be a marked decrease in H with molar volume, i.e., the more chlorinated PCDFs have lower air-water partition coefficients. The slope of about 0.0072 units per cm³/mol or a factor of about 1.4 per 20 cm³/mol arises because increasing molar volumes causes a more rapid decrease in vapor pressure than in solubility.

In summary, a 20 cm³/mol increase in LeBas molar volume (corresponding to an additional chlorine atom) causes approximately
    (1) a decrease in log solubility of 0.72 units (a factor of 5.2);
    (2) a decrease in log vapor pressure of 0.86 units (a factor of 7.2);
    (3) a decrease in log H of 0.14 units (a factor of 1.4);
    (4) an increase in log K_{OW} of 0.44 units (a factor of 2.75).

The net result is a consistent increase in hydrophobicity with molecular weight and also an increased tendency, but of reduced magnitude, to partition from air into water. The PCDFs are quite similar in this respect to the structurally similar "dioxins".

Selected Values

The scarcity of information on the physical-chemical properties and reaction rate data of the dibenzofuran series is evident in the preceding compiled data sheets. The selected values are often based on a single measurement, or estimated, or correlated values. The physical-chemical properties listed for each chemical were inspected, examined in the light of the QSPR plots and values selected as reported in Tables 4.1 and 4.2. Since the PCDFs are structurally similar to dioxins, their half-life classes are suggested to be similar to those of dioxins given in Table 4.3. The air half-lives range from 55 hours (class 3) to 550 hours (class 5) for PCDFs and higher congeners. Water phase half-lives are generally longer, with the soil half-lives believed to be about a factor of 30 or 3 classes longer still. Sediment half-lives are believed to be longer still. The higher congeners are well-preserved in sediments.

Again, these values should be used only for approximate general assessments. For more detailed site-specific assessments estimates should be made of partitioning or reactivity under local conditions, preferably by direct experimental measurement.

Evaluative Calculations

Level I, II and III calculations are shown for five selected dibenzofurans, the aim being to cover the range of properties from the nonchlorine substituted dibenzofuran (DF) to octachlorodibenzofuran (OCDF). It is possible to infer the likely environmental behavior of other congeners from those of similar structure and hence properties.

The Level I calculation shows that the soil is the primary medium of accumulation for the PCDFs, but air accounts for some 10% and water some 5% in the case of DF. Increasing chlorination reduces the air-water partition coefficient and more importantly increases sorption to organic carbon thus there is an increased tendency to accumulate in soil. Sediments generally account for about 2% because of their smaller volume.

The Level II calculations suggest that for DF, the 10% in air is sufficient to cause 80% of the removal to be from air with about equal advective and reaction loss. As chlorination increases and the proportion in air decreases, air becomes less important contributing some 35% of the losses for dichlorodibenzfuran (DCDF), and 8% for the tetrachlorodibenzofuran (TCDF). Soil becomes the primary medium of loss, and environmental persistence is eventually controlled entirely by the rate of reaction in soil. The persistence thus becomes very long corresponding to the slow reaction rate.

The Level III calculations are viewed as the most realistic since they take into account the medium of discharge. When discharge is to air (in the first row of the Table in the Level III diagrams) much of the DF and DCDF is found in air and is removed form air. As chlorination increases, there is a tendency for increased deposition to soil, because of the low vapor pressure, and soil becomes the dominant medium. There is a corresponding increase in persistence as the chemical migrates into a medium from which there is only slow removal. When discharge is to water (in the second row in the same Table in the Level III diagrams), there is appreciable loss by reaction and advection from water of the lower chlorine-number congeners, but the rate of transport to sediments becomes appreciable for the higher congeners. The result is that sediments become the primary medium of accumulation and persistence is correspondingly longer. When discharge is to soil, eventually all the chemical remains there, it reacts slowly, and it has a long residence time. Volatilization is very slow, even for the unsubstituted dibenzofuran (DF), in which of the 1000 kg/h discharged, 991 kg/h reacts, 6 kg/h volatilizes and 3 kg/h is transported to water in run-off.

As expected, the PCDFs behave similarly to the "dioxins". The congeners with zero to two chlorines show some volatility and the opportunity to be advected or reacted. But as chlorination increases, the combination of lower vapor pressure and higher K_{ow} drives the chemicals into the solid phases of aerosols (resulting in rapid deposition to soil and water), suspended matter in the water column (resulting in deposition to sediments) and soils and sediments. Since the reaction rates in these solid media are believed to be slow, the reaction persistence is correspondingly long. Once the PCDFs have reached soils and sediments, they have little tendency to migrate from there, i.e., they remain indefinitely until degraded. It is thus very important to establish accurate values of the rates of degradation of this series of chemicals in soils, sediments and water.

4.5 REFERENCES

Andren, A.W. (1986) Personal Communication.

Atkinson, R. (1987a) Estimation of OH radical reaction rate constants and atmospheric lifetimes for polychlorobiphenyls, dibenzo-p-dioxins, and dibenzofurans. *Environ. Sci. Technol.* 21, 305-307.

Atkinson, R. (1987b) A structure-activity relationship for the estimation of rate constants for the gas-phase reactions of OH radicals with organic compounds. *Int'l. J. Chem. Kinet.* 19, 799-828.

Banerjee, S., Yalkowsky, S.H., Valvani, S.C. (1980) Water solubility and octanol/water partition coefficient of organics. Limitations of solubility-partition coefficient correlation. *Environ. Sci. Technol.* 14, 1227-1229.

Birnbaum, L.S. (1985) The role of structure in the disposition of halogenated aromatic xenobiotics. *Environ. Health Perspectives* 61, 11-20.

Birnbaum, L.S., Decad, G.M., Matthews, H.B. (1980) Disposition and excretion of 2,3,7,8-tetrachlorodibenzofuran in the rat. *Toxicol. Appl. Pharmacol.* 55, 342-352.

Birnbaum, L.S., Decad, G.M., Matthews, H.B., McConnell, E.E. (1981) Fate of 2,3,7,8-tetrachlorodibenzofuran in the monkey. *Toxicol. Appl. Pharmacol.* 57, 189-196.

Boublik, T., Fried, V., Hala, E. (1973) *The Vapour Pressure of Pure Substances.* Elsevier, Amsterdam.

Boublik, T., Fried, V., Hala, E. (1984) *The Vapour Pressures of Pure Substances.* (second revised edition), Elsevier, Amsterdam.

Broman, D., Näf, C., Rolff, C., Zebühr, Y. (1991) Occurence and dynamics of polychlorinated dibenzo-p-dioxins, dibenzofurans and polycyclic aromatic hydrocarbons in the mixed surface layer of remote coastal and offshore waters of the Baltic. *Environ. Sci. Technol.* 25(11), 1850-1864.

Burkhard, L.P., Kuehl, D.W. (1986) n-Octanol/water partition coefficients by reverse phase liquid chromatography/mass spectrometry for eight tetrachlorinated planar molecules. *Chemosphere* 15, 163-167.

Buser, H.R. (1988) Rapid photolytic decomposition of brominated and brominated/chlorinated dibenzodioxins and dibenzofurans. *Chemosphere* 17, 889-903.

Choudhary, G.G., Hutzinger, O. (1982) Photochemical formation and degradation of polychlorinated dibenzofurans and dibenzo-p-dioxins. *Residue Rev.* 84, 113-161.

Choudhary, G.G., Webster, G.R.B. (1985a) Environmental photochemistry of PCDDs. Part I. Kinetics and quantum yields of the photodegradation of 1,2,3,4,7-penta- and 1,2,3,4,7,8-hexachlorodibenzo-p-dioxin in aqueous acetonitrile. *Chemosphere* 14, 9-26.

Choudhary, G.G., Webster, G.R.B. (1985b) Quantum yields for the photodecomposition of polychlorinated dibenzo-p-dioxins (PCDDs) in water-acetonitrile solution. *Chemosphere* 14, 893-896.

Clark, K.E., Mackay, D. (1991) Dietary uptake and biomagnification of four chlorinated hydrocarbons by guppies. *Environ. Toxicol. Chem.* 10, 1205-1217.

Crosby, D.G., Moilanen, K.W. (1973) Photodecomposition of chlorinated biphenyls and dibenzofurans. *Bull. Environ. Contam. Toxicol.* 6, 372.

Dean, J.D., Editor (1979) *Lange's Handbook of Chemistry.* 12th ed., McGraw-Hill, Inc., New York.

Dean, J.D., Editor (1985) *Lange's Handbook of Chemistry.* 13th ed. McGraw-Hill, Inc., New York.

Decad, G.M., Birnbaum, L.S., Matthews, H.B. (1981a) 2,3,7,8-tetrachlorodibenzofuran tissue distribution and excretion in guinea pigs. *Toxicol. Appl. Pharmacol.* 57, 231-240.

Decad, G.M., Birnbaum, L.S., Matthews, H.B. (1981b) Distribution and excretion of 2,3,7,8-tetrachlorodibenzofuran in C57BL/6J and DBA/2J mice. *Toxicol. Appl. Pharmacol.* 59, 564-573.

Doucette, W.J. (1985) *Measurement and Estimation of Octanol/Water Partition Coefficients and Aqueous Solubilities for Halogenated Aromatic Hydorcarbons.* Ph.D. thesis, University of Wisconsin-Madison.

Doucette, W.J., Andren, A.W. (1987) Correlation of octanol/water partition coefficients and total molecular surface area for highly hydrophobic aromatic compounds. *Environ. Sci. Technol.* 21, 821-824.

Doucette, W.J., Andren, A.W. (1988a) Aqueous solubility of selected biphenyl, furan and dioxin congeners. *Chemosphere* 17, 243-252.

Doucette, W.J., Andren, A.W. (1988b) Estimation of octanol/water partition coefficients: Evaluation of six methods for highly hydorphobic aromatic hydrocarbons. *Chemosphere* 17, 345-359.

Dunn III, W.J., Koehler, M., Stalling, D.L., Schwartz, T.R. (1986) Relationship between gas chromatographic retention of polychlorinated dibenzofurans and calculated molecular surface area. *Anal. Chem.* 58, 1835-1838.

Eitzer, B.D., Hites, R.A. (1988) Vapor pressures of chlorinated dioxins and dibenzofurans. *Environ. Sci. Technol.* 22, 1362-1364.

Eitzer, B.D., Hites, R.A. (1989) Polychlorinated dibenzo-p-dioxins and dibenzofurans in the ambient atmosphere of Bloomington, Indiana. *Environ. Sci. Technol.* 23, 1389-1395.

Fiedler, H., Schramm, K.-W. (1990) QSAR generated octanol-water partition coefficients of selected mixed halogenated dibenzodioxins and dibenzofurans. *Chemosphere* 20, 1597-1602.

Friesen, K.J., Vilk, J., Muir, D.C.G. (1990) Aqueous solubilities of selected 2,3,7,8-substituted polychlorinated debenzofurans (PCDFs). *Chemosphere* 20, 27-32.

Geyer, H.J., Scheunert, I., Korte, F. (1987) Correlation between the bioconcentration potential of organic environmental chemicals in humans and their n-octanol/water partition coefficients. *Chemosphere* 16, 239-252.

Gobas, F.A.P.C, Lahittete, J.M., Garofalo, G., Shiu, W.Y., Mackay, D. (1988) A novel method for measuring menbrane-water partition coefficients of hydrophobic organic chemicals: comparison with 1-octanol-water partitioning. *J. Pharm. Sci.* 77, 265-272.

Gobas, F.A.P.C., Schrap, S.M. (1990) Bioconcentration fo some polychlorinated dibenzo-p-dioxins and octachlorodibenzofuran in the guppy (*Poecilia reticulata*). *Chemosphere* 20, 495-512.

Gray, A.P., Cepa, S.P., Solomon, I.J., Anilin, O. (1976) *J. Org. Chem.* 41, 2435.

Howard, P.H., Boethling, R.S., Jarvis, W.F., Meylan, W.M., Michalenko, E.M. (1991) *Handbook of Environmental Degradation Rates*. Lewis Publishers, Inc., Chelsea, Michigan.

Ioannou, Y.M., Birnbaum, L.S., Matthews, H.B. (1983) Toxicity and distribution of 2,3,7,8-tetrachlorodibenzofuran in male guinea pigs. *J. Toxicol. Environ. Health* 12, 541-553.

Isnard, P., Lambert, S. (1988) Estimating bioconcentraion factors from octanol-water partition coefficient and aqueous solubility. *Chemosphere* 17, 21-34.

Isnard, P., Lambert, S. (1989) Aqueous solubility and n-octanol/water partition coefficient correlations. *Chemosphere* 18, 1837-1853.

Kearney, P.C., Isensee, A.R., Helling, C.S., Woolsen, E.A., Plimmer, J.R. (1971) Environmental significance of chlorodioxins. In: *Chlorodioxins-Origin and Fate. Adv. Chem. Ser.* 120, 105-111.

Kearney, P.C., Woolson, E.A., Ellington, Jr., C.P. (1972) Persistence and metabolism of chlorodioxins in soils. *Environ. Sci. Technol.* 6, 1017-1019.

Kearney, P.C., Woolson, E.A., Isensee, A.R., Helling, C.S. (1973) Tetrrachlorodibenzodioxin in the environment: Sources, fate and decontamination. *Envirn. Health Perspectives* (No. 5), 273-277.

Kuehl, D.W., Cook, P.M., Batterman, A.R. (1986) Uptake and depuration studies of PCDDs and PCDFs in freshwater fish. *Chemosphere* 15(9-12), 2023-2026.

Kuehl, D.W., Cook, P.M., Batterman, A.R., Lothenbach, D., Butterworth, B.C. (1987) Bioavailability of polychlorinated dibenzo-p-dioxins and dibenzofurans from contaminated Wisconsin river sediment to carp. *Chemosphere* 16(4), 667-679.

Kuroki, H., Haraguchi, K., Masuda, Y. (1984) Synthesis of polychlorinated dibenzofuran isomers and their gas chromatographic profiles. *Chemosphere* 13(4), 561-573.

Lee, M.D., Wilson, J.T., Ward, C.H. (1984) Microbial degradation of selected aromatics in a hazardous waste site. *Devel. Indust. Microbiol.* 25, 557-565.

Loonen, H., Parsons, J.R., Govers, H.A.J. (1991) Dietry accumulation of PCDDs and PCDFs in guppies. *Chemosphere* 23, 1349-1357.

Lu, P.Y., Metcalf, R.L., Carlson, E.M. (1978) *Environ. Health Perspectives* 24, 201.

Ma, K.C., Shiu, W.Y., Mackay, D. (1990) *A Critically Reviewed Compilation of Physical and Chemical and Persistence Data for 110 Selected EMPPL Substances*. A report prepared for the Ontario Ministry of Environment, Water Resources Branch, Toronto, Ontario.

Mabey, W., Mill, T. (1978). Critical review of hydrolysis of organic compounds in water under environmental conditions. *J. Phys. Chem. Ref Data* 7, 383-415.

Mackay, D. (1982) Correlation of bioconcentration factors. *Environ. Sci. Technol.* 16, 274-278.

Mackay, D. (1991) *Multimedia Environmental Models*. Lewis Publishers, Inc., Chelsea, Michigan.

563

Mehrle, P.M., Buckler, D.R., Little, E.E., Smith, L.M., Petty, J.D., Peterman, P.H., Stalling, D.L., De Graeve, G.M.,Coyle, J.J., Adams, W.J. (1988) Toxicity and bioconcentration of 2,3,7,8-tetrachlorodibenzodioxin and 2,3,7,8-tetrachloro-dibenzofuran in rainbow trout. *Environ. Toxicol. & Chem.* 7, 47-62.

Muir, D.C.G., Yarechewski, A.L., Metner, D.A., Lockhart, W.L., Webster, G.R.B., Friesen, K.J. (1990) Dietary accumulation and sustained hepatic mixed function oxidase enzyme induction by 2,3,7,8-pentachlorodibenzofuran in rainbow trout. *Environ. Toxicol. & Chem.* 9, 1463-1472.

Niimi, A.J., Oliver, B.G. (1986) Biological half-lives of chlorinated dibenzo-p-dioxins and dibenzofurans in rainbow trout (*salmo gairdneri*). *Environ. Toxicol. & Chem.* 5, 49-53.

Olling, M., Derks, H.J.G.M., Berende, P.L.M., Liem, A.K.D., de Jong, A.P.J.M. (1991) Toxicokinetics of eight ^{13}C-labelled polychlorinated dibenzo-p-dioxins and -furans in lactating cows. *Chemosphere* 23, 1377-1385.

Olsen, J.R., Gasiewicz, T.A., Neal, R.A. (1980) Tissue distribution, excretion and metabolism of 2,3,7,8-tetrachlorodibenzo-p-dioxin (TCDD) in the golden Syrian hamster. *Toxico. Appl. Pharmacol.* 56, 78-85.

Opperhuizen, A. (1986) Bioconcentration of hydrophobic chemicals in fish. In: *Aquatic Toxicology and Environmental Fate.* 9th Vol., Poston, T.M., Purdy, R., Editors, ASTM STP 921, pp.304-315., American Society for Testing and Materials, Philadephia.

Opperhuizen, A., Vander Velde, E.W., Gobas, F.A.P.C., Liem, D.A.K., Van der Steen, J.M., Hutzinger, O. (1985) Relationship between bioconcentration in fish and steric factors of hydrophobic chemicals. *Chemosphere* 14, 1871-1896.

Opperhuizen, A., Wagenaar, W.J., Van der Wielen, F.W.M., Van den Berg, M., Olie, K., O. Hutzinger, O., Gobas, F.A.P.C. (1986) Uptake and elimination of PCDD/PCDF congeners by fish after aqueous exposure to a fly ash extract from a municipal incinerator. *Chemosphere* 15, 2049-2053.

Opperhuizen, A., Sijm, D.T.H.M. (1990) Bioaccumulation and biotransformation of polychlorinated dibenzo-p-dioxins and dibenzofurans in fish. *Environ. Toxicol. & Chem.* 9, 175-186.

Palauschek, N., Scholz, S. (1987) Destruction of polychlorinated dibenzo-p-dioxins and dibenzofurans in contaminated water samples using ozone. *Chemosphere* 16, 1857-1863.

Pearlman, R.S., Yalkowsky, S.H., Banerjee, S. (1984) Water solubilities of polynuclear aromatic and heteroaromatic compounds. *J. Phys. Chem. Ref. Data* 13, 555-562.

Rordorf, B.F. (1986b) Thermal properties of dioxins, furans and related compounds. *Chemosphere* 15, 1325-1332.

Rordorf, B.F. (1989) Prediction of vapor pressures, boiling points and enthalpies of fusion for twenty-nine halogenated dibenzo-p-dioxins and fifty-five dibenzofurans by a vapor pressure correlation method. *Chemosphere* 18, 783-788.

Rordorf, B.F., Sarna, L.P., Webster, G.R.B., Safe, S.H., Safe, L.M., Lenoir, D., Schwind, K.H., Hutzinger, O. (1990) Vapor pressure measurements on halogenated dibenzo-p-dioxins and dibenzofurans. An extended data set for a correlation method. *Chemosphere* 20, 1603-1609.

Sabljic, A. (1987) Nonempirical modelling of environmental distribution and toxicity of major organic pollutants. In: *QSAR in Environmental Toxicology-II*. Kaiser, K.L.E., Editor, pp. 309-322., D. Reidel Publ. Co., Dordrecht, Netherlands.

Sarna, L.P., Hodge, P.E., Webster, G.R.B. (1984) Octanol/water partition coefficients of chlorinated dioxins and dibenzofurans by reversed-phase HPLC using several C_{18} columns. *Chemosphere* 13, 975-983.

Servos, M.R., Muir, C.G., Webster, G.R.B. (1992a) Environmental fate of polychlorinated dibenzo-p-dioxins in lake enclosures. *Can. J. Fish Aquat. Sci.* 49, 722-734.

Servos, M.R., Muir, C.G., Webster, G.R.B. (1992b) Bioavailability of polychlorinated dibenzo-p-dioxins in lake enclosures. *Can. J. Fish Aquat. Sci.* 49, 735-742.

Shiu, W.Y., Mackay, D. (1992) Temperature dependence of aqueous solubility and octanol/water partition coefficient of dibenzofuran. (to be submitted to *J. Chem. Eng. Data*)

Sijm, D.T.H.M., Wever, H., de Vries, P.J., Opperhuizen, A. (1989a) Octan-1-ol/water partition coefficients of polychlorinated dibenzo-p-dioxins and dibenzofurans: experimantal values determined with a stirring method. *Chemosphere* 19, 263-266.

Sijm, D.T.H.M., Yarechewski, A.L., Muir, D.C.G., Webster, G.R.B., Seinen, W., Opperhuizen, A. (1990) Biotransformation and tissue distribution of 1,2,3,7-tetrachlorodibenzo-p-dioxin, 1,2,3,7-pentachlorodibenzo-p-dioxin and 2,3,4,7,8-pentachlorodibenzofuran in rainbow trout. *Chemosphere* 21, 845-866.

Stephenson, R.M., Malanowski, A. (1987) *Handbook of the Thermodynamics of Organic Compounds*. Elsevier, New York.

The Merck Index (1989) *An Encyclopedia of Chemicals, Drugs and Biologicals*. Budavari, S., Editor, Merck & Co., Inc., Rahway, N.J., 11th edition.

The Merck Index (1983) *An Encyclopedia of Chemicals, Drugs and Biologicals*. Windholz, M. Editor, Merck and Co., Inc., Rahway, N.J., U.S.A., 10th edition.

Tysklind, M., Rappe, C. (1991) Photolytic transformation of polychlorinated dioxins and dibenzofurans in fly ash. *Chemosphere* 23, 1365-1375.

Van den Berg, M., Olie, K. (1985) Polychlorinated dibenzofurans (PCDFs). Environmental occurance and physical, chemical and biological properties. *Toxicol. & Environ. Chem.* 9, 171-217.

Van den Berg, M., De Jongh, J., Eckhart, P., Van der Wielen, F.W.M. (1989) The elimination and absence of pharmacokinetic interaction of some polychlorinated dibenzofurans (PCDFs) in the liver of the rat. *Chemosphere* 18(1-6), 665-675.

Veith, G.D., Austin, N.M., Morris, R.T. (1979a) A rapid method for estimation log P for organic chemicals. *Water Res.* 13, 43-47.

Veith, G.D., Defor, D.L. Bergstedt, B.V. (1979b) Measuring and estimating the bioconcentration factor of chemicals in fish. *J. Fish Res. Board Can.* 26, 1040-1048.

Veith, G.D., Call, D.J., Brooke, L.T. (1983) Structure toxicity relationships for the fathead minnow, Pimephales promelas, narcotic industrial chemicals. *Can. J. Fish Aquat. Sci.* 40, 743-748.

Veith, G.D., Macek, K.J., Petrocelli, S.R., Carroll, J. (1980) An evaluation of using partition coefficients and water solubility to estimate bioconcentration factors for organic chemicals in fish. In: *Aquatic Toxicology. ASTM STP* 707, Eaton, J.G., Parrish, P.R., Hendricks, A.C., Eds., pp 116-129, Amer. Soc. for Testing and Materials, Philadelphia.

Ward, C.H., Tomson, M.B., Bedient, P.B., Lee, M.D. (1986) Transport and fate processes in the subsurface. *Water Res. Symp.* 13, 19-39.

Weast, R. (1984) *Handbook of Chemistry and Physics.* 64th ed., CRC Press, Boca Raton, Florida.

Weast, R. (1972-73) *Handbook of Chemistry and Physics.* 53rd ed., CRC Press, Cleveland.

Weast, R. (1977) *Handbook of Chemistry and Physics.* 57th ed., CRC Press, Boca Raton, Florida.

Weast, R. (1982-83) *Handbook of Chemistry and Physics.* 63rd. ed., CRC Press, Boca Raton, Florida.

Weast, R.C., Ed. (1983-84). *Handbook of Chemistry and Physics.* 64th ed., CRC Press, Boca Raton, Florida.

Zitko, V., Choi, P.M.K. (1973) Oral toxicity of chlorinated dibenzofurans to juvenile Atlantic salmon. *Bull. Environ. Contam. Toxicol.* 10, 120-122.

List of Symbols and Abbreviations:

A_i	area of phase i, m^2
ALPM	Automated Log-P Measurement
BCF	bioconcentration factor
C	molar concentration, mol/L or mmol/m^3
C^S	saturated aqueous concentration, mol/L or mmol/m^3
C_L	subcooled liquid concentration, mol/L or mmol/m^3
C_S	solid molar concentration, mol/L or mmol/m^3
C_A	concentration in air phase, mol/L or mmol/m^3
C_W	concentration in water phase, mol/L or mmol/m^3
^{14}C	radioactive labelled carbon-14 compound
D	D values, mol/Pa·h
D_A	D values for advection, mol/Pa·h
D_{Ai}	D values for advective loss in phase i, mol/Pa·h
D_R	D value for reaction, mol/Pa·h
D_{Ri}	D value for reaction loss in phase i, mol/Pa·h
D_{ij}	intermedia D values, mol/Pa·h
D_{VW}	intermedia D value for air-water diffusion (absorption), mol/Pa·h
D_{RW}	intermedia D value for air-water dissolution, mol/Pa·h
D_{QW}	D value for total particle transport (dry and wet), mol/Pa·h
D_{RS}	D value for rain dissolution (air-soil), mol/Pa·h
D_{QS}	D value for wet and dry deposition (air-soil), mol/Pa·h
D_{VS}	D value for total soil-air transport, mol/Pa·h
D_S	D value for air-soil boundary layer diffusion, mol/Pa·h
D_{SW}	D value for water transport in soil, mol/Pa·h
D_{SA}	D value for air transport in soil, mol/Pa·h
D_{Ti}	total transport D value in bulk phase i, mol/Pa·h
DOC	dissolved organic carbon
E	emission rate, mol/h or kg/h

EPICS	Equilibrium Partitioning In Closed System
F	Fugacity ratio
f	fugacity, Pa
f_i	fugacity in pure phase i, Pa
f-const.	fragmental constants
fluo.	fluorescence method
G	advective inflow, mol/h
G_B	advective inflow to bottom sediment mol/h
ΔG_v	Gibbs's free energy of vaporization kJ/mol or kcal/mol
GC	gas chromatography
GC/FID	GC analysis with flame ionization detector
GC/ECD	GC analysis with electron capture detector
GC-RT	GC retention time
gen. col.	generator-column
H, HLC	Henry's law constant, Pa m^3/mol
ΔH_{fus}	enthalpy of fusion, kcal/mol
ΔH_v	enthalpy of vaporization, kJ/mol or kcal/mol
HPLC	high pressure liquid chromatography
HPLC/UV	HPLC analysis with UV detector
HPLC/fluo.	HPLC analysis with fluorescence detector
HPLC-k'	HPLC-capacity factor correlation
HPLC-RI	HPLC-retention index correlation
HPLC-RT	HPLC-retention time correlation
HPLC-RV	HPLC-retention volume correlation
IP	ionization potential
J	intermediate quantities for fugacity calculation
k	first-order rate constant, h^{-1} (hour^{-1})
k_i	first-order rate constant in phase i, h^{-1}
k_A	air-water mass transfer coefficient, air-side, m/h

k_W	air-water mass transfer coefficient, water-side, m/h
K_{AW}	dimensionless air/water partition coefficient
K_B	bioconcentration factor
K_h	association coefficient
K_{OC}	organic-carbon sorption partition coefficient
K_{OM}	organic-matter sorption partition coefficient
K_{OW}	octanol-water partition coefficient
K_p	sorption coefficient
k_1	uptake/accumulation rate constant, d^{-1} (day^{-1})
k_2	elimination/clearance/depuration rate constant, d^{-1}
k_b	biodegradation rate constant, d^{-1}
k_h	hydrolysis rate constant, d^{-1}
k_p	photolysis rate constant, d^{-1}
L	lipid content of fish
LSC	Liquid Scintillation Counting
m_i	amount of chemical in phase i, mol or kg
M	total amount of chemical, mol or kg
MO	molecular orbital calculation
M.P.	melting point, °C
MR	molar refraction
MW	molecular weight, g/mol
n_C	number of carbon atoms
n_{Cl}	number of chlorine atoms
P	vapor pressure, Pa (Pascal)
P_L	liquid or subcooled liquid vapor pressure, Pa
P_S	solid vapor pressure, Pa
Q	scavenge ratio
QSPR	Quantitative Structure-Property Relationship
QSAR	Quantitative Structural-Activity Relationship

RP-HPLC	Reversed Phase High Pressure Liqud Chromatography
RP-TLC	Reversed Phase Thin Layer Chromatography
S	water solubility, mg/L or g/m^3
ΔS_{fus}	entropy of fusion, $J/mol \cdot K$ or $cal/mol \cdot K$ (e.u.)
$S_{octanol}$	solubility in octanol
t	residence time, h (hour)
t_o	overall residence time, h
t_A	advection persistence time, h
t_B	sediment burial residence time, h
t_R	reaction persistence time, h
$t_{1/2}$	half-life, h
T_{ij}	intermedia transport rate, mol/h or kg/h
T	system temperature, K
T_B	boiling point, K
T_M	melting point, K
TLC	thin-layer chromatography
TMV	total molecular volume per molecule, $Å^3$ (Angstrom^3)
TSA	total surface area per molecule, $Å^2$
U_1	air side, air-water MTC (same as k_A), m/h
U_2	water side, air-water MTC (same as k_W), m/h
U_3	rain rate (same as U_R), m/h
U_4	aerosol deposition rate, m/h
U_5	soil-air phase diffusion MTC, m/h
U_6	soil-water phase diffusion MTC, m/h
U_7	soil-air boundary layer MTC, m/h
U_8	sediment-water MTC, m/h
U_9	sediment deposition rate, m/h
U^{10}	sediment resuspension rate, m/h
U_{11}	soil-water run-off rate, m/h

U_{12}	soil-solids run-off rate, m/h
U_R	rain rate, m/h
U_Q	dry deposition velocity, m/h
U_B	sediment burial rate, m/h
UV	UV spectrometry
UNIFAC	UNIQUAC Functional Group Activity Coefficients
V_i	volume of pure phase i, m^3
V_S	volume of bottom sediment, m^3
V_{Bi}	volume of bulk phase i, m^3
V_I	intrinsic molar volume, cm^3/mol
V_M	molar volume, cm^3/mol
v_i	volume fraction of phase i
v_Q	volume fraction of aerosol
VOC	volatile organic chemicals
W	molecular mass, g/mol
Z_i	fugacity capacity of phase i, mol/m^3 Pa
Z_{Bi}	fugacity capacity of bulk phase i, mol/m^3 Pa

Greek characters:

π-const.	substituent constants
γ	solute activity coefficient
γ_o	solute activity coefficient in octanol phase
γ_W	solute activity coefficient in water phase
ρ_i	density of pure phase i, kg/m^3
ρ_{Bi}	density of bulk phase i, kg/m^3
χ	molecular connectivity indices
ϕ_{OC}	organic carbon fraction
ϕ_i	organic carbon fraction in phase i

APPENDICES:

A1. BASIC COMPUTER PROGRAM FOR FUGACITY CALCULATIONS

```
10 REM Fugacity Level I,II and III program, 6 compartments,(LEWIS)
20 REM Select condensed print
30 WIDTH "lpt1:",132
40 LPRINT CHR$(15)
50 DIM N$(9),V(9),Z(9),C(9),F(9),M(9),P(9),CG(9),CU(9),
DEN(9),ORG(9),VZ(9),DR(9),DA(9),CB(9),A(9),PA(9),PR(9),RK(9),GA(9)
60 DIM NR(9),NA(9),I(9),GD(9,9),D(9,9),N(9,9),GRA(9),TD(9,9),
HL(9),U(20),UY(20),U$(20)
70 REM N$  = six phases : air, water, soil, sediment, susp sedt and fish
80 REM V   = volume of the six phases  (m3)
90 REM DEN = density of the six phases (kg/m3)
100 REM HT  = depth of air, water, soil and sediment (m)
110 REM AR  = area of air, water, soil and sediment  (m2)
120 REM ORG = the fraction of organic carbons in sediment and susp sedt
130 REM Z   = Z values for each phase (mol/m3.Pa)
140 REM VZ  = VZ values for each phase (mol/Pa)
150 REM F   = fugacity for each phase (Pa)
160 REM C   = concentration of chemical in each phase (mol/m3)
170 REM CG  = concentration of chemical in each phase (g/m3)
180 REM CU  = concentration of chemical in each phase (ug/g)
190 REM M   = the total amount of chemical in each phase (mol)
200 REM MK  = the total amount of chemical in each phase (kg)
210 REM P   = the mole percent of chemical in each phase (%)
212 REM DR  = reaction D values
214 REM DA  = advection D values
216 REM U   = transport velocities
220 N$(1)="Air      "
230 N$(2)="Water    "
240 N$(3)="Soil     "
250 N$(4)="Sediment "
260 N$(5)="Susp sedt"
270 N$(6)="Fish     "
280 ART=100000!*1000000!:FAR(2)=.1:FAR(3)=1-FAR(2)'total area fractions
290 AR(2)=FAR(2)*ART:AR(3)=FAR(3)*ART:AR(1)=ART:AR(4)=AR(2)'areas m2
300 HT(1)=1000:HT(2)=20:HT(3)=.1:HT(4)=.01
310 V(1)=AR(1)*HT(1):V(2)=AR(2)*HT(2):V(3)=AR(3)*HT(3):V(4)=AR(4)*HT(4)
320 V(5)=.000005*V(2):V(6)=.000001*V(2)
330 REM input properties
340 PRINT "Select chemical ,user-spec =1, benzene= 2, HCB= 3, 123TCB= 4, Type 10 to
exit program"
350 INPUT QC
```

```
360 ON QC GOTO 370,510,520,530,4360
370 INPUT "Name of chemical ",CHEM$
380 INPUT "Temperature eg 25 deg C";TC
390 INPUT "Melting point temperature or data temperature if chemical is liquid eg 80 deg
C";TM
400 INPUT "Molecular mass eg 200 g/mol";WM
410 INPUT "Vapor pressure eg 2 Pa ";P
420 INPUT "Water solubility eg 50 g/m3 ";SG
430 INPUT "Log octanol water coefficient eg 4.0 ";LKOW
440 PRINT "Input overall reaction rate half-lives eg 100 h "
450 PRINT "For zero reaction enter a fictitiously long half life eg 1E11 h"
460 INPUT "Half-life in air       ";HL(1)
470 INPUT "Half-life in water     ";HL(2)
480 INPUT "Half-life in soil      ";HL(3)
490 INPUT "Half-life in sediment   ";HL(4)
500 GOTO 590
510 CHEM$ = "Benzene":TC = 25!:WM = 78.11:P = 12700:SG = 1780:LKOW = 2.13:TM = 5.53:
HL(1) = 17:HL(2) = 170:HL(3) = 550:HL(4) = 1700:GOTO 590
520 CHEM$ = "Hexachlorobenzene(HCB)":TC = 25!:WM = 284.8:P = .0015:SG = .005:
LKOW = 5.47:TM = 230!:HL(1) = 17000:HL(2) = 55000!:HL(3) = 55000!:HL(4) = 55000!:
GOTO 590
530 CHEM$ = "1,2,3-Triclorobenzene,(123TCB)":TC = 25!:WM = 181.45:P = 28:SG = 21:
LKOW = 4.10:TM = 53:HL(1) = 550:HL(2) = 1700!:HL(3) = 5500!:HL(4) = 17000!:GOTO 590
550 CHEM$ = "NAPHTHALENE":TC = 25!:WM = 128.19:P = 10.4:SG = 31:LKOW = 3.37:
TM = 80.5:HL(1) = 17:HL(2)170:HL(3) = 1700:HL(4) = 5500:GOTO 590
570 CHEM$ = "2,3,7,8-Tetrachlorodibenzo-p-dioxin,(2378TCDD)":TC = 25!:
WM = 322.0:P = 0.0000002:SG = 0.0000193:LKOW = 6.80:TM = 305:HL(1) = 170:HL(2) = 550:
HL(3) = 5500:HL(4) = 17000:GOTO 590
590 MTK = 100000!
600 MT = MTK*1000/WM
610 REM     Input for Fugacity Level II program
620 EK = 1000
630 E = EK*1000/WM
640 PRINT "Input emission rates of chemical for Level III calculation kg/h"
650 INPUT "Emission into air     ";IK(1)
660 INPUT "Emission into water    ";IK(2)
670 INPUT "Emission into soil     ";IK(3)
680 GRA(1) = 100
690 GRA(2) = 1000
700 GRA(4) = 50000!
710 S = SG/WM 'solubility mol/m3
720 H = P/S   'Henry's law constant Pa.m3/mol
730 KOW = 10^LKOW 'Octanol-water partition coefficient
```

```
740 KOC=.41*KOW  'Organic carbon-water partition coefficient
750 KFW=.05*KOW  'Fish-water bioconcentration factor
760 TK=TC+273.15 'Temperature K
770 RG=8.314     'Gas constant

780 IF TM>TC GOTO 790 ELSE GOTO 810
790 FR=EXP(6.79*(1-(TM+273.15)/TK))
800 GOTO 820
810 FR=1
820 PL=P/FR
830 ORG(3)=.02:ORG(4)=.04:ORG(5)=.2'Organic carbon contents g/g
840 DEN(1)=.029*101325!/RG/TK:DEN(2)=1000:DEN(3)=2400'Densities kg/m3
850 DEN(4)=2400:DEN(5)=1500:DEN(6)=1000:DEN(7)=2000'Densities kg/m3
860 REM calculate Z values
870 Z(1)=1/RG/TK 'Z values
880 Z(2)=1/H
890 Z(3)=Z(2)*DEN(3)*ORG(3)*KOC/1000
900 Z(4)=Z(2)*DEN(4)*ORG(4)*KOC/1000
910 Z(5)=Z(2)*DEN(5)*ORG(5)*KOC/1000
920 Z(6)=Z(2)*DEN(6)*KFW/1000
930 K71=6000000!/PL
940 Z(7)=Z(1)*K71
950 K12=Z(1)/Z(2) 'Partition coefficients
960 K32=Z(3)/Z(2)
970 K42=Z(4)/Z(2)
980 K52=Z(5)/Z(2)
990 K62=Z(6)/Z(2)
1000 REM calculate distribution
1010 VZT=0
1020 FOR N= 1 TO 6
1030 VZ(N)=V(N)*Z(N)
1040 VZT=VZT+VZ(N)
1050 NEXT N
1060 F1=MT/VZT 'fugacity
1070 FOR N=1 TO 6
1080 F(N)=F1
1090 C(N)=F(N)*Z(N) 'concentration mol/m3
1100 M(N)=C(N)*V(N) 'amount mol
1110 MK(N)=M(N)*WM/1000
1120 P(N)=100*M(N)/MT 'percentages
1130 CG(N)=C(N)*WM  'concentration g/m3
1140 CU(N)=CG(N)*1000/DEN(N)'concentration ug/g
1150 NEXT N
```

1160 REM print out results
1170 LPRINT " PROGRAM 'LEWIS':SIX COMPARTMENT FUGACITY LEVEL I CALCULATION "
1180 LPRINT " "
1190 LPRINT "Properties of "CHEM$
1200 LPRINT " "
1210 LPRINT "Temperature deg C   ";TC
1220 LPRINT "Molecular mass g/mol   ";WM
1230 LPRINT "Melting point deg C   ";TM
1240 LPRINT "Fugacity ratio   ";FR
1250 LPRINT "Vapor pressure Pa   ";P
1260 LPRINT "Sub-cooled liquid vapor press Pa ";PL
1270 LPRINT "Solubility g/m3   ";SG
1280 LPRINT "Solubility mol/m3   ";S
1290 LPRINT "Henry's law constant Pa.m3/mol   ";H
1300 LPRINT "Log octanol-water p-coefficient ";LKOW
1310 LPRINT "Octanol-water partn-coefficient ";KOW
1320 LPRINT "Organic C-water ptn-coefficient ";KOC
1330 LPRINT "Fish-water partition coefficient ";KFW
1340 LPRINT "Air-water partition coefficient ";K12
1350 LPRINT "Soil-water partition coefficient ";K32
1360 LPRINT "Sedt-water partition coefficient ";K42
1370 LPRINT "Susp sedt-water partn coeffnt   ";K52
1380 LPRINT "Aerosol-air partition coeff   ";K71
1390 LPRINT "Aerosol Z value   ";Z(7)
1400 LPRINT "Aerosol density kg/m3   ";DEN(7)
1410 LPRINT " "
1420 LPRINT "Amount of chemical moles   ";MT
1430 LPRINT "Amount of chemical kilograms   ";MTK
1440 LPRINT "Fugacity Pa   ";F1
1450 LPRINT "Total of VZ products   ";VZT
1460 LPRINT " "
1470 LPRINT "Phase properties and compositions"
1480 LPRINT " "
1490 LPRINT "Phase   "TAB(15) N$(1) TAB(30) N$(2) TAB(45) N$(3) TAB(60) N$(4) TAB(75) N$(5) TAB(90) N$(6)
1500 LPRINT "Volume m3   "TAB(15) V(1) TAB(30) V(2) TAB(45) V(3) TAB(60) V(4) TAB(75) V(5) TAB(90) V(6)
1510 LPRINT "Density kg/m3"TAB(15) DEN(1) TAB(30) DEN(2) TAB(45) DEN(3) TAB(60) DEN(4) TAB(75) DEN(5) TAB(90) DEN(6)
1520 LPRINT "Depth m   "TAB(15) HT(1) TAB(30) HT(2) TAB(45) HT(3) TAB(60) HT(4)
1530 LPRINT "Area m2   "TAB(15) AR(1) TAB(30) AR(2) TAB(45) AR(3) TAB(60)AR(4)
1540 LPRINT "Frn org carb " TAB(45) ORG(3) TAB(60) ORG(4) TAB(75) ORG(5)

576

1550 LPRINT "Z mol/m3.Pa "TAB(15) Z(1) TAB(30) Z(2) TAB(45) Z(3) TAB(60) Z(4) TAB(75) Z(5) TAB(90) Z(6)

1560 LPRINT "VZ mol/Pa   "TAB(15) VZ(1) TAB(30) VZ(2) TAB(45) VZ(3) TAB(60) VZ(4)) TAB(75) VZ(5) TAB(90) VZ(6)

1570 LPRINT "Fugacity Pa "TAB(15) F(1) TAB(30) F(2) TAB(45) F(3) TAB(60) F(4) TAB(75) F(5) TAB(90) F(6)

1580 LPRINT "Conc mol/m3 "TAB(15) C(1) TAB(30) C(2) TAB(45) C(3) TAB(60) C(4) TAB(75) C(5) TAB(90) C(6)

1590 LPRINT "Conc g/m3   "TAB(15) CG(1) TAB(30) CG(2) TAB(45) CG(3) TAB(60) CG(4) TAB(75) CG(5) TAB(90) CG(6)

1600 LPRINT "Conc ug/g   "TAB(15) CU(1) TAB(30) CU(2) TAB(45) CU(3) TAB(60) CU(4) TAB(75) CU(5) TAB(90) CU(6)

1610 LPRINT "Amount mol "TAB(15) M(1) TAB(30) M(2) TAB(45) M(3) TAB(60) M(4) TAB(75) M(5) TAB(90) M(6)

1620 LPRINT "Amount kg   "TAB(15) MK(1) TAB(30) MK(2) TAB(45) MK(3) TAB(60) MK(4) TAB(75) MK(5) TAB(90) MK(6)

1630 LPRINT "Amount %   "TAB(15) P(1) TAB(30) P(2) TAB(45) P(3) TAB(60) P(4) TAB(75) P(5) TAB(90) P(6)

1640 LPRINT CHR$(12)

1650 REM Fugacity Level II program, 6 compartments

1660 REM calculate total inflows

1670 GA(1) = V(1)/GRA(1)

1680 GA(2) = V(2)/GRA(2)

1690 GA(4) = V(4)/GRA(4)

1700 REM calculate D values

1710 NRT=0:NRTK=0:NAT=0:NATK=0:VZT=0:MT=0:DT=0:DTA=0:DTR=0 'set totals to zero

1720 FOR N= 1 TO 4

1730 RK(N)=.693/HL(N) 'rate constants from half-lives

1740 DR(N)=V(N)*Z(N)*RK(N):DA(N)=GA(N)*Z(N) 'reaction and advection D values

1750 DTR=DTR+DR(N):DTA=DTA+DA(N) 'total D values

1760 NEXT N

1770 DT=DTR+DTA 'total D value

1780 F2=E/DT 'fugacity

1790 FOR N=1 TO 6

1800 F(N)=F2

1810 C(N)=F(N)*Z(N) 'concentration mol/m3

1820 M(N)=C(N)*V(N) 'amount mol

1830 MK(N)=M(N)*WM/1000

1840 MT=MT+M(N) 'total amount

1850 CG(N)=C(N)*WM 'concentration g/m3

1860 CU(N)=CG(N)*1000/DEN(N) 'concentration ug/g

1870 NR(N)=V(N)*C(N)*RK(N):NRK(N)=NR(N)*WM/1000 'reaction rates mol/h and kg/h

```
1880 NA(N)=GA(N)*C(N):NAK(N)=NA(N)*WM/1000 'advection rates mol/h and kg/h
1890 NRT=NRT+NR(N):NAT=NAT+NA(N) 'total rates mol/h
1900 NRTK=NRTK+NRK(N):NATK=NATK+NAK(N) 'total rates kg/h
1910 NEXT N
1920 NT=NRT+NAT:NTK=NRTK+NATK
1930 MTK=MT*WM/1000
1940 FOR N=1 TO 6
1950 P(N)=100*M(N)/MT 'percentages of amount
1960 PR(N)=100*NR(N)/NT 'percentages of reaction rate
1970 PA(N)=100*NA(N)/NT 'percentages of advection rate
1980 NEXT N
1990 IF NRT=0 THEN TR=0 ELSE TR=MT/NRT
2000 IF NAT=0 THEN TA=0 ELSE TA=MT/NAT
2010 TOV=MT/NT 'overall residence time h
2020 REM print out results
2030 LPRINT
2040 LPRINT  "SIX COMPARTMENT FUGACITY LEVEL II CALCULATION  ";CHEM$
2050 LPRINT " "
2060 LPRINT "Emission rate of chemical mol/h  ";E
2070 LPRINT "Emission rate of chemical  kg/h  ";EK
2080 LPRINT "Fugacity Pa                  ";F2
2090 LPRINT "Total amount of chemical mol     ";MT
2100 LPRINT "Total amount of chemical kg      ";MTK
2110 LPRINT " "
2120 LPRINT "Phase properties,compositions and rates"
2130 LPRINT " "
2140 LPRINT "Phase       "TAB(15) N$(1) TAB(30) N$(2) TAB(45) N$(3) TAB(60) N$(4)
TAB(75) N$(5) TAB(90) N$(6)
2150 LPRINT "Adv.flow m3/h"TAB(15) GA(1) TAB(30) GA(2) TAB(45) GA(3) TAB(60)
GA(4)
2160 LPRINT "Adv.restime h"TAB(15) GRA(1) TAB(30) GRA(2) TAB(45) GRA(3) TAB(60)
GRA(4)
2170 LPRINT "Rct halflife h"TAB(15) HL(1) TAB(30) HL(2) TAB(45) HL(3) TAB(60)
HL(4)
2180 LPRINT "Rct rate c.h-1"TAB(15) RK(1) TAB(30) RK(2) TAB(45) RK(3) TAB(60)
RK(4)
2190 LPRINT "Fugacity Pa "TAB(15) F(1) TAB(30) F(2) TAB(45) F(3) TAB(60) F(4)
TAB(75) F(5) TAB(90) F(6)
2200 LPRINT "Conc mol/m3 "TAB(15) C(1) TAB(30) C(2) TAB(45) C(3) TAB(60) C(4)
TAB(75) C(5) TAB(90) C(6)
2210 LPRINT "Conc g/m3   "TAB(15) CG(1) TAB(30) CG(2) TAB(45) CG(3) TAB(60)
CG(4 ) TAB(75) CG(5) TAB(90) CG(6)
2220 LPRINT "Conc ug/g   "TAB(15) CU(1) TAB(30) CU(2) TAB(45) CU(3) TAB(60)
```

CU(4) TAB(75) CU(5) TAB(90) CU(6)
2230 LPRINT "Amount mol "TAB(15) M(1) TAB(30) M(2) TAB(45) M(3) TAB(60) M(4)
TAB(75) M(5) TAB(90) M(6)
2240 LPRINT "Amount kg "TAB(15) MK(1) TAB(30) MK(2) TAB(45) MK(3) TAB(60)
MK(4) TAB(75) MK(5) TAB(90) MK(6)
2250 LPRINT "Amount % "TAB(15) P(1) TAB(30) P(2) TAB(45) P(3) TAB(60) P(4)
TAB(75) P(5) TAB(90) P(6)
2260 LPRINT "D rct mol/Pa.h"TAB(15) DR(1) TAB(30) DR(2) TAB(45) DR(3) TAB(60)
DR(4)
2270 LPRINT "D adv mol/Pa.h"TAB(15) DA(1) TAB(30) DA(2) TAB(45) DA(3) TAB(60)
DA(4)
2280 LPRINT "Rct rate mol/h"TAB(15) NR(1) TAB(30) NR(2) TAB(45) NR(3) TAB(60)
NR(4)
2290 LPRINT "Adv rate mol/h"TAB(15) NA(1) TAB(30) NA(2) TAB(45) NA(3) TAB(60)
NA(4)
2300 LPRINT "Rct rate kg/h "TAB(15) NRK(1) TAB(30) NRK(2) TAB(45) NRK(3) TAB(60)
NRK(4)
2310 LPRINT "Adv rate kg/h "TAB(15) NAK(1) TAB(30) NAK(2) TAB(45) NAK(3)
TAB(60) NAK(4)
2320 LPRINT "Reaction % "TAB(15) PR(1) TAB(30) PR(2) TAB(45) PR(3) TAB(60)
PR(4)
2330 LPRINT "Advection % "TAB(15) PA(1) TAB(30) PA(2) TAB(45) PA(3) TAB(60)
PA(4)
2340 LPRINT " "
2350 LPRINT "Total advection D value ";DTA
2360 LPRINT "Total reaction D value ";DTR
2370 LPRINT "Total D value ";DT
2380 LPRINT " "
2390 LPRINT "Output by reaction mol/h ";NRT
2400 LPRINT "Output by advection mol/h ";NAT
2410 LPRINT "Total output by reaction and advection mol/h ";NT
2420 LPRINT" "
2430 LPRINT "Output by reaction kg/h ";NRTK
2440 LPRINT "Output by advection kg/h ";NATK
2450 LPRINT "Total output by reaction and advection kg/h ";NTK
2460 LPRINT" "
2470 LPRINT "Overall residence time h ";TOV
2480 LPRINT "Reaction residence time h ";TR
2490 LPRINT "Advection residence time h ";TA
2500 LPRINT CHR$(12)
2510 LPRINT
2520 REM Fugacity Level III Program
2530 REM Set bulk phase volumes, densities and Z values

579

```
2540 VB(1)=V(1):VB(2)=V(2):VB(3)=1.8E+10:VB(4)=5E+08
2550 VA(1)=1:VQ(1)=2E-11 'volume fractions
2560 VW(2)=1:VP(2)=.000005:VF(2)=.000001
2570 VA(3)=.2:VW(3)=.3:VE(3)=.5
2580 VW(4)=.8:VS(4)=.2
2590 DENB(1)=VA(1)*DEN(1)+VQ(1)*DEN(7)
2600 DENB(2)=VW(2)*DEN(2)+VP(2)*DEN(5)+VF(2)*DEN(6)
2610 DENB(3)=VW(3)*DEN(2)+VA(3)*DEN(1)+VE(3)*DEN(3)
2620 DENB(4)=VW(4)*DEN(2)+VS(4)*DEN(4)
2630 ZB(1)=VA(1)*Z(1)+VQ(1)*Z(7)
2640 ZB(2)=VW(2)*Z(2)+VP(2)*Z(5)+VF(2)*Z(6)
2650 ZB(3)=VA(3)*Z(1)+VW(3)*Z(2)+VE(3)*Z(3)
2660 ZB(4)=VW(4)*Z(2)+VS(4)*Z(4)
2670 KB12=ZB(1)/ZB(2)
2680 KB32=ZB(3)/ZB(2)
2690 KB42=ZB(4)/ZB(2)
2700 REM Parameters
2710 U(1)=5          :U$(1)="air side air-water MTC          "
2720 U(2)=.05        :U$(2)="water side air-water MTC        "
2730 U(3)=.0001      :U$(3)="rain rate                       "
2740 U(4)=6E-10      :U$(4)="aerosol deposition velocity     "
2750 U(5)=.02        :U$(5)="soil air phase diffusion MTC     "
2760 U(6)=.00001     :U$(6)="soil water phase diffusion MTC   "
2770 U(7)=5          :U$(7)="soil air boundary layer MTC      "
2780 U(8)=.0001      :U$(8)="sediment-water diffusion MTC     "
2790 U(9)=.0000005   :U$(9)="sediment deposition velocity     "
2800 U(10)=.0000002:U$(10)="sediment resuspension velocity   "
2810 U(11)=.00005   :U$(11)="soil water runoff rate           "
2820 U(12)=1E-08    :U$(12)="soil solids runoff rate          "
2830 'Calculate D values
2840 DRW=AR(2)*U(3)*Z(2)
2850 DQW=AR(2)*U(4)*Z(7)
2860 DVWA=AR(2)*U(1)*Z(1)
2870 DVWW=AR(2)*U(2)*Z(2)
2880 DVW=1/(1/DVWA+1/DVWW)
2890 D(2,1)=DVW
2900 D(1,2)=DVW+DQW+DRW
2910 DVSB=AR(3)*U(7)*Z(1)
2920 DVSA=AR(3)*U(5)*Z(1)
2930 DVSW=AR(3)*U(6)*Z(2)
2940 DRS=AR(3)*U(3)*Z(2)
2950 DQS=AR(3)*U(4)*Z(7)
2960 DVS=1/(1/DVSB+1/(DVSW+DVSA))
```

```
2970 D(3,1)=DVS
2980 D(1,3)=DVS+DRS+DQS
2990 DSWD=AR(2)*U(8)*Z(2)
3000 DSD=AR(2)*U(9)*Z(5)
3010 DSR=AR(4)*U(10)*Z(4)
3020 D(2,4)=DSWD+DSD
3030 D(4,2)=DSWD+DSR
3040 DSWW=AR(3)*U(11)*Z(2)
3050 DSWS=AR(3)*U(12)*Z(3)
3060 D(3,2)=DSWW+DSWS
3070 D(2,3)=0
3080 REM calculate total chemical inflows
3090 IN=0:INK=0
3100 FOR N=1 TO 4
3110 I(N)=IK(N)*1000/WM
3120 IN=IN+I(N):INK=INK+IK(N)
3130 NEXT N
3140 REM calculate reaction and advection D values for bulk phases
3150 GAB(1)=VB(1)/GRA(1)
3160 GAB(2)=VB(2)/GRA(2)
3170 GAB(4)=VB(4)/GRA(4)
3180 VZBT=0
3190 FOR N= 1 TO 4
3200 RK(N)=.693/HL(N)
3210 DR(N)=VB(N)*ZB(N)*RK(N):DA(N)=GAB(N)*ZB(N)
3220 VZB(N)=VB(N)*ZB(N)
3230 VZBT=VZBT+VZB(N)
3240 NEXT N
3250 FOR N=1 TO 4
3260 FOR NN=1 TO 4
3270 GD(N,NN)=D(N,NN)/ZB(N)
3280 IF GD(N,NN)=0 GOTO 3300 ELSE GOTO 3290
3290 TD(N,NN)=.693*VB(N)/GD(N,NN)
3300 NEXT NN
3310 NEXT N
3320 DT(1)=DR(1)+DA(1)+D(1,2)+D(1,3)
3330 DT(2)=DR(2)+DA(2)+D(2,1)+D(2,3)+D(2,4)
3340 DT(3)=DR(3)+DA(3)+D(3,1)+D(3,2)
3350 DT(4)=DR(4)+DA(4)+D(4,2)
3360 J1=I(1)/DT(1)+I(3)*D(3,1)/DT(3)/DT(1)
3370 J2=D(2,1)/DT(1)
3380 J3=1-D(3,1)*D(1,3)/DT(1)/DT(3)
3390 J4=D(1,2)+D(3,2)*D(1,3)/DT(3)
```

```
3400 F(2)=(I(2)+J1*J4/J3+I(3)*D(3,2)/DT(3)+I(4)*D(4,2)/DT(4))/(DT(2)-J2*J4/J3
-D(2,4)*D(4,2)/DT(4))
3410 F(1)=(J1+F(2)*J2)/J3
3420 F(3)=(I(3)+F(1)*D(1,3))/DT(3)
3430 F(4)= (I(4)+F(2)*D(2,4))/DT(4)
3440 NRT=0:NAT=0:MT=0
3450 FOR N=1 TO 4
3460 C(N)=F(N)*ZB(N)
3470 M(N)=C(N)*VB(N)
3480 MK(N)=M(N)*WM/1000
3490 MT=MT+M(N)
3500 CG(N)=C(N)*WM
3510 CU(N)=CG(N)*1000/DENB(N)
3520 NR(N)=F(N)*DR(N):NRK(N)=NR(N)*WM/1000
3530 NA(N)=F(N)*DA(N):NAK(N)=NA(N)*WM/1000
3540 NRT=NRT+NR(N):NAT=NAT+NA(N)
3550 NEXT N
3560 MTK=MT*WM/1000
3570 NRTK=NRT*WM/1000:NATK=NAT*WM/1000
3580 NT=NRT+NAT:NTK=NT*WM/1000
3590 FOR N=1 TO 4
3600 P(N)=100*M(N)/MT
3610 PR(N)=100*NR(N)/NT
3620 PA(N)=100*NA(N)/NT
3630 NEXT N
3640 N(1,2)=D(1,2)*F(1):NK(1,2)=N(1,2)*WM/1000
3650 N(1,3)=D(1,3)*F(1):NK(1,3)=N(1,3)*WM/1000
3660 N(2,1)=D(2,1)*F(2):NK(2,1)=N(2,1)*WM/1000
3670 N(2,4)=D(2,4)*F(2):NK(2,4)=N(2,4)*WM/1000
3680 N(3,1)=D(3,1)*F(3):NK(3,1)=N(3,1)*WM/1000
3690 N(3,2)=D(3,2)*F(3):NK(3,2)=N(3,2)*WM/1000
3700 N(4,2)=D(4,2)*F(4):NK(4,2)=N(4,2)*WM/1000
3710 TR=MT/(NRT+.0000001)
3720 TA=MT/(NAT+.0000001)
3730 TOV=MT/NT
3740 TOVD=TOV/24
3750 REM print out results
3760 LPRINT  " FOUR COMPARTMENT FUGACITY LEVEL III
CALCULATION",CHEM$
3770 LPRINT
3780 LPRINT "Bulk phase properties,compositions and rates"
3790 LPRINT " "
3800 LPRINT "Phase        "TAB(15) N$(1) TAB(30) N$(2) TAB(45) N$(3) TAB(60) N$(4)
```

TAB(75) "Total"
3810 LPRINT "Bulk vol m3 "TAB(15) VB(1) TAB(30) VB(2) TAB(45) VB(3) TAB(60) VB(4)
3820 LPRINT "Density kg/m3"TAB(15) DENB(1) TAB(30) DENB(2) TAB(45) DENB(3) TAB(60) DENB(4)
3830 LPRINT "Bulk Z value"TAB(15) ZB(1) TAB(30) ZB(2) TAB(45) ZB(3) TAB(60) ZB(4)
3840 LPRINT "Bulk VZ   "TAB(15) VZB(1) TAB(30) VZB(2) TAB(45) VZB(3) TAB(60) VZB(4) TAB(75) VZBT
3850 LPRINT "Emission mol/h"TAB(15) I(1) TAB(30) I(2) TAB(45) I(3) TAB(60) I(4) TAB(75) IN
3860 LPRINT "Emission kg/h "TAB(15) IK(1) TAB(30) IK(2) TAB(45) IK(3) TAB(60) IK(4) TAB(75) INK
3870 LPRINT "Fugacity Pa "TAB(15) F(1) TAB(30) F(2) TAB(45) F(3) TAB(60) F(4)
3880 LPRINT "Conc mol/m3 "TAB(15) C(1) TAB(30) C(2) TAB(45) C(3) TAB(60) C(4)
3890 LPRINT "Conc g/m3   "TAB(15) CG(1) TAB(30) CG(2) TAB(45) CG(3) TAB(60) CG(4)
3900 LPRINT "Conc ug/g   "TAB(15) CU(1) TAB(30) CU(2) TAB(45) CU(3) TAB(60) CU(4)
3910 LPRINT "Amount mol   "TAB(15) M(1) TAB(30) M(2) TAB(45) M(3) TAB(60) M(4) TAB(75) MT
3920 LPRINT "Amount kg   "TAB(15) MK(1) TAB(30) MK(2) TAB(45) MK(3) TAB(60) MK(4) TAB(75) MTK
3930 LPRINT "Amount %   "TAB(15) P(1) TAB(30) P(2) TAB(45) P(3) TAB(60) P(4)
3940 LPRINT "Adv.flow m3/h"TAB(15) GAB(1) TAB(30) GAB(2) TAB(45) GAB(3) TAB(60) GAB(4)
3950 LPRINT "D rct mol/Pa.h"TAB(15) DR(1) TAB(30) DR(2) TAB(45) DR(3) TAB(60) DR(4)
3960 LPRINT "D adv mol/Pa.h"TAB(15) DA(1) TAB(30) DA(2) TAB(45) DA(3) TAB(60) DA(4)
3970 LPRINT "Rct rate mol/h"TAB(15) NR(1) TAB(30) NR(2) TAB(45) NR(3) TAB(60) NR(4) TAB(75) NRT
3980 LPRINT "Rct rate kg/h "TAB(15) NRK(1) TAB(30) NRK(2) TAB(45) NRK(3) TAB(60) NRK(4) TAB(75) NRTK
3990 LPRINT "Adv rate mol/h"TAB(15) NA(1) TAB(30) NA(2) TAB(45) NA(3) TAB(60) NA(4) TAB(75) NAT
4000 LPRINT "Adv rate kg/h "TAB(15) NAK(1) TAB(30) NAK(2) TAB(45) NAK(3) TAB(60) NAK(4) TAB(75) NATK
4010 LPRINT "Reaction %   "TAB(15) PR(1) TAB(30) PR(2) TAB(45) PR(3) TAB(60) PR(4)
4020 LPRINT "Advection %   "TAB(15) PA(1) TAB(30) PA(2) TAB(45) PA(3) TAB(60) PA(4)
4030 LPRINT " "

```
4040 LPRINT "Overall residence time   h   ";TOV
4050 LPRINT "Reaction residence time  h   ";TR;
4060 LPRINT "    Advection residence time h   ";TA
4070 LPRINT
4080 LPRINT "Intermedia Data.   Half times  Equiv flows  D values    Rates  of transport "
4090 LPRINT "                      h        m3/h      mol/Pa.h  mol/h    kg/h"
4100 LPRINT "Air to water     ";:LPRINT USING " ##.####^^^^ ";TD(1,2);GD(1,2);D(1,2)
;N(1,2);NK(1,2)
4110 LPRINT "Air to soil      ";:LPRINT USING " ##.####^^^^ ";TD(1,3);GD(1,3);D(1,3)
;N(1,3);NK(1,3)
4120 LPRINT "Water to air     ";:LPRINT USING " ##.####^^^^ ";TD(2,1);GD(2,1);D(2,1)
;N(2,1);NK(2,1)
4130 LPRINT "Water to sediment";:LPRINT USING " ##.####^^^^
";TD(2,4);GD(2,4);D(2,4) ;N(2,4);NK(2,4)
4140 LPRINT "Soil to air      ";:LPRINT USING " ##.####^^^^ ";TD(3,1);GD(3,1),D(3,1)
,N(3,1);NK(3,1)
4150 LPRINT "Soil to water    ";:LPRINT USING " ##.####^^^^ ";TD(3,2);GD(3,2);D(3,2)
;N(3,2);NK(3,2)
4160 LPRINT "Sediment to water";:LPRINT USING " ##.####^^^^
";TD(4,2);GD(4,2),D(4,2) ,N(4,2);NK(4,2)
4170 LPRINT "  Transport velocity parameters          m/h           m/year  "
4180 FOR I=1 TO 12
4190 UY(I)=U(I)*8760
4200 LPRINT TAB(5) I TAB(10) U$(I) TAB(45) U(I) TAB(60) UY(I)
4210 NEXT I
4220 LPRINT "Individual process D values "
4230 LPRINT "Air-water diffusion (air-side)       ";DVWA TAB(50);
4240 LPRINT "Air-water diffusion (water-side)     ";DVWW
4250 LPRINT "Air-water diffusion (overall)        ";DVW
4260 LPRINT "Rain dissolution to water            ";DRW  TAB(50);
4270 LPRINT "Aerosol deposition to water          ";DQW
4280 LPRINT "Rain dissolution to soil             ";DRS  TAB(50);
4290 LPRINT "Aerosol deposition to soil           ";DQS
4300 LPRINT "Soil-air diffusion (air-phase)       ";DVSA TAB(50);
4310 LPRINT "Soil-air diffusion (water-phase)     ";DVSW
4320 LPRINT "Soil-air diffusion (bndry layer)     ";DVSB TAB(50);
4330 LPRINT "Soil-air diffusion (overall)         ";DVS
4340 LPRINT "Water-sediment diffusion             ";DSWD
4350 LPRINT "Water-sediment deposition            ";DSD TAB(50);
4360 LPRINT "Sediment-water resuspension          ";DSR
4370 LPRINT "Soil-water runoff (water)            ";DSWW TAB(50);
4380 LPRINT "Soil-water runoff (solids)           ";DSWS
```

584

PROGRAM 'LEWIS':SIX COMPARTMENT FUGACITY LEVEL I CALCULATION

Properties of Naphthalene

Temperature deg C	25
Molecular mass g/mol	128.18
Melting point deg C	80.5
Fugacity ratio	.2825374
Vapor pressure Pa	10.4
Sub-cooled liquid vapor press Pa	36.80929
Solubility g/m3	31
Solubility mol/m3	.2418474
Henry's law constant Pa.m3/mol	43.00232
Log octanol-water p-coefficient	3.37
Octanol-water partn-coefficient	2344.229
Organic C-water ptn-coefficient	961.1338
Fish-water partition coefficient	117.2114
Air-water partition coefficient	.0173479
Soil-water partition coefficient	46.13442
Sedt-water partition coefficient	92.26883
Susp sedt-water partn coeffnt	288.3402
Aerosol-air partition coeff	163002.4
Aerosol Z value	65.75806
Aerosol density kg/m3	2000

Amount of chemical moles	780153
Amount of chemical kilograms	100000
Fugacity Pa	1.42182E-05
Total of VZ products	5.487004E+10

Phase properties and compositions

Phase	Air	Water	Soil	Sediment	Susp sedt	Fish
Volume m3	1E+14	2E+11	8.999999E+09	1E+08	999999.9	200000
Density kg/m3	1.185413	1000	2400	2400	1500	1000
Depth m	1000	20	.1	.01		
Area m2	1E+11	1E+10	9E+10	1E+10		
Frn org carb			.02	.04	.2	
Z mol/m3.Pa	4.034179E-04	2.325456E-02	1.072836	2.145671	6.705223	2.725701
VZ mol/Pa	4.034179E+10	4.650912E+09	9.655519E+09	2.145671E+08	6705223	545140.1
Fugacity Pa	1.42182E-05	1.42182E-05	1.42182E-05	1.42182E-05	1.42182E-05	1.42182E-05
Conc mol/m3	5.735875E-09	3.306379E-07	1.525379E-05	3.050757E-05	9.533618E-05	3.875455E-05
Conc g/m3	7.352243E-07	4.238117E-05	1.95523E-03	3.910461E-03	1.222019E-02	4.967558E-03
Conc ug/g	6.202262E-04	4.238117E-05	8.146793E-04	1.629359E-03	8.146794E-03	4.967558E-03
Amount mol	573587.5	66127.58	137284.1	3050.757	95.33617	7.750909
Amount kg	73522.43	8476.232	17597.07	391.0461	12.22019	.9935114
Amount %	73.52244	8.476232	17.59707	.3910461	1.222019E-02	9.935114E-04

SIX COMPARTMENT FUGACITY LEVEL II CALCULATION Naphthalene

Emission rate of chemical mol/h   7801.53
Emission rate of chemical kg/h   1000
Fugacity Pa   3.758397E-06
Total amount of chemical mol   206223.4
Total amount of chemical kg   26433.71

Phase properties,compositions and rates

Phase	Air	Water	Soil	Sediment	Susp sedt	Fish
Adv.flow m3/h	1E+12	2E+08	0	2000		
Adv.restime h	100	1000	0	50000		
Rct halflife h	17	170	1700	550		
Rct rate c.h-1	4.076471E-02	4.076471E-03	4.076471E-04	.00126		
Fugacity Pa	3.758397E-06	3.758397E-06	3.758397E-06	3.758397E-06	3.758397E-06	3.758397E-06
Conc mol/m3	1.516204E-09	8.739986E-08	4.032142E-06	8.064283E-06	2.520089E-05	1.024426E-05
Conc g/m3	1.943471E-07	1.120291E-05	5.168399E-04	1.03368E-03	3.23025E-03	1.31311E-03
Conc ug/g	1.639488E-04	1.120291E-05	2.153499E-04	4.306999E-04	.0021535	1.31311E-03
Amount mol	151620.4	17479.97	36289.27	806.4283	25.20089	2.048853
Amount kg	19434.71	2240.583	4651.559	103.368	3.230249	.2626219
Amount %	73.52244	8.476234	17.59707	.3910461	1.222019E-02	9.935113E-04
D rct mol/Pa.h	1.644521E+09	1.89593E+07	3936044	270354.6		
D adv mol/Pa.h	4.034179E+08	4650912	0	4291.343		
Rct rate mol/h	6180.763	71.2566	14.79322	1.0161		
Adv rate mol/h	1516.204	17.47997	0	1.612857E-02		
Rct rate kg/h	792.2501	9.133669	1.896194	.1302437		
Adv rate kg/h	194.3471	2.240583	0	2.067359E-03		
Reaction %	79.22501	.913367	.1896195	1.302437E-02		
Advection %	19.43471	.2240583	0	2.06736E-04		

Total advection D value   4.080731E+08
Total reaction D value   1.667687E+09
Total D value   2.07576E+09

Output by reaction   mol/h 6267.829
Output by advection   mol/h 1533.701
Total output by reaction and advection mol/h 7801.53

Output by reaction   kg/h   803.4102
Output by advection   kg/h   196.5897
Total output by reaction and advection kg/h 999.9999

Overall residence time   h   26.43371
Reaction residence time h   32.90188
Advection residence time h   134.4613

FOUR COMPARTMENT FUGACITY LEVEL III CALCULATION   Naphthalene

Bulk phase properties,compositions and rates

Phase	Air	Water	Soil	Sediment	Total
Bulk vol m3	1E+14	2E+11	1.8E+10	5E+08	
Density kg/m3	1.185413	1000.009	1500.237	1280	
Bulk Z value	4.034192E-04	2.329081E-02	.5434749	.4477379	
Bulk VZ	4.034192E+10	4.658162E+09	9.782548E+09	2.238689E+08	5.50065E+10
Emission mol/h	7801.53	0	0	0	7801.53
Emission kg/h	1000	0	0	0	1000
Fugacity Pa	3.797488E-06	9.07406E-07	7.511529E-07	8.554287E-07	
Conc mol/m3	1.53198E-09	2.113422E-08	4.082327E-07	3.830078E-07	
Conc g/m3	1.963691E-07	2.708984E-06	5.232727E-05	4.909394E-05	
Conc ug/g	1.656546E-04	2.708961E-06	3.487933E-05	3.835464E-05	
Amount mol	153198	4226.844	7348.189	191.5039	164964.5
Amount kg	19636.91	541.7969	941.8908	24.54697	21145.15
Amount %	92.86723	2.562276	4.454407	.116088	
Adv.flow m3/h	1E+12	2E+08	0	10000	
D rct mol/Pa.h	1.644527E+09	1.898886E+07	3987827	28207.49	
D adv mol/Pa.h	4.034192E+08	4658162	0	4477.379	
Rct rate mol/h	6245.07	17.23061	2.995468	2.412949E-02	6265.32
Rct rate kg/h	800.493	2.208619	.383959	3.092918E-03	803.0886
Adv rate mol/h	1531.98	4.226844	0	3.830079E-03	1536.21
Adv rate kg/h	196.3691	.5417968	0	4.909394E-04	196.9114
Reaction %	80.0493	.2208619	.0383959	3.092918E-04	
Advection %	19.63692	5.417969E-02	0	4.909395E-05	

Overall residence time   h   21.14515
Reaction residence time   h   26.32978   Advection residence time h   107.3841

Intermedia Data.	Half times	Equiv flows	D values	Rates of transport	
	h	m3/h	mol/Pa.h	mol/h	kg/h
Air to water	3.7783E+03	1.8341E+10	7.3993E+06	2.8099E+01	3.6017E+00
Air to soil	2.9217E+04	2.3719E+09	9.5686E+05	3.6337E+00	4.6576E-01
Water to air	4.3767E+02	3.1668E+08	7.3757E+06	6.6927E+00	8.5787E-01
Water to sediment	5.6852E+04	2.4379E+06	5.6781E+04	5.1523E-02	6.6042E-03
Soil to air	9.1117E+03	1.3690E+06	7.4402E+05	5.5887E-01	7.1636E-02
Soil to water	6.4191E+04	1.9433E+05	1.0561E+05	7.9330E-02	1.0169E-02
Sediment to water	5.6321E+03	6.1522E+04	2.7546E+04	2.3564E-02	3.0204E-03

	Transport velocity parameters		m/h	m/year
1	air side air-water MTC	5	43800
2	water side air-water MTC	.05	438
3	rain rate	.0001	.876
4	aerosol deposition velocity	6E-10	5.256E-06
5	soil air phase diffusion MTC	.02	175.2
6	soil water phase diffusion MTC	.00001	.0876
7	soil air boundary layer MTC	5	43800
8	sediment-water diffusion MTC	.0001	.876
9	sediment deposition velocity	.0000005	.00438
10	sediment resuspension velocity	.0000002	.001752
11	soil water runoff rate	.00005	.438
12	soil solids runoff rate	1E-08	.0000876

Individual process D values

Air-water diffusion (air-side)	2.017089E+07	Air-water diffusion (water-side)	1.162728E+07
Air-water diffusion (overall)	7375664		
Rain dissolution to water	23254.56	Aerosol deposition to water	394.5484
Rain dissolution to soil	209291.1	Aerosol deposition to soil	3550.935
Soil-air diffusion (air-phase)	726152.2	Soil-air diffusion (water-phase)	20929.1
Soil-air diffusion (bndry layer)	1.815381E+08	Soil-air diffusion (overall)	744019.5
Water-sediment diffusion	23254.56		
Water-sediment deposition	33526.12	Sediment-water resuspension	4291.343
Soil-water runoff (water)	104645.5	Soil-water runoff (solids)	965.552

FOUR COMPARTMENT FUGACITY LEVEL III CALCULATION   Naphthalene

Bulk phase properties,compositions and rates

Phase	Air	Water	Soil	Sediment	Total
Bulk vol m3	1E+14	2E+11	1.8E+10	5E+08	
Density kg/m3	1.185413	1000.009	1500.237	1280	
Bulk Z value	4.034192E-04	2.329081E-02	.5434749	.4477379	
Bulk VZ	4.034192E+10	4.658162E+09	9.782548E+09	2.238689E+08	5.50065E+10
Emission mol/h	0	7801.53	0	0	7801.53
Emission kg/h	0	1000	0	0	1000
Fugacity Pa	9.013497E-07	2.512742E-04	1.782893E-07	4.542377E-05	
Conc mol/m3	3.636218E-10	5.852381E-06	9.689574E-08	2.033794E-05	
Conc g/m3	4.660903E-08	7.501581E-04	1.24201E-05	2.606917E-03	
Conc ug/g	3.931881E-05	7.501517E-04	8.278756E-06	2.036654E-03	
Amount mol	36362.18	1170476	1744.123	10168.97	1218751
Amount kg	4660.904	150031.6	223.5617	1303.459	156219.5
Amount %	2.98356	96.03896	.1431074	.8343762	
Adv.flow m3/h	1E+12	2E+08	0	10000	
D rct mol/Pa.h	1.644527E+09	1.898886E+07	3987827	282074.9	
D adv mol/Pa.h	4.034192E+08	4658162	0	4477.379	
Rct rate mol/h	1482.293	4771.412	.7109868	12.8129	6267.229
Rct rate kg/h	190.0004	611.5995	9.113428E-02	1.642358	803.3334
Adv rate mol/h	363.6218	1170.476	0	.2033794	1534.301
Adv rate kg/h	46.60904	150.0316	0	2.606917E-02	196.6667
Reaction %	19.00003	61.15995	9.113428E-03	.1642358	
Advection %	4.660903	15.00316	0	2.606917E-03	

Overall residence time   h   156.2195
Reaction residence time   h   194.4642   Advection residence time h   794.3365

Intermedia Data.	Half times	Equiv flows	D values	Rates	of transport
	h	m3/h	mol/Pa.h	mol/h	kg/h
Air to water	3.7783E+03	1.8341E+10	7.3993E+06	6.6694E+00	8.5488E-01
Air to soil	2.9217E+04	2.3719E+09	9.5686E+05	8.6247E-01	1.1055E-01
Water to air	4.3767E+02	3.1668E+08	7.3757E+06	1.8533E+03	2.3756E+02
Water to sediment	5.6852E+04	2.4379E+06	5.6781E+04	1.4268E+01	1.8288E+00
Soil to air	9.1117E+03	1.3690E+06	7.4402E+05	1.3265E-01	1.7003E-02
Soil to water	6.4191E+04	1.9433E+05	1.0561E+05	1.8829E-02	2.4135E-03
Sediment to water	5.6321E+03	6.1522E+04	2.7546E+04	1.2512E+00	1.6038E-01

	Transport velocity parameters		m/h	m/year
1	air side air-water MTC		5	43800
2	water side air-water MTC		.05	438
3	rain rate		.0001	.876
4	aerosol deposition velocity		6E-10	5.256E-06
5	soil air phase diffusion MTC		.02	175.2
6	soil water phase diffusion MTC		.00001	.0876
7	soil air boundary layer MTC		5	43800
8	sediment-water diffusion MTC		.0001	.876
9	sediment deposition velocity		.0000005	.00438
10	sediment resuspension velocity		.0000002	.001752
11	soil water runoff rate		.00005	.438
12	soil solids runoff rate		1E-08	.0000876

Individual process D values

Air-water diffusion (air-side)	2.017089E+07	Air-water diffusion (water-side)	1.162728E+07
Air-water diffusion (overall)	7375664		
Rain dissolution to water	23254.56	Aerosol deposition to water	394.5484
Rain dissolution to soil	209291.1	Aerosol deposition to soil	3550.935
Soil-air diffusion (air-phase)	726152.2	Soil-air diffusion (water-phase)	20929.1
Soil-air diffusion (bndry layer)	1.815381E+08	Soil-air diffusion (overall)	744019.5
Water-sediment diffusion	23254.56		
Water-sediment deposition	33526.12	Sediment-water resuspension	4291.343
Soil-water runoff (water)	104645.5	Soil-water runoff (solids)	965.552

FOUR COMPARTMENT FUGACITY LEVEL III CALCULATION   Naphthalene

Bulk phase properties,compositions and rates

Phase	Air	Water	Soil	Sediment	Total
Bulk vol m3	1E+14	2E+11	1.8E+10	5E+08	
Density kg/m3	1.185413	1000.009	1500.237	1280	
Bulk Z value	4.034192E-04	2.329081E-02	.5434749	.4477379	
Bulk VZ	4.034192E+10	4.658162E+09	9.782548E+09	2.238689E+08	5.50065E+10
Emission mol/h	0	0	7801.53	0	7801.53
Emission kg/h	0	0	1000	0	1000
Fugacity Pa	6.037597E-07	5.629077E-06	1.612853E-03	5.306637E-06	
Conc mol/m3	2.435683E-10	1.311058E-07	8.76545E-04	2.375982E-06	
Conc g/m3	3.122058E-08	1.680513E-05	.1123555	3.045534E-06	
Conc ug/g	2.63373E-05	1.680499E-05	7.489185E-02	2.379323E-04	
Amount mol	24356.83	26221.15	1.577781E+07	1187.991	1.582958E+07
Amount kg	3122.058	3361.027	2022400	152.2767	2029035
Amount %	.1538691	.1656466	99.67298	7.504883E-03	
Adv.flow m3/h	1E+12	2E+08	0	10000	
D rct mol/Pa.h	1.644527E+09	1.898886E+07	3987827	28207.49	
D adv mol/Pa.h	4.034192E+08	4658162	0	4477.379	
Rct rate mol/h	992.8988	106.8898	6431.779	.1496869	7531.717
Rct rate kg/h	127.2698	13.70113	824.4253	1.918686E-02	965.4154
Adv rate mol/h	243.5683	26.22115	0	2.375982E-02	269.8132
Adv rate kg/h	31.22058	3.361027	0	3.045534E-03	34.58465
Reaction %	12.72698	1.370113	82.44253	1.918686E-03	
Advection %	3.122058	.3361027	0	3.045534E-04	

Overall residence time   h   2029.035
Reaction residence time h   2101.722   Advection residence time h   58668.66

Intermedia Data.	Half times	Equiv flows	D values	Rates	of transport	
	h	m3/h	mol/Pa.h	mol/h	kg/h	
Air to water	3.7783E+03	1.8341E+10	7.3993E+06	4.4674E+00	5.7263E-01	
Air to soil	2.9217E+04	2.3719E+09	9.5686E+05	5.7771E-01	7.4051E-02	
Water to air	4.3767E+02	3.1668E+08	7.3757E+06	4.1518E+01	5.3218E+00	
Water to sediment	5.6852E+04	2.4379E+06	5.6781E+04	3.1962E-01	4.0969E-02	
Soil to air	9.1117E+03	1.3690E+06	7.4402E+05	1.2000E+03	1.5382E+02	
Soil to water	6.4191E+04	1.9433E+05	1.0561E+05	1.7034E+02	2.1834E+01	
Sediment to water	5.6321E+03	6.1522E+04	2.7546E+04	1.4618E-01	1.8737E-02	

	Transport velocity parameters	m/h	m/year
1	air side air-water MTC	5	43800
2	water side air-water MTC	.05	438
3	rain rate	.0001	.876
4	aerosol deposition velocity	6E-10	5.256E-06
5	soil air phase diffusion MTC	.02	175.2
6	soil water phase diffusion MTC	.00001	.0876
7	soil air boundary layer MTC	5	43800
8	sediment-water diffusion MTC	.0001	.876
9	sediment deposition velocity	.0000005	.00438
10	sediment resuspension velocity	.0000002	.001752
11	soil water runoff rate	.00005	.438
12	soil solids runoff rate	1E-08	.0000876

Individual process D values

Air-water diffusion (air-side)	2.017089E+07	Air-water diffusion (water-side)	1.162728E+07
Air-water diffusion (overall)	7375664		
Rain dissolution to water	23254.56	Aerosol deposition to water	394.5484
Rain dissolution to soil	209291.1	Aerosol deposition to soil	3550.935
Soil-air diffusion (air-phase)	726152.2	Soil-air diffusion (water-phase)	20929.1
Soil-air diffusion (bndry layer)	1.815381E+08	Soil-air diffusion (overall)	744019.5
Water-sediment diffusion	23254.56		
Water-sediment deposition	33526.12	Sediment-water resuspension	4291.343
Soil-water runoff (water)	104645.5	Soil-water runoff (solids)	965.552

FOUR COMPARTMENT FUGACITY LEVEL III CALCULATION   Naphthalene

Bulk phase properties,compositions and rates

Phase	Air	Water	Soil	Sediment	Total
Bulk vol m3	1E+14	2E+11	1.8E+10	5E+08	
Density kg/m3	1.185413	1000.009	1500.237	1280	
Bulk Z value	4.034192E-04	2.329081E-02	.5434749	.4477379	
Bulk VZ	4.034192E+10	4.658162E+09	9.782548E+09	2.238689E+08	5.50065E+10
Emission mol/h	4680.918	2340.459	780.153	0	7801.53
Emission kg/h	600	300	100	0	1000
Fugacity Pa	2.609457E-06	7.654061E-05	1.617895E-04	7.215628E-05	
Conc mol/m3	1.052705E-09	1.782693E-06	8.792853E-05	3.23071E-05	
Conc g/m3	1.349357E-07	2.285056E-04	1.127068E-02	4.141124E-03	
Conc ug/g	1.138301E-04	2.285036E-04	7.512598E-03	3.235253E-03	
Amount mol	105270.5	356538.6	1582714	16153.55	2060676
Amount kg	13493.57	45701.12	202872.2	2070.562	264137.5
Amount %	5.108541	17.30202	76.80555	.7838956	
Adv.flow m3/h	1E+12	2E+08	0	10000	
D rct mol/Pa.h	1.644527E+09	1.898886E+07	3987827	28207.49	
D adv mol/Pa.h	4.034192E+08	4658162	0	4477.379	
Rct rate mol/h	4291.321	1453.419	645.1886	2.035347	6391.963
Rct rate kg/h	550.0614	186.2993	82.70026	.2608908	819.3218
Adv rate mol/h	1052.705	356.5386	0	.323071	1409.567
Adv rate kg/h	134.9357	45.70112	0	4.141124E-02	180.6782
Reaction %	55.00614	18.62993	8.270026	2.608908E-02	
Advection %	13.49357	4.570112	0	4.141124E-03	

Overall residence time   h   264.1375
Reaction residence time   h   322.3855   Advection residence time h   1461.922

Intermedia Data.	Half times h	Equiv flows m3/h	D values mol/Pa.h	Rates of transport mol/h	kg/h
Air to water	3.7783E+03	1.8341E+10	7.3993E+06	1.9308E+01	2.4749E+00
Air to soil	2.9217E+04	2.3719E+09	9.5686E+05	2.4969E+00	3.2005E-01
Water to air	4.3767E+02	3.1668E+08	7.3757E+06	5.6454E+02	7.2362E+01
Water to sediment	5.6852E+04	2.4379E+06	5.6781E+04	4.3460E+00	5.5707E-01
Soil to air	9.1117E+03	1.3690E+06	7.4402E+05	1.2037E+02	1.5430E+01
Soil to water	6.4191E+04	1.9433E+05	1.0561E+05	1.7087E+01	2.1902E+00
Sediment to water	5.6321E+03	6.1522E+04	2.7546E+04	1.9876E+00	2.5477E-01

	Transport velocity parameters	m/h	m/year
1	air side air-water MTC	5	43800
2	water side air-water MTC	.05	438
3	rain rate	.0001	.876
4	aerosol deposition velocity	6E-10	5.256E-06
5	soil air phase diffusion MTC	.02	175.2
6	soil water phase diffusion MTC	.00001	.0876
7	soil air boundary layer MTC	5	43800
8	sediment-water diffusion MTC	.0001	.876
9	sediment deposition velocity	.0000005	.00438
10	sediment resuspension velocity	.0000002	.001752
11	soil water runoff rate	.00005	.438
12	soil solids runoff rate	1E-08	.0000876

Individual process D values

Air-water diffusion (air-side)	2.017089E+07	Air-water diffusion (water-side)	1.162728E+07
Air-water diffusion (overall)	7375664		
Rain dissolution to water	23254.56	Aerosol deposition to water	394.5484
Rain dissolution to soil	209291.1	Aerosol deposition to soil	3550.935
Soil-air diffusion (air-phase)	726152.2	Soil-air diffusion (water-phase)	20929.1
Soil-air diffusion (bndry layer)	1.815381E+08	Soil-air diffusion (overall)	744019.5
Water-sediment diffusion	23254.56		
Water-sediment deposition	33526.12	Sediment-water resuspension	4291.343
Soil-water runoff (water)	104645.5	Soil-water runoff (solids)	965.552

A2. Listing of Fugacity calculations for Lotus 123 spreadsheet program.

Fugacity calculations: Naphthalene
LEVEL I, II and III

*	Amount of chemicals, moles	780152.90 moles	100000 kg
*	Emission rate of chemicals, E =	7801.5290 mol/h	1000 kg/h
	Gas constant, Pa m3/mol K, R=	8.314	
	System temperature, K, T= (t + 273.15)	298.15	
*	Molecular weight, g/mol MW =	128.18	
*	Melting point, t C M.P.	80.55	
#	If M.P. > 25 C enter Tm (mp+273.15) or else system temp. K	353.7	* input data
*	Fugacity ratio = exp(6.79(1-Tm/T)) for solid comp'ds	0.2822	* input data
*	Solubility, g/m3 or mg/L	31	* input
	molar solubility, mol/L, c=S/MW S =	2.42E-01	
*	Vapor pressure, Pa P =	10.4	* input data
	Vap. pressure, subcooled liquid, Pa	36.851	
	Henry's law constant, Pa m3/mol, H=p/c	43.002322	
*	Octanol/water partition coefficient, log Kow	3.37	* input data
	Kow =	2344	
	Partition coefficient, organic C, Koc = 0.41*Kow*y	961.13381	
	for soil (mole fraction organic C), y(3) =	0.02	
	suspended sediment, y(5) =	0.2	
	bottom sediment, y(6) =	0.04	
	Kp(3) =0.41*Kow*y(3)	19	
	Kp(4) =0.41*Kow*y(4)	192	
	Kp(5) = 0.41*Kow*y(5)	38	
	Bioconcentraion factor, K(6) = 0.050*Kow	117	
	Air/water partition coeff., Z(air)/Z(water)	0.0173479	
	Soil/water partition coeff., Z(soil)/Z(water)	46.134423	
	Sediment/water partition coeff., Z(sediment)/Z(water)	92.268846	
	Sus. sediment/water partition coeff., Z(ss)/Z(water)	288.34014	
	Aerosol/water partition coeff., Z(aerosol)/Z(air)	162816.84	
	Densities, g/cm3 or kg/L		
	air, d(1) = (0.029*101325/RT)	0.0011854	
	water, d(2)	1	
	soil, d(3) =	2.4	
	bottom sediment, d(4) =	2.4	
	suspended sediment, d(5) =	1.5	
	biota, d(6) =	1	
	aerosol, d(7)	2	
	Fugacity capacities, Z:		
	Z(1) or Z(air) = 1/RT	4.034E-04	
	Z(2) or Z(water) = 1/H = c/p	2.325E-02	
	Z(3) or Z(soil) = Kp(s)*Z(water)*d(s)	1.073E+00	
	Z(4) or Z(bottom sediment) = Kp(bs)*Z(water)*d(bs)	2.146E+00	
	Z(5) or Z(suspended sediment) = Kp(ss)*Z(water)*d(ss)	6.705E+00	
	Z(6) or Z(biota) = K *Z(water)*d(B)	2.726E+00	
	Z(7) or Z(aerosol) = 6*E6/p(L)RT	6.568E+01	
	Fugacity (Level I), f = total no. of moles/sum(ViZi)	1.422E-05	
	Fugacity (Level II), f = emission/sum(D values)	3.759E-06	

* input

	Adv. flow, G(air)	1.000E+12 mol/hr		
	Adv. flow G(water)	5.00E+08 mol/hr		
	Adv. flow, G(sed.)	10000 mol/hr		
	Half-lives, hours			
*	t(air)	17 h		
*	t(water)	170 h		
*	t(soil)	1700 h		
*	t(sediment)	5500 h		

Emission, kg/h :

	air	water	soil
E	600	300	100
E(A)	1000	0	0
E(B)	0	1000	0
E(C)	0	0	1000

Residence time: h

air	100	h
water	1000	h
sediment	50000	h

Tansport parameters:

Parameter	m/h, k	m/yr, k/8760
air-water MTC, air side, U(1)	5	43800
air-water MTC, Water side, U(2)	0.05	438
rain rate, U(3) = 0.85/8760	0.0001	0.876
aerosol deposition velocity, U(4)	6.00E-10	0.0000052
soil air phase diffusion MTC, U(5)	0.02	175.2
soil water phase diffusion MTC, U(6)	1.00E-05	0.0876
soil air boundary layer MTC, U(7)	5	43800
sediment-water MTC, U(8)	0.0001	0.876
sediment deposition rate, U(9)	5.00E-07	0.00438
sediment resuspended rate, U(10)	2.00E-07	0.001752
soil water runoff rate, U(11)	5.00E-05	0.438
soil solids runoff rate, U(12)	1.00E-08	0.0000876
sediment burial rate, U(13)	0	0
diffusion to stratosphere, U(14)	0	0

Define unit world:

Compartment	Volume, Vi, m3	Depth, h, m	Area, A, m2	Density d, kg/m3	Fugacity cap., Zi mol/Pa m3	VZ	*input Advective flow, G mol/h	Residence time, V/G t(R), h	*input Reaction half-life t(1/2), h	*input Rate const. 0.693/t k, 1/hr	Emission rate, E mol/h	E(A) mol/h	E(B) mol/h	E(C) mol/h
Air (1)	1.00E+14	1000	1.000E+11	1.1854132	4.034E-04	4.03E+10	1.000E+12	100	17	4.076E-02	4680.92	7801.53	0	0
Water (2)	2.00E+11	20	1.000E+10	1000	2.325E-02	4.65E+09	2.000E+08	1000	170	4.076E-03	2340.46	0	7801.53	0
Soil (3)	9.00E+09	0.1	9.000E+10	2400	1.073E+00	9.66E+09	0		1700	4.076E-04	780.15	0	0	7801.53
bottom sediment (4)	1.00E+08	0.01	1.000E+10	2400	2.146E+00	2.15E+08	2000	50000	5500	1.260E-04	0.00	0	0	0
Sus. sediment (5)	1.00E+06			1500	6.705E+00	6.71E+06			1.00E+11	6.930E-12				
Biota (fish) (6)	2.00E+05			1000	2.726E+00	5.45E+05			1.00E+11	6.930E-12				
Aerosol (7)	2000			2000	6.568E-01	5.49E+10								

Level I calculation:

Compartment	Volume, Vi, m3	Fugacity capacity Zi	VZ	Conc'n, c c = f*z mol/m3	Amount m= ciVi mol	Amount w=m*MW/1E3 kg	Amount %	Conc'n, S mg/L (or g/m3)	Conc'n (S/d)*1000 ug/g
Air (1)	1.00E+14	4.034E-04	4.034E+10	5.736E-09	5.74E+05	7.352E-04	7.352E+01	7.35E-07	6.20E-04
Water (2)	2.00E+11	2.325E-02	4.651E+09	3.306E-07	6.61E+04	8.476E+03	8.48E+00	4.24E-05	4.24E-05
Soil (3)	9.00E+09	1.073E+00	9.656E+09	1.525E-05	1.37E+05	1.760E+04	1.76E+01	1.96E-03	8.15E-04
Bottom sediment (4)	1.00E+08	2.146E+00	2.146E+08	3.051E-05	3.05E+03	3.910E+02	3.91E-01	3.91E-03	1.63E-03
Sus. sediment (5)	1.00E+06	6.705E+00	6.705E+06	9.534E-05	9.53E+01	1.222E+01	1.22E-02	1.22E-02	8.15E-03
Biota (fish) (6)	2.00E+05	2.726E+00	5.451E+05	3.875E-05	7.75E+00	9.935E-01	9.94E-04	4.97E-03	4.97E-03
			5.487E+10		780152.90	100000	100		

Level II phase properties and rates:

Compartment	Rate const. k, 1/hr	D(reaction) VZk	D(advec'n) GZ	conc'n c=f*Z mol/m3	Amount m=ciVi mol	Amount m*MW/1000 kg	conc'n mg/L (or g/m3)	Conc'n ug/g (S/d)*1000	Loss Reaction mol/h Vck	Loss Advection mol/h Gc	% Loss reaction	% Loss advection	Removal %
Air (1)	0.040764705	1.645E+09	4.034E+08	1.516E-09	1.52E+05	1.94E+04	1.944E-07	1.64E-04	6.182E+03	1.516E+03	7.92E+01	19.44	98.67
Water (2)	0.004076470	1.896E+07	4.651E+06	8.741E-08	1.75E+04	2.24E+03	1.120E-05	1.12E-05	7.127E+01	1.748E+01	9.13E-01	0.224	1.138
Soil (3)	0.00407647	3.936E+06		4.033E-06	4.65E+04	4.65E+03	5.169E-04	2.15E-04	1.479E+01		1.90E-01		1.90E-01
Bottom sediment (4)	0.000126	2.704E+04	4.291E+03	8.065E-06	8.07E+02	1.03E+02	1.034E-03	4.31E-04	1.016E+01	1.613E-02	1.30E-03	2.07E-04	1.51E-03
Sus. sediment (5)	6.9300E-12	4.647E-05		2.520E-05	2.52E+01	3.23E+00	3.231E-03	2.15E-03	1.747E-10		2.24E-12		2.24E-12
Biota (fish) (6)	6.9300E-12	3.778E-06		1.025E-05	2.05E+00	2.63E-01	1.313E-03	1.31E-03	1.420E-11		1.82E-13		1.82E-13
Total R + A		1.667E+09	4.081E+08		206247.96	26436.864			6267.66	1533.88	80.34	19.66	100
			2.076E+09							7801.54523			

```
Total amount of chemicals,          206247.96 moles        26437 kg
Total reaction D value              1.67E+09
Total advection D value             40806813
Total D value                       2.08E+09
Fugacity, E/sum D values            3.759E-06
Output by reaction, mol/h           6267.6617              803.38888 kg/h
Output by advection, mol/h          1533.8834              123.55431 kg/h
Total output, mol/h                 7801.5452              2759.4065 kg/h
Overall resistence time, h          26.436864
Reaction resistence time, h         32.906684
Advection resistence time, h        134.46129
```

Compartment	Subcomp't	volume fraction vi	Fugacity capacity Zi	Bulk vol. VB, m3	Bulk ZB(i) sum(viZi)	Bulk partition coeff. (Zi/Zw)	Bulk VZ=VB*Zi	Bulk den. sum(vidi) kg/m3	Bulk Adv. flow GAB=VB/Gi
Air (1)	Air	1	0.0004034	1.00E+14	0.0004034	1.73E-02	4.034E+10	1.18541328	1.00E+12
	Aerosol	2.0000E-11	65.68						
Water (2)	Water	1	0.0232545	2.00E+11	0.0232908	46.134423	4.658E+09	1000.0085	2.00E+08
	Particulate	0.000005	6.7052225						
	Biota(fish)	0.000001	2.7257002						
Soil (3)	Air	0.2	0.0004034	1.80E+10	5.43E-01	92.268846	9.783E+09	1500.23708	
	Water	0.3	0.0232545						
	Solids	0.5	1.0728356						
Bottom sediment (4)	Water	0.8	0.0232545	5.00E+08	0.4477378	117.21144	2.239E+08	1280	10000
	Solids	0.2	2.1456712						
			1.0149069				5.501E+10		

Level III Intermedia Data:

	D values Dij mol/Pa h	Eq. flows Dij/Z(i) GDij, m3/h	half-life .693Vi/G t(1/2), h	Rate of transport Dij*f(i) N, mol/h	N*MW/1000 Nk, kg/h
Air to water (D12)	7.3993E+06	1.834E+10	3.778E+03	1.931E+01	2.47E+00
Air to soil (D13)	9.5686E+05	2.372E+09	2.922E+04	2.497E+00	3.20E-01
Water to air (D21)	7.3757E+06	3.167E+08	4.377E+02	5.645E+02	7.24E+01
Water to sed. (D24)	5.6781E+04	2.438E+06	5.685E+04	4.346E+00	5.57E-01
Soil to air (D31)	7.4402E+05	1.369E+06	4.556E+03	1.204E+02	1.54E+01
Soil to water (D32)	1.0561E+05	1.943E+06	3.210E+04	1.709E+01	2.19E+00
Sed. to water (D42)	2.7546E+04	6.152E+04	1.126E+03	1.988E+00	2.55E-01
Water to soil (D23)	0				
Sed. burial DL(4)	0.0000E+00				
Stratosphere DL(1)	0.0000E+00				

Equations for Dij values:

$$D12 = A(2)*(1/(1/U(1)*Z(1)+1/(U(2)*Z(2)+U(4)*Z(7)))$$
$$D13 = A(3)*(1/(1/U(5)*Z(1)+U(6)*Z(2)))+U(3)*Z(2)+U(4)*Z(7))$$
$$D21 = A(2)*(1/(1/U(1)*Z(1)+1/(U(2)*Z(2)))$$
$$D24 = A(2)*(U(8)*Z(2)+U(9)*Z(5))$$
$$D31 = A(3)*(1/(1/U(5)*Z(1)+U(6)*Z(2)+1/(U(7)*Z(1))+U(3)*Z(2)+U(4)*Z(7))$$
$$D32 = A(3)*(U(11)*Z(2)+U(12)*Z(3))$$
$$D42 = A(4)*(U(8)*Z(2)+U(10)*Z(4))$$

	D values Dij mol/Pa h	Eq. flows Dij/Z(i) GDij, m3/h	half-life .693Vi/G t(1/2), h	E(A) Rate of transport Dij*f(i) N, mol/h	N*MW/1000 Nk, kg/h	E(B) Rate of transport Dij*f(i) N, mol/h	N*MW/1000 Nk, kg/h	E(C) Rate of transport Dij*f(i) N, mol/h	N*MW/1000 Nk, kg/h
Air to water (D12)	7.3993E+06	1.834E+10	3.778E+03	2.810E+01	3.60E+00	6.674E+00	8.55E-01	4.467E+00	5.726E-01
Air to soil (D13)	9.5686E+05	2.372E+09	2.922E+04	3.634E+00	4.66E-01	8.630E-01	1.11E-01	5.777E-01	7.405E-02
Water to air (D21)	7.3757E+06	3.167E+08	4.377E+02	6.693E+00	8.58E-01	1.855E+03	2.38E+02	4.152E+01	5.322E+00
Water to sed. (D24)	5.6781E+04	2.438E+06	5.685E+04	5.152E-02	6.60E-03	1.428E+01	1.83E+00	3.196E-01	4.097E-02
Soil to air (D31)	7.4402E+05	1.369E+06	4.556E+03	5.589E-01	7.16E-02	1.327E-01	1.70E-02	1.200E+03	1.538E+02
Soil to water (D32)	1.0561E+05	1.943E+06	3.210E+04	7.933E-02	1.02E-02	1.884E-02	2.42E-03	1.703E+02	2.183E+01
Sed. to water (D42)	2.7546E+04	6.152E+04	1.126E+03	2.356E-02	3.02E-03	6.530E+00	8.37E-01	1.462E-01	1.874E-02
Water to soil (D23)	0								
Sed. burial DL(4)	0.0000E+00								
Stratosphere DL(1)	0.0000E+00								

Phase properties and rates:

Compartment	Rate const. k, 1/hr	D Values Reaction VB*ZB*k mol/Pa h	Advection GAB*ZB mol/Pa h	DTs	Js	Js, E(A)	Js, E(B)	Js, E(C)
Air (1)	0.040764705	1.645E+09	4.034E+08	2.056E+09	2.335E-06	3.794E-06	0.000E+00	5.835E-07
Water (2)	0.00407647	1.899E+07	4.658E+06	3.108E+07	3.587E-03	3.587E-03	3.587E-03	3.587E-03
Soil (3)	0.000407647	3.988E+06	0	4.837E+06	9.999E-01	9.999E-01	9.999E-01	9.999E-01
Bottom sediment (4)	0.000126	2.821E+04	4.477E+03	6.023E+04	7.420E+06	7.420E+06	7.420E+06	7.420E+06
Total R + A		1.668E+09	4.081E+08	2.076E+09				

Level III calculations:
For E(1,2,3), i.e., E(1)=600, E(2)=300 and E(3)=100 kg/h

Compartment	Fugacity f/s Pa	Concentration C mol/m3	S mg/L	ug/g	amount m, mol	amount %	Loss Reaction mol/h (Ci*DRi)	Loss Advection mol/h (Ci*DAi)	Loss rate mol/h (Ci*DL)	Reaction %	Advection %	Removal %
Air (1)	2.609E-06	1.053E-09	1.349E-07	1.138E-04	1.05E+05	5.10854	4.29E+03	1052.7048	0.00E+00 0	55.0061	13.4936	68.50
Water (2)	7.654E-05	1.783E-06	2.285E-04	2.285E-04	3.57E+05	17.30202	1.45E+03	356.5386	0	18.6299	4.5701	23.20
Soil (3)	1.618E-04	8.793E-05	1.127E-02	7.513E-03	1.58E+06	76.80554	6.45E+02	0.0000	0	8.2700	0.0000	8.27
Bottom sediment (4)	7.216E-05	3.231E-05	4.141E-03	3.235E-03	1.62E+04	0.78390	2.04E+00	0.3231	0.00E+00 0.0000	0.0261	0.0041	3.02E-02
Total					2.06E+06	100	6.39E+03	1409.5664 7.80E+03	0.0000			100

For E(A), i.e., E(1) = 1000 kg/h conditions:

Compartment	Fugacity f Pa	Concentration mol/m3	mg/L	ug/g	amount m, mol	amount %	Loss Reaction mol/h (Ci*DRi)	Loss Advection mol/h (Ci*DAi)	Loss rate mol/h (Ci*DL)	Reaction %	Advection %	Removal %
Air (1)	3.797E-06	1.532E-09	1.964E-07	1.657E-04	1.53E+05	92.86725	6245.07	1531.9794	0.00E+00 0	80.0493	19.6369	99.69
Water (2)	9.074E-07	2.113E-08	2.709E-06	2.709E-06	4.23E+03	2.56228	17.2306	4.2268	0	0.2209	0.0542	0.28
Soil (3)	7.511E-07	4.082E-07	5.233E-05	3.488E-05	7.35E+03	4.45439	2.9955	0.0000	0	0.0384	0.0000	0.0384
Bottom sediment (4)	8.554E-07	3.830E-07	4.909E-05	3.835E-05	1.92E+02	1.161E-01	0.0241	3.830E-03	0.00E+00 0.0000	3.09E-04	4.91E-05	3.58E-04
Total					1.65E+05	100	6.27E+03	1536.2100 7802	0.0000			100

For E(B), i.e., E(2) = 1000 kg/h conditions:

Compartment	Fugacity f Pa	Concentration mol/m3	mg/L	ug/g	amount m, mol	amount %	Reaction mol/h	Advection mol/h	Loss rate mol/h	Reaction %	Advection %	Removal %
Air (1)	9.020E-07	3.639E-10	4.664E-08	3.935E-05	36387	2.88220	1483.3000	363.8687	0.00E+00 0	19.0129	4.6641	23.68
Water (2)	2.584E-04	5.856E-06	7.507E-04	7.507E-04	1.17E+06	92.77618	4774.65	1171.2711	0	61.2013	15.0133	76.21
Soil (3)	1.784E-07	9.696E-08	1.243E-05	8.284E-06	1745.30	0.13824	0.7115	0.0000	0	0.0091	0.0000	0.0091
Bottom sediment (4)	2.370E-04	1.061E-04	1.360E-02	1.063E-02	53066	4.20337	6.6864	1.0613	0.00E+00 0.0000	0.0857	1.36E-02	0.0993
Total					1262469.5		6265.35	1536.2011 7802	0.0000			100

For E(C), i.e., E(3) = 1000 kg/h conditions:

Compartment	Fugacity f Pa	Concentration mol/m3	mg/L	ug/g	amount m, mol	amount %	Reaction mol/h	Advection mol/h	Loss rate mol/h	Reaction %	Advection %	Removal %
Air (1)	6.038E-07	2.436E-10	3.122E-08	2.634E-05	2.44E+04	0.15387	992.90	243.5682	0.00E+00 0	12.7270	3.1221	15.85
Water (2)	5.629E-06	1.311E-07	1.681E-05	1.680E-05	2.62E+04	0.16565	106.8897	26.2211	0	1.3701	0.3361	1.71
Soil (3)	1.613E-03	8.765E-04	1.12E-01	7.489E-02	1.58E+07	99.67298	6431.78	0.0000	0	82.4425	0.0000	82.44
Bottom sediment (4)	5.307E-06	2.376E-06	3.046E-04	2.379E-04	1.19E+03	0.00750	0.1497	2.376E-02	0.00E+00 0.0000	1.92E-03	3.05E-04	2.22E-03
Total					15829575.		7531.72	269.8131 7802	0.0000			100

Level III summary

	E	E(A)	E(B)	E(C)
Total emission rate mol/h	7801.5290	7801.5290	7801.5290	7801.5290
Total VZ products	5.501E+10	5.501E+10	5.501E+10	5.501E+10
Total amount of chemicals	206676.0	164964.44	1262469.5	15829575.
Total advection D value	1.668E+09	1.67E+09	1.67E+09	1.67E+09
Total reaction D value	408081854	408081854	408081854	408081854
Total D value	2.076E+09	2.08E+09	2.08E+09	2.08E+09
Output by reaction, mol/h	6391.9627	6265.3190	6265.3498	7531.7159
Output by advection, mol/h	1409.5663	1536.2100	1536.2010	269.81314
Output by losses	0	0	0	0
Overall residence time, h	264.13745	21.145141	161.82289	2029.0350
Reaction residence time, h	322.38549	26.329774	201.50024	2101.7223
Advection residence time, h	1461.9219	107.38403	821.81268	58668.659

Output for fugacity Level I, II and III calculations:

Naphthalene

Level I calculation:

Compartment	Z	Concentration			Amount	Amount
	mol/m3 Pa	mol/m3	mg/L (or g/m3)	ug/g	kg	%
Air	4.034E-04	5.736E-09	7.352E-07	6.202E-04	73522	73.522
Water	2.325E-02	3.306E-07	4.238E-05	4.238E-05	8476.23	8.476
Soil	1.073E+00	1.525E-05	1.955E-03	8.147E-04	17597	17.597
Biota (fish)	2.726E+00	3.875E-05	4.968E-03	4.968E-03	0.9935	9.94E-04
Suspended sediment	6.705E+00	9.534E-05	1.222E-02	8.147E-03	12.220	1.22E-02
Bottom sediment	2.146E+00	3.051E-05	3.910E-03	1.629E-03	391.05	0.3910
Total					100000	100

f = 1.422E-05 Pa

Level II Calculation:

Compartment	Half-life h	D Values		Conc'n mol/m3	Loss Reaction kg/h	Loss Advection kg/h	Removal %
		D(reaction)	D(advec'n)				
Air	17	1.64E+09	4.03E+08	1.52E-09	792.34	194.37	98.671
Water	170	1.90E+07	4.65E+06	8.74E-08	9.135	2.241	1.138
Soil	1700	3.94E+06		4.03E-06	1.896		0.190
Biota (fish)				1.02E-05			
Suspended sediment				2.52E-05			
Bottom sediment	5500	2.70E+04	4.29E+03	8.07E-06	1.30E-02	2.07E-03	1.51E-03
Total R + A		1.67E+09	4.08E+08		803.39	196.61	100
		2.08E+09				1000	

f = 3.759E-06 Pa

Total amount = 26437 kg

Overall residence time = 26.44 h
Reaction time = 32.91 h
Advection time = 134.46 h

596

Level III Calculation: Naphthalene

Phase Properties and Rates:

Compartment	Bulk Z mol/m3 Pa	Half-life h	D Values Reaction mol/Pa h	D Values Advection mol/Pa h
Air (1)	4.034E-04	17	1.64E+09	4.03E+08
Water (2)	2.329E-02	170	1.90E+07	4.66E+06
Soil (3)	5.435E-01	1700	3.99E+06	
Bottom sediment (4)	4.477E-01	5500	2.82E+04	4.48E+03

Intermedia D values:

diff., DL	mol/Pa h	0
air/water D12	7.399E+06	
air/soil, D13	9.569E+05	
water/air, D21	7.376E+06	
water/sed., D24	5.678E+04	
soil/air, D31	7.440E+05	
soil/water, D32	1.056E+05	
sed./water, D42	2.755E+04	
sed., burial, DL	0	

	E(1)=1000	E(2)=1000	E(3)=1000	E(1,2,3)
Overall residence time =	21.15	161.82	2029.04	264.14 h
Reaction time =	26.33	201.50	2101.72	322.39 h
Advection time =	107.38	821.81	58668.66	1461.92 h

Phase Properties, Compositions, Transport and Transformation Rates:

Emission, kg/h E(1)	E(2)	E(3)	Fugacity, Pa f(1)	f(2)	f(3)	f(4)	Concentration, g/m3 C(1)	C(2)	C(3)	C(4)
1000	0	0	3.797E-06	9.074E-07	7.511E-07	8.554E-07	1.964E-07	2.709E-06	5.233E-05	4.909E-05
0	1000	0	9.020E-07	2.514E-04	1.784E-07	2.370E-04	4.664E-08	7.507E-04	1.243E-05	1.360E-02
0	0	1000	6.038E-07	5.629E-06	1.613E-03	5.307E-06	3.122E-08	1.681E-05	1.124E-01	3.046E-04
600	300	100	2.609E-06	7.654E-05	1.618E-04	7.216E-05	1.349E-07	2.285E-04	1.127E-02	4.141E-03

Amounts, kg

m(1)	m(2)	m(3)	m(4)	amount, kg
1.964E+04	5.418E+02	9.419E+02	2.455E+01	2.115E+04
4.664E+03	1.501E+05	2.237E+02	6.802E+03	1.618E+05
3.122E+03	3.361E+03	2.022E+06	1.523E+02	2.029E+06
1.349E+04	4.570E+04	2.029E+05	2.071E+03	2.641E+05

Emission, kg/h E(1)	E(2)	E(3)	Loss, Reaction, kg/h R(1)	R(2)	R(3)	R(4)	Loss, Advection, kg/h A(1)	A(2)	A(4)
1000	0	0	8.005E+02	2.209E+00	3.84E-01	3.093E-03	1.964E+02	5.418E+01	4.909E-04
0	1000	0	1.901E+02	6.120E+02	9.12E-02	8.571E-01	4.664E+01	1.501E+02	1.360E-01
0	0	1000	1.273E+02	1.370E+02	8.24E+02	1.919E-02	3.122E+01	3.361E+00	3.046E-03
600	300	100	5.501E+02	1.863E+02	8.27E+01	2.609E-01	1.349E+02	4.570E+01	4.141E-02

Intermedia Rate of Transport, kg/h

T12 air-water	T21 water-air	T13 air-soil	T31 soil-air	T24 water-sed	T32 soil-water	T42 sed-water
3.602E+00	8.579E-01	4.658E-01	7.164E-02	6.604E-03	1.017E-02	3.020E-03
8.555E-01	2.377E+02	1.106E-01	1.701E-02	1.830E+00	2.415E-03	8.370E-01
5.726E-01	5.322E+00	7.405E-02	1.538E+02	4.097E-02	2.183E+01	1.874E-02
2.475E+00	7.236E+01	3.200E-01	1.543E+01	5.571E-01	2.190E+00	2.548E-01